Basic Proof Theory

Cambridge Tracts in Theoretical Computer Science 43

Titles in the series

1. G. Chaitin *Algorithmic Information Theory*
2. L. C. Paulson *Logic and Computation*
3. M. Spivey *Understanding Z*
5. A. Ramsey *Formal Methods in Artificial Intelligence*
6. S. Vickers *Topology via Logic*
7. J.-Y. Girard, Y. Lafont & P. Taylor *Proofs and Types*
8. J. Clifford *Formal Semantics & Progmatics for Natural Language Processing*
9. M. Winslett *Updating Logical Databases*
10. K. McEvoy & J. V. Tucker (eds) *Theoretical Foundations of VLSI Design*
11. T. H. Tse *A Unifying Framework for Structured Analysis and Design Models*
12. G. Brewka *Nonmonotonic Reasoning*
14. S. G. Hoggar *Mathematics for Computer Graphics*
15. S. Dasgupta *Design Theory and Computer Science*
17. J. C. M. Baeten (ed) *Applications of Process Algebra*
18. J. C. M. Baeten & W. P. Weijland *Process Algebra*
19. M. Manzano *Extensions of First Order Logic*
21. D. A. Wolfram *The Clausal Theory of Types*
22. V. Stoltenberg-Hansen, I. Lindström & E. Griffor *Mathematical Theory of Domains*
23. E.-R. Olderog *Nets, Terms and Formulas*
26. P. D. Mosses *Action Semantics*
27. W. H. Hesselink *Programs, Recursion and Unbounded Choice*
28. P. Padawitz *Deductive and Declarative Programming*
29. P. Gärdenfors (ed) *Belief Revision*
30. M. Anthony & N. Biggs *Computational Learning Theory*
31. T. F. Melham *Higher Order Logic and Hardware Verification*
32. R. L. Carpenter *The Logic of Typed Feature Structures*
33. E. G. Manes *Predicate Transformer Semantics*
34. F. Nielson & H. R. Nielson *Two Level Functional Languages*
35. L. Feijs & H. Jonkers *Formal Specification and Design*
36. S. Mauw & G. J. Veltink (eds) *Algebraic Specification of Communication Protocols*
37. V. Stavridou *Formal Methods in Circuit Design*
38. N. Shankar *Metamathematics, Machines and Gödel's Proof*
39. J. B. Paris *The Uncertain Reasoner's Companion*
40. J. Dessel & J. Esparza *Free Choice Petri Nets*
41. J.-J. Ch. Meyer & W. van der Hoek *Epistemic Logic for AI and Computer Science*
42. J. R. Hindley *Basic Simple Type Theory*
43. A. Troelstra & H. Schwichtenberg *Basic Proof Theory*
44. J. Barwise & J. Seligman *Information Flow*
45. A. Asperti & S. Guerrini *The Optimal Implementation of Functional Programming Languages*
46. R. M. Amadio & P.-L. Curien *Domains and Lambda-Calculi*
47. W.-P. de Roever & K. Engelhardt *Data Refinement*
48. H. Kleine Büning & T. Lettman *Propositional Logic*
49. L. Novak & A. Gibbons *Hybrid Graph Theory and Network Analysis*

Basic Proof Theory

Second Edition

A.S. Troelstra
University of Amsterdam

H. Schwichtenberg
University of Munich

PUBLISHED BY THE PRESS SYNDICATE OF THE UNIVERSITY OF CAMBRIDGE
The Pitt Building, Trumpington Street, Cambridge, United Kingdom

CAMBRIDGE UNIVERSITY PRESS
The Edinburgh Building, Cambridge CB2 2RU, UK http://www.cup.cam.ac.uk
40 West 20th Street, New York, NY 10011–4211, USA http://www.cup.org
10 Stamford Road, Oakleigh, Melbourne 3166, Australia
Ruiz de Alarcón 13, 28014 Madrid, Spain

First published 1996
Second edition 2000

Printed in the United Kingdom at the University Press, Cambridge

Typeset by the author in Computer Modern 10/13pt, in LaTeX 2ε [EPC]

A catalogue record of this book is available from the British Library

Library of Congress Cataloguing in Publication data

ISBN 0 521 77911 1 paperback

Contents

Preface

Preface to the first edition

The discovery of the set-theoretic paradoxes around the turn of the century, and the resulting uncertainties and doubts concerning the use of high-level abstractions among mathematicians, led D. Hilbert to the formulation of his programme: to prove the consistency of axiomatizations of the essential parts of mathematics by methods which might be considered as evident and reliable because of their elementary combinatorial ("finitistic") character.

Although, by Gödel's incompleteness results, Hilbert's programme could not be carried out as originally envisaged, for a long time variations of Hilbert's programme have been the driving force behind the development of proof theory. Since the programme called for a complete formalization of the relevant parts of mathematics, including the logical steps in mathematical arguments, interest in proofs as combinatorial structures in their own right was awakened. This is the subject of structural proof theory; its true beginnings may be dated from the publication of the landmark-paper Gentzen [1935].

Nowadays there are more reasons, besides Hilbert's programme, for studying structural proof theory. For example, automated theorem proving implies an interest in proofs as combinatorial structures; and in logic programming, formal deductions are used in computing.

There are several monographs on proof theory (Schütte [1960,1977], Takeuti [1987], Pohlers [1989]) inspired by Hilbert's programme and the questions this engendered, such as "measuring" the strength of subsystems of analysis in terms of provable instances of transfinite induction for definable well-orderings (more precisely, ordinal notations). Pohlers [1989] is particularly recommended as an introduction to this branch of proof theory.

Girard [1987b] presents a wider panorama of proof theory, and is not easy reading for the beginner, though recommended for the more experienced.

The present text attempts to fill a lacuna in the literature, a gap which exists between introductory books such as Heindorf [1994], and textbooks on mathematical logic (such as the classic Kleene [1952a], or the recent van Dalen [1994]) on the one hand, and the more advanced monographs mentioned above on the other hand.

Our text concentrates on the structural proof theory of first-order logic and its applications, and compares different styles of formalization at some length. A glimpse of the proof theory of first-order arithmetic and second-order logic is also provided, illustrating techniques in relatively simple situations which are applied elsewhere to far more complex systems.

As preliminary knowledge on the part of the reader we assume some familiarity with first-order logic as may be obtained from, for example, van Dalen [1994]. A slight knowledge of elementary recursion theory is also helpful, although not necessary except for a few passages. Locally, other preliminary knowledge will be assumed, but this will be noted explicitly.

Several short courses may be based on a suitable selection of the material. For example, chapters 1, 2, 6 and 10 develop the theory of natural deduction and lead to a proof of the "classical" result of Gentzen on the relation between the ordinal ε_0 and first-order arithmetic. A course based on the first five chapters concentrates on Gentzen systems and cut elimination with (elementary) applications.

There are many interconnections between the present text and Hindley [1997]; the latter concentrates on type-assignment systems (systems of rules for assigning types to untyped lambda terms) which are not treated here. In our text we only consider theories with "rigid typing", where each term and all of its subterms carry along a fixed type. Hindley's book may be regarded as a companion volume providing a treatment of deductions as they appear in type-assignment systems.

We have been warned by colleagues from computer science that references to sources more than five years old will make a text look outdated. For readers inclined to agree with this we recommend contemplation of the following platitudes: (1) a more recent treatment of a topic is not automatically an improvement over earlier treatments; (2) if a subject is worthwhile, it will in due time acquire a history going back more than five years; (3) results of lasting interest do exist; (4) limiting the horizon to five years entails a serious lack of historical perspective.

Numbered exercises are scattered throughout the text. These are immediately recognizable as such, since they have been set in smaller type and have been marked with the symbol ♠.

Many of these exercises are of a routine character ("complete the proof of this lemma"). We believe that (a) such exercises are very helpful in familiarizing the student with the material, and (b) listing these routine exercises explicitly makes it easy for a course leader to assign definite tasks to the students.

At the end of each chapter, except the first, there is a section called "Notes". There we have collected historical credits and suggestions for further reading; also we mention other work related to the topic of the chapter. These notes do not pretend to give a history of the subject, but may be of help in gaining some historical perspective, and point the way to the sources. There is no attempt at completeness; with the subject rapidly expanding this has become well-nigh impossible.

The references in the index to names of persons concern in the majority of cases a citation of a publication. In case of publications with more than one author, only the first author's name is indexed. Occurrences of author names in the bibliography have not been indexed. There is a separate list, where symbols and notations of more than local significance have been indexed.

The text started as a set of course notes for part of a course "Introduction to Constructivism and Proof Theory" for graduate students at the University of Amsterdam. When the first author decided to expand these notes into a book, he felt that at least some of the classical results on the proof theory of first-order arithmetic ought to be included; hence the second author was asked to become coauthor, and more particularly, to provide a chapter on the proof theory of first-order arithmetic. The second author's contribution did not restrict itself to this; many of his suggestions for improvement and inclusion of further results have been adopted, and a lot of material from his course notes and papers has found its way into the text.

We are indebted for comments and information to K. R. Apt, J. F. A. K. van Benthem, H. C. Doets, J. R. Hindley, G. E. Mints, V. Sanchez, S. V. Solovjov, A. Weiermann; the text was prepared with the help of some useful Latex macros for the typesetting of prooftrees by S. Buss and for the typesetting of ordinary trees by D. Roorda. M. Behrend of the Cambridge University Press very carefully annotated the near-final version of the text, expunging many blemishes and improving typographical consistency.

Amsterdam/München A. S. Troelstra
Spring 1996 H. Schwichtenberg

Preface to the second edition

In preparing this revised edition we used the opportunity to correct many errata in the first edition. Moreover certain sections were rewritten and some new material inserted, especially in chapters 3–6. The principal changes are the following.

Chapter 1: section 1.3 has been largely rewritten. Chapter 2: the material in 2.1.10 is new. Chapter 3: more prominence has been given to a Kleene-style variant of the G3-systems (3.5.11), and multi-succedent versions of **G3**[**mi**] are defined in the body of the text (3.5.10). A general definition of systems with local rules (3.4) is also new. Chapter 4: the proof of cut elimination for G3-systems (4.1.5) has been completely rewritten, and a sketch of cut elimination for the systems m-**G3**[**mi**] has been added (4.1.10). There are new sections on cut elimination for extensions of G3-systems, with applications to predicate logic with equality and the intuitionistic theory of apartness. Chapter 5: a result on the growth of size of proofs in propositional logic under cut elimination (5.2) has been included. Chapter 6: extensions of N-systems with extra rules, with an application to E-logic, are new; the section on E-logic replaces an inadequate treatment of the same results in chapter 4 of the first edition. Chapter 11: a new proof of strong normalization for **λ2**. New are also the "Solutions to selected exercises"; these are intended as a help to those readers who study the text on their own. Exercises for which a (partial) solution is provided are marked with *. The updating of the bibliography primarily concerns the parts of the text which have been revised.

In a review of the first edition it has been noted that complexity-theoretic aspects are largely absent. We felt that this area is so vast that it would require a separate monograph of its own, to be written by an expert in the area. Another complaint was that our account was lacking in motivation and philosophical background. This has not been remedied in the present edition, although a few words of extra explanation have been added here and there. A typically philosophical problem we did not deal with is the question: when are two proofs to be considered equal? We doubt, however, whether this question will ever have a simple answer; it may well be that there are many answers, depending on aims and points of view.

One terminological change deserves to be noted: we changed "contraction" (in the sense of a step in transforming terms of type theory and proofs in natural deduction), into "conversion". Similarly, "contracts to" and "contractum" have been replaced by "converts to" and "conversum" repectively. On the other hand "contraction" as the name of a structural rule in Gentzen systems is maintained. See under the remarks at the end of 1.2.5 for a motivation.

We are again indebted to many people for comments and corrections, in particular H. van Ditmarsch, L. Gordeev, J. R. Hindley, R. Matthes, G. E. Mints, and the students in our classes. Special thanks are due to Sara Negri, who unselfishly offered to read carefully a large part of the text; in this she has been assisted by Jan von Plato. Their comments led to many improvements. For the remaining defects of the text the authors bear sole responsibility.

Amsterdam/München A. S. Troelstra
March 2000 H. Schwichtenberg

Chapter 1

Introduction

Proof theory may be roughly divided into two parts: structural proof theory and interpretational proof theory. Structural proof theory is based on a combinatorial analysis of the structure of formal proofs; the central methods are cut elimination and normalization.

In interpretational proof theory, the tools are (often semantically motivated) syntactical translations of one formal theory into another. We shall encounter examples of such translations in this book, such as the Gödel–Gentzen embedding of classical logic into minimal logic (2.3), and the modal embedding of intuitionistic logic into the modal logic **S4** (9.2). Other well-known examples from the literature are the formalized version of Kleene's realizability for intuitionistic arithmetic and Gödel's Dialectica interpretation (see, for example, Troelstra [1973]).

The present text is concerned with the more basic parts of structural proof theory. In the first part of this text (chapters 2–7) we study several formalizations of standard logics. "Standard logics", in this text, means minimal, intuitionistic and classical first-order predicate logic. Chapter 8 describes the connection between cartesian closed categories and minimal conjunction–implication logic; this serves as an example of the applications of proof theory in category theory. Chapter 9 illustrates the extension to other logics (namely the modal logic **S4** and linear logic) of the techniques introduced before in the study of standard logics. The final two chapters deal with first-order arithmetic and second-order logic respectively.

The first section of this chapter contains notational conventions and definitions, to be consulted only when needed, so a quick scan of the contents will suffice to begin with. The second section presents a concise introduction to simple type theory, with rigid typing; the parallel between (extensions of) simple type theory and systems of natural deduction, under the catch-phrase "formulas-as-types", is an important theme in the sequel. Then follows a brief informal introduction to the three principal types of formalism we shall encounter later on, the N-, H- and G-systems, or Natural deduction, Hilbert systems, and Gentzen systems respectively. Formal definitions of these systems will be given in chapters 2 and 3.

1.1 Preliminaries

The material in this section consists primarily of definitions and notational conventions, and may be skipped until needed.

Some very general abbreviations are "iff" for "if and only if", "IH" for "induction hypothesis", "w.l.o.g." for "without loss of generality". To indicate *literal identity* of two expressions, we use $\equiv$. (In dealing with expressions with bound variables, this is taken to be literal identity modulo renaming of bound variables; see 1.1.2 below.)

The symbol $\boxtimes$ is used to mark the end of proofs, definitions, stipulations of notational conventions.

$\mathbb{N}$ is used for the natural numbers, zero included. Set-theoretic notations such as $\in$, $\subset$ are standard.

1.1.1. *The language of first-order predicate logic*

The standard language considered contains $\vee, \wedge, \rightarrow, \bot, \forall, \exists$ as primitive logical operators ($\bot$ being the degenerate case of a zero-place logical operator, i.e. a logical constant), countably infinite supplies of individual variables, n-place relation symbols for all $n \in \mathbb{N}$, symbols for n-ary functions for all $n \in \mathbb{N}$. 0-place relation symbols are also called proposition letters or proposition variables; 0-argument function symbols are also called (individual) constants. The language will *not*, unless stated otherwise, contain $=$ as a primitive.

Atomic formulas are formulas of the form $Rt_1 \dots t_n$, R a relation symbol, $t_1, \dots, t_n$ individiual terms, $\bot$ is not regarded as atomic. For formulas which are either atomic or $\bot$ we use the term *prime* formula

We use certain categories of letters, possibly with sub- or superscripts or primed, as metavariables for certain syntactical categories (locally different conventions may be introduced):

- x, y, z, u, v, w for individual variables;

- f, g, h for arbitrary function symbols;

- c, d for individual constants;

- t, s, r for arbitrary terms;

- P, Q for atomic formulas;

- R for relation symbols of the language;

- A, B, C, D, E, F for arbitrary formulas in the language.

Notation. For the countable set of *proposition variables* we write $\mathcal{PV}$. We introduce abbreviations:

$$\begin{aligned} A \leftrightarrow B &:= (A \to B) \wedge (B \to A), \\ \neg A &:= A \to \bot, \\ \top &:= \bot \to \bot. \end{aligned}$$

In this text, $\top$ ("truth") is sometimes added as a primitive. If Γ is a finite sequence $A_1, \ldots, A_n$ of formulas, $\bigwedge \Gamma$ is the iterated conjunction $(\ldots(A_1 \wedge A_2) \wedge \ldots A_n)$, and $\bigvee \Gamma$ the iterated disjunction $(\ldots(A_1 \vee A_2) \vee \ldots A_n)$. If Γ is empty, we identify $\bigvee \Gamma$ with $\bot$, and $\bigwedge \Gamma$ with $\top$. ⊠

Notation. (*Saving on parentheses*) In writing formulas we save on parentheses by assuming that $\forall, \exists, \neg$ bind more strongly than $\vee, \wedge$, and that in turn $\vee, \wedge$ bind more strongly than $\to, \leftrightarrow$. Outermost parentheses are also usually dropped. Thus $A \wedge \neg B \to C$ is read as $((A \wedge (\neg B)) \to C)$. In the case of iterated implications we sometimes use the short notation

$$A_1 \to A_2 \to \ldots A_{n-1} \to A_n \quad \text{for} \quad A_1 \to (A_2 \to \ldots (A_{n-1} \to A_n) \ldots).$$

We also save on parentheses by writing e.g. $Rxyz$, $Rt_0t_1t_2$ instead of $R(x, y, z)$, $R(t_0, t_1, t_2)$, where R is some predicate letter. Similarly for a unary function symbol with a (typographically) simple argument, so fx for $f(x)$, etc. In this case no confusion will arise. But readability requires that we write in full $R(fx, gy, hz)$, instead of $Rfxgyhz$. ⊠

1.1.2. *Substitution, free and bound variables*

Expressions $\mathcal{E}, \mathcal{E}'$ which differ only in the names of bound variables will be regarded by us as identical. This is sometimes expressed by saying that $\mathcal{E}$ and $\mathcal{E}'$ are α-equivalent. In other words, we are only interested in certain equivalence classes of (the concrete representations of) expressions, expressions "modulo renaming of bound variables". There are methods of finding unique representatives for such equivalence classes, for example the namefree terms of de Bruijn [1972]. See also Barendregt [1984, Appendix C].

For the human reader such representations are less convenient, so we shall stick to the use of bound variables. But it should be realized that the issues of handling bound variables, renaming procedures and substitution are essential and non-trivial when it comes to implementing algorithms.

In the definition of "substitution of expression $\mathcal{E}'$ for variable x in expression $\mathcal{E}$", either one requires that *no* variable free in $\mathcal{E}'$ becomes bound by a variable-binding operator in $\mathcal{E}$, when the free occurrences of x are replaced by $\mathcal{E}'$ (also expressed by saying that there must be no "clashes of variables"), "*$\mathcal{E}'$ is free for x in $\mathcal{E}$*", or the substitution operation is taken to involve a systematic renaming operation for the bound variables, avoiding clashes. Having stated that we are only interested in expressions modulo renaming bound variables, we can without loss of generality assume that substitution is always possible.

Also, it is never a real restriction to assume that distinct quantifier occurrences are followed by distinct variables, and that the sets of bound and free variables of a formula are disjoint.

NOTATION. "FV" is used for the (set of) free variables of an expression; so FV(t) is the set of variables free in the term t, FV(A) the set of variables free in formula A etc.

$\mathcal{E}[x/t]$ denotes the result of substituting the term t for the variable x in the expression $\mathcal{E}$. Similarly, $\mathcal{E}[\vec{x}/\vec{t}]$ is the result of *simultaneously* substituting the terms $\vec{t} = t_1, \dots, t_n$ for the variables $\vec{x} = x_1, \dots, x_n$ respectively.

For substitutions of predicates for predicate variables (predicate symbols) we use essentially the same notational conventions. If in a formula A, containing an n-ary relation variable X^n, X^n is to be replaced by a formula B, seen as an n-ary predicate of n of its variables $\vec{x} \equiv x_1, \dots, x_n$, we write $A[X^n/\lambda\vec{x}.B]$ for the formula which is obtained from A by replacing every occurrence $X^n\vec{t}$ by $B[\vec{x}/\vec{t}]$ (neither individual variables nor relation variables of $\forall\vec{x}\,B$ are allowed to become bound when substituting).

Note that B may contain other free variables besides $\vec{x}$, and that the "$\lambda\vec{x}$" is needed to indicate which terms are substituted for which variables.

Locally we shall adopt the following convention. In an argument, once a formula has been introduced as $A(x)$, i.e., A with a designated free variable x, we write $A(t)$ for $A[x/t]$, and similarly with more variables. ⊠

1.1.3. *Subformulas*

DEFINITION. (*Gentzen subformula*) Unless stated otherwise, the notion of *subformula* we use will be that of a subformula in the sense of Gentzen. (Gentzen) subformulas of A are defined by

(i) A is a subformula of A;

(ii) if $B \circ C$ is a subformula of A then so are B, C, for $\circ = \vee, \wedge, \rightarrow$;

(iii) if $\forall xB$ or $\exists xB$ is a subformula of A, then so is $B[x/t]$, for all t free for x in B.

If we replace the third clause by:

(iii)′ if $\forall xB$ or $\exists xB$ is a subformula of A then so is B,

we obtain the notion of *literal* subformula. ⊠

Definition. The notions of *positive, negative, strictly positive* subformula are defined in a similar style:

(i) A is a positive and a stricly positive subformula of itself;

(ii) if $B \wedge C$ or $B \vee C$ is a positive [negative, strictly positive] subformula of A, then so are B, C;

(iii) if $\forall x B$ or $\exists x B$ is a positive [negative, strictly positive] subformula of A, then so is $B[x/t]$ for any t free for x in B;

(iv) if $B \to C$ is a positive [negative] subformula of A, then B is a negative [positive] subformula of A, and C is a positive [negative] subformula of A;

(v) if $B \to C$ is a strictly positive subformula of A then so is C.

A strictly positive subformula of A is also called a *strictly positive part (s.p.p.)* of A. Note that the set of subformulas of A is the union of the positive and the negative subformulas of A.

Literal positive, negative, strictly positive subformulas may be defined in the obvious way by restricting the clause for quantifiers. ⊠

Example. $(P \to Q) \to R \vee \forall x R'(x)$ has as s.p.p.'s the whole formula, $R \vee \forall x R'(x)$, R, $\forall x R'(x)$, $R'(t)$. The positive subformulas are the s.p.p.'s and in addition P; the negative subformulas are $P \to Q$, Q.

1.1.4. *Contexts and formula occurrences*

Formula occurrences (f.o.'s) will play an even more important role than the formulas themselves. An f.o. is nothing but a formula with a position in another structure (prooftree, sequent, a larger formula etc.). If no confusion is to be feared, we shall permit ourselves a certain "abus de langage" and talk about formulas when really f.o.'s are meant.

The notion of a (sub)formula occurrence in a formula or sequent is intuitively obvious, but for formal proofs of metamathematical properties it is sometimes necessary to use a rigorous formal definition. This may be given via the notion of a *context*. Roughly speaking, a context is nothing but a formula with an occurrence of a special propositional variable, a "placeholder". Alternatively, a context is sometimes described as a formula with a hole in it.

Definition. We define *positive* ($\mathcal{P}$) and *negative (formula-)contexts* ($\mathcal{N}$) simultaneously by an inductive definition given by the three clauses (i)–(iii) below. The symbol "$*$" in clause (i) functions as a special proposition letter (not in the language of predicate logic), a *placeholder* so to speak.

(i) $* \in \mathcal{P}$;

and if $B^+ \in \mathcal{P}$, $B^- \in \mathcal{N}$, and A is any formula, then

(ii) $A \wedge B^+$, $B^+ \wedge A$, $A \vee B^+$, $B^+ \vee A$, $A \to B^+$, $B^- \to A$, $\forall x B^+$, $\exists x B^+ \in \mathcal{P}$;

(iii) $A \wedge B^-$, $B^- \wedge A$, $A \vee B^-$, $B^- \vee A$, $A \to B^-$, $B^+ \to A$, $\forall x B^-$, $\exists x B^- \in \mathcal{N}$.

The set of *formula contexts* is the union of $\mathcal{P}$ and $\mathcal{N}$. Note that a context contains always only a single occurrence of $*$. We may think of a context as a formula in the language extended by $*$, in which $*$ occurs only once. In a positive [negative] context, $*$ is a positive [negative] subformula. Below we give a formal definition of (sub)formula occurrence *via* the notion of context.

For arbitrary contexts we sometimes write $F[*]$, $G[*]$, ... Then $F[A]$, $G[A]$, ... are the formulas obtained by replacing $*$ by A (literally, without renaming variables).

The notion of context may be generalized to a context with several placeholders $*_1, \ldots, *_n$, which are treated as extra proposition variables, each of which may occur only once in the context.

The *strictly positive* contexts $\mathcal{SP}$ are defined by

(iv) $* \in \mathcal{SP}$; and if $B \in \mathcal{SP}$, then

(v) $A \wedge B$, $B \wedge A$, $A \vee B$, $B \vee A$, $A \to B$, $\forall x B$, $\exists x B \in \mathcal{SP}$.

An alternative style of presentation of this definition is

$$\begin{aligned}
\mathcal{P} &= * \mid A \wedge \mathcal{P} \mid \mathcal{P} \wedge A \mid A \vee \mathcal{P} \mid \mathcal{P} \vee A \mid A \to \mathcal{P} \mid \mathcal{N} \to A \mid \forall x \mathcal{P} \mid \exists x \mathcal{P}, \\
\mathcal{N} &= A \wedge \mathcal{N} \mid \mathcal{N} \wedge A \mid A \vee \mathcal{N} \mid \mathcal{N} \vee A \mid A \to \mathcal{N} \mid \mathcal{P} \to A \mid \forall x \mathcal{N} \mid \exists x \mathcal{N}, \\
\mathcal{SP} &= * \mid A \wedge \mathcal{SP} \mid \mathcal{SP} \wedge A \mid A \vee \mathcal{SP} \mid \mathcal{SP} \vee A \mid A \to \mathcal{SP} \mid \forall x \mathcal{SP} \mid \exists x \mathcal{SP}.
\end{aligned}$$

A *formula occurrence* (*f.o.* for short) in a formula B is a literal subformula A together with a context indicating the place where A occurs (so B may be obtained by replacing $*$ in the context by A). In the obvious way we can now define *positive, strictly positive* and *negative occurrence.* ⊠

1.1.5. *Finite multisets*

Finite *multisets*, i.e. "sets with multiplicity", or to put it otherwise, finite sequences modulo the ordering, will play an important role in this text.

NOTATION. If Δ is a multiset, we use $|\Delta|$ for the number of its elements. For the multiset union of Γ and Δ we write $\Gamma \cup \Delta$ or in certain situations simply Γ, Δ or even $\Gamma\Delta$ (namely when writing sequents, which will be introduced later). The notation Γ, A or ΓA then designates a multiset which is the union of Γ and the singleton multiset containing only A.

If "c" is some unary operator and $\Gamma \equiv A_1, \ldots, A_n$ is a finite multiset of formulas, we write $\mathrm{c}\Gamma$ for the multiset $\mathrm{c}A_1, \ldots, \mathrm{c}A_n$.

Finite sets may be regarded as special cases of finite multisets: a multiset where each element occurs with multiplicity one represents a finite set. For the set underlying a multiset Γ, we write $\mathrm{Set}(\Gamma)$; this multiset contains the formulas of Γ with multiplicity one. ⊠

NOTATION. We shall use the notations $\bigwedge \Gamma$, $\bigvee \Gamma$ also in case Γ is a multiset. $\bigwedge \Gamma$, $\bigvee \Gamma$ are then the conjunction, respectively disjunction of Γ' for some sequence Γ' corresponding to Γ. $\bigwedge \Gamma$, $\bigvee \Gamma$ are then well-defined modulo logical equivalence, as long as in our logic $\wedge, \vee$ obey the laws of symmetry and associativity. ⊠

DEFINITION. The notions of *(positive, negative) formula occurrence* may be defined for sequents, i.e., expressions of the form $\Gamma \Rightarrow \Delta$, with Γ, Δ finite multisets, as (positive, negative) formula occurrences in the corresponding formulas $\bigwedge \Gamma \to \bigvee \Delta$. ⊠

1.1.6. *Deducibility and deduction from hypotheses, conservativity*

NOTATION. In our formalisms, we derive either formulas or sequents (as introduced in the preceding definition). For sequents derived in a formalism **S** we write

$$\mathbf{S} \vdash \Gamma \Rightarrow \Delta \quad \text{or} \quad \vdash_{\mathbf{S}} \Gamma \Rightarrow \Delta,$$

and for formulas derived in **S**

$$\mathbf{S} \vdash A \quad \text{or} \quad \vdash_{\mathbf{S}} A.$$

If we want to indicate that a deduction $\mathcal{D}$ derives $\Gamma \Rightarrow \Delta$, we can write $\mathcal{D} \vdash_{\mathbf{S}} \Gamma \Rightarrow \Delta$ (or $\mathcal{D} \vdash \Gamma \Rightarrow \Delta$ if **S** is evident).

For formalisms based on sequents, $\mathbf{S} \vdash A$ will coincide with $\mathbf{S} \vdash \Rightarrow A$ (sequent $\Gamma \Rightarrow A$ with Γ empty).

If a formula A is derivable from a finite multiset Γ of hypotheses or assumptions, we write

$$\Gamma \vdash_{\mathbf{S}} A.$$

In systems with sequents this is equivalent to $\mathbf{S} \vdash \Gamma \Rightarrow A$. (N.B. In the literature $\Gamma \vdash A$ is sometimes given a slightly different definition for which the deduction theorem does not hold; cf. remark in 9.1.2. Moreover, some authors use $\vdash$ instead of our sequent-arrow $\Rightarrow$.)

A *theory* is a set of sentences (closed formulas); with each formalism is associated a theory of deducible sentences. Since for the theories associated with the formalisms in this book, it is always true that the set of deducible *formulas* and the set of pairs $\{(\Gamma, A) \mid \Gamma \vdash A\}$ are uniquely determined by the theory, we shall also speak of formulas belonging to a theory, and use the expression "A is deducible from Γ in a theory".

In particular, we write

$$\Gamma \vdash_{\mathrm{m}} A, \qquad \Gamma \vdash_{\mathrm{i}} A, \qquad \Gamma \vdash_{\mathrm{c}} A$$

for deducibility in our standard logical theories **M**, **I**, **C** respectively (cf. the next subsection). ⊠

DEFINITION. A system **S** is *conservative* over a system $\mathbf{S}' \subset \mathbf{S}$, if for formulas A in the language of $\mathbf{S}'$ we have that if $\mathbf{S} \vdash A$, then $\mathbf{S}' \vdash A$. For systems with sequents, conservativity similarly means: if $\vdash \Gamma \Rightarrow A$ in **S**, with $\Gamma \Rightarrow A$ in the language of $\mathbf{S}'$, then $\vdash \Gamma \Rightarrow A$ in $\mathbf{S}'$. Similarly for theories. ⊠

1.1.7. *Names for theories and systems*

Where we are only interested in the logics as theories, i.e. as *sets of theorems*, we use **M**, **I** and **C** for minimal, intuitionistic and classical predicate calculus respectively; **Mp**, **Ip** and **Cp** are the corresponding propositional systems. If we are interested only in formulas constructed from a set of operators $\mathcal{A}$ say, we write $\mathcal{A}$-**S** or $\mathcal{A}$**S** for the system **S** restricted to formulas with operators from $\mathcal{A}$. Thus $\to\wedge$-**M** is **M** restricted to formulas in $\wedge, \to$ only.

On the other hand, where the notion of formal deduction is under investigation, we have to distinguish between the various formalisms characterizing the same theory. In choosing designations, we use some mnemonic conventions:

- We use "**N**", "**H**", "**G**" for "Natural Deduction", "Hilbert system" and "Gentzen system" respectively. "**GS**" (from "Gentzen–Schütte") is used as a designation for a group of calculi with one-sided sequents (always classical).

- We use "**c**" for "classical", "**i**" for "intuitionistic", "**m**" for "minimal", "**s**" for "**S4**", "**p**" for "propositional", "**e**" for "E-logic". If **p** is absent, the system includes quantifiers. The superscript "^{2}" is used for second-order systems.

- Variants may be designated by additions of extra boldface capitals, numbers, superscripts such as "*" etc. Thus, for example, **G1c** is close to the original sequent calculus LK of Gentzen (and **G1i** to Gentzen's LJ), **G2c** is a variant with weakening absorbed into the logical rules, **G3c** a system with weakening *and* contraction absorbed into the rules, **GK** (from Gentzen–Kleene) refers to Gentzen systems very close to the system G3 of Kleene, etc.

- In order to indicate several formal systems at once, without writing down the exhaustive list, we use the following type of abbreviation: **S**[**abc**] refers to **Sa**, **Sb**, **Sc**; **S**[**ab**][**cd**] refers to **Sac**, **Sbc**, **Sad**, **Sbd**, etc.; [**mic**] stands for "**m**, or **i** or **c**"; [**mi**] for "**m** or **i**"; [**123**] for "**1**, **2** or **3**", etc. In such contracted statements an obvious parallelism is maintained, e.g. "**G**[**123**]**c** satisfies $\mathcal{A}$ iff **G**[**123**]**i** satisfies $\mathcal{B}$" is read as: "**G1c** (respectively **G2c**, **G3c**) satisfies $\mathcal{A}$ iff **G1i** (respectively **G2i**, **G3i**) satisfies $\mathcal{B}$".

1.1.8. *Finite trees*

DEFINITION. (*Terminology for trees*) Trees are partially ordered sets $(X, \leq)$ with a lowest element and all sets $\{y : y \leq x\}$ for $x \in X$ linearly ordered. The elements of X are called the *nodes* of the tree; *branches* are maximal linearly ordered subsets of X (i.e. subsets which cannot be extended further).

Trees are supposed to grow upwards; the single node at the bottom is called the *root* or *bottom node* of the tree. If a *branch* of a tree is finite, it ends in a *leaf* or *top node* of the tree. If n, m are nodes of a tree with partial ordering $\prec$, and $n \prec m$, then m is a *successor* of n, n a *predecessor* of m. If $n \prec m$ and there are no nodes properly between n and m, then n is an *immediate* predecessor of m, and m an *immediate* successor of n.

A tree is said to be k-*branching* (*strictly* k-*branching*), if each node has at most k (exactly k) immediate successors.

We also consider labelled trees, with a function assigning objects (e.g. formulas) to the nodes. The terminology for trees is also applied to labelled trees. ⊠

1.1.9. DEFINITION. The *length* or *size* of a finite tree is the number of nodes in the tree. We write $\mathrm{s}(\mathcal{T})$ for the size of $\mathcal{T}$.

The *depth (of a tree)* or *height (of a tree)* $|\mathcal{T}|$ of a tree $\mathcal{T}$ is the maximum length of the branches in the tree, where the *length* of a branch is the number of nodes in the branch minus 1.

The *leafsize* $\mathrm{ls}(\mathcal{T})$ of a tree $\mathcal{T}$ is the number of top nodes of the tree. ⊠

For future use we note: Let $\mathcal{T}$ be a tree which is at most k-branching, i.e. each node has at most k ($k \geq 1$) immediate successors. Then

$$\mathrm{s}(\mathcal{T}) \leq k^{|\mathcal{T}|+1}, \quad \mathrm{ls}(\mathcal{T}) \leq \mathrm{s}(\mathcal{T}).$$

For strictly 2-branching trees $\mathrm{s}(\mathcal{T}) = 2\mathrm{ls}(\mathcal{T}) - 1$.

Formulas may also be regarded as (labelled) trees. The definitions of size and depth specialized to formulas yield the following definition.

DEFINITION. The *depth* $|A|$ of a formula A is the maximum length of a branch in its construction tree. In other words, we define recursively $|P| = 0$ for atomic P, $|\bot| = 0$, $|A \circ B| = \max(|A|, |B|) + 1$ for binary operators $\circ$, $|\circ A| = |A| + 1$ for unary operators $\circ$.

The *size* or *length* $\mathrm{s}(A)$ of a formula A is the number of occurrences of logical symbols and atomic formulas (parentheses not counted) in A: $\mathrm{s}(P) = 1$ for P atomic, $\mathrm{s}(\bot) = 0$, $\mathrm{s}(A \circ B) = \mathrm{s}(A) + \mathrm{s}(B) + 1$ for binary operators $\circ$, $\mathrm{s}(\circ A) = \mathrm{s}(A) + 1$ for unary operators $\circ$. ⊠

For formulas we therefore have

$$\mathrm{s}(A) \leq 2^{|A|+1}.$$

1.2 Simple type theories

This section briefly describes typed combinatory logic and typed lambda calculus, and may be skipped by readers already familiar with simple type theories. For more detailed information on type theories, see Barendregt [1992], Hindley [1997]. Below, we consider only formalisms with *rigid typing*, i.e. systems where every term and all subterms of a term carry a fixed type. Hindley [1997] deals with systems of *type assignment*, where untyped terms are assigned types according to certain rules. The untyped terms may possess many different types, or no type at all. There are many parallels between rigidly typed systems and type-assignment systems, but in the theory of type assignment there is a host of new questions, sometimes very subtle, to study. But theories of type assignment fall outside the scope of this book.

1.2.1. DEFINITION. (*The set of simple types*) The set of *simple types* $\mathcal{T}_\to$ is constructed from a countable set of *type variables* $P_0, P_1, P_2, \ldots$ by means of a type-forming operation (*function-type constructor*) $\to$. In other words, simple types are generated by two clauses:

(i) type variables belong to $\mathcal{T}_\to$;

(ii) if $A, B \in \mathcal{T}_\to$, then $(A \to B) \in \mathcal{T}_\to$.

A type of the form $A \to B$ is called a *function type*. "Generated" means that nothing belongs to $\mathcal{T}_\to$ except on the basis of (i) and (ii). Since the types have the form of propositional formulas, we can use the same abbreviations in writing types as in writing formulas (cf. 1.1.1). ⊠

Intuitively, types denote special sets. We may think of the type variables as standing for arbitrary, unspecified sets, and given types A, B, the type $A \to B$ is a set of functions from A to B.

1.2.2. DEFINITION. (*Terms of the simply typed lambda calculus* $\boldsymbol{\lambda}_{\to}$) All terms appear with a type; for terms of type A we use t^A, s^A, r^A, possibly with extra sub- or superscripts. The terms are generated by the following three clauses:

(i) For each $A \in \mathcal{T}_{\to}$ there is a countably infinite supply of variables of type A; for arbitrary variables of type A we use $u^A, v^A, w^A, x^A, y^A, z^A$ (possibly with extra sub- or superscripts);

(ii) if $t^{A\to B}$, s^A are terms of types $A \to B$, A, then $\mathrm{App}(t^{A\to B}, s^A)^B$ is a term of type B;

(iii) if t^B is a term of type B and x^A a variable of type A, then $(\lambda x^A.t^B)^{A\to B}$ is a term of type $A \to B$. ⊠

NOTATION. For $\mathrm{App}(t^{A\to B}, s^A)^B$ we usually write simply $(t^{A\to B}s^A)^B$.

There is a good deal of redundancy in the typing of terms; provided the types of x, t, s are known, the types of $(\lambda x.t)$, (ts) are known and need not be indicated by a superscript. In general, we shall omit type-indications whenever possible without creating confusion. When writing ts it is always assumed that this is a meaningful application, and hence that for suitable A, B the term t has type $A \to B$, s type A.

If the type of the whole term is omitted, we usually simplify (ts) by dropping the outer parentheses and writing simply ts. The abbreviation $t_1t_2\dots t_n$ is defined by recursion on n as $(t_1t_2\dots t_{n-1})t_n$, i.e. $t_1t_2\dots t_n$ is $(\dots((t_1t_2)t_3\dots)t_n)$.

For $\lambda x_1.(\lambda x_2.(\dots(\lambda x_n.t)\dots))$ we write $\lambda x_1x_2\dots x_n.t$. Application binds more strongly than $\lambda x.$, so $\lambda x.tt'$ is $\lambda x.(tt')$, not $(\lambda x.t)t'$.

A frequently used alternative notation for x^A, t^B is $x{:}\,A, t{:}\,B$ respectively. The notations t^A and $t{:}\,A$ are used interchangeably and may occur mixed; readability determines the choice. ⊠

EXAMPLES. $\mathbf{k}_\lambda^{A,B} := \lambda x^A y^B.x^A$, $\mathbf{s}_\lambda^{A,B,C} := \lambda x^{A\to(B\to C)}y^{A\to B}z^A.xz(yz)$.

1.2.3. DEFINITION. The set $\mathrm{FV}(t)$ of variables free in t is specified by:

$$
\begin{aligned}
\mathrm{FV}(x^A) &:= x^A,\\
\mathrm{FV}(ts) &:= FV(t) \cup \mathrm{FV}(s),\\
\mathrm{FV}(\lambda x.t) &:= \mathrm{FV}(t) \setminus \{x\}.
\end{aligned}
$$
⊠

1.2.4. DEFINITION. (*Substitution*) The operation of substitution of a term s for a variable x in a term t (notation $t[x/s]$) may be defined by recursion on the complexity of t, as follows.

$$\begin{array}{ll} x[x/s] & := s, \\ y[x/s] & := y \text{ for } y \not\equiv x, \\ (t_1t_2)[x/s] & := t_1[x/s]t_2[x/s], \\ (\lambda x.t)[x/s] & := \lambda x.t, \\ (\lambda y.t)[x/s] & := \lambda y.t[x/s] \text{ for } y \not\equiv x;\ \text{w.l.o.g. } y \notin \mathrm{FV}(s). \end{array}$$

A similar definition may be given for simultaneous substitution $t[\vec{x}/\vec{s}]$. ⊠

LEMMA. *(Substitution lemma) If $x \not\equiv y$, $x \notin \mathrm{FV}(t_2)$, then*

$$t[x/t_1][y/t_2] \equiv t[y/t_2][x/t_1[y/t_2]].$$

PROOF. By induction on the depth of t. ⊠

1.2.5. DEFINITION. (*Conversion, reduction, normal form*) Let T be a set of terms, and let conv be a binary relation on T, written in infix notation: t conv s. If $t \operatorname{conv} s$, we say that t *converts to* s; t is called a *redex* or *convertible* term, and s the *conversum* of t. The replacement of a redex by its conversum is called a *conversion*. We write $t \succ_1 s$ (t *reduces in one step to* s) if s is obtained from t by replacement of (an occurrence of) a redex t' of t by a conversum t'' of t', i.e. by a single conversion. The relation $\succ$ ("*properly reduces to*") is the transitive closure of $\succ_1$ and $\succeq$ ("*reduces to*") is the reflexive and transitive closure of $\succ_1$. The relation $\succeq$ is said to be the notion of reduction *generated* by cont. $\prec_1, \prec, \preceq$ are the relations converse to $\succ_1, \succ, \succeq$ respectively.

With the notion of reduction generated by conv we associate a relation on T called *conversion equality*: $t =_{\mathrm{conv}} s$ (t *is equal by conversion to* s) if there is a sequence $t_0, \ldots, t_n$ with $t_0 \equiv t$, $t_n \equiv s$, and $t_i \preceq t_{i+1}$ or $t_i \succeq t_{i+1}$ for each i, $0 \leq i < n$. The subscript "conv" is usually omitted when clear from the context.

A term t is *in normal form*, or t is *normal*, if t does not contain a redex. t *has a normal form* if there is a normal s such that $t \succeq s$.

A *reduction sequence* is a (finite or infinite) sequence of pairs (t_0, δ_0), (t_1, δ_1), (t_2, δ_2), ... with δ_i an (occurrence of a) redex in t_i and $t_i \succ t_{i+1}$ by conversion of δ_i, for all i. This may be written as

$$t_0 \overset{\delta_0}{\succ_1} t_1 \overset{\delta_1}{\succ_1} t_2 \overset{\delta_2}{\succ_1} \ldots.$$

We often omit the δ_i, simply writing $t_0 \succ_1 t_1 \succ_1 t_2 \ldots$.

Finite reduction sequences are partially ordered under the initial part relation ("sequence σ is an initial part of sequence τ"); the collection of finite reduction sequences starting from a term t forms a tree, the *reduction tree* of t. The branches of this tree may be identified with the collection of all infinite and all terminating finite reduction sequences.

A term is *strongly normalizing* (is SN) if its reduction tree is finite. ⊠

REMARKS. (i) As to the terminology, in the literature on lambda calculus and combinatory logic, writers use mostly "contraction", "contracts", "contractum", instead of "conversion", "converts", "conversum". In the lambda calculus literature "conversion" is used for a more general notion: there t converts to s if t and s can be shown to be equal by reduction steps (going in both directions). On the other hand, there is a tradition, deriving from Prawitz [1965], of using "conversion" instead of "contraction" for the corresponding notion applied to natural deductions.

Moreover, "contraction" is also widely used in the literature on Gentzen systems (to be discussed later) for a specific deduction rule, whereas the notion of "conversion" of the lambda calculus literature is hardly used here. Therefore after prolonged hesitation we have chosen the terminology adopted here.

(ii) Usually it is more convenient to think of the reduction tree of a term t as a tree with its nodes labelled with terms; t is put at the root, and if s is the label of the node ν, there is, for each pair (s', δ) such that $s \overset{\delta}{\succ}_1 s'$, an immediate successor ν' to ν, with label s'.

Instead of the notion defined above, we may also consider a less refined notion of reduction sequence by disregarding the redexes; that is to say, we identify sequences

$$t_0 \overset{\delta_0}{\succ}_1 t_1 \overset{\delta_1}{\succ}_1 t_2 \overset{\delta_2}{\succ}_1 \ldots \quad \text{and} \quad t'_0 \overset{\epsilon_0}{\succ}_1 t'_1 \overset{\epsilon_1}{\succ}_1 t'_2 \overset{\epsilon_2}{\succ}_1 \ldots$$

if $t_i = t'_i$ for all i. The notion of reduction tree is then changed accordingly. The arguments in this book using reduction sequences hold with both notions of reduction sequence.

NOTATION. We shall distinguish different conversion relations by subscripts; so we have, for example, cont_β, $\text{cont}_{\beta\eta}$ (to be defined below). Similarly for the associated relations of one-step reduction: $\succ_{\beta,1}$, $\succ_\beta$, $\succeq_\beta$, etc. We write $=_\beta$ instead of $=_{\text{cont}_\beta}$ etc. ⊠

1.2.6. EXAMPLES. For us, the most important reduction is the one induced by β-conversion:

$$(\lambda x^A.t^B)s^A \ \ \text{cont}_\beta \ \ t^B[x^A/s^A].$$

η-conversion is given by

$$\lambda x^A.tx \ \ \text{cont}_\eta \ \ t \qquad (x \notin \text{FV}(t)).$$

$\beta\eta$-conversion $\text{cont}_{\beta\eta}$ is $\text{cont}_\beta \cup \text{cont}_\eta$.

It is to be noted that in defining $\succ_{\beta,1}, \succ_{\beta\eta,1}$ conversion of redexes occurring within the scope of a λ-abstraction operator is permitted. However, no free variables may become bound when executing the substitution in a β-conversion. An example of a reduction sequence is the following:

$$\begin{array}{ll}(\lambda xyz.xz(yz))(\lambda uv.u)(\lambda u'v'.u') & \succ_{\beta,1} \\ (\lambda yz.(\lambda uv.u)z(yz))(\lambda u'v'.u') & \succ_{\beta,1} \\ (\lambda yz.(\lambda v.z)(yz))(\lambda u'v'.u') & \succ_{\beta,1} \\ (\lambda yz.z)(\lambda u'v'.u') & \succ_{\beta,1} \\ \lambda z.z. & \end{array}$$

Relative to $\text{cont}_{\beta\eta}$ conversion of different redexes may yield the same result: $(\lambda x.yx)z \succ_1 yz$ either by converting the β-redex $(\lambda x.yx)z$ or by converting the η-redex $\lambda x.yx$. So here the crude and the more refined notion of reduction sequence, mentioned above, differ.

DEFINITION. A relation R is said to be *confluent*, or to have the *Church–Rosser property (CR)*, if, whenever $t_0 \, R \, t_1$ and $t_0 \, R \, t_2$, then there is a t_3 such that $t_1 \, R \, t_3$ and $t_2 \, R \, t_3$. A relation R is said to be *weakly confluent*, or to have the *weak Church–Rosser property (WCR)*, if, whenever $t_0 \, R \, t_1, t_0 \, R \, t_2$ then there is a t_3 such that $t_1 \, R^* \, t_3$, $t_2 \, R^* \, t_3$, where R^* is the reflexive and transitive closure of R. ⊠

1.2.7. THEOREM. *For a confluent reduction relation $\succeq$ the normal forms of terms are unique. Furthermore, if $\succeq$ is a confluent reduction relation we have: $t = t'$ iff there is a term t'' such that $t \succeq t''$ and $t' \succeq t''$.*

PROOF. The first claim is obvious. The second claim is proved as follows. If $t = t'$ (for the equality induced by $\succeq$), then by definition there is a chain $t \equiv t_0, t_1, \ldots, t_n \equiv t'$, such that for all $i < n$ $t_i \succeq t_{i+1}$ or $t_{i+1} \succeq t_i$. The existence of the required t'' is now established by induction on n. Consider the step from n to $n+1$. By induction hypothesis there is an s such that $t_0 \succeq s$, $t_n \succeq s$. If $t_{n+1} \succeq t_n$, take $t'' = s$; if $t_n \succeq t_{n+1}$, use the confluence to find a t'' such that $s \succeq t''$ and $t_{n+1} \succeq t''$. ⊠

1.2.8. THEOREM. *(Newman's lemma) Let $\succeq$ be the transitive and reflexive closure of $\succ_1$, and let $\succ_1$ be weakly confluent. Then the normal form w.r.t. $\succ_1$ of a strongly normalizing t is unique. Moreover, if all terms are strongly normalizing w.r.t. $\succ_1$, then the relation $\succeq$ is confluent.*

PROOF. Assume WCR, and let us write $s \in \text{UN}$ to indicate that s has a unique normal form. If a term is strongly normalizing, then so are all terms occurring in its reduction tree. In order to show that a strongly normalizing t has a unique normal form (and hence satisfies CR), we argue by contradiction.

We show that if $t \in$ SN, $t \notin$ UN, then we can find a $t_1 \prec t$ with $t_1 \notin$ UN. Repeating this construction leads to an infinite sequence $t \succ t_1 \succ t_2 \succ \ldots$ contradicting the strong normalizability of t.

So let $t \in$ SN, $t \notin$ UN. Then there are two reduction sequences $t \succ_1 t'_1 \succ t'_2 \succ \ldots \succ_1 t'$ and $t \succ_1 t''_1 \succ_1 t''_2 \succ \ldots \succ_1 t''$ with t', t'' distinct normal terms. Then either $t'_1 = t''_1$, or $t'_1 \neq t''_1$. In the first case we can take $t_1 := t'_1 = t''_1$. In the second case, by WCR we can find a t^* such that $t^* \prec t'_1, t''_1$; $t \in$ SN, hence $t^* \succeq t'''$ for some normal t'''. Since $t' \neq t'''$ or $t'' \neq t'''$, either $t'_1 \notin$ UN or $t''_1 \notin$ UN; so take $t_1 := t'_1$ if $t' \neq t'''$, $t_1 := t''_1$ otherwise. The final statement of the theorem follows immediately. ⊠

1.2.9. DEFINITION. The *simple typed lambda calculus* $\boldsymbol{\lambda}_\to$ is the calculus of β-reduction and β-equality on the set of terms of $\boldsymbol{\lambda}_\to$ defined in 1.2.2. More explicitly, $\boldsymbol{\lambda}_\to$ has the term system as described, with the following axioms and rules for $\preceq$ (is $\preceq_\beta$) and $=$ (is $=_\beta$):

$$t \succeq t \qquad (\lambda x^A.t^B)s^A \succeq t^B[x^A/s^A]$$

$$\frac{t \succeq s}{rt \succeq rs} \qquad \frac{t \succeq s}{tr \succeq sr} \qquad \frac{t \succeq s}{\lambda x.t \succeq \lambda x.s} \qquad \frac{t \succeq s \quad s \succeq r}{t \succeq r}$$

$$\frac{t \succeq s}{t = s} \qquad \frac{t = s}{s = t} \qquad \frac{t = s \quad s = r}{t = r}$$

The *extensional simple typed lambda calculus* $\boldsymbol{\lambda\eta}_\to$ is the calculus of $\beta\eta$-reduction and $\beta\eta$-equality $=_{\beta\eta}$ and the set of terms $\boldsymbol{\lambda}_\to$; in addition to the axioms and rules already stated for the calculus $\boldsymbol{\lambda}_\to$ there is the axiom

$$\lambda x.tx \succeq t \quad (x \notin \mathrm{FV}(t)).$$ ⊠

1.2.10. LEMMA. *(Substitutivity of $\succeq_\beta$ and $\succeq_{\beta\eta}$) For $\succeq$ either $\succeq_\beta$ or $\succeq_{\beta\eta}$ we have*

if $s \succeq s'$ then $s[y/s''] \succeq s'[y/s'']$.

PROOF. By induction on the depth of a proof of $s \succeq s'$. It suffices to check the crucial basis step, where s is $(\lambda x.t)t'$, and s' is $t[x/t']$: $(\lambda x.t)t'[y/s''] = (\lambda x.(t[y/s''])t'[y/s''] = t[y/s''][x/t'[y/s'']] = t[x/t'][y/s'']$ using (1.2.4). Here it is assumed that $x \not\equiv y$, $x \notin \mathrm{FV}(s'')$ (if not, rename x). ⊠

1.2.11. PROPOSITION. *$\succ_{\beta,1}$ and $\succ_{\beta\eta,1}$ are weakly confluent.*

PROOF. By distinguishing cases. If the conversions leading from t to t' and from t to t'' concern disjoint redexes, then t''' is simply obtained by converting both redexes. More interesting are the cases where the redexes are nested.

If $t \equiv \dots(\lambda x.s)s' \dots$, $t' \equiv \dots s[x/s'] \dots$, and $t'' \equiv \dots(\lambda x.s)s'' \dots$, $s' \succ_1 s''$, then $t''' \equiv \dots s[x/s''] \dots$, and $t' \succeq t'''$ in as many steps as there are occurrences of x in s, $t'' \succeq t'''$ in a single step.

If $t \equiv \dots(\lambda x.s)s' \dots$, $t' \equiv \dots s[x/s'] \dots$, and $t'' \equiv \dots(\lambda x.s'')s' \dots$, $s \succ_1 s''$, then $t''' \equiv \dots s''[x/s'] \dots$. Here we have to use the fact that if $s \succeq s''$, then $s[x/s'] \succeq s''[x/s']$, i.e. the compatibility of reduction with substitution.

The cases involving η-conversion we leave to the reader. ⊠

1.2.12. Theorem. *The terms of* $\boldsymbol{\lambda}_{\to}$, $\boldsymbol{\lambda\eta}_{\to}$ *are strongly normalizing for* $\succeq_\beta$ *and* $\succeq_{\beta\eta}$ *respectively, and hence the* β*- and* $\beta\eta$*-normal forms are unique.*

Proof. For $\succeq_\beta$ and $\succeq_{\beta\eta}$ see sections 6.8 and 8.3 respectively. ⊠

From the preceding theorem it follows that the reduction relations are confluent. This can also be proved directly, without relying on strong normalization, by the following method, due to W. W. Tait and P. Martin-Löf (see Barendregt [1984, 3.2]) which also applies to the untyped lambda calculus. The idea is to prove confluence for a relation $\succeq_{\mathrm{p}}$ which intuitively corresponds to conversion of a finite set of redexes such that in case of nesting the inner redexes are converted before the outer ones.

1.2.13. Definition. $\succeq_{\mathrm{p}}$ on $\boldsymbol{\lambda}_{\to}$ is generated by the axiom and rules

(id) $x \succeq_{\mathrm{p}} x$

$$(\lambda\text{mon})\ \frac{t \succeq_{\mathrm{p}} t'}{\lambda x.t \succeq_{\mathrm{p}} \lambda x.t'} \qquad (\text{appmon})\ \frac{t \succeq_{\mathrm{p}} t' \quad s \succeq_{\mathrm{p}} s'}{ts \succeq_{\mathrm{p}} t's'}$$

$$(\beta\text{par})\ \frac{t \succeq_{\mathrm{p}} t' \quad s \succeq_{\mathrm{p}} s'}{(\lambda x.t)s \succeq_{\mathrm{p}} t'[x/s']} \qquad (\eta\text{par})\ \frac{t \succeq_{\mathrm{p}} t'}{\lambda x.tx \succeq_{\mathrm{p}} t'}\ (x \notin \mathrm{FV}(t))$$ ⊠

We need some lemmas.

1.2.14. Lemma. *(Substitutivity of* $\succeq_{\mathrm{p}}$*) If* $t \succeq_{\mathrm{p}} t'$, $s \succeq_{\mathrm{p}} s'$, *then* $t[x/s] \succeq_{\mathrm{p}} t'[x/s']$.

Proof. By induction on $|t|$. Assume, without loss of generality, $x \notin \mathrm{FV}(s)$. We consider one case and leave others to the reader. Let $t \equiv (\lambda y.t_1)t_2$ and assume (induction hypothesis):

if $t_1 \succeq_{\mathrm{p}} t_1'$ and $s \succeq_{\mathrm{p}} s'$, then $t_1[x/s] \succeq_{\mathrm{p}} t_1'[x/s']$,
if $t_2 \succeq_{\mathrm{p}} t_2'$ and $s \succeq_{\mathrm{p}} s'$, then $t_2[x/s] \succeq_{\mathrm{p}} t_2'[x/s']$.

Then

$t \succeq_{\mathrm{p}} t_1'[y/t_2']$, and
$t[x/s] \equiv (\lambda y.t_1[x/s])t_2[x/s] \succeq_{\mathrm{p}} t_1'[x/s'][y/t_2'[x/s']]$

by the IH, and by 1.2.4 this last expression is $(t_1'[y/t_2'])[x/s']$. ⊠

1.2.15. LEMMA. $\succeq_p$ *is confluent.*

PROOF. By induction on $|t|$ we show: for all t', t'', if $t \succeq_p t', t''$ then there is a t''' such that $t', t'' \succeq_p t'''$.
Case 1. If $t \succeq_p t', t''$ by application of the same clause in the definition of $\succeq_p$, the claim follows immediately from the IH, using 1.2.14 in the case of βpar.
Case 2. Let

$$t \equiv \lambda x.t_0 x \succeq_p \lambda x.t'_0 x, \text{ where } t_0 \succeq_p t'_0 \ (\lambda\text{mon}), \text{ and}$$
$$t \succeq_p t''_0, \text{ where } t_0 \succeq_p t''_0 \ (\eta\text{par}).$$

Apply the IH to $t_0 \succeq_p t'_0, t''_0$ to find t'''_0 such that $t'_0, t''_0 \succeq_p t'''_0$. We can then take $t''' \equiv t'''_0$.
Case 3. Let

$$t \equiv \lambda x.(\lambda x.t_0)x \succeq_p \lambda x.t'_0, \text{ where } t_0 \succeq_p t'_0 \ (\beta\text{par}, \lambda\text{mon}), \text{ and}$$
$$t \succeq_p t''_0, \text{ where } \lambda x.t_0 \succeq_p t''_0 \ (\eta\text{par}).$$

Then $\lambda x.t_0 \succeq_p \lambda x.t'_0$; since $|\lambda x.t'_0| < |t|$, the IH applies and there is t'''_0 such that $\lambda x.t'_0, t''_0 \succeq_p t'''_0$. Then we can put $t''' \equiv t'''_0$.
Case 4. Let

$$t \equiv (\lambda x.t_0)t_1 \succeq_p (\lambda x.t'_0)t'_1,$$
$$\text{where } t_0 \succeq_p t'_0, \ \ t_1 \succeq_p t'_1 \ (\lambda\text{mon, appmon}), \text{ and}$$
$$t \succeq_p t''_0[x/t''_1], \text{ where } t_0 \succeq_p t''_0, \ \ t_1 \succeq_p t''_1 \ (\beta\text{par}).$$

By the IH we find t'''_0, t'''_1 such that

$$t'_0, t''_0 \succeq_p t'''_0, \quad t'_1, t''_1 \succeq_p t'''_1;$$

then

$$(\lambda x.t'_0)t'_1 \succeq_p t'''_0[x/t'''_1] \ (\beta\text{par}) \text{ and}$$
$$t''_0[x/t''_1] \succeq_p t'''_0[x/t'''_1] \ (\text{substitutivity of } \succeq_p).$$

Take $t''' \equiv t'''_0[x/t'''_1]$.
Case 5. Let

$$t \equiv (\lambda x.t_0 x)t_1 \succeq_p t'_0 t'_1,$$
$$\text{where } t_0 \succeq_p t'_0, \ \ t_1 \succeq_p t'_1, \ \ x \notin \mathrm{FV}(t_0) \ (\eta\text{par, appmon}), \text{ and}$$
$$t \succeq_p t''_0[x/t''_1], \text{ where } t_0 x \succeq_p t''_0, \ \ t_1 \succeq_p t''_1 \ (\beta\text{par}).$$

Also $t_0 x \succeq_p t'_0 x$. Apply the IH to $t_0 x$ (which is possible since $|t_0 x| < |t|$) to find t'''_0 with

$$t'_0 x, t''_0 \succeq_p t'''_0,$$

and apply the IH to t_1 to find t_1''' such that

$$t_1', t_1'' \succeq_{\mathrm{p}} t_1'''.$$

Then

$$t_0' t_1 \equiv (t_0' x)[x/t_1'] \succeq_{\mathrm{p}} (t_0''' x)[x/t_1'''] \equiv t_0''' t_1''' \text{ and}$$
$$t_0''[x/t_1''] \succeq_{\mathrm{p}} t_0'''[x/t_1''']$$

(both by substitutivity of $\succeq_{\mathrm{p}}$). ⊠

1.2.16. THEOREM. *β- and βη-reduction are confluent.*

PROOF. The reflexive closure of $\succ_1$ for $\beta\eta$-reduction is contained in $\succeq_{\mathrm{p}}$, and $\succeq$ is therefore the transitive closure of $\succeq_{\mathrm{p}}$. Write $t \succeq_{\mathrm{p},n} t'$ if there is a chain $t \equiv t_0 \succeq_{\mathrm{p}} t_1 \succeq_{\mathrm{p}} t_2 \succeq_{\mathrm{p}} \ldots \succeq_{\mathrm{p}} t_n \equiv t'$. Then we show by induction on $n + m$, using the preceding lemma, that if $t \succeq_{\mathrm{p},n} t'$, $t \succeq_{\mathrm{p},m} t''$ then there is a t''' such that $t' \succeq_{\mathrm{p},m} t'''$, $t'' \succeq_{\mathrm{p},n} t'''$. ⊠

1.2.17. *Typed combinatory logic*

We now turn to the description of (simple) typed combinatory logic, which is an analogue of $\boldsymbol{\lambda}_{\to}$ without bound variables.

DEFINITION. (*Terms of typed combinatory logic* $\mathbf{CL}_{\to}$) The terms are inductively defined as in the case of $\boldsymbol{\lambda}_{\to}$, but now with the clauses

(i) For each $A \in \mathcal{T}_{\to}$ there is a countably infinite supply of variables of type A; for arbitrary variables of type A we use $u^A, v^A, w^A, x^A, y^A, z^A$ (possibly with extra sub- or superscripts)

(ii) for all $A, B, C \in \mathcal{T}$ there are constant terms

$$\mathbf{k}^{A,B} \in A \to (B \to A),$$
$$\mathbf{s}^{A,B,C} \in (A \to (B \to C)) \to ((A \to B) \to (A \to C));$$

(iii) if $t^{A \to B}$, s^A are terms of the types shown, then $\mathrm{App}(t^{A\to B}, s^A)^B$ is a term of type B.

Conventions for notation remain as before. Free variables are defined as in $\boldsymbol{\lambda}_{\to}$, but of course we put $\mathrm{FV}(\mathbf{k}) = \mathrm{FV}(\mathbf{s}) = \emptyset$. ⊠

DEFINITION. The *weak reduction* relation $\succeq_{\rm w}$ on the terms of $\mathbf{CL}_\to$ is generated by a conversion relation $\mathrm{cont}_{\rm w}$ consisting of the following pairs:

$$\mathbf{k}^{A,B}x^Ay^B \ \mathrm{cont}_{\rm w}\ x, \quad \mathbf{s}^{A,B,C}x^{A\to(B\to C)}y^{A\to B}z^A \ \mathrm{cont}_{\rm w}\ xz(yz).$$

In other words, $\mathbf{CL}_\to$ is the term system defined above with the following axioms and rules for $\succeq_{\rm w}$ and $=_{\rm w}$ (abbreviated to $\succeq, =$):

$$t \succeq t \qquad \mathbf{k}xy \succeq x \qquad \mathbf{s}xyz \succeq xz(yz)$$

$$\frac{t \succeq s}{rt \succeq rs} \qquad \frac{t \succeq s}{tr \succeq sr} \qquad \frac{t \succeq s \quad s \succeq r}{t \succeq r}$$

$$\frac{t \succeq s}{t = s} \qquad \frac{t = s}{s = t} \qquad \frac{t = s \quad s = r}{t = r}$$

⊠

1.2.18. THEOREM. *The weak reduction relation in $\mathbf{CL}_\to$ is confluent and strongly normalizing, so normal forms are unique.*

PROOF. Similar to the proof for $\boldsymbol{\lambda}_\to$, but easier (cf. 6.8.6). ⊠

1.2.19. The effect of lambda-abstraction can be achieved to some extent in $\mathbf{CL}_\to$, as shown by the following theorem.

THEOREM. *To each term t in $\mathbf{CL}_\to$ there is another term $\lambda^*x^A.t$ such that*

(i) $x^A \notin \mathrm{FV}(\lambda^*x^A.t)$,

(ii) $(\lambda^*x^A.t)s^A \succ_{\rm w} t[x^A/s^A]$.

PROOF. We define $\lambda^*x^A.t$ by recursion on the construction of t:

$$\lambda^*x^A.x := \mathbf{s}^{A,A\to A,A}\mathbf{k}^{A,A\to A}\mathbf{k}^{A,A},$$

$$\lambda^*x^A.y^B := \mathbf{k}^{B,A}y^B \text{ for } y \not\equiv x,$$

$$\lambda^*x^A.t_1^{B\to C}t_2^B := \mathbf{s}^{A,B,C}(\lambda^*x.t_1)(\lambda^*x.t_2).$$

The properties stated in the theorem are now easily verified by induction on the complexity of t. ⊠

COROLLARY. $\mathbf{CL}_\to$ *is combinatorially complete, i.e. for every applicative combination t of* $\mathbf{k}$, $\mathbf{s}$ *and variables $x_1, x_2, \ldots, x_n$ there is a closed term s such that in $\mathbf{CL}_\to \vdash sx_1\ldots x_n =_{\rm w} t$, in fact even $\mathbf{CL}_\to \vdash sx_1\ldots x_n \succeq_{\rm w} t$.* ⊠

REMARK. Note that the defined abstraction operator $\lambda^* x$ fails to have an important property of λx: it is not true that, if $t = t'$, then $\lambda^* x.t = \lambda^* x.t'$. Counterexample (dropping all type indications): $\mathbf{k}x\mathbf{k} = x$, but $\lambda^* x.\mathbf{k}x\mathbf{k} = \mathbf{s}(\mathbf{s}(\mathbf{kk})(\mathbf{skk}))(\mathbf{kk})$, $\lambda^* x.x = \mathbf{skk}$. The latter two terms are both in weak normal form, but distinct; hence by the theorem of the uniqueness of normal form, they cannot be proved to be equal in $\mathbf{CL}_{\rightarrow}$.

1.2.19A. ♠ Consider the following variant $\lambda^\circ x.t$ of $\lambda^* x.t$: $\lambda^\circ x.x := \mathbf{skk}$, $\lambda^\circ x.t := \mathbf{k}t$ if $x \notin \mathrm{FV}(t)$, $\lambda^\circ x.tx := t$ if $x \notin \mathrm{FV}(t)$, $\lambda^\circ x.ts := \mathbf{s}(\lambda^\circ x.t)(\lambda^\circ x.s)$ if the preceding clauses do not apply. Show that this alternative defined abstraction operator has the properties mentioned in the theorem above, and in addition

(iii) $\lambda^\circ x.tx \succeq_{\mathrm{w}} t$ if $x \notin t$,

(iv) $(\lambda^\circ x.t)[y/s] = \lambda^\circ x.t[y/s]$ if $y \not\equiv x$, $x \notin \mathrm{FV}(s)$.

Show by examples that $\lambda^* x.t$ does not have these properties in general. Also, verify that it is still not true that if $t = t'$, then $\lambda^\circ x.t = \lambda^\circ x.t'$.

1.2.20. *Computational content*

In the remainder of this section we shall show that there is some "computational content" in simple type theory: for a suitable representation of the natural numbers we can represent certain number-theoretic functions. This will be utilized in 6.9.2 and 11.2.2.

DEFINITION. The *Church numerals* of type A are β-normal terms $\overline{n}_A$ of type $(A \rightarrow A) \rightarrow (A \rightarrow A)$, $n \in \mathbb{N}$, defined by

$$\overline{n}_A := \lambda f^{A \rightarrow A} \lambda x^A . f^n(x),$$

where $f^0(x) := x$, $f^{n+1}(x) := f(f^n(x))$. N_A is the set of all the $\overline{n}_A$. ⊠

N.B. If we want to use $\beta\eta$-normal terms, we must use $\lambda f^{A \rightarrow A}.f$ instead of $\lambda fx.fx$ for $\overline{1}_A$.

DEFINITION. A function $f \colon \mathbb{N}^k \rightarrow \mathbb{N}$ is said to be *A-representable* if there is a term F of $\boldsymbol{\lambda}_{\rightarrow}$ such that (abbreviating $\overline{n}_A$ as $\overline{n}$)

$$F\overline{n}_1 \ldots \overline{n}_k = \overline{f(n_1 \ldots n_k)}$$

for all $n_1, \ldots, n_k \in \mathbb{N}$, $\overline{n}_i = (\overline{n}_i)_A$. ⊠

DEFINITION. *Polynomials, extended polynomials*

(i) The n-argument *projections* $\mathbf{p}_i^n$ are given by $\mathbf{p}_i^n(x_1, \ldots, x_n) = x_i$, the unary constant functions $\mathbf{c}_m$ by $\mathbf{c}_m(x) = m$, and sg, $\overline{\text{sg}}$ are unary functions which satisfy $\text{sg}(Sn) = 1$, $\text{sg}(0) = 0$, $\overline{\text{sg}}(Sn) = 0$, $\overline{\text{sg}}(0) = 1$, where S is the successor function.

(ii) The n-argument function f is the *composition* of m-argument g, n-argument $h_1, \ldots, h_m$ if f satisfies $f(\vec{x}) = g(h_1(\vec{x}), \ldots, h_m(\vec{x}))$.

(iii) The *polynomials* in n variables are generated from $\mathbf{p}_i^n$, $\mathbf{c}_m$, addition and multiplication by closure under composition. The *extended polynomials* are generated from $\mathbf{p}_i^n$, $\mathbf{c}_m$, sg, $\overline{\text{sg}}$, addition and multiplication by closure under composition. ⊠

1.2.20A. ♠ Show that all terms in β-normal form of type $(P \to P) \to (P \to P)$, P a propositional variable, are either of the form $\overline{n}_P$ or of the form $\lambda f^{P \to P}.f$.

1.2.21. THEOREM. *All extended polynomials are representable in* $\boldsymbol{\lambda}_{\to}$.

PROOF. Abbreviate $\mathbb{N}_A$ as N. Take as representing terms for addition, multiplication, projections, constant functions, sg, $\overline{\text{sg}}$:

$$
\begin{aligned}
F_+ &:= \lambda x^N y^N f^{A \to A} z^A . x f(y f z), \\
F_\times &:= \lambda x^N y^N f^{A \to A} . x(y f), \\
F_{\mathbf{p}_i^k} &:= \lambda x_1^N \ldots x_k^N . x_i, \\
F_{\mathbf{c}_n} &:= \lambda x^N . \overline{n}, \\
F_{\text{sg}} &:= \lambda x^N f^{A \to A} z^A . x(\lambda u^A . f z) z, \\
F_{\overline{\text{sg}}} &:= \lambda x^N f^{A \to A} z^A . x(\lambda u^A . z)(f z).
\end{aligned}
$$

It is easy to verify that F_+, $F_\times$ represent addition and multiplication respectively, by showing that

$$(\overline{n}_A f^{A \to A}) \circ (\overline{m}_A f^{A \to A}) =_\beta (\overline{n+m})_A (f^{A \to A}), \quad \overline{n}_A \circ \overline{m}_A =_\beta (\overline{nm})_A,$$

where $f \circ g := \lambda z. f(g(z))$. The proof that the representable functions are closed under composition is left to the reader. ⊠

A proof of the converse of this theorem (in the case where A is a proposition variable) may be found in Schwichtenberg [1976].

REMARK. Extended polynomials are of course majorized (bounded above) by polynomials.

However, if we permit ourselves the use of Church numerals of different types, and in particular liberalize the notion of representation of a function by permitting numerals of different types for the input and the output, we can represent more than extended polynomials. In particular we can express exponentiation

$$\overline{n}_{A\to A}\overline{m}_A = \overline{(m^n)}_A \quad (n > 0).$$

1.2.21A. ♠ Complete the proof of the theorem and verify the remark.

1.3 Three types of formalism

The greater part of this text deals with the theory of the "standard" logics, that is minimal, intuitionistic and classical logic. In this section we introduce the three styles of formalization: natural deduction, Gentzen systems and Hilbert systems. (On the names and history of these types of formalism, see the notes to chapters 2 and 3.) The first two will play a leading role in the sequel; the Hilbert systems are well known and widely used in logic, but less important from the viewpoint of structural proof theory. Each of these formalization styles will be illustrated for implication logic.

Deductions will be presented as trees; the nodes will be labelled with formulas (in the case of natural deduction and Hilbert systems) or with sequents (for the Gentzen system); the labels at the immediate successors of a node ν are the premises of a rule application, the label at ν the conclusion. At the root of the tree we find the conclusion of the whole deduction.

The word *proof* will as a rule be reserved for the meta-level; for formal arguments we preferably use *deduction* or *derivation*. But *prooftree* will mean the same as *deduction tree* or *derivation tree*, and a "natural deduction proof" will be a formal deduction in one of the systems of natural deduction. Rules are schemas; an *instance of a rule* is also called a *rule-application* or *inference.*

If a node C in the underlying tree with say two predecessors and one successor looks like the tree on the left, we represent this more compactly as on the right:

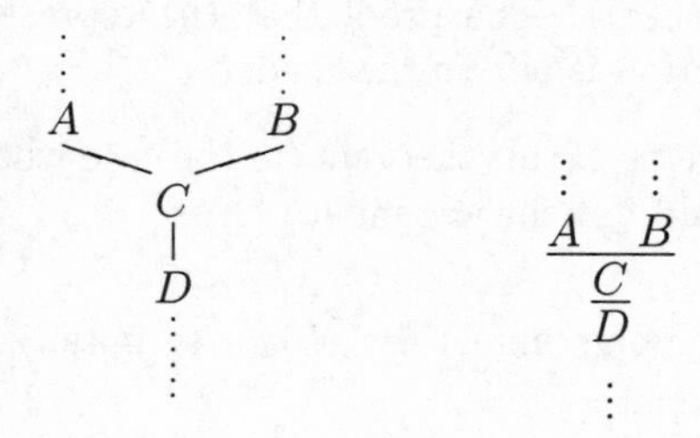

We use script $\mathcal{D}$, $\mathcal{E}$, possibly sub- and/or superscripted, for deductions.

1.3.1. *The BHK-interpretation*

Minimal logic and intuitionistic logic differ only in the treatment of negation, or (equivalently) falsehood, and minimal implication logic is the same as intuitionistic implication logic. The informal interpretation underlying intuitionistic logic is the so-called Brouwer–Heyting–Kolmogorov interpretation (BHK-interpretation for short); this interpretation tells us what it means to prove a compound statement such as $A \to B$ in terms of what it means to prove the components A and B (cf. classical logic, where the truthvalue of $A \to B$ is defined relative to the truthvalues of A and B). As primitive notions in the BHK-interpretation there appear "construction" and "(constructive, informal) proof". These notions are admittedly imprecise, but nevertheless one may convincingly argue that the usual laws of intuitionistic logic hold for them, and that, for our understanding of these primitives, certain principles of classical logic are not valid for the interpretation. We here reproduce the clause for implication only:

> A construction p proves $A \to B$ if p transforms any possible proof q of A into a proof $p(q)$ of B.

A *logical law* of implication logic, according to the BHK-interpretation, is a formula for which we can give a proof, no matter how we interpret the atomic formulas. A *rule* is valid for this interpretation if we know how to construct a proof for the conclusion, given proofs of the premises.

The following two rules for $\to$ are obviously valid on the basis of the BHK-interpretation:

(a) If, starting from a hypothetical (unspecified) proof u of A, we can find a proof $t(u)$ of B, then we have in fact given a proof of $A \to B$ (without the assumption that u proves A). This proof may be denoted by $\lambda u.t(u)$.

(b) Given a proof t of $A \to B$, and a proof s of A, we can apply t to s to obtain a proof of B. For this proof we may write $\mathrm{App}(t, s)$ or ts (t applied to s).

1.3.2. *A natural deduction system for minimal implication logic*

Characteristic for natural deduction is the use of assumptions which may be *closed* at some later step in the deduction. Assumptions are formula occurrences appearing at the top nodes (leaves) of the prooftree; they may be open or closed. Assumptions are provided with markers (a type of label). Any kind of symbol may be used for the markers, but below we suppose the markers to be certain symbols for variables, such as u, v, w, possibly sub- or superscripted.

The assumptions in a deduction which are occurrences of the same formula with the same marker form together an *assumption class*. The notations

$$\begin{array}{c}[A]^u\\ \mathcal{D}\\ B\end{array} \qquad \begin{array}{c}A^u\\ \mathcal{D}\\ B\end{array} \qquad \begin{array}{c}\mathcal{D}'\\ [A]\\ \mathcal{D}\\ B\end{array} \qquad \begin{array}{c}\mathcal{D}'\\ A\\ \mathcal{D}\\ B\end{array}$$

have the following meaning, from left to right: (1) a deduction $\mathcal{D}$ with conclusion B and a set $[A]$ of open assumptions, consisting of all occurrences of the formula A at top nodes of the prooftree $\mathcal{D}$ with marker u (*note*: both B and the $[A]$ are part of $\mathcal{D}$, and we do not talk about the *multi*set $[A]^u$ since we are dealing with formula *occurrences*); (2) a deduction $\mathcal{D}$ with conclusion B and a single assumption of the form A marked u occurring at some top node; (3) deduction $\mathcal{D}$ with a deduction $\mathcal{D}'$, with conclusion A, substituted for the assumptions $[A]^u$ of $\mathcal{D}$; (4) the same, but now for a single assumption occurrence A in $\mathcal{D}$. Under (3) the formula A shown is the conclusion of $\mathcal{D}'$ as well as the formula in an assumption class of $\mathcal{D}$.

In cases (3) and (4) this metamathematical notation may be considered imprecise, since we have not indicated the label of A before substitution. But in practice this will not cause confusion. Note that the marker u disappears by the substitution: only topformulas bear markers.

We now consider a system $\rightarrow$**Nm** for the minimal theory of implication. Prooftrees are constructed according to the following principles.

A single formula occurrence A labelled with a marker is a single-node prooftree, representing a deduction with conclusion A from open assumption A.

There are two rules for constructing new prooftrees from given ones, which correspond precisely to the two principles (a), (b) valid for the BHK-interpretation, mentioned above, and which may be rendered schematically as follows:

$$\dfrac{\begin{array}{c}[A]^u\\ \mathcal{D}\\ B\end{array}}{A \rightarrow B}\rightarrow\mathrm{I},u \qquad\qquad \dfrac{\begin{array}{c}\mathcal{D}\\ A \rightarrow B\end{array} \quad \begin{array}{c}\mathcal{D}'\\ A\end{array}}{B}\rightarrow\mathrm{E}$$

By application of the rule $\rightarrow$I of *implication introduction*, a new prooftree is formed from $\mathcal{D}$ by adding at the bottom the conclusion $A \rightarrow B$ while *closing* the set of open assumptions A marked by u. All other open assumptions remain open in the new prooftree.

The rule $\rightarrow$E of *implication elimination* (also known as *modus ponens*) constructs from two deductions $\mathcal{D}, \mathcal{D}'$ with conclusions $A \rightarrow B, A$ a new combined deduction with conclusion B, which has as open assumptions the open assumptions of $\mathcal{D}$ and $\mathcal{D}'$ combined.

Two occurrences α, β of the same formula belong to the same *assumption class* if they bear the same label and either are both open or have both been closed at the same inference.

It should be noted that in the rule →I the "degenerate case", where $[A]^u$ is empty, is permitted; thus for example the following is a correct deduction:

$$\dfrac{\dfrac{A^u}{B \to A}\,v}{A \to (B \to A)}\,u$$

At the first inference an empty class of occurrences is discharged; we have assigned this "invisible class" a label v, for reasons of uniformity of treatment, but obviously the choice of label is unimportant as long as it differs from all other labels in use; in practice the label at the inference may be omitted in such cases.

In applying the rule →I, we do not assume that $[A]$ consists of *all* open assumptions of the form A occurring above the inference. Consider for example the following two distinct (inefficient) deductions of $A{\to}(A{\to}A)$:

$$\dfrac{\dfrac{\dfrac{\dfrac{A^u}{A \to A}\,v \qquad A^w}{A}}{A \to A}\,u}{A \to (A \to A)}\,w \qquad\qquad \dfrac{\dfrac{\dfrac{\dfrac{A^u}{A \to A}\,u \qquad A^v}{A}}{A \to A}\,v}{A \to (A \to A)}\,w$$

The formula tree in these deductions is the same, but the pattern of closing assumptions differs. In the second deduction *all* assumptions of the given form which are still open before application of an inference →I are closed simultaneously. Deductions which have this property are said to obey the *Complete Discharge Convention*; we shall briefly return to this in 2.1.9. But, no matter how natural this convention may seem if one is interested in deducible formulas, for deductions as combinatorial structures it is an undesirable restriction, as we shall see later.

1.3.3. EXAMPLE.

$$\dfrac{\dfrac{\dfrac{\dfrac{\dfrac{A \to (B \to C)^u \qquad A^w}{B \to C} \qquad \dfrac{A \to B^v \qquad A^w}{B}}{C}}{A \to C}\,w}{(A \to B) \to (A \to C)}\,v}{(A \to (B \to C)) \to ((A \to B) \to (A \to C))}\,u$$

We have not indicated the rules used, since these are evident.

1.3.4. *Formulas-as-types*

As already suggested by the notation, the BHK-valid principles (a) and (b) correspond to function abstraction, and application of a function to an argument respectively. Starting from variables u, v, w associated with assumption formulas, these two principles precisely generate the terms of simple type theory $\boldsymbol{\lambda}_{\rightarrow}$.

Transferring these ideas to the formal rules constructing prooftrees, we see that parallel to the construction of the prooftree, we may write next to each formula occurrence the term describing the proof obtained in the subdeduction with this occurrence as conclusion.

(i) To assumptions A correspond variables of type A; more precisely, formulas with the same marker get the same variable. If we have already used variable symbols as markers, we can use these same variables for the correspondence.

(ii) For the rules →I and →E the assignment of terms to the conclusion, constructed from term(s) for the premise(s), is shown below.

$$\dfrac{\begin{matrix}[u\colon A]\\ \mathcal{D}\\ t\colon B\end{matrix}}{\lambda u^A.t^B\colon A \to B}\,u \qquad\qquad \dfrac{\begin{matrix}\mathcal{D}\\ t\colon A\to B\end{matrix}\quad \begin{matrix}\mathcal{D}'\\ s\colon A\end{matrix}}{(t^{A\to B}s^A)\colon B}$$

Thus there is a very close relationship between $\boldsymbol{\lambda}_{\rightarrow}$ and →**Nm**, which at first comes as a surprise. In fact, the terms of $\boldsymbol{\lambda}_{\rightarrow}$ are nothing else but an alternative notation system for deductions in →**Nm**. That is to say, if we consider just the term assigned to the conclusion of a deduction, and assuming not only the whole term to carry its type, but also all its subterms, the prooftree may be unambiguously reconstructed from this term. This is the basic observation of the formulas-as-types isomorphism, an observation which has proved very fruitful, since it is capable of being extended to many more complicated logical systems on the one hand, and more complicated type theories on the other hand, and permits us to lift results and methods of type theory to logic and vice versa.

By way of illustration, we repeat our previous example, but now at each node of the prooftree we also exhibit the corresponding terms. We have not shown the types of subterms, since these follow readily from the construction of the tree. We have dropped the superscript markers at the assumptions, as well as the repetition of markers at the line where an assumption class is discharged, since these are now redundant.

1.3.5. EXAMPLE.

$$\dfrac{\dfrac{\dfrac{\dfrac{\dfrac{u\colon A\to(B\to C)\quad w\colon A}{uw\colon B\to C}\quad\dfrac{v\colon A\to B\quad w\colon A}{vw\colon B}}{uw(vw)\colon C}}{\lambda w.uw(vw)\colon A\to C}}{\lambda vw.uw(vw)\colon (A\to B)\to(A\to C)}}{\lambda uvw.uw(vw)\colon (A\to(B\to C))\to((A\to B)\to(A\to C))}$$

1.3.6. *Identity of prooftrees.* When are two prooftrees to be regarded as identical? Taking the formulas-as-types isomorphism as our guideline, we can say that two prooftrees are the same, if the corresponding terms of simple type theory are the same (modulo renaming bound variables). Thus the following two prooftrees are to be regarded as identical:

$$\dfrac{\dfrac{\dfrac{(A \to A)^u \quad A^v}{A}}{A \to A}\,v \qquad A^w}{\dfrac{A}{A \to A}\,w} \qquad\qquad \dfrac{\dfrac{\dfrac{(A \to A)^u \quad A^v}{A}}{A \to A}\,v \qquad A^v}{\dfrac{A}{A \to A}\,v}$$

since the first one corresponds to the term $\lambda w.(\lambda v.uv)w$, and the second to $\lambda v.(\lambda v.uv)v$, and these terms are the same modulo renaming of bound variables. On the other hand, the two deductions at the end of 1.3.2 correspond to $\lambda wu.(\lambda v.u)w$ and $\lambda wv.(\lambda u.u)v$ respectively, which are distinct terms.

In the right hand tree, the upper $\to$I closes only the *upper* occurrence A^v; the lower $\to$I only the *lower* occurrence of A^v (since at that place the upper occurrence has already been closed). In other words, the two A^v-occurrences belong to distinct assumption classes, since they are closed at different places.

Without loss of generality we may assume that the labels for distinct assumption classes of the same formula are always distinct, as in the first of the two prooftrees above.

However, there is more to the formulas-as-types isomorphism than just another system of notation. The notion of β-reduction is also meaningful for prooftrees. A β-conversion

$$(\lambda x^A.t^B)s^A \ \text{cont}_\beta \ t^B[x^A/s^A]$$

corresponds to a transformation on prooftrees:

$$\dfrac{\dfrac{\begin{matrix}[A]^u \\ \mathcal{D} \\ B\end{matrix}}{A \to B}\,u \qquad \begin{matrix}\mathcal{D}' \\ A\end{matrix}}{B} \qquad \mapsto \qquad \begin{matrix}\mathcal{D}' \\ [A] \\ \mathcal{D} \\ B\end{matrix}$$

Here the prooftree on the right is the prooftree obtained from $\mathcal{D}$ by replacing all occurrences of A in the class $[A]$ in $\mathcal{D}$ by $\mathcal{D}'$. Note that the f.o. $A \to B$ in the left deduction is a local maximum of complexity, being first introduced, only to be removed immediately afterwards. The conversion may be said to remove a "detour" in the proof. A proof without detours is said to be a *normal* proof. Normal deductions may be said to embody an idea of "direct" proof.

In a normal proof the left (major) premise of $\to$E is never the conclusion of $\to$I. One can show that such normal deductions have the subformula property: if a normal deduction $\mathcal{D}$ derives A from open assumptions Γ, then all formulas occurring in the deduction are subformulas of formulas in Γ, A.

The term notation for deductions is compact and precise, and tells us exactly how we should manage open and closed assumptions when we substitute one prooftree into another one. Using distinct markers for distinct closed assumption classes corresponds to the use of separate variables for each occurrence of a binding operator. The tree notation on the other hand gives us some geometric intuition. It is not so compact, and although in principle we can treat the trees with the same rigour as the terms, it is not always feasible to do so; one is led to the use of suggestive, but not always completely precise, notation. In our discussion of natural deduction we shall extend the term notation to full predicate logic (2.2) and give a notion of reduction for the full system in chapter 6.

1.3.7. *Gentzen systems*

There are two motivations leading to Gentzen systems, which will be discussed below. The first one views a Gentzen system as a metacalculus for natural deduction; this applies in particular to systems for minimal and intuitionistic logic. The second motivation is semantical: Gentzen systems for classical logic are obtained by analysing truth conditions for formulas. This also applies to intuitionistic and minimal logic if we use Kripke semantics instead of classical semantics.

A Gentzen system as a metacalculus. Let us first consider a Gentzen system obtained as a metacalculus for the system $\to$**Nm**. Consider the following four construction steps for prooftrees.

1. The single-node tree with label A, marker u is a prooftree.
2. Add at the bottom of a prooftree an application of $\to$I, discharging an assumption class.
3. Given a prooftree $\mathcal{D}$ with open assumption class $[B]^u$ and a prooftree $\mathcal{D}_1$ deriving A, replace all occurrences of B in $[B]^u$ by

$$\frac{A \to B^v \qquad \begin{matrix}\mathcal{D}_1\\ A\end{matrix}}{B}$$

4. Substitute a deduction of A for the occurrences of an (open) assumption class $[A]^u$ of another deduction.

These construction (or generation) principles suffice to obtain any prooftree of $\to$**Nm**, for the first construction rule gives us the single-node prooftree which derives A from assumption A, the second rule corresponds to applications of $\to$I, and closure under $\to$E is seen as follows: in order to obtain the tree

$$\dfrac{\begin{matrix}\mathcal{D}_1 & \mathcal{D}_2 \\ A \to B & A\end{matrix}}{B}$$

from the prooftrees $\mathcal{D}_1, \mathcal{D}_2$, we first combine the first and third construction principles to obtain

$$\dfrac{\begin{matrix} & \mathcal{D}_2 \\ A \to B^u & A\end{matrix}}{B}$$

and then use the fourth (substitution) principle to obtain the desired tree.

Let $\Gamma \Rightarrow A$ express that A is deducible in $\to$**Nm** from assumptions in Γ. Then the four construction principles correspond to the following axiom and rules for obtaining statements $\Gamma \Rightarrow A$:

$$\Gamma \cup \{A\} \Rightarrow A \text{ (Axiom)}$$

$$\dfrac{\Gamma \cup \{A\} \Rightarrow B}{\Gamma \Rightarrow A \to B}\,\text{R}\!\to \qquad \dfrac{\Gamma \Rightarrow A \qquad \Delta \cup \{B\} \Rightarrow C}{\Gamma \cup \Delta \cup \{A \to B\} \Rightarrow C}\,\text{L}\!\to$$

$$\dfrac{\Gamma \Rightarrow A \qquad \Delta \cup \{A\} \Rightarrow B}{\Gamma \cup \Delta \Rightarrow B}\,\text{Cut}$$

Call the resulting system $\mathcal{S}$ (an ad hoc designation). Here in the sequents $\Gamma \Rightarrow A$ the Γ is treated as a (finite) set. For bookkeeping reasons it is often convenient to work with multisets instead; multisets are "(finite) sets with repetitions", or equivalently, finite sequences modulo the order of the elements. If we rewrite the system above with multisets, we get the Gentzen system described below, which we shall designate ad hoc by $\mathcal{S}'$, and where in the sequents $\Gamma \Rightarrow A$ the Γ is now a multiset. The rules and axioms of $\mathcal{S}'$ are

$$A \Rightarrow A \quad \text{(Axiom)}$$

$$\dfrac{\Gamma \Rightarrow A \qquad \Delta, B \Rightarrow C}{\Gamma, \Delta, A \to B \Rightarrow C}\,\text{L}\!\to \qquad \dfrac{\Gamma, A \Rightarrow B}{\Gamma \Rightarrow A \to B}\,\text{R}\!\to$$

$$\dfrac{\Gamma \Rightarrow A}{\Gamma, B \Rightarrow A}\,\text{LW} \qquad \dfrac{\Gamma, B, B \Rightarrow A}{\Gamma, B \Rightarrow A}\,\text{LC}$$

$$\dfrac{\Gamma \Rightarrow A \qquad A, \Delta \Rightarrow B}{\Gamma, \Delta \Rightarrow B}\,\text{Cut}$$

R$\to$ and L$\to$ are called the logical rules, LW, LC and Cut the structural rules. LC is called the rule of (left-)*contraction*, LW the rule of (left-)*weakening*. Due to the presence of LC and LW, derivability of $\Gamma \Rightarrow A$ is equivalent to derivability of $\text{Set}(\Gamma) \Rightarrow A$ where $\text{Set}(\Gamma)$ is the set underlying the multiset

Γ. We have simplified the axiom, since some applications of LW produce $\Gamma, A \Rightarrow A$ from $A \Rightarrow A$.

If one uses sequences instead of sets, in order to retain equivalence of derivability of $\Gamma \Rightarrow A$ and derivability of $\mathrm{Set}(\Gamma) \Rightarrow A$, an extra rule of exchange then has to be added:

$$\frac{\Gamma, A, B, \Delta \Rightarrow C}{\Gamma, B, A, \Delta \Rightarrow C}\,\mathrm{LE}$$

Example. (Of a deduction in $\mathcal{S}$ and $\mathcal{S}'$)

$$\dfrac{\dfrac{\dfrac{A \Rightarrow A \qquad B \Rightarrow B}{A \to B, A \Rightarrow B}\,\mathrm{L}{\to}}{A \Rightarrow (A \to B) \to B}\,\mathrm{R}{\to}}{\Rightarrow A \to ((A \to B) \to B)}\,\mathrm{R}{\to}$$

The natural deduction of $(A \to (B \to C)) \to ((A \to B) \to (A \to C))$, which we gave earlier in example 1.3.3 may be obtained by repeated use of the generation principles 1–3 (not 4) as follows:

$$\dfrac{\dfrac{\dfrac{\dfrac{\dfrac{3{:}\, A \to (B \to C)^u \qquad 3{:}\, A^w}{1{:}\, B \to C} \qquad \dfrac{2{:}\, A \to B^v \qquad 2{:}\, A^w}{1{:}\, B}}{0{:}\, C}}{4{:}\, A \to C}\,w}{5{:}\, (A \to B) \to (A \to C)}\,v}{6{:}\, (A \to (B \to C)) \to ((A \to B) \to (A \to C))}\,u$$

In the displayed tree, the numbers 0–6 indicate the seven steps in the construction of the tree. The number 0 corresponds to an application of construction principle 1, the numbers 4–6 to applications of principle 2, the numbers 1 and 3 to applications of principle 3, and number 2 to an application of principle 3 in the construction of the subtree with conclusion B. We can now readily transform this into a sequent deduction in $\mathcal{S}$:

$$\dfrac{\dfrac{\dfrac{\dfrac{A \Rightarrow A \qquad \dfrac{A \Rightarrow A \qquad \dfrac{B \Rightarrow B \quad C \Rightarrow C\,(0)}{B, B \to C \Rightarrow C\,(1)}}{A, A \to B, B \to C \Rightarrow C\,(2)}}{A, A \to B, A \to (B \to C) \Rightarrow C\,(3)}}{A \to B, A \to (B \to C) \Rightarrow A \to C\,(4)}}{A \to (B \to C) \Rightarrow (A \to B) \to (A \to C)\,(5)}}{\Rightarrow (A \to (B \to C)) \to ((A \to B) \to (A \to C))\,(6)}$$

where the lines 1–6 correspond to the steps 1–6 above; the only axiom application appearing as a right premise for L$\to$, namely $C \Rightarrow C$, corresponds to step 0.

Only a slight change is necessary to formulate the deduction in the calculus with multisets:

$$\dfrac{\dfrac{\dfrac{\dfrac{\dfrac{A \Rightarrow A \quad \dfrac{A \Rightarrow A \quad \dfrac{B \Rightarrow B \quad C \Rightarrow C}{B, B \to C \Rightarrow C}}{A, A \to B, B \to C \Rightarrow C}}{A, A, A \to B, A \to (B \to C) \Rightarrow C}\text{LC}}{A, A \to B, A \to (B \to C) \Rightarrow C}}{A \to B, A \to (B \to C) \Rightarrow A \to C}}{A \to (B \to C) \Rightarrow (A \to B) \to (A \to C)}}{\Rightarrow (A \to (B \to C)) \to ((A \to B) \to (A \to C))}$$

The appearance of two occurrences of A just before the LC-inference corresponds to the two occurrences of A^w in the original deduction in $\to$**Nm**.

It is not hard to convince oneself that, as long as only the principles 1-3 for the construction of prooftrees are applied, the resulting proof will always be *normal.* Conversely, it may be proved that all normal prooftrees can be obtained using construction principles 1–3 only. Thus we see that normal prooftrees in $\to$**Nm** correspond to deduction in the sequent calculus without Cut; and since every proof in natural deduction may be transformed into a normal proof of the same conclusion, using (at most) the same assumptions, it also follows for the sequent calculus that every deducible sequent $\Gamma \Rightarrow A$ must have a deduction without Cut.

Deductions in $\mathcal{S}$ without the rule Cut have a very nice property, which is immediately obvious: the *subformula property*: all formulas occurring in a deduction of $\Gamma \Rightarrow A$ are subformulas of the formulas in Γ, A.

A point worth noting is that the correspondence between sequent calculus deductions and natural deductions is usually not one-to-one. For example, in our transformation of the example 1.3.3 above, the steps 2 and 3 might have been interchanged, resulting in a different deduction in $\mathcal{S}$.

Another remark concerns construction principle 3: it follows in an indirect way from the rule $\to$E. Instead of $\to$E we might take the following:

$$\to\text{E}^*, u \; \dfrac{\begin{array}{c} \mathcal{D}_0 \\ A \to B \end{array} \quad \begin{array}{c} \mathcal{D}_1 \\ A \end{array} \quad \begin{array}{c} [B]^u \\ \mathcal{D}_2 \\ C \end{array}}{C}$$

which closely corresponds to construction principle 3 (cf. 6.12.4).

1.3.7A. ♠ There are other possible choices for the construction principles for prooftrees. For example, we might replace principle 3 by the following principle 3′:

Given a prooftree $\mathcal{D}$ with open assumption class $[B]^u$, replace all occurrences of B in $[B]^u$ by

$$\dfrac{A \to B^v \quad A}{B}$$

Show that this also generates all natural deduction prooftrees for implication logic; what sequent rules do these modified principles give rise to?

1.3.7B. ♠ Show that the following prooftree requires an application of construction principle 4:

$$\dfrac{\dfrac{\dfrac{\dfrac{A \to A \to B \quad A^u}{A \to B} \quad A^u}{B}}{A \to B}\,u \qquad A^v}{B}$$

1.3.8. *Semantical motivation of Gentzen systems*

For classical logic, we may arrive in a very natural way at a Gentzen system by semantical considerations. Here we use sequents $\Gamma \Rightarrow \Delta$, with Γ and Δ finite sets; the intuitive interpretation is that $\Gamma \Rightarrow \Delta$ is valid iff $\bigwedge \Gamma \to \bigvee \Delta$ is true. Now suppose we want to find out if there is a valuation making all of Γ true and all of Δ false. We can break down this problem by means of two rules, one for reducing $A \to B$ on the left, another for reducing $A \to B$ on the right:

$$\dfrac{\Gamma \Rightarrow A, \Delta \qquad \Gamma, B \Rightarrow \Delta}{\Gamma, A {\to} B \Rightarrow \Delta}\,\text{L}{\to} \qquad \dfrac{\Gamma, A \Rightarrow B, \Delta}{\Gamma \Rightarrow A {\to} B, \Delta}\,\text{R}{\to}$$

The problem of finding the required solution for the sequent at the bottom is equivalent to finding the solution(s) for (each of) the sequent(s) at the upper line. Thus starting at the bottom, we may work our way upwards; along each branch, the possibility of applying the rules stops, if all components have been reduced to atoms. If all branches terminate in sequents of the form $\Gamma', P \Rightarrow P, \Delta'$, there is no valuation for the sequent $\Gamma \Rightarrow \Delta$ making Γ true and Δ false. Taking sequents $\Gamma', P \Rightarrow P, \Delta'$ as axioms, the search tree for the valuation has then in fact become a derivation of the sequent $\Gamma \Rightarrow \Delta$ from axioms and L$\to$, R$\to$. This very simple argument constitutes also a completeness proof for classical propositional logic, relative to a Gentzen system without Cut. This idea for a completeness proof may also be adapted (in a not entirely trivial way) to intuitionistic and minimal logic, with Kripke semantics as the intended semantics.

The reader may be inclined to ask, why consider Gentzen systems at all? They do look more involved than natural deduction. There are two reasons for this. First of all, for certain logics Gentzen systems may be justified by semantical arguments in cases where it is not obvious how to construct an appropriate natural deduction system. Secondly, given the fact that there is a special interest in systems with the subformula property (on which many elementary proof-theoretic applications rest), we note that the condition of normality, guaranteeing the subformula property for natural deduction, is a *global* property of the deduction involving the order in which the rules are applied, whereas for Gentzen systems this is simply achieved by excluding the Cut rule.

1.3.9. *A Hilbert system*

A third type of formalism, extensively used in the logical literature, is the Hilbert system. Here there is a notion of deduction from assumptions, as for natural deductions, but *assumptions are never closed.* In Hilbert systems, the number of rules is reduced at the expense of introducing formulas as axioms. In most systems of this type, modus ponens ($\to$E) is in fact the only rule for propositional logic.

The Hilbert system $\to$**Hm** for minimal implication logic has as axioms all formulas of the forms:

$$A \to (B \to A) \quad (\mathbf{k}\text{-axioms}),$$
$$(A \to (B \to C)) \to ((A \to B) \to (A \to C)) \quad (\mathbf{s}\text{-axioms}),$$

for arbitrary A, B and C, and has $\to$E as the only rule. A deduction of B from assumptions Γ is then a tree with formulas from Γ and axioms at the top nodes, and the conclusion B at the root. (Usually, one finds K and S instead of **k** and **s** in the literature, since K and S are standard notations in combinatory logic. However, in modal logic one also encounters an axiom schema K, and we wish to avoid confusion.)

Example. A deduction $\mathcal{D}_A$ of $A \to A$:

$$\dfrac{\dfrac{[A\to((A\to A)\to A)] \to [(A\to(A\to A))\to(A\to A)] \qquad A\to((A\to A)\to A)}{(A\to(A\to A)) \to (A\to A)} \qquad A\to(A\to A)}{A \to A}$$

Deductions in Hilbert systems are often presented in linear format. Thus, in the case of implication logic, we may define a deduction of a formula A as a sequence $A_1, A_2, \dots, A_n$ such that $A \equiv A_n$, and moreover for each k ($1 \le k \le n$) either A_k is an axiom, or there are A_i, A_j with $i, j < k$ such that $A_j \equiv A_i \to A_k$. For example, the prooftree above may be represented by the following sequence:

(1)	$[A\to((A\to A)\to A)] \to [(A\to(A\to A))\to(A\to A)]$	**s**-axiom
(2)	$A\to((A\to A)\to A)$	**k**-axiom
(3)	$(A\to(A\to A)) \to (A\to A)$	(1), (2)
(4)	$A \to (A \to A)$	**k**-axiom
(5)	$A \to A$	(3), (4)

In fact, it is also possible to present natural deduction proofs and Gentzen system deductions in such a linear form. Where the primary aim is to discuss the actual construction of deductions, this is common practice in the literature on natural deduction. The disadvantage of the tree format, when compared with the linear format, is that the width of prooftrees for somewhat more complicated deductions soon makes it impracticable to exhibit them. On the other hand, as we shall see, the tree format for natural deductions has decided advantages for meta-theoretical considerations, since it provides an element of geometrical intuition.

There is also a formulas-as-types isomorphism for $\rightarrow$**Hm**, but this time the corresponding term system is $\mathbf{CL}_{\rightarrow}$, where the constants **k** and **s** represent the axioms (cf. 1.2.17)

$$\begin{aligned}&\mathbf{k}^{A,B} : A \rightarrow (B \rightarrow A),\\&\mathbf{s}^{A,B,C} : (A \rightarrow (B \rightarrow C)) \rightarrow ((A \rightarrow B) \rightarrow (A \rightarrow C)),\end{aligned}$$

and application corresponds to $\rightarrow$E as for natural deduction.

EXAMPLES. We write AB as an abbreviation for $A \rightarrow B$.

$$\frac{\mathbf{k}^{B(CB),A} : (B(CB))(A(B(CB))) \qquad \mathbf{k}^{B,C} : B(CB)}{\mathbf{k}^{B(CB),A}\mathbf{k}^{B,C} : A(B(CB))}$$

The prooftree exhibited before, establishing $A \rightarrow A$, corresponds to a term

$$\mathbf{s}^{A,A\rightarrow A,A}\mathbf{k}^{A,A\rightarrow A}\mathbf{k}^{A,A}.$$

The notion of weak reduction of course transfers from terms of $\mathbf{CL}_{\rightarrow}$ to $\rightarrow$**Hm**, but is of far less interest than β-reduction for $\rightarrow$**Nm**. However, the construction of an "abstraction-surrogate" λ^*x in 1.2.19 plays a role in proving the equivalence (w.r.t. derivable formulas) between systems of natural deduction and Hilbert systems, since it corresponds to a deduction theorem (see 2.4.2), and thereby provides us with a method for translating natural deduction proofs into Hilbert system proofs.

Chapter 2

N-systems and H-systems

Until we come to chapter 9, we shall concentrate on our three standard logics: classical logic **C**, intuitionistic logic **I** and minimal logic **M**. The informal interpretation (semantics) for **C** needs no explanation here. The logic **I** was originally motivated by L. E. J. Brouwer's philosophy of mathematics (more information in Troelstra and van Dalen [1988, chapter 1]); the informal interpretation of the intuitionistic logical operators, in terms of the primitive notions of "construction" and "constructive proof", is known as the "Brouwer–Heyting–Kolmogorov interpretation" (see 1.3.1, 2.5.1). Minimal logic **M** is a minor variant of **I**, obtained by rejection of the principle "from a falsehood follows whatever you like" (Latin: "ex falso sequitur quodlibet", hence the principle is often elliptically referred to as "ex falso"), so that, in **M**, the logical symbol for falsehood $\perp$ behaves like some unprovable propositional constant, not playing a role in the axioms or rules.

This chapter opens with a precise description of N-systems for the full first-order language with proofs in the form of deduction trees, assumptions appearing at top nodes. After that we present in detail the corresponding term system for the intuitionistic N-system, an extension of simple type theory. Once a precise formalism has been specified, we are ready for a section on the Gödel–Gentzen embedding of classical logic into minimal logic. This section gives some insight into the relations between **C** on the one hand and **M**, **I** on the other hand. Finally we introduce Hilbert systems for our standard logics and prove their deductive equivalence to the corresponding N-systems.

2.1 Natural deduction systems

We use script $\mathcal{D}$, $\mathcal{E}$, possibly sub- and/or superscripted, for deductions, and adopt the notational conventions for prooftrees with assumptions and conclusion adopted at the beginning of 1.3.2.

2.1.1. Definition. (*The systems* **Nm**, **Ni**, **Nc**) Assumptions are formula occurrences always appearing at the top of a branch (assumptions are "leaves" of the tree), and are supposed to be labelled by markers (e.g. natural numbers, or variable symbols). The set of assumptions of the same form with the same marker forms an *assumption class*. Distinct formulas must have distinct markers. We permit empty assumption classes!

Assumptions may be closed; assumption classes are always closed "en bloc", that is to say, at each inference, either all assumptions in a class are closed, or they are all left open. Closure is indicated by repeating the marker(s) of the class(es) at the inference.

For ease in the exposition, we shall reserve u, v, w for assumption markers, and x, y, z for individual variables.

Deductions in the system of natural deduction are generated as follows.

Basis. The single-node tree with label A (i.e. a single occurrence of A) is a (natural) *deduction* from the open assumption A; there are no closed assumptions.

Inductive step. Let $\mathcal{D}_1, \mathcal{D}_2, \mathcal{D}_3$ be deductions. A (natural) *deduction* $\mathcal{D}$ may be constructed according to one of the rules below. The classes $[A]^u$, $[B]^v$, $[\neg A]^u$ below contain open assumptions of the deductions of the premises of the final inference, but are closed in the whole deduction.

For $\wedge$, $\vee$, $\to$, $\forall$, $\exists$ we have *introduction rules (I-rules)* and *elimination rules (E-rules)*.

$$\frac{\begin{matrix}\mathcal{D}_1 & \mathcal{D}_2 \\ A & B\end{matrix}}{A \wedge B}\wedge\text{I} \qquad \frac{\begin{matrix}\mathcal{D}_1 \\ A \wedge B\end{matrix}}{A}\wedge\text{E}_\text{R} \qquad \frac{\begin{matrix}\mathcal{D}_1 \\ A \wedge B\end{matrix}}{B}\wedge\text{E}_\text{L}$$

$$\frac{\begin{matrix}[A]^u \\ \mathcal{D}_1 \\ B\end{matrix}}{A \to B}\to\text{I}, u \qquad \frac{\begin{matrix}\mathcal{D}_1 & \mathcal{D}_2 \\ A \to B & A\end{matrix}}{B}\to\text{E}$$

$$\frac{\begin{matrix}\mathcal{D}_1 \\ A\end{matrix}}{A \vee B}\vee\text{I}_\text{R} \qquad \frac{\begin{matrix}\mathcal{D}_1 \\ B\end{matrix}}{A \vee B}\vee\text{I}_\text{L} \qquad \frac{\begin{matrix} & [A]^u & [B]^v \\ \mathcal{D}_1 & \mathcal{D}_2 & \mathcal{D}_3 \\ A \vee B & C & C\end{matrix}}{C}\vee\text{E}, u, v$$

$$\frac{\begin{matrix}\mathcal{D}_1 \\ A[x/y]\end{matrix}}{\forall x A}\forall\text{I}$$

In $\forall$I: $y \equiv x$ or $y \notin \text{FV}(A)$, and y not free in any assumption open in $\mathcal{D}_1$.

$$\frac{\begin{matrix}\mathcal{D}_1 \\ \forall x A\end{matrix}}{A[x/t]}\forall\text{E}$$

$$\frac{\begin{matrix}\mathcal{D}_1 \\ A[x/t]\end{matrix}}{\exists x A}\exists\text{I} \qquad \frac{\begin{matrix} & [A[x/y]]^u \\ \mathcal{D}_1 & \mathcal{D}_2 \\ \exists x A & C\end{matrix}}{C}\exists\text{E}, u$$

In $\exists$E: $y \equiv x$ or $y \notin \text{FV}(A)$, and y not free in C nor in any assumption open in $\mathcal{D}_2$ except in $[A[x/y]]^u$.

This completes the description of the rules for minimal logic **Nm**. Note that $\perp$ has not been mentioned in any of the above rules, and therefore it behaves in minimal logic as an arbitrary unprovable propositional constant.

To obtain the intuitionistic and classical systems **Ni** and **Nc** we add the *intuitionistic absurdity rule* $\perp_i$ and the more general *classical absurdity rule* $\perp_c$ respectively:

$$\dfrac{\begin{array}{c}\mathcal{D}_1\\ \perp\end{array}}{A}\,\perp_i \qquad\qquad \dfrac{\begin{array}{c}[\neg A]^u\\ \mathcal{D}_1\\ \perp\end{array}}{A}\,\perp_c, u$$

($\perp_c$ is more general than $\perp_i$ since $[\neg A]^u$) may be empty.) In an E-rule application, the premise containing the occurrence of the logical operator being eliminated is called the *major* premise; the other premise(s) are called the *minor* premise(s) of the rule application. As a standard convention in displaying prooftrees, we place the major premises of elimination rule applications in leftmost position.

To spell out the *open* and *closed* assumptions for the rules exhibited above:

- When $\to$I is applied, the set $[A]^u$ of open assumptions of the form A in $\mathcal{D}$ becomes closed; when $\vee$E is applied, the set $[A]^u$ of open assumptions of the form A in $\mathcal{D}_2$ and the set $[B]^v$ of open assumptions of the form B in $\mathcal{D}_3$ become closed; when $\exists$E is applied, the set $[A[x/y]]^u$ of open assumptions of the form A in $\mathcal{D}_2$ becomes closed; when $\perp_c$ is applied, the set $[\neg A]^u$ of open assumptions of the form $\neg A$ in $\mathcal{D}_1$ becomes closed. All other assumptions, not covered by the cases just mentioned, stay open.

As to the individual variables which are considered to be free in a deduction, we stipulate

- The deduction consisting of assumption A only has $\mathrm{FV}(A)$ as free variables;

- at each rule application, the free individual variables are inherited from the immediate subdeductions, except that

- in an application of $\exists$E the occurrences of the free variable y in $\mathcal{D}_2$ become bound, and in an application of $\forall$I the occurrences of variable y in $\mathcal{D}_1$ become bound, and

- in $\to$I the variables in $\mathrm{FV}(A)$ have to be added in case $[A]^u$ is empty, in $\vee\mathrm{I_R}$ those in $\mathrm{FV}(B)$ have to be added, and in $\vee\mathrm{I_L}$ those in $\mathrm{FV}(A)$ have to be added.

The individual variable becoming bound in an application α of $\forall$I or $\exists$E is said to be the *proper* variable of α.

Instead of closed (assumption) one also finds in the literature the terminology *eliminated* or *cancelled* or *discharged*. Because of the correspondence of closed assumptions with bound variables in the term calculus (to be explained in detail in the next section) we also sometimes use "bound" for "closed". If A is among the open assumptions of a deduction $\mathcal{D}$ with conclusion B, the conclusion B in $\mathcal{D}$ is said to *depend* on A in $\mathcal{D}$. From now on we regard "assumption of $\mathcal{D}$" and "open assumption of $\mathcal{D}$" as synonymous. ⊠

2.1.2. Definition. A convenient global assumption in the presentation of a deduction is the *variable convention*. A deduction is said to satisfy the variable convention if the proper variables of the applications of $\exists$E and $\forall$I are kept distinct. That is to say, the proper variable of an application α of $\exists$E or $\forall$I occurs in the deduction only above α.

If moreover the bound and free variables are kept distinct, the deduction is said to be a *pure-variable* deduction. Henceforth we shall usually assume that the pure-variable condition is satisfied. ⊠

2.1.3. Remarks. (i) Since in our notation for prooftrees, $[A]^u$ refers to all assumptions A labelled u, it is tacitly understood that in $\vee$E the label u occurs in $\mathcal{D}_2$ only, and v in $\mathcal{D}_3$ only. Similarly, in $\exists$E the marker u occurs in $\mathcal{D}_2$ only. This restriction may be relaxed, at the expense of a much more clumsy formulation of $\exists$E and $\vee$E.

(ii) The rules of (extensions of) natural deduction systems are often presented in a more informal style. Instead of using inductive clauses "If $\mathcal{D}_0$, $\mathcal{D}_1, \ldots$ are correct deductions, then so is $\mathcal{D}$" as we did above, we can also describe the rules by exhibiting premises and assumptions to be discharged (if any), where the deductions between assumptions and premises are indicated by vertical dots. Thus $\to$E and $\vee$E are given by the schemas

$$\frac{A \to B \quad A}{B}\to\text{E} \qquad\qquad \frac{A \vee B \quad \begin{matrix}[A]^u\\ \vdots\\ C\end{matrix} \quad \begin{matrix}[B]^v\\ \vdots\\ C\end{matrix}}{C}\vee\text{E},u,v$$

2.1.4. Examples. The first example is in the classical system **Nc**:

$$\dfrac{\dfrac{\dfrac{\neg\forall x\neg A(x)^{w} \qquad \dfrac{\dfrac{\dfrac{\neg\exists xA(x)^{u} \quad \dfrac{A(x)^{v}}{\exists xA(x)}\,\exists\mathrm{I}}{\bot}\,{\to}\mathrm{I}}{\neg A(x)}\,{\to}\mathrm{I},v}{\forall x\neg A(x)}\,\forall\mathrm{I}}{\bot}\,{\to}\mathrm{E}}{\exists xA(x)}\,\bot_{\mathrm{c}},u}{\neg\forall x\neg A(x)\to\exists xA(x)}\,{\to}\mathrm{I},w$$

The next example is in **Ni**:

$$\dfrac{\dfrac{\dfrac{\begin{array}{c}[A]^{u'}\\ \mathcal{D}\\ Px\vee\neg Px\end{array} \qquad \dfrac{\dfrac{\begin{array}{c}[A]^{u'}\\ \mathcal{D}'\\ Px\to\exists yQy\end{array} \quad Px^{u}}{\exists yQy}\,{\to}\mathrm{E} \quad \dfrac{\dfrac{Qy^{w}}{Px\to Qy}\,{\to}\mathrm{I}}{\exists y(Px\to Qy)}\,\exists\mathrm{I}}{\exists y(Px\to Qy)}\,\exists\mathrm{E},w \qquad \dfrac{\dfrac{\dfrac{\dfrac{\neg Px^{v} \quad Px^{w'}}{\bot}\,{\to}\mathrm{E}}{Qy}\,\bot_{\mathrm{i}}}{Px\to Qy}\,{\to}\mathrm{I},w'}{\exists y(Px\to Qy)}\,\exists\mathrm{I}}{\exists y(Px\to Qy)}\,\vee\mathrm{E},u,v}{\forall x\exists y(Px\to Qy)}\,\forall\mathrm{I}}{A\to\forall x\exists y(Px\to Qy)}\,{\to}\mathrm{I},u'$$

where $A \equiv \forall x(Px \vee \neg Px) \wedge \forall x(Px \to \exists yQy)$, and where $\mathcal{D}$, $\mathcal{D}'$ are

$$\dfrac{\dfrac{A^{u'}}{\forall x(Px\vee\neg Px)}\,\wedge\mathrm{E}}{Px\vee\neg Px}\,\forall\mathrm{E} \qquad\qquad \dfrac{\dfrac{A^{u'}}{\forall x(Px\to\exists yQy)}\,\wedge\mathrm{E}}{Px\to\exists yQy}\,\forall\mathrm{E}$$

We also give two examples of incorrect deductions, violating the conditions on variables in ∀I, ∃E. The conclusions are obviously incorrect, since they are not generally valid for the standard semantics for classical logic. We have marked the incorrect assumption discharges with an exclamation mark.

$$\dfrac{\dfrac{\dfrac{Px^{u}}{\forall y\,Py}\,\forall\mathrm{I},\,!}{Px\to\forall y\,Py}\,{\to}\mathrm{I},u}{\forall x(Px\to\forall y\,Py)}\,\forall\mathrm{I} \qquad\qquad \dfrac{\dfrac{\dfrac{\dfrac{\exists x\,Px^{w} \quad \dfrac{Py\to Q^{u} \quad Py^{v}}{Q}\,{\to}\mathrm{E}}{Q}\,\exists\mathrm{E},v,!}{\exists x\,Px\to Q}\,{\to}\mathrm{I},w}{(Py\to Q)\to(\exists x\,Px\to Q)}\,{\to}\mathrm{I},u}{\forall y((Py\to Q)\to(\exists x\,Px\to Q))}\,\forall\mathrm{I}$$

2.1.5. DEFINITION. The theories (sets of theorems) generated by **Nm**, **Ni** and **Nc** are denoted by **M** (minimal logic), **I** (intuitionistic logic) and **C** (classical logic) respectively.

$\Gamma \vdash_{\mathbf{S}} A$ for $\mathbf{S} = \mathbf{M}, \mathbf{I}, \mathbf{C}$ iff A is derivable from the set of assumptions Γ in the N-system for $\mathbf{S}$. ⊠

2.1.6. *Identity of proof trees.* (i) Prooftrees are completely determined if we indicate at every node which is not a top node which rule has been applied to obtain the formula at the node from the formulas at the nodes immediately above it, plus the assumption classes discharged, if any.

(ii) From 1.3.6 we recall that two prooftrees are regarded as (essentially) the same, if (1) the underlying (unlabelled) trees are isomorphic, (2) nodes corresponding under the isomorphism get assigned the same formulas, (3) again modulo the isomorphism, the partitioning of assumptions into assumption classes is the same, and (4) corresponding assumption classes are discharged at corresponding nodes. Moreover, under the isomorphism, corresponding *open* assumptions should get the same marker.

Needless to say, in many cases it is not really necessary to indicate the rule which has been used to arrive at a particular node, since this is already determined by the form of the formulas at the nodes; but in a few cases the rule applied cannot be unambiguously reconstructed from the formulas alone. Nor is it essential to indicate a variable for a discharged assumption class at a rule application, if the assumption class happens to be empty. The reason why we insisted, in the definition of deduction above, that in principle this variable should always be present, is that this convention leads to the most straightforward correspondence between deductions and terms of a typed lambda calculus, discussed in the next section.

2.1.7. Remarks. (i) The absurdity rules $\bot_{\mathrm{i}}$ and $\bot_{\mathrm{c}}$ might be called elimination rules for $\bot$, since they eliminate the constant $\bot$; this suggests the designations $\bot\mathrm{E}_{\mathrm{i}}$, $\bot\mathrm{E}_{\mathrm{c}}$ for these rules. However, they behave rather differently from the other E-rules, since neither has the conclusion a direct connection with the premise, nor is there an assumption directly related to the premise, as in $\vee$E, $\exists$E. Therefore we have kept the customary designation for these rules. This anomalous behaviour suggests another possibility: taking **Nm** as the basic system, **Ni** and **Nc** are regarded as **Nm** with extra axioms added. For **Ni** one adds $\forall\vec{x}(\bot \to A)$, with $\vec{x} \equiv \mathrm{FV}(A)$, and for **Nc** one adds stability axioms $\forall\vec{x}(\neg\neg A \to A)$ (see 2.3.6).

(ii) Sometimes it is more natural to write $\forall$E and $\exists$I as two-premise rules, with the individual term as second premise (a *minor* premise in case of $\forall$E):

$$\frac{\forall x A \qquad t}{A[x/t]} \qquad\qquad \frac{A[x/t] \qquad t}{\exists x A}$$

This emphasizes the analogies between $\to$E and $\forall$E, and between $\wedge$I and $\exists$I. The mixing of deduction with term construction might seem strange at first sight, but becomes less so if one keeps in mind that writing down a term implicitly entails a proof that by the rules of term construction the term denotes something which is in the domain of the variables. Such extra premises become indispensable if we consider logics where terms do not always denote; see 6.5. This convention is also utilized in chapter 10. If one wants to stress the relation to type theory, one writes $t: D$ (D domain of individuals) for the second premise.

(iii) The statement of the rules ∀I and ∃E may be simplified somewhat if we rely on our convention that formulas differing only in the naming of bound variables are equal. These rules may then be written as:

$$\dfrac{\begin{matrix}\mathcal{D}_1\\ A\end{matrix}}{\forall x A}\,\forall\text{I} \qquad\qquad \dfrac{\begin{matrix}\mathcal{D}_1\\ \exists x A\end{matrix} \quad \begin{matrix}[A]^u\\ \mathcal{D}_2\\ C\end{matrix}}{C}\,\exists\text{E},u$$

where in ∀I x is not free in any assumption open in $\mathcal{D}_1$, and in ∃E x is not free in C nor in any assumption open in $\mathcal{D}_2$ except in $[A]^u$.

(iv) In theories based on logic, we may accommodate axioms as rules without premises; so an axiom appears in a prooftree as a top node with a line over it (in practice we often drop this line).

2.1.8. *Natural deductions in sequent style*

In the format described above, the assumptions open at any node ν in a deduction tree $\mathcal{D}$ are found by looking at the top nodes above ν; the ones bearing a label not yet discharged between the top node and ν are still open at ν. Less economical in writing, but for metamathematical treatment sometimes more convenient, is a style of presentation where the open assumptions are carried along and exhibited at each node. We call the set of open assumptions at a node the *context*. A context is a set

$$u_1: A_1, u_2: A_2, \ldots, u_n: A_n$$

where the u_i are pairwise distinct; the A_i need not be distinct. The deductions now become trees where each node is labelled with a sequent of the form $\Gamma \Rightarrow B$, Γ a context. Below, when writing a union of contexts such as $\Gamma\Delta$ (short for $\Gamma \cup \Delta$), it will always be assumed that the union is *consistent*, that is to say, again forms a context. In this form the rules and axioms now read as follows:

$$u : A \Rightarrow A \quad \text{(Axiom)}$$

$$\frac{\Gamma[u: A] \Rightarrow B}{\Gamma \Rightarrow A \to B}\to\text{I} \qquad \frac{\Gamma \Rightarrow A \to B \qquad \Delta \Rightarrow A}{\Gamma\Delta \Rightarrow B}\to\text{E}$$

$$\frac{\Gamma \Rightarrow A \qquad \Delta \Rightarrow B}{\Gamma\Delta \Rightarrow A \wedge B}\wedge\text{I} \qquad \frac{\Gamma \Rightarrow A_0 \wedge A_1}{\Gamma \Rightarrow A_i}\wedge\text{E}$$

$$\frac{\Gamma \Rightarrow A_i}{\Gamma \Rightarrow A_0 \vee A_1}\vee\text{I} \qquad \frac{\Gamma \Rightarrow A \vee B \quad \Delta[u: A] \Rightarrow C \quad \Delta'[v: B] \Rightarrow C}{\Gamma\Delta\Delta' \Rightarrow C}\vee\text{E}$$

$$\frac{\Gamma[x\colon \neg A] \Rightarrow \bot}{\Gamma \Rightarrow A}\bot_c \qquad \frac{\Gamma \Rightarrow \bot}{\Gamma \Rightarrow A}\bot_i$$

$$\frac{\Gamma \Rightarrow A[x/y]}{\Gamma \Rightarrow \forall x A}\forall\mathrm{I} \qquad \frac{\Gamma \Rightarrow \forall x A}{\Gamma \Rightarrow A[x/t]}\forall\mathrm{E}$$

$$\frac{\Gamma \Rightarrow A[x/t]}{\Gamma \Rightarrow \exists x A}\exists\mathrm{I} \qquad \frac{\Gamma \Rightarrow \exists y A[x/y] \qquad \Delta[u\colon A] \Rightarrow C}{\Gamma\Delta \Rightarrow C}\exists\mathrm{E}$$

Here $[u\colon C]$ means that the assumption $u\colon C$ in the context may be present or absent. Moreover, in $\to$I $u\colon A$ does not occur in Γ, in $\vee$E $u\colon A$ and $v\colon B$ do not occur in $\Gamma\Delta\Delta'$, in $\bot_c$ $u\colon \neg A$ does not occur in Γ, and in $\exists$E $u\colon A$ does not occur in $\Gamma\Delta$.

The correspondence is now such that at any node *precisely* the inhabited assumption classes which are not yet closed at that node are listed.

2.1.8A. ♠ Give proofs in **Nm** or **Ni** of

$A \to (B \to A)$;
$A \to A \vee B,\ \ B \to (A \vee B)$;
$(A \to C) \to [(B \to C) \to (A \vee B \to C)]$;
$A \wedge B \to A,\ \ A \wedge B \to B,\ \ A \to (B \to A \wedge B)$;
$\bot \to A$;
$\forall x A \to A[x/t];\ \ \ A[x/t] \to \exists x A$;
$\forall x(B \to A) \leftrightarrow (B \to \forall y A[x/y])\ \ (x \notin \mathrm{FV}(B),\ y \equiv x \text{ or } y \notin \mathrm{FV}(A))$;
$\forall x(A \to B) \leftrightarrow (\exists y A[x/y] \to B)\ \ (x \notin \mathrm{FV}(B),\ y \equiv x \text{ or } y \notin \mathrm{FV}(A))$.

2.1.8B. ♠* Give proofs in **Nm** of

$A \to \neg\neg A$;
$\neg\neg\neg A \leftrightarrow \neg A$;
$\neg\neg(A \to B) \to (\neg\neg A \to \neg\neg B)$;
$\neg\neg(A \wedge B) \leftrightarrow (\neg\neg A \wedge \neg\neg B)$;
$\neg(A \vee B) \leftrightarrow (\neg A \wedge \neg B)$;
$\neg\neg\forall x A \to \forall x \neg\neg A$.

2.1.8C. ♠ Give proofs in **Nm** of

$(B \to C) \to (A \to B) \to A \to C$ (**b**-axioms),
$(A \to B \to C) \to B \to A \to C$ (**c**-axioms),
$(A \to A \to B) \to A \to B$ (**w**-axioms).

2.1.8D. ♠ Prove in **Ni** $(\neg\neg A \to \neg\neg B) \to \neg\neg(A \to B)$. *Hint.* First construct deductions of $\neg\neg A$ and of $\neg B$ from the assumption $\neg(A \to B)$.

2.1.8E. ♠* Prove in **Nc**

$$A \vee B \leftrightarrow \neg(\neg A \wedge \neg B),$$
$$\exists x A \leftrightarrow \neg \forall x \neg A,$$
$$((A \to B) \to A) \to A \text{ (Peirce's law).}$$

2.1.8F. ♠* Construct in $\to$**Nm** a proof of

$$((A \to B) \to C) \to (A \to C) \to C$$

from two instances of Peirce's law as assumptions: $((A \to B) \to A) \to A$ and $((C \to A) \to C) \to C$.

2.1.8G. ♠* Derive in $\to$**Nm** $\mathrm{P}_{A,B\wedge C}$ from $\mathrm{P}_{A,B}$ and $\mathrm{P}_{A,C}$, where $\mathrm{P}_{X,Y}$ is $((X \to Y) \to X) \to X$, i.e., Peirce's law for X and Y.

2.1.8H. ♠* Let $F[*]$, $G[*]$ be a positive and negative context respectively. Prove in **Nm** that

$$\vdash \forall \vec{x}(A \to B) \to (F[A] \to F[B]),$$
$$\vdash \forall \vec{x}(A \to B) \to (G[B] \to G[A]),$$

where $\vec{x}$ consists of the variables in $A \to B$ becoming bound by substitution of A and B into $F[*]$ in the first line, and into $G[*]$ in the second line.

2.1.9. *The Complete Discharge Convention*

One possibility left open by the definition of deductions in the preceding section is to discharge always *all* open assumptions of the same form, whenever possible.

Thus in $\to$I we can take $[A]^x$ to represent *all* assumptions of the form A which are still open at the premise B of the inference and occur above B; in an application of $\vee$E $[A]^u$, $[B]^v$ represent all assumptions of the form A still open at C in the second subdeduction, and all assumptions of the form B still open at C in the third subdeduction respectively; in an application of $\exists$E $[A[x/y]]^u$ represents all assumptions of this form in the second subdeduction still open at C.

It is easy to see that a deduction remains correct, if we modify the discharge of assumptions according to this convention. We call this convention the "*Complete Discharge Convention*", or *CDC* for short.

Note that the use of markers, and the repetition of markers at inferences where assumption classes are being discharged, is redundant if one adopts CDC (although still convenient as a bookkeeping device).

From the viewpoint of deducibility, both versions of the notion of deduction are acceptable; CDC has the advantage of simplicity. But as we shall discover, the general notion is much better-behaved when it comes to studying normalization of deductions. In particular, the so-called "formulas-as-types" analogy ("isomorphism"), which has strong motivating power and permits us to transfer techniques from the study of the lambda calculus to the study of natural deduction, applies only to the general notion of deduction, not to deductions based on CDC.

2.1.10. *Digression: representing CDC natural deduction with sequents*

Let $\mathbf{N}_0$ be intuitionistic natural deduction for implication under CDC. $\mathbf{N}_0$ can be presented as a calculus in sequent notation, $\mathbf{N}_1$, in the following (obvious) way:

$$A \Rightarrow A \qquad \frac{\Gamma \Rightarrow B}{\Gamma \setminus \{A\} \Rightarrow A \to B} \qquad \frac{\Gamma \Rightarrow A \to B \qquad \Gamma' \Rightarrow A}{\Gamma \cup \Gamma' \Rightarrow B}$$

Here the antecedents are regarded as *sets*, not multisets.

$\mathbf{N}_1$-deductions are obtained from $\mathbf{N}_0$-deductions by replacing the formula A at node ν by the sequent $\Gamma \Rightarrow A$, where Γ is the set of assumptions open at ν.

At first sight one might think that the following calculus $\mathbf{N}_2$ –

$$\Gamma, A \Rightarrow A \qquad \frac{\Gamma, A \Rightarrow B}{\Gamma \Rightarrow A \to B} \qquad \frac{\Gamma \Rightarrow A \to B \qquad \Gamma \Rightarrow A}{\Gamma \Rightarrow B}$$

– where the antecedents of the sequents are treated as *multisets*, represents a step towards the standard $\to$**Ni** (without CDC), since it looks as if distinct occurrences of the same formula in the antecedent might be used to represent differently labelled assumption classes in $\mathbf{N}_0$-deductions. But this impression is mistaken; in fact, if we strip the dummy assumptions from deductions in $\mathbf{N}_2$, there is a one-to-one correspondence with the deductions in $\mathbf{N}_1$.

Let us write $\mathcal{D}[\Gamma' \Rightarrow]$ for the deduction in $\mathbf{N}_2$ obtained from $\mathcal{D}$ by replacing at each node ν of $\mathcal{D}$ the sequent $\Gamma \Rightarrow A$ at that node with $\Gamma\Gamma' \Rightarrow A$.

We show how to associate to each deduction $\mathcal{D}$ of $\Gamma \Rightarrow A$ in $\mathbf{N}_2$ a deduction $\mathcal{D}'$ of $\Gamma' \Rightarrow A$, $\Gamma' \subset \Gamma$ with the same tree structure, such that

(i) all $A \in \Gamma'$ occur as conclusion of an axiom and Γ' is a set,

(ii) $\mathcal{D}$ is $\mathcal{D}'[(\Gamma \setminus \Gamma') \Rightarrow]$, that is to say $\mathcal{D}$ is obtained from $\mathcal{D}'$ by weakening the sequents throughout with the same multiset,

(iii) if $\mathcal{D}_\nu$ is the subdeduction of $\mathcal{D}$ associated with node ν, and we replace everywhere in $\mathcal{D}$ the conclusion of $\mathcal{D}_\nu$ by the conclusion of $(\mathcal{D}')_\nu$, then the result is a deduction $\phi(\mathcal{D})$ in $\mathbf{N}_1$.

The construction is by induction on the depth of $\mathcal{D}$, and the properties just listed are verified by induction on the depth of $\mathcal{D}$.

Case 1. To an axiom $\Gamma, A \Rightarrow A$ we associate the axiom $A \Rightarrow A$.

Case 2. Let the proof $\mathcal{D}$ end with →I:

$$\dfrac{\begin{array}{c}\mathcal{D}_0\\ \Gamma, A \Rightarrow B\end{array}}{\Gamma \Rightarrow A \to B}$$

To $\mathcal{D}_0$ we have already assigned, by IH, a $\mathcal{D}'_0$ with conclusion $\Gamma' \Rightarrow B$. There are two cases: A does not occur in Γ, or Γ is of the form Γ'', A, A not in Γ''. In the first case, $\mathcal{D}'$ is as on the left, in the second case as on the right below:

$$\dfrac{\begin{array}{c}\mathcal{D}'_0[A \Rightarrow]\\ \Gamma', A \Rightarrow B\end{array}}{\Gamma' \Rightarrow A \to B} \qquad \dfrac{\begin{array}{c}\mathcal{D}'_0\\ \Gamma'', A \Rightarrow B\end{array}}{\Gamma'' \Rightarrow A \to B}$$

Case 3. Let $\mathcal{D}$ end with →E:

$$\dfrac{\begin{array}{c}\mathcal{D}_0\\ \Gamma \Rightarrow A \to B\end{array} \quad \begin{array}{c}\mathcal{D}_1\\ \Gamma \Rightarrow A\end{array}}{\Gamma \Rightarrow B}$$

The IH produces two deductions

$$\begin{array}{c}\mathcal{D}'_0\\ \Gamma'_0 \Rightarrow A \to B\end{array} \qquad \begin{array}{c}\mathcal{D}'_1\\ \Gamma'_1 \Rightarrow A\end{array}$$

and we take for $\mathcal{D}'$

$$\dfrac{\begin{array}{c}\mathcal{D}'_0[(\Gamma'_1 \setminus \Gamma'_0) \Rightarrow]\\ \Gamma'_0 \cup \Gamma'_1 \Rightarrow A \to B\end{array} \quad \begin{array}{c}\mathcal{D}'_1[(\Gamma'_0 \setminus \Gamma'_1) \Rightarrow]\\ \Gamma'_0 \cup \Gamma'_1 \Rightarrow A\end{array}}{\Gamma' \equiv \Gamma'_0 \cup \Gamma'_1 \Rightarrow B}$$

We leave it to the reader to construct a map ψ from $\mathbf{N}_1$ to $\mathbf{N}_2$ which is inverse to ϕ.

2.1.10A. ♠ Define the map ψ mentioned above and show that it is inverse to ϕ.

2.2 Ni as a term calculus

2.2.1. Extending the term notation for implication logic, described in section 1.3, we can also identify the full calculi **Ni**, **Nm** with a system of typed terms in a very natural way. The typed terms serve as an *alternative notation system.* In a sense, this makes the use of calligraphic $\mathcal{D}, \mathcal{E}$ for deductions in the case of N-systems redundant; we might as well use metavariables for terms in a type theory (say s, t) for deductions. Nevertheless we shall use both notations: $\mathcal{D}, \mathcal{E}$ if we wish to emphasize the prooftrees, and ordinary term notation if we wish to exploit the formulas-as-types parallel and study computational aspects.

2.2.2. Definition. (*Term calculus for the full system* **Ni**) The variables with formula type are distinct from the individual variables occurring in the types (formulas), and the sets of variables for distinct types are disjoint. We exhibit the generation of terms in parallel to the rules. To each rule corresponds a specific operator. For example, the first term-labelled rule $\wedge$I corresponds to a clause: if $t_0\colon A_0$ and $t_1\colon A_1$ are terms, then $\mathbf{p}(t_0{:}A_0, t_1{:}A_1)\colon A_0 \wedge A_1$, or $\mathbf{p}(t_0^{A_0}, t_1^{A_1})^{A_0\wedge A_1}$, is a term. Together with the listing of the clauses for the generation of the terms, we specify variable conditions, the free *assumption* variables ($\mathrm{FV_a}$) and the free *individual* variables ($\mathrm{FV_i}$). As in type theory, we abbreviate $\mathrm{App}(t, s)$ as ts.

$$u\colon A \qquad \mathrm{FV_i}(u) := \mathrm{FV}(A),\ \ \mathrm{FV_a}(u) := \{u\}.$$

$$\frac{t\colon A \qquad s\colon B}{\mathbf{p}(t^A, s^B)\colon A \wedge B}\wedge\mathrm{I} \qquad \begin{array}{l}\mathrm{FV_i}(\mathbf{p}(t,s)) := \mathrm{FV_i}(t) \cup \mathrm{FV_i}(s);\\ \mathrm{FV_a}(\mathbf{p}(t,s)) := \mathrm{FV_a}(t) \cup \mathrm{FV_a}(s).\end{array}$$

$$\frac{t\colon A_0 \wedge A_1}{\mathbf{p}_j(t^{A_0\wedge A_1})\colon A_j}\wedge\mathrm{E}\ \ (j{\in}\{0,1\}) \qquad \begin{array}{l}\mathrm{FV_i}(\mathbf{p}_j(t)) := \mathrm{FV_i}(t);\\ \mathrm{FV_a}(\mathbf{p}_j(t)) := \mathrm{FV_a}(t).\end{array}$$

$$\frac{t\colon A_j}{\mathbf{k}_j(t^{A_j})\colon A_0 \vee A_1}\vee\mathrm{I}\ \ (j{\in}\{0,1\}) \qquad \begin{array}{l}\mathrm{FV_i}(\mathbf{k}_j(t)) := \mathrm{FV_i}(t) \cup \mathrm{FV}(A_{1-j}),\ \text{and}\\ \mathrm{FV_a}(\mathbf{k}_j(t)) := \mathrm{FV_a}(t).\end{array}$$

$$\frac{t\colon A \vee B \qquad \begin{array}{c}[u\colon A]\\ \mathcal{D}_0\\ s\colon C\end{array} \qquad \begin{array}{c}[v\colon B]\\ \mathcal{D}_1\\ s'\colon C\end{array}}{\mathrm{E}^{\vee}_{u,v}(t^{A\vee B}, s^C, s'^C)\colon C}\vee\mathrm{E} \qquad \begin{array}{l}u \notin \mathrm{FV_a}(t, s'),\ v \notin \mathrm{FV_a}(t, s),\\ \mathrm{FV_i}(\mathrm{E}^{\vee}_{u,v}(t,s,s')) := \mathrm{FV_i}(t,s,s');\\ \mathrm{FV_a}(\mathrm{E}^{\vee}_{u,v}(t,s,s')) :=\\ \quad \mathrm{FV_a}(t) \cup (\mathrm{FV_a}(s) \setminus \{u\}) \cup (\mathrm{FV_a}(s') \setminus \{v\}).\end{array}$$

$$\frac{\begin{array}{c}[u\colon A]\\ \mathcal{D}\\ t\colon B\end{array}}{(\lambda u^A.t^B)\colon A \to B}\to\mathrm{I} \qquad \begin{array}{l}\mathrm{FV_i}(\lambda u^A.t) := \mathrm{FV_i}(t) \cup \mathrm{FV}(A);\\ \mathrm{FV_a}(\lambda u^A.t) := \mathrm{FV_a}(t) \setminus \{u\}.\end{array}$$

$$\frac{t\colon A{\to}B \qquad s\colon A}{t^{A\to B}s^A\colon B}\to\mathrm{E} \qquad \begin{array}{l}\mathrm{FV_i}(ts) := \mathrm{FV_i}(t) \cup \mathrm{FV_i}(s);\\ \mathrm{FV_a}(ts) := \mathrm{FV_a}(t) \cup \mathrm{FV_a}(s).\end{array}$$

$$\frac{t[x/y]\colon A[x/y]}{\lambda x.t^A\colon \forall x A}\forall\text{I}$$

$y \equiv x$ or $y \notin \mathrm{FV}(A)$, and
if $u^B \in \mathrm{FV_a}(t)$, then $x \notin \mathrm{FV}(B)$;
$\mathrm{FV_i}(\lambda x.t) := \mathrm{FV_i}(t) \setminus \{x\}$;
$\mathrm{FV_a}(\lambda x.t) := \mathrm{FV_a}(t)$.

$$\frac{t\colon \forall x A}{t^{\forall x A} s\colon A[x/s]}\forall\text{E}$$

$\mathrm{FV_i}(ts) := \mathrm{FV_i}(t) \cup \mathrm{FV}(s)$;
$\mathrm{FV_a}(ts) := \mathrm{FV_a}(t)$.

$$\frac{t\colon A[x/s]}{\mathbf{p}(t^{A[x/s]}, s)\colon \exists x A}\exists\text{I}$$

$\mathrm{FV_i}(\mathbf{p}(t,s)) := \mathrm{FV_i}(t)$,
$\mathrm{FV_a}(\mathbf{p}(t,s)) := \mathrm{FV_a}(t)$.

$$\frac{t\colon \exists x A \qquad \begin{array}{c}[u\colon A[x/y]]\\ \mathcal{D}\\ s\colon C\end{array}}{\mathrm{E}^{\exists}_{u,y}(t^{\exists x A}, s^C)\colon C}\exists\text{E}$$

$y \equiv x$ or $y \notin \mathrm{FV}(A)$,
$u \notin \mathrm{FV_a}(t)$, $y \notin \mathrm{FV}(C)$, and
if $v^B \in \mathrm{FV_a}(s) \setminus \{u\}$, then $y \notin \mathrm{FV}(B)$;
$\mathrm{FV_i}(\mathrm{E}^{\exists}_{u,y}(t,s) := \mathrm{FV_i}(t) \cup (\mathrm{FV_i}(s) \setminus \{y\})$;
$\mathrm{FV_a}(\mathrm{E}^{\exists}_{u,y}(t,s) := \mathrm{FV_a}(t) \cup (\mathrm{FV_a}(s) \setminus \{u\})$.

$$\frac{t\colon \bot}{\mathrm{E}^{\bot}_A(t^{\bot})\colon A}\bot_{\mathrm{i}}$$

$\mathrm{FV_i}(\mathrm{E}^{\bot}_A(t)) := \mathrm{FV_i}(t) \cup \mathrm{FV}(A)$;
$\mathrm{FV_a}(\mathrm{E}^{\bot}_A(t)) := \mathrm{FV_i}(t)$.

Finally, we put $\mathrm{FV}(t^A) := \mathrm{FV_i}(t^A) \cup \mathrm{FV_a}(t^A)$. ⊠

REMARKS. (i) Dropping the terms and retaining the formulas in the schemas above produces ordinary prooftrees, provided we keep assumptions labelled by variables, and indicate where they are discharged.

(ii) The term assigned to the conclusion describes in fact the complete prooftree, i.e. the deduction can unambiguously be reconstructed from this term.

(iii) Since the variables are always supposed to have a definite type (we could say that individual variables have a type I), it would have been possible to define FV straight away, instead of $\mathrm{FV_i}$ and $\mathrm{FV_a}$ separately, but the resulting definition would not have been very perspicuous.

(iv) We may assume that proper parameters of applications of ∃E and ∀I are always kept distinct and are used only in the subdeduction terminating in the rule application concerned; this would have resulted in slight simplifications in the stipulations for free variables above. Similarly, assuming that all bound assumption variables are kept distinct permits slight simplifications.

(v) The conditions $u \notin \mathrm{FV_a}(t, s')$, $v \notin \mathrm{FV_a}(t, s)$ in ∨E, and the condition $u \notin \mathrm{FV_a}(t)$ in ∃E may be dropped, but this would introduce an imperfection in the correlation between deduction trees as described earlier in 2.1.1 and the term calculus. The conditions just mentioned correspond to the conditions in 2.1.1 that the u in $\vee$E occurs in $\mathcal{D}_2$ only etc.

(vi) There is considerable redundancy in the typing of terms and subterms, and in practice we shall drop types whenever we can do so without creating confusion.

(vii) Instead of the use of subscripted variables for the operators, we can use alternative notations, such as $\mathrm{E}^{\exists}(t, (y, z)s)$, $\mathrm{E}^{\vee}(t, (y)s, (z)s')$. Here variables are bound by "( )", so as not to cause confusion with the λ which is associated with $\to$I and $\forall$I.

(viii) As noted before, in the rules $\forall$E and $\exists$I the term s may appear as a second premise; in certain situations this is a natural thing to do (analogy with type theories).

(ix) Extra axioms may be represented by addition of constants of the appropriate types.

2.2.2A. ♠* Give proofs in **Nm** of $(A \vee B \to C) \to ((A \to C) \wedge (B \to C))$ $(A, B, C$ arbitrary), and of $\forall x(Rx \to R'y) \to (\exists x Rx \to R'y)$ $(R, R'$ unary relation symbols) and label the nodes with the appropriate terms; compute $\mathrm{FV_a}$ and $\mathrm{FV_i}$ for the terms assigned to the conclusions.

2.3 The relation between C, I and M

In this section we discuss some embeddings of **C** into **M** or **I**, via the so-called "negative translation". This translation exists in a number of variants.

2.3.1. DEFINITION. A formula A in a first-order language is said to belong to the *negative fragment* (or "*A is negative*") if atomic formulas P occur only negated (i.e. in a context $P \to \bot$) in A, and A does not contain $\vee, \exists$. ⊠

2.3.2. LEMMA. *For A negative,* $\mathbf{M} \vdash A \leftrightarrow \neg\neg A$.

PROOF. As seen by inspection of exercise 2.1.8B, the following are all provable in **Nm**:

$$A \to \neg\neg A, \quad \neg\neg\neg A \leftrightarrow \neg A;$$
$$\neg\neg(A \wedge B) \to \neg\neg A \wedge \neg\neg B;$$
$$\neg\neg(A \to B) \to (\neg\neg A \to \neg\neg B), \quad (\neg\neg A \to \neg\neg B) \leftrightarrow (A \to \neg\neg B);$$
$$\neg\neg\forall x A \to \forall x \neg\neg A.$$

Using these implications, we establish the lemma by induction on the depth of A; A has one of the forms $\neg P$ (P atomic), $\bot$, $B \wedge C$, $B \to C$, $\forall x A$. Consider e.g. the case $A \equiv B \to C$. Then $\neg\neg A \equiv \neg\neg(B \to C)$ which implies $(B \to \neg\neg C)$, and by IH $(B \to C)$; this finally yields $\neg\neg(B \to C)$. We leave the other cases to the reader. ⊠

2.3.2A. ♠ Do the remaining cases.

2.3.3. Definition. For all formulas of predicate logic the *(Gödel–Gentzen) negative translation* ^{g} is defined inductively by

(i) $P^{\mathrm{g}} := \neg\neg P$ for atomic P;
(ii) $\bot^{\mathrm{g}} := \bot$;
(iii) $(A \wedge B)^{\mathrm{g}} := A^{\mathrm{g}} \wedge B^{\mathrm{g}}$;
(iv) $(A \to B)^{\mathrm{g}} := A^{\mathrm{g}} \to B^{\mathrm{g}}$;
(v) $(\forall x A)^{\mathrm{g}} := \forall x A^{\mathrm{g}}$;
(vi) $(A \vee B)^{\mathrm{g}} := \neg(\neg A^{\mathrm{g}} \wedge \neg B^{\mathrm{g}})$;
(vii) $(\exists x A)^{\mathrm{g}} := \neg\forall x \neg A^{\mathrm{g}}$.

Inessential variants are obtained by dropping clause (ii) and applying the first clause to $\bot$ as well, or by adding a process of systematically replacing $\neg\neg\neg$ by $\neg$. ⊠

2.3.4. Theorem. *For all A*

(i) $\mathbf{C} \vdash A \leftrightarrow A^{\mathrm{g}}$;

(ii) $\Gamma \vdash_{\mathrm{c}} A \Leftrightarrow \Gamma^{\mathrm{g}} \vdash_{\mathrm{m}} A^{\mathrm{g}}$,

where $\Gamma^{\mathrm{g}} := \{B^{\mathrm{g}} : B \in \Gamma\}$.

Proof. The proof shows by induction on the length of deductions in **Nc** that whenever $\Gamma \vdash A$, then $\Gamma^{\mathrm{g}} \vdash A^{\mathrm{g}}$. The rules for $\vee, \exists$ are in **Nc** derivable from the other rules, if we use the classical definitions $A \vee B := \neg(\neg A \wedge \neg B)$, $\exists x A := \neg\forall x \neg A$. So we may restrict attention to **Nc** for the language $\wedge\forall\!\to\!\bot$.

All applications of rules, except applications of $\bot_{\mathrm{c}}$, translate into applications of the corresponding rules of **Nm**, e.g.

$$\dfrac{\begin{matrix}[A]^x\\ \mathcal{D}\\ B\end{matrix}}{A \to B}\,x \qquad \text{translates as} \qquad \dfrac{\begin{matrix}[A^{\mathrm{g}}]^x\\ \mathcal{D}^{\mathrm{g}}\\ B^{\mathrm{g}}\end{matrix}}{(A \to B)^{\mathrm{g}}}\,x$$

For the translation of $\bot_{\mathrm{c}}$ we need lemma 2.3.2:

$$\dfrac{\begin{matrix}[\neg A]^x\\ \mathcal{D}\\ \bot\end{matrix}}{A}\,x \qquad \text{translates as} \qquad \dfrac{\begin{matrix}\mathcal{D}_A\\ \neg\neg A^{\mathrm{g}} \to A^{\mathrm{g}}\end{matrix} \qquad \dfrac{\begin{matrix}[\neg A^{\mathrm{g}}]^x\\ \mathcal{D}^{\mathrm{g}}\\ \bot\end{matrix}}{\neg\neg A^{\mathrm{g}}}\to\mathrm{I}, x}{A^{\mathrm{g}}}$$

where $\mathcal{D}_A$ is a standard proof of $\neg\neg A^{\mathrm{g}} \to A^{\mathrm{g}}$ as given by lemma 2.3.2. ⊠

2.3.5. Corollary. *For negative A,* $\mathbf{C} \vdash A$ *iff* $\mathbf{M} \vdash A$, *i.e.* **C** *is conservative over* **M** *w.r.t. negative formulas.*

2.3.5A. ♠ Derive the rules for defined $\vee, \exists$ from the other rules in **Nc**.

In a very similar way we obtain the following:

2.3.6. THEOREM. *Let Γ, A be formulas without $\vee, \exists$, and let* $\mathbf{Nc} \vdash \Gamma \Rightarrow A$. *Then there is a proof of* $\mathbf{M} \vdash \Gamma, \Delta \Rightarrow A$ *where Δ consists of assumptions* $\forall \vec{x}\,(\neg\neg R\vec{x} \to R\vec{x}\,)$, *$R$ a relation symbol occurring in Γ, A. (Such assumptions are called stability assumptions.)*

PROOF. Since $\vee, \exists$ are classically definable, we may assume that the whole proof of $\mathbf{Nc} \vdash \Gamma \Rightarrow A$ is carried out in the language without $\vee, \exists$. For this fragment, all instances of $\bot_c$ are reducible to instances with atomic conclusion relative to the rules of **Nm** (exercise). For the rest, the proof proceeds straightforwardly by induction on the length of classical deductions in the language without $\vee, \exists$. ⊠

2.3.6A. ♠ Show that in **Ni** all instances of $\bot_i$ are derivable from the instances of $\bot_i$ with atomic conclusion. Show that in **Nc**, for the language without $\vee, \exists$, all instances of $\bot_c$ are derivable from instances $\bot_c$ with atomic conclusion. (For a hint, see 6.1.11.)

2.3.7. *Other versions of the negative translation*

One of the best known variants is *Kolmogorov's negative translation* ^{k}. A^k is obtained by simultaneously inserting double negations in front of all subformulas of A, including A itself, but excepting $\bot$, which is left unchanged. Inductively we may define ^{k} by:

$$
\begin{array}{ll}
P^k & := \neg\neg P \text{ for } P \text{ atomic;} \\
\bot^k & := \bot; \\
(A \circ B)^k & := \neg\neg(A^k \circ B^k) \text{ for } \circ \in \{\wedge, \vee, \to\}; \\
(QxA)^k & := \neg\neg(Qx)A^k \text{ for } Q \in \{\forall, \exists\}.
\end{array}
$$

Another variant A^q (*Kuroda's negative translation*) ("q" from "quantifier") is obtained as follows: insert $\neg\neg$ *after* each occurrence of $\forall$, and in front of the whole formula.

2.3.8. PROPOSITION. $\mathbf{M} \vdash A^g \leftrightarrow A^k, \quad \mathbf{I} \vdash A^g \leftrightarrow A^q$.

2.3.8A. ♠* Prove the proposition.

COROLLARY. *For formulas A not containing $\forall$, $\mathbf{C} \vdash \neg A$ iff $\mathbf{I} \vdash \neg A$.*

PROOF. Let $\mathbf{C} \vdash \neg A$, then $\mathbf{C} \vdash \neg A^g$, hence $\mathbf{M} \vdash \neg A^g$. Now $(\neg A)^g \equiv \neg A^g$, and by the proposition $\mathbf{I} \vdash (\neg A)^g \leftrightarrow (\neg A)^q$. But $(\neg A)^q \equiv \neg\neg\neg A$, and $\mathbf{I} \vdash \neg\neg\neg A \to \neg A$. ⊠

2.4 Hilbert systems

Hilbert systems, H-systems for short, are very convenient in proofs of many metamathematical properties established by induction on lengths of deductions. But the main theme of this text contains cutfree and normalizable systems, so we shall not return to Hilbert systems after this section, which is mainly devoted to a proof of equivalence of Hilbert systems with other systems studied here.

By a Hilbert system we mean an axiomatization with axioms and as sole rules →E and ∀I; so there are no rules which close hypotheses (= assumptions). In a more liberal concept of Hilbert formalism one can permit other rules besides or instead of →E and ∀I, provided that no rule closes assumptions. (For example, we could allow a rule: from A, B derive $A \wedge B$.)

2.4.1. DEFINITION. (*Hilbert systems* **Hc, Hm, Hi** *for* **C, M** *and* **I**) The axioms for **Hm** are

$$
\begin{array}{l}
A \to (B \to A), \quad (A \to (B \to C)) \to ((A \to B) \to (A \to C)); \\
A \to A \vee B, \quad B \to A \vee B; \\
(A \to C) \to ((B \to C) \to (A \vee B \to C)); \\
A \wedge B \to A, \quad A \wedge B \to B, \quad A \to (B \to (A \wedge B)); \\
\forall x A \to A[x/t], \quad A[x/t] \to \exists x A; \\
\forall x(B \to A) \to (B \to \forall y A[x/y]) \quad (x \notin \mathrm{FV}(B),\ y \equiv x \text{ or } y \notin \mathrm{FV}(A)); \\
\forall x(A \to B) \to (\exists y A[x/y] \to B) \quad (x \notin \mathrm{FV}(B),\ y \equiv x \text{ or } y \notin \mathrm{FV}(A)).
\end{array}
$$

Hi has in addition the axiom $\bot \to A$, and **Hc** is **Hi** plus an additional axiom schema $\neg\neg A \to A$ (*law of double negation*). Instead of the law of double negation, one can also take the *law of the excluded middle* $A \vee \neg A$.
Rules for deductions from a *set* of assumptions Γ:

Ass If $A \in \Gamma$, then $\Gamma \vdash A$.

→E If $\Gamma \vdash A \to B$, $\Gamma \vdash A$, then $\Gamma \vdash B$;

∀I If $\Gamma \vdash A$, then $\Gamma \vdash \forall y A[x/y]$, $(x \notin \mathrm{FV}(\Gamma),\ y \equiv x \text{ or } y \notin \mathrm{FV}(A))$.

→E is also known as *Modus Ponens* (MP), and ∀I as the rule of *Generalization* (G).

Deductions from assumptions Γ may be exhibited as prooftrees, where axioms and assumptions from Γ appear at the top nodes, and lower nodes are formed either by the single-premise rule ∀I or by the two-premise rule →E. As observed already in 1.3.9, quite often the notion of deduction is presented in linear format, as follows:

$A_1, \ldots, A_n$ is said to be a deduction of A from Γ, if $A_n \equiv A$, and each A_i is either an element of Γ, or an instance of a logical axiom, or follows from A_j, $j < i$, by ∀I, or follows from A_j and A_k with j and $k < i$, by →E. ⊠

2.4.2. THEOREM. **H[mic]** *and* **N[mic]** *are equivalent, i.e.* $\Gamma \vdash A$ *in* **H[mic]** *iff* $\vdash \Gamma \Rightarrow A$ *in* **N[mic]**.

PROOF. We concentrate on the intuitionistic case. The direction from left to right is straightforward: we only have to check that all axioms of **Hi** are in fact derivable in **Ni** (exercise 2.1.8A, example 1.3.3); the rules $\to$E, $\forall$I are also available in **Ni**.

Now as to the direction from right to left. First we show how to transform a deduction in **Ni** into a deduction in the intermediate system with axioms of **Hi** and rules $\to$I, $\forall$I and $\to$E only, by induction on the height of deductions. There are as many cases as there are rules. On the left hand side we show a derivation of height $k+1$ terminating in $\wedge$I, $\vee$E, $\exists$E; on the right we indicate the transformation into a deduction with axioms and $\to$I and $\to$E only. By induction hypothesis, $\mathcal{D}_1, \mathcal{D}_2, \mathcal{D}_3$ have already been transformed into $\mathcal{D}_1', \mathcal{D}_2', \mathcal{D}_3'$ respectively.

$$\dfrac{\begin{array}{c}\mathcal{D}_1\\A\end{array} \quad \begin{array}{c}\mathcal{D}_2\\B\end{array}}{A\wedge B} \quad \overset{h}{\mapsto} \quad \dfrac{\dfrac{A\to(B\to(A\wedge B)) \quad \begin{array}{c}\mathcal{D}_1'\\A\end{array}}{B\to(A\wedge B)} \quad \begin{array}{c}\mathcal{D}_2'\\B\end{array}}{A\wedge B}$$

$$\dfrac{\begin{array}{c}\mathcal{D}_1\\A\vee B\end{array} \quad \begin{array}{c}[A]\\\mathcal{D}_2\\C\end{array} \quad \begin{array}{c}[B]\\\mathcal{D}_3\\C\end{array}}{C} \quad \overset{h}{\mapsto} \quad \dfrac{\dfrac{\dfrac{(A{\to}C){\to}((B{\to}C){\to}(A{\vee}B{\to}C)) \quad \dfrac{\begin{array}{c}[A]\\\mathcal{D}_2'\\C\end{array}}{A\to C}}{(B{\to}C){\to}(A{\vee}B{\to}C)} \quad \dfrac{\begin{array}{c}[B]\\\mathcal{D}_3'\\C\end{array}}{B\to C}}{A{\vee}B{\to}C} \quad \begin{array}{c}\mathcal{D}_1'\\A\vee B\end{array}}{C}$$

$$\dfrac{\begin{array}{c}\mathcal{D}_1\\\exists xA\end{array} \quad \begin{array}{c}[A[x/y]]\\\mathcal{D}_2\\C\end{array}}{C} \quad \overset{h}{\mapsto} \quad \dfrac{\dfrac{\forall x(A{\to}C){\to}(\exists xA{\to}C) \quad \dfrac{\dfrac{\begin{array}{c}[A[x/y]]\\\mathcal{D}_2'\\C\end{array}}{A[x/y]{\to}C}}{\forall x(A{\to}C)}}{\exists xA\to C} \quad \begin{array}{c}\mathcal{D}_1'\\\exists xA\end{array}}{C}$$

In the second prooftree it is tacitly assumed that $x \notin \mathrm{FV}(C)$; if this is not the case, we must do some renaming of variables. Alternatively, we may rely on our convention that formulas and prooftrees are identified if differing only in the names of bound variables. The remaining cases are left as an exercise.

Now we shall show, by induction on the height of a deduction in the intermediate system, how to remove all applications of $\to$I. First we recall that $A \to A$ can be proved in **Hi** by the standard deduction $\mathcal{D}_A$ exhibited in subsection 1.3.9.

Consider any deduction $\mathcal{D}$ with axioms, open hypotheses and the rules $\to$I and $\to$E. Suppose $\mathcal{D}$ ends with $\to$I, say $\mathcal{D}$ is of the form

$$\dfrac{\begin{array}{c}[A]^x\\ \mathcal{D}_1\\ B\end{array}}{A \to B}\,x$$

and let $\mathcal{D}_1'$ be the result of eliminating $\to$I from $\mathcal{D}_1$ (induction hypothesis). We have to show how to eliminate the final $\to$I from

$$\dfrac{\begin{array}{c}[A]^x\\ \mathcal{D}_1'\\ B\end{array}}{A \to B}\,x$$

We do this by showing how to transform each subdeduction $\mathcal{D}_0$ (with conclusion C say) of $\mathcal{D}_1'$ into a deduction

$$\begin{array}{c}\mathcal{D}_0^*\\ A \to C\end{array}$$

by induction on the height of $\mathcal{D}_0$.

Basis. A top formula occurrence C not in the class indicated by $[A]^x$ in $\mathcal{D}_1'$ is replaced by

$$\dfrac{C \to (A \to C) \qquad C}{A \to C}$$

and the top formula occurrences in $[A]^x$ are replaced by $\mathcal{D}_A$.

Induction step. $\mathcal{D}_0$ ends with $\to$E or with $\forall$I. We indicate the corresponding step in the construction of $\mathcal{D}_0^*$ below.

$$\dfrac{\begin{array}{c}\mathcal{D}_2\\ D \to C\end{array} \quad \begin{array}{c}\mathcal{D}_3\\ D\end{array}}{C} \quad \text{goes to} \quad \dfrac{\dfrac{(A{\to}D{\to}C){\to}(A{\to}D) \to A{\to}C \qquad \begin{array}{c}\mathcal{D}_2^*\\ A{\to}D{\to}C\end{array}}{(A{\to}D){\to}A \to C} \qquad \begin{array}{c}\mathcal{D}_3^*\\ A \to D\end{array}}{A \to C}$$

and for $C \equiv \forall x B$

$$\dfrac{\begin{array}{c}\mathcal{D}_2\\ B[x/y]\end{array}}{\forall x B} \quad \text{goes to} \quad \dfrac{\forall x(A{\to}B){\to}(A \to \forall x B) \qquad \dfrac{\begin{array}{c}\mathcal{D}_2^*\\ A \to B[x/y]\end{array}}{\forall x(A \to B)}}{A \to \forall x B}$$

By induction hypothesis, $\mathcal{D}_2^*, \mathcal{D}_3^*$ have already been constructed for $\mathcal{D}_2, \mathcal{D}_3$. The case for minimal logic is contained in the argument above. The extension to the classical case we leave as an exercise. ⊠

REMARKS. (i) This argument may also be read as showing that **Hi** itself is closed under $\rightarrow$I, i.e. it shows how to construct a proof of $\Gamma \vdash A \rightarrow B$ from a proof of $\Gamma, A \vdash B$. This is called the *deduction theorem* for **Hi**.

(ii) The compactness of term notation is well illustrated by the following example. Suppose we want to show that the rule $\vee$E can be replaced by instances of the axiom $(A{\rightarrow}C) \rightarrow (B{\rightarrow}C) \rightarrow (A \vee B \rightarrow C)$. Let $s_0(x^A){:}\,C$, $s_1(y^B){:}\,C$, $t{:}\,A{\vee}B$ be given, and let $\mathbf{d}$ be a constant for the axiom. Then $\mathbf{d}(\lambda x^A.s_0^C)(\lambda y^B.s_1^C)t^{A\vee B}{:}\,C$ takes the place of the voluminous prooftree occurring at the relevant place in the argument above.

(iii) If we concentrate on implication logic, we see that the method of eliminating an open assumption A, in the proof of the deduction theorem, is *exactly* the same, step for step, as the definition of the abstraction operator $\lambda^* x^A$ in combinatory logic (1.2.19). Consider, for example, the induction step in this construction. By induction hypothesis, we have constructed on the prooftree side deductions of $\Gamma \vdash A \rightarrow (C \rightarrow D)$, corresponding to a term $\lambda^* x^A.t^{C\rightarrow D}$, and of $\Gamma \vdash A \rightarrow C$, corresponding to a term $\lambda^* x^A.s^C$. The deduction of $A \rightarrow D$ constructed from this corresponds to

$$\dfrac{\dfrac{\mathbf{s}^{A,C,D}{:}\,(A(CD))((AC)(AD)) \qquad \lambda^* x^A.t^{C\rightarrow D}{:}\,A(CD)}{\mathbf{s}(\lambda^* x^A.t){:}\,(AC)(AD)} \qquad \lambda^* x^A.s^C{:}\,AC}{\mathbf{s}(\lambda^* x.t)(\lambda^* x.s){:}\,AD}$$

where $\rightarrow$ has been dropped, that is to say, EF is short for $E \rightarrow F$.

2.4.2A. ♠ Do the remaining cases of the transformation of a natural deduction in the intermediate system.

2.4.2B. ♠ Show the equivalence of **Hc** and **Nc**.

2.4.2C. ♠ Show that **Hi** with $\neg$ as primitive operator may be axiomatized by replacing the axiom schema $\bot \rightarrow A$ by $A \rightarrow (\neg A \rightarrow B)$ and $(A \rightarrow B) \rightarrow (\neg B \rightarrow \neg A)$.

2.4.2D. ♠* An alternative axiomatization for $\rightarrow$**Hm** is obtained by taking as rule modus ponens, and as axiom schemas $A \rightarrow B \rightarrow A$, $(A \rightarrow A \rightarrow B) \rightarrow A \rightarrow B$ (contraction), $(A \rightarrow B \rightarrow C) \rightarrow (B \rightarrow A \rightarrow C)$ (permutation), $(A \rightarrow B) \rightarrow (C \rightarrow A) \rightarrow (C \rightarrow B)$ (when combined with permutation this is just transitivity of implication). Prove the equivalence.

2.4.2E. ♠ Show that a Hilbert system for $\forall\rightarrow\wedge$**M** is obtained by taking as the only rule $\rightarrow$E, and as axioms $\forall\vec{x}F$, where F is a formula of one of the following forms: $A \rightarrow B \rightarrow A$; $(A \rightarrow B \rightarrow C) \rightarrow (A \rightarrow B) \rightarrow A \rightarrow C$; $A \rightarrow B \rightarrow A \wedge B$; $A_0 \wedge A_1 \rightarrow A_i$ $(i \in \{0,1\})$; $B \rightarrow \forall y B$ $(y \notin \mathrm{FV}(B))$; $\forall y(B[x/y] \rightarrow C[x/y]) \rightarrow$

$\forall xB \to \forall xC$ ($y \notin \mathrm{FV}(\forall x(B \to C))$; $\forall xA \to A[x/t]$. Can you extend this to full **M**?

2.4.2F. ♠ Show that the following axiom schemas and rules yield a Hilbert system for **Ip**: (1) $A \to A$, (2) if A, $A \to B$ then B, (3) if $A \to B$ and $B \to C$ then $A \to C$, (4) $A \wedge B \to A$, $A \wedge B \to B$, (5) $A \to A \vee B$, $B \to A \vee B$, (6) if $A \to C$, $B \to C$ then $A \vee B \to C$, (7) if $A \to B$, $A \to C$ then $A \to B \wedge C$, (8) if $A \wedge B \to C$ then $A \to (B \to C)$, (9) if $A \to (B \to C)$ then $A \wedge B \to C$, (10) $\bot \to A$. This is an example of a Hilbert system for propositional logic with more rules than just modus ponens. (Spector [1962], Troelstra [1973].)

2.5 Notes

2.5.1. *The Brouwer–Heyting–Kolmogorov interpretation.* This interpretation (BHK-interpretation for short) of intuitionistic logic explains what it means to prove a logically compound statement in terms of what it means to prove its components; the explanations use the notions of *construction* and *constructive proof* as unexplained primitive notions. For atomic formulas the notion of proof is supposed to be given. For propositional logic the clauses of BHK are

- p proves $A \wedge B$ iff p is a pair (p_0, p_1) and p_0 proves A, p_1 proves B,
- p proves $A \vee B$ iff p is either of the form $(0, p_1)$, and p_1 proves A, or of the form $(1, p_1)$ and p_1 proves B,
- p proves $A \to B$ iff p is a construction transforming any proof c of A into a proof $p(c)$ of B,
- $\bot$ is a proposition without proof.

It will be clear that $A \vee \neg A$ (that is to say, $A \vee (A \to \bot)$) is not generally valid in this interpretation, since the validity of $A \vee \neg A$ would mean that for *every* proposition A we can either prove A or refute A, that is to say we can either prove A or give a construction which obtains a contradiction from any alleged proof of A. (The only way of making $A \vee \neg A$ generally valid would be to give "proof" and "construction" a non-standard, obviously unintended interpretation.) More information on the BHK-interpretation and its history may be found in Troelstra and van Dalen [1988, 1.3, 1.5.3], Troelstra [1983, 1990].

For predicate logic we may add clauses:

- p proves $(\forall x{\in}D)A$ if p is a construction such that for all $d \in D$, $p(d)$ proves $A[x/d]$,

- p proves $(\exists x{\in}D)A$ if p is of the form (d, p') with d an element of D, and p' a proof of $A[x/d]$.

2.5.2. *Natural deduction.* Gentzen [1935] introduced natural deduction systems NJ and NK for intuitionistic and classical logic respectively. NJ is like **Ni**, except that $\neg$ is treated as a primitive with two rules

$$\dfrac{\begin{matrix}[A]^u \\ \mathcal{D} \\ \bot\end{matrix}}{\neg A}\,\neg\mathrm{I},u \qquad\qquad \dfrac{\begin{matrix}\mathcal{D}_1 \\ A\end{matrix} \quad \begin{matrix}\mathcal{D}_2 \\ \neg A\end{matrix}}{\bot}\,\neg\mathrm{E}$$

which reduce to instances of $\to$I, $\to$E if $\neg A$ is defined. Gentzen's NK is obtained from NJ by adding axioms $A \vee \neg A$. Gentzen was not the first to introduce this type of formalism; before him, Jaśkowski [1934] gave such a formalism (in linear, not in tree format) for classical logic (cf. Curry [1963, p. 249]).

Gentzen's examples, and his description of the handling of open and closed assumptions, may be interpreted as referring to the N-systems as described here, but are also compatible with CDC. In the latter case, however, Gentzen's marking of discharged assumptions would be redundant. On the other hand, Gentzen's examples are all compatible with CDC.

Subsection 2.1.10 improves the discussion of Troelstra [1999, p. 99], where the system $\mathbf{N}_3$ is misstated; $\mathbf{N}_1$ as defined here is the correct version.

In Curry [1950,1963] natural deduction is treated in the same manner as done by Gentzen. For classical systems he considers several formulations, taking as his basic one the intuitionistic system with a rule already considered by Gentzen:

$$\frac{\neg\neg A}{A}$$

However, *in the absence of negation* Curry includes for the classical systems the *Peirce rule* P:

$$\dfrac{\begin{matrix}[A \to B]^u \\ \mathcal{D} \\ A\end{matrix}}{A}\,\mathrm{P},u$$

Beth [1962b,1962a] also considers natural deduction for **C** with the Peirce rule. Our **Nm**, **Ni** and **Nc** coincide with the systems for **M**, **I** and **C** in Prawitz [1965] respectively, except that Prawitz adopts CDC as his standard convention. The more liberal convention concerning the closure of open assumptions is also described by Prawitz, but actually used by him only in his discussion of the normalization for a natural deduction system for the modal logic **S4**.

Many different presentation styles for systems of natural deduction are considered in the literature. For example, Jaśkowski [1934] and Smullyan [1965,1966] present the proofs in linear style, with nested "boxes"; when a new assumption is introduced, all formulas derived under that assumption are placed in a rectangular box, which is closed when the assumption is discharged. See also Prawitz [1965, appendix C].

Although natural deduction systems were not the exclusive discovery of Gentzen, they certainly became widely known and used as a result of Gentzen [1935]. However, we have reserved the name "Gentzen system" for the calculi with left- and right-introduction rules (discussed in the next chapter), since not only are they exclusively due to Gentzen, but it was also for these formalisms, not for the N-systems, that Gentzen formulated a basic metamathematical result, namely cut elimination. The words used by Gentzen in the preamble to Gentzen [1935] indicate that he had something like normalization for NJ, but not for NK; however, he did not present his proof for NJ.

2.5.3. *Hilbert systems.* Kleene [1952a] uses the term "Hilbert-type system"; this was apparently suggested by Gentzen [1935], who speaks of "einem dem Hilbertschen Formalismus angeglichenen Kalkül". Papers and books such as Hilbert [1926,1928], Hilbert and Ackermann [1928], Hilbert and Bernays [1934] have made such formalisms widely known, but they date from long before Hilbert; already Frege [1879] introduced a formalism of this kind (if one disregards the enormous notational differences). We have simplified the term "Hilbert-type system" to "Hilbert system". There is one aspect in which our system differs from the systems used by Hilbert and Frege: they stated the axioms, not as schemas, but with proposition variables and/or relation variables for A, B, C, and added a *substitution rule.* As far as we know, von Neumann [1927] was the first to use axiom schemas. Extensive historical notes may be found in Church [1956, section 29].

Many different Hilbert systems for **I** and **C** appear in the literature. Our systems are fairly close to the formalisms used in Kleene [1952a]. The axioms for implication in exercise 2.4.2D are the ones adopted by Hilbert [1928].

Kolmogorov [1925] gave a Hilbert system for minimal logic for the fragment $\neg$, $\to$, $\forall$, $\exists$. The name "minimal calculus" (German: Minimalkalkül) or "minimal logic" was coined by Johansson [1937], who was the first to give a formalization w.r.t. all operators in the form of a Gentzen system.

Glivenko [1929] contains a partial axiomatization for the intuitionistic logic of $\to, \wedge, \vee, \neg$. The first (Hilbert-type) axiomatization for the full system **I** is in Heyting [1930a,1930b]. Heyting [1930b] attempts unsuccessfully to use a formal substitution operator, and makes an incompletely realized attempt to take "partial terms" into account, i.e. terms which need not always denote something. Heyting's propositional rule "If A, B, then $A \wedge B$" was shown to be redundant by Bernays (see his letter to Heyting, reproduced in Troelstra [1990]). Bernays also considered the problem of formalizing intuitionistic logic, and noted that a suitable formalism could be obtained by dropping $\neg\neg A \to A$ from Hilbert's formalism, but these results were not published.

The *deduction theorem*, which is crucial to our proof of equivalence of natural deduction and Hilbert systems, was discovered several times independently. The first published proof appears to be the one by Herbrand [1930] (already announced in Herbrand [1928]). Tarski [1956, p.32, footnote] claims earlier, unpublished discovery in 1921. For more historical information see Curry [1963, p. 249].

2.5.4. *Rule of detachment.* Interesting variants of Hilbert systems for propositional logics are systems based on the so-called *Condensed Detachment rule* (CD). These systems are based on axioms and the rule

$$\mathrm{CD}\,\frac{A \to B \qquad C}{\mathrm{cd}(A \to B, C)}$$

where $\mathrm{cd}(A \to B, C)$ is defined as follows. Let $\mathrm{PV}(F)$ be the set of propositional variables in F. If A and C have a common substitution instance, let $D = A\sigma_1 = C\sigma_2$ (σ_1, σ_2 substitutions defined on (part of) the propositional variables of A and C respectively) be a *most general* common substitution instance such that $\mathrm{PV}(D) \cap (\mathrm{PV}(B) \setminus \mathrm{PV}(A)) = \emptyset$; then take $\mathrm{cd}(A \to B, C)$ to be $B\sigma_1$. (There is an algorithm for finding a most general common substitution instance, namely the unification algorithm discussed in 7.2.11, where $\to$ is treated as a binary function constant.) It can be shown that, for example, $\to$-**M** is complete for the system based on the axioms obtained from the schemas **k, s** (1.3.9) and **b, c, w** (2.1.8C) by choosing distinct propositional variables P, Q, R for A, B, C, plus the rule CD. For more information on this see Hindley and Meredith [1990], Hindley [1997, chapter 6].

2.5.5. *Negative translation.* See Kolmogorov [1925], Gödel [1933b], Gentzen [1933a], Kuroda [1951]. More on the negative translation and its variants, as well as stronger conservativity results for **C** relative to **I** and **M**, may be found in Troelstra and van Dalen [1988, section 2.3].

2.5.6. *Formulas-as-types.* For intuitionistic implication logic, the idea is clearly present in Curry and Feys [1958, sections 9E–F]; in embryonic form already in Curry [1942, p. 60, footnote 28]; the first hint is perhaps found in Curry [1934, p. 588]. The idea was not elaborated and/or exploited by Curry, possibly for the following two reasons: the parallel presents itself less forcefully in the setting of type-assignment systems than in the case of type theories with rigid typing, and, related to this, the parallel did not seem relevant to the problems Curry was working on. In any case it is a fact that the parallel is not even mentioned in Curry [1963].

In Howard [1980] (informally circulating since 1969), the parallel is made explicit for all the logical operators (with some credit to P. Martin-Löf). N. G. de Bruijn has been developing a language AUTOMATH for the writing and checking of mathematical proofs since 1967; he independently arrived at formulas-as-types to deal with logic in his language. The logical community at large became only slowly aware of this work. There is now an excellent account in Nederpelt et al. [1994], with an introduction and reproduction of the more important papers on AUTOMATH, many of which had not been widely accessible before.

Chapter 3

Gentzen systems

Gentzen [1935] introduced his calculi LK, LJ as formalisms more amenable to metamathematical treatment than natural deduction. For these systems he developed the technique of cut elimination. Even if nowadays normalization as an "equivalent" technique is widely used, there are still many reasons to study calculi in the style of LK and LJ (henceforth to be called *Gentzen calculi* or *Gentzen systems*, or simply *G-systems*):

- Where *normal* natural deductions are characterized by a restriction on the form of the proof – more precisely, a restriction on the order in which certain rules may succeed each other – cutfree Gentzen systems are simply characterized by the absence of the Cut rule.

- Certain results are more easily obtained for cutfree proofs in G-systems than for normal proofs in N-systems.

- The treatment of classical logic in Gentzen systems is more elegant than in N-systems.

The Gentzen systems for **M**, **I** and **C** have many variants. There is no reason for the reader to get confused by this fact. Firstly, we wish to stress that in dealing with Gentzen systems, no particular variant is to be preferred over all the others; one should choose a variant suited to the purpose at hand. Secondly, there is some method in the apparent confusion.

As our basic system we present in the first section below a slightly modified form of Gentzen's original calculi LJ and LK for intuitionistic and classical logic respectively: the G1-calculi. In these calculi the roles of the logical rules and the so-called structural rules are kept distinct.

It is possible to absorb the structural rules into the logical rules; this leads to the formulation of the G3-calculi (section 3.5) with the G2-calculi (not very important in their own right) as an intermediate step. Finally we formulate (section 3.6) for classical logic the Gentzen–Schütte systems (GS-systems), exploiting the De Morgan dualities. The use of one-sided sequents practically halves the number of rules. In later chapters we shall encounter the G4-

and G5-systems, designed for special purposes. Two sections respectively introduce the Cut rule, and establish deductive equivalence between N- and G-systems.

3.1 The G1- and G2-systems

The Gentzen systems **G1c**, **G1i** below (for classical and intuitionistic logic) are almost identical with the original Gentzen calculi LK and LJ respectively. The systems derive *sequents*, that is to say expressions $\Gamma \Rightarrow \Delta$, with Γ, Δ finite multisets (not sequences, as for Gentzen's LJ, LK); for the notational conventions in connection with finite multisets, see 1.1.5.

3.1.1. DEFINITION. (*The Gentzen systems* **G1c**,**G1m**,**G1i**) Proofs or deductions are labelled finite trees with a single root, with axioms at the top nodes, and each node-label connected with the labels of the (immediate) successor nodes (if any) according to one of the rules. The rules are divided into *left- (L-)* and *right- (R-) rules.* For a logical operator $\circledast$ say, L$\circledast$, R$\circledast$ indicate the rules where a formula with $\circledast$ as main operator is introduced on the left and on the right respectively. The axioms and rules for **G1c** are:

Axioms

$$\text{Ax}\ \ A \Rightarrow A \qquad\qquad \text{L}\bot\ \ \bot \Rightarrow$$

Rules for weakening (W) and contraction (C)

$$\text{LW}\ \frac{\Gamma \Rightarrow \Delta}{A, \Gamma \Rightarrow \Delta} \qquad \text{RW}\ \frac{\Gamma \Rightarrow \Delta}{\Gamma \Rightarrow \Delta, A}$$

$$\text{LC}\ \frac{A, A, \Gamma \Rightarrow \Delta}{A, \Gamma \Rightarrow \Delta} \qquad \text{RC}\ \frac{\Gamma \Rightarrow \Delta, A, A}{\Gamma \Rightarrow \Delta, A}$$

Rules for the logical operators

$$\text{L}\wedge\ \frac{A_i, \Gamma \Rightarrow \Delta}{A_0 \wedge A_1, \Gamma \Rightarrow \Delta}\ (i = 0, 1) \qquad \text{R}\wedge\ \frac{\Gamma \Rightarrow \Delta, A \qquad \Gamma \Rightarrow \Delta, B}{\Gamma \Rightarrow \Delta, A \wedge B}$$

$$\text{L}\vee\ \frac{A, \Gamma \Rightarrow \Delta \qquad B, \Gamma \Rightarrow \Delta}{A \vee B, \Gamma \Rightarrow \Delta} \qquad \text{R}\vee\ \frac{\Gamma \Rightarrow \Delta, A_i}{\Gamma \Rightarrow \Delta, A_0 \vee A_1}\ (i = 0, 1)$$

$$\text{L}\!\rightarrow\ \frac{\Gamma \Rightarrow \Delta, A \qquad B, \Gamma \Rightarrow \Delta}{A \rightarrow B, \Gamma \Rightarrow \Delta} \qquad \text{R}\!\rightarrow\ \frac{A, \Gamma \Rightarrow \Delta, B}{\Gamma \Rightarrow \Delta, A \rightarrow B}$$

$$\text{L}\forall\ \frac{A[x/t], \Gamma \Rightarrow \Delta}{\forall x A, \Gamma \Rightarrow \Delta} \qquad \text{R}\forall\ \frac{\Gamma \Rightarrow \Delta, A[x/y]}{\Gamma \Rightarrow \Delta, \forall x A}$$

$$\text{L}\exists\ \frac{A[x/y], \Gamma \Rightarrow \Delta}{\exists x A, \Gamma \Rightarrow \Delta} \qquad \text{R}\exists\ \frac{\Gamma \Rightarrow \Delta, A[x/t]}{\Gamma \Rightarrow \Delta, \exists x A}$$

where in L$\exists$, R$\forall$, y is not free in the conclusion.

The variable y in an application α of R$\forall$ or L$\exists$ is called the *proper* variable of α. The proper variable of α occurs only above α.

In the rules the Γ, Δ are called the *side formulas* or the *context*. In the conclusion of each rule, the formula not in the context is called the *principal* or *main* formula. In a sequent $\Gamma \Rightarrow \Delta$ Γ is called the *antecedent*, and Δ the *succedent*. The formula(s) in the premise(s) from which the principal formula derives (i.e. the formulas not belonging to the context) are the *active* formulas. (Gentzen calls such formulas "side formulas", which rather suggests an element of the context.) In the axiom Ax, both occurrences of A are principal; in L$\bot$ the occurrence of $\bot$ is principal.

The intuitionistic system **G1i** is the subsystem of **G1c** obtained by restricting all axioms and rules to sequents with *at most one* formula on the right, and replacing L$\to$ by

$$\text{L}\!\to\ \frac{\Gamma \Rightarrow A \qquad B, \Gamma \Rightarrow \Delta}{A \to B, \Gamma \Rightarrow \Delta}$$

G1m, the system for **M**, is **G1i** minus L$\bot$. Note that, due to the absence of L$\bot$, every sequent derivable in **G1m** must have a non-empty succedent, i.e., the succedent consists of a single formula. (This is straightforward by a simple induction on the depth of deductions.)

For the possibility of restricting the active formulas in Ax to prime A, see 3.1.9. ⊠

3.1.2. DEFINITION. As for N-systems, a convenient global assumption for deductions is that the proper variables of applications of L$\exists$ and R$\forall$ are kept distinct; this is called the *variable convention.*

If moreover the free and bound variables in a deduction are kept disjoint, the deduction is said to be a *pure-variable* deduction.We shall usually assume our deductions to satisfy the pure-variable condition ⊠

3.1.2A. ♠ Show that each deduction may be transformed into a pure-variable deduction.

3.1.3. EXAMPLES. (Some proofs in **G1c**,**G1m**) We have not explicitly indicated the rules used. The following two deductions are in **G1m**:

$$\dfrac{\dfrac{\dfrac{B \Rightarrow B}{A \wedge B \Rightarrow B} \quad \dfrac{A \Rightarrow A}{A \wedge B \Rightarrow A}}{A \wedge B \Rightarrow B \wedge A}}{\Rightarrow (A \wedge B) \to (B \wedge A)} \qquad \dfrac{\dfrac{\dfrac{A \Rightarrow A \quad \dfrac{A \Rightarrow A \quad \dfrac{B \Rightarrow B}{A, B \Rightarrow B}}{A, A \to B \Rightarrow B}}{A, A \to (A \to B) \Rightarrow B}}{A \to (A \to B) \Rightarrow A \to B}}{\Rightarrow (A \to (A \to B)) \to (A \to B)}$$

The fact that more than one formula may occur on the right enters essentially into certain classical proofs, for example the following two deductions in **G1c** (the left deduction derives *Peirce's Law*, in the right deduction $x \notin \mathrm{FV}(B)$):

$$\dfrac{\dfrac{\dfrac{\dfrac{P \Rightarrow P}{P \Rightarrow P, Q}}{\Rightarrow P, P \to Q} \quad P \Rightarrow P}{(P \to Q) \to P \Rightarrow P}}{\Rightarrow ((P \to Q) \to P) \to P} \qquad \dfrac{\dfrac{\dfrac{\dfrac{\dfrac{Ax \Rightarrow Ax}{Ax \Rightarrow Ax, B}}{\Rightarrow Ax, Ax \to B}}{\Rightarrow Ax, \exists x(Ax \to B)}}{\Rightarrow \forall x Ax, \exists x(Ax \to B)} \quad \dfrac{\dfrac{\dfrac{B \Rightarrow B}{B, Ax \Rightarrow B}}{B \Rightarrow Ax \to B}}{B \Rightarrow \exists x(Ax \to B)}}{(\forall x Ax \to B) \Rightarrow \exists x(Ax \to B)}$$

It not difficult to see that there are no proofs of the conclusions if we admit only sequents with at most one formula on the right.

The use of the contraction rule in the following two deductions (the left one is a deduction in **G1m**, the right one is a deduction both in **G1c** and **G1i**) cannot be avoided:

$$\dfrac{\dfrac{\dfrac{\dfrac{\dfrac{\dfrac{\dfrac{P \Rightarrow P}{P \Rightarrow P \vee \neg P} \quad \dfrac{\bot \Rightarrow \bot}{P, \bot \Rightarrow \bot}}{P, (P \vee \neg P) \to \bot \Rightarrow \bot}}{(P \vee \neg P) \to \bot \Rightarrow \neg P}}{(P \vee \neg P) \to \bot \Rightarrow P \vee \neg P} \quad \dfrac{\bot \Rightarrow \bot}{(P \vee \neg P) \to \bot, \bot \Rightarrow \bot}}{(P \vee \neg P) \to \bot, (P \vee \neg P) \to \bot \Rightarrow \bot}}{(P \vee \neg P) \to \bot \Rightarrow \bot}\,\mathrm{LC}}{\Rightarrow ((P \vee \neg P) \to \bot) \to \bot} \qquad \dfrac{\dfrac{\dfrac{\dfrac{\dfrac{P \Rightarrow P \quad \dfrac{\bot \Rightarrow}{P, \bot \Rightarrow}}{P, P \to \bot \Rightarrow}}{P \wedge \neg P, \neg P \Rightarrow}}{P \wedge \neg P, P \wedge \neg P \Rightarrow}}{P \wedge \neg P \Rightarrow}\,\mathrm{LC}}{P \wedge \neg P \Rightarrow \bot}$$

3.1.3A. ♠ Prove in **G1m** $A \Rightarrow A$ for arbitrary A from atomic instances $P \Rightarrow P$.

3.1.3B. ♠ Give sequent calculus proofs in **G1m** of

$A \wedge B \to A,\ \ A \wedge B \to B,\ \ A \to (B \to (A \wedge B)),$
$A \to A \vee B,\ \ B \to A \vee B,$
$(A \vee B \to C) \to ((A \to C) \wedge (B \to C)),$
$A \to (B \to A),$
$(A \to (B \to C)) \to ((A \to B) \to (A \to C)),$
$\forall x A \to A[x/t],\ \ A[x/t] \to \exists x A.$

3.1.3C. ♠* Prove in **G1i** that $(\neg\neg A \to \neg\neg B) \to \neg\neg(A \to B)$.

3.1.3D. ♠* Give sequent calculus proofs in **G1c** of

$$(A \to \exists x B) \to \exists x(A \to B) \quad (x \notin \mathrm{FV}(A))$$
$$\exists x(Ax \to \forall y Ay),$$
$$(A \to B) \vee (B \to A).$$

3.1.4. NOTATION. Some notational conventions in exhibiting deductions in sequent calculi:

- Double lines indicate some (possibly zero) applications of structural rules.
- In prooftrees the union of finite multisets $\Gamma, \Gamma', \Gamma'', \ldots$ of formulas is indicated simply by juxtaposition: $\Gamma\Gamma'\Gamma''$, or the multisets are separated by commas for greater readability: $\Gamma, \Gamma', \Gamma''$. The union of a multiset Γ with a singleton multiset $\{A\}$ is written ΓA or Γ, A.
- In prooftrees A^n stands for a multiset consisting of n copies of A; so A^0 is the empty multiset. ⊠

3.1.5. REMARK. *Context-sharing and contextfree rules.* In the two-premise rules, the contexts in both premises are the same (exception: succedent of intuitionistic L→, because of the restriction to at most one formula in the succedent). Rules with such a treatment of contexts are called *context-sharing.* But because of the presence of the structural rules of contraction and weakening, equivalent systems are obtained if some or all of these context-sharing rules are replaced by *context-independent* (*non-sharing, context-free*) versions, where the contexts of both premises are simply joined together. For example, the context-independent versions of L→ and R∧ are

$$\frac{\Gamma \Rightarrow A, \Delta \qquad \Gamma', B \Rightarrow \Delta'}{\Gamma, \Gamma', A \to B \Rightarrow \Delta, \Delta'} \qquad \frac{\Gamma \Rightarrow A, \Delta \qquad \Gamma' \Rightarrow B, \Delta'}{\Gamma, \Gamma' \Rightarrow A \wedge B, \Delta, \Delta'}$$

To see that the two versions of say R∧ are equivalent, consider

$$\frac{\dfrac{\Gamma \Rightarrow A\Delta}{\Gamma\Gamma' \Rightarrow A\Delta\Delta'}\,\mathrm{W} \qquad \dfrac{\Gamma' \Rightarrow B\Delta'}{\Gamma\Gamma' \Rightarrow B\Delta\Delta'}\,\mathrm{W}}{\Gamma\Gamma' \Rightarrow (A \wedge B)\Delta\Delta'} \qquad \frac{\dfrac{\Gamma \Rightarrow A\Delta \quad \Gamma \Rightarrow B\Delta}{\Gamma\Gamma \Rightarrow (A \wedge B)\Delta\Delta}}{\Gamma \Rightarrow (A \wedge B)\Delta}\,\mathrm{C}$$

If we replace in **G1c** the rule L→ by its non-sharing version, the intuitionistic version may be obtained by simply restricting attention to sequents with at most one formula on the right everywhere.

For "non-sharing" and "sharing" sometimes the terms *multiplicative* and *additive* respectively are used. This terminology derives from linear logic, where the distinction between sharing and non-sharing versions of the rules is crucial; the terminology was suggested by consideration of a particular type-theoretic model of linear logic (Girard domains). However, "context-free" and "context-sharing" as defined above apply to rules with more than one premise only, whereas "multiplicative" and "additive" also apply to rules with a single premise (see 9.3.1); hence, in chapter 9 the meaning of "context-free" and "context-sharing" will be extended and equated with "multiplicative" and "additive" respectively.

3.1.6. *The systems* **G2[mic]**. Due to the weakening rules, we obtain an equivalent system if we replace the axioms by the more general versions:

$$\Gamma, A \Rightarrow A, \Delta \qquad \bot, \Gamma \Rightarrow \Delta$$

This suggests consideration of:

DEFINITION. **G2c** is the system obtained from **G1c** by taking the generalized axioms and leaving out the weakening rules. The intuitionistic system **G2i** is the subsystem of **G2c** obtained by restricting all axioms and rules to sequents with *at most one* formula on the right. **G2m** is obtained from **G2i** by dropping the rule L⊥. ⊠

As for **G1m**, all sequents derivable in **G2m** have a single formula in the succedent.

3.1.7. PROPOSITION. *(Depth-preserving weakening, equivalence of G1- and G2-systems) Let us write* $\vdash_n \Gamma \Rightarrow \Delta$ *if there is a deduction of depth at most* n. *In* **G2[mic]**,

$$\textit{if } \vdash_n \Gamma \Rightarrow \Delta \textit{ then } \vdash_n \Gamma\Gamma' \Rightarrow \Delta\Delta'$$

where $|\Delta\Delta'| \leq 1$ *for* **G2[mi]**). *As a consequence,*

$$\mathbf{G1[mic]} \vdash \Gamma \Rightarrow \Delta \textit{ iff } \mathbf{G2[mic]} \vdash \Gamma \Rightarrow \Delta.$$

PROOF. By induction on the length of derivations. Or, starting at the bottom conclusion, and working our way upwards, we add Γ' and Δ' to the left and right side of sequents respectively, except when we encounter an application of intuitionistic L→, where we add Γ' in both premises, but Δ' in the right premise only. ⊠

(Essentially) the same proof works for the other systems we shall consider in the sequel. The proof of the next lemma is left to the reader.

3.1.8. LEMMA. *(Elimination of empty succedents) If $\Gamma \Rightarrow A$ is provable in* **G2**[**mi**], *then there is a proof which exclusively contains sequents with a single formula on the right. If $\Gamma \Rightarrow$ is provable, then there is also a proof of $\Gamma \Rightarrow A$, for any A.*

If $\Gamma \Rightarrow A$ is provable in **G1i**, *it is also provable in* **G1i***, *a system obtained from* **G1i** *by replacing the axiom* L$\bot$ *by the set of axioms $\bot \Rightarrow B$. The formula B may be restricted to being prime (in fact, even to atomic, since $\bot, \Gamma \Rightarrow \bot$ is an instance of* Ax*).*

3.1.8A. ♠ Carry out the proof of the preceding lemma.

3.1.9. PROPOSITION. *(Restriction to prime instances of axioms) The A in the axioms of* **G**[**12**][**mic**] *may be restricted to prime formulas, while 3.1.7 stays true under this restriction. In the systems* **G2**[**mic**] *the formulas in Γ, Δ in the axioms may also be assumed to be prime, but with these restrictions proposition 3.1.7 does not hold.* ⊠

3.1.9A. ♠ Prove the proposition.

3.1.10. REMARK. We can also restrict the axiom Ax to the case that A is atomic instead of just prime. The special case $\Gamma, \bot \Rightarrow \bot, \Delta$ is also an instance of L$\bot$ in **G2**[**ic**], and is derivable by weakening from L$\bot$ in **G1**[**ic**]. In the systems for minimal logic, we have to add $\bot \Rightarrow \bot$ or $\Gamma, \bot \Rightarrow \bot$.

3.2 The Cut rule

The systems introduced so far, namely **G**[**12**][**mic**], all obey the *subformula property*: in any deduction of a sequent $\Gamma \Rightarrow \Delta$, only subformulas of Γ and Δ occur. A consequence of this fact is the *separation property* for **G**[**12**][**mic**]: a proof of a sequent $\Gamma \Rightarrow \Delta$ requires logical rules only for the logical operators ($\bot$ is regarded as a 0-place operator) actually occurring in this sequent. This is no longer the case when the so-called Cut rule is added. The Cut rule

$$\text{Cut}\ \frac{\Gamma \Rightarrow \Delta, A \qquad A, \Gamma' \Rightarrow \Delta'}{\Gamma\Gamma' \Rightarrow \Delta\Delta'}$$

expresses a form of transitivity of $\Rightarrow$. The A in the instance exhibited is called the *cutformula*; an application of the rule Cut is called a *cut*. We may add this rule to our systems, but then the subformula property is no longer valid; the cutformula A in the premises is not necessarily a subformula of a formula in $\Gamma\Gamma'\Delta\Delta'$.

There is also a *context-sharing* version of the *Cut* rule, also called *additive Cut* (left classical, right intuitionistic):

$$\text{Cut}_{\text{cs}}\frac{\Gamma \Rightarrow \Delta A \qquad A\Gamma \Rightarrow \Delta}{\Gamma \Rightarrow \Delta} \qquad \text{Cut}_{\text{cs}}\frac{\Gamma \Rightarrow A \qquad A\Gamma \Rightarrow B}{\Gamma \Rightarrow B}$$

It is easy to see that due to the presence of weakening and contraction the addition of Cut_{cs} is equivalent to the addition of Cut.

Do new formulas become derivable by adding Cut? The answer is: not if we treat formulas modulo renaming of bound variables. The following example shows that the possibility of renaming bound variables is essential in predicate logic: without renaming bound variables, we cannot derive $\forall x\forall y(Ry \wedge Qx) \Rightarrow Qy$ (R, Q unary relation variables) in **G1c** or **G2c** without Cut. To see this, note that such a proof, say in **G2c**, ought to have the following structure:

$$\cfrac{\cfrac{\cfrac{Qt \Rightarrow Qy}{Rs \wedge Qt \Rightarrow Qy}}{\forall y(Ry \wedge Qt) \Rightarrow Qy}}{\forall x\forall y(Ry \wedge Qx) \Rightarrow Qy}$$

The top can only be an axiom if $t \equiv y$, but this is a forbidden substitution: t is not free for x in $\forall y(Ry \wedge Qx)$. Note that introducing contractions would not help in finding a proof. On the other hand, with Cut we can give a proof:

$$\cfrac{\cfrac{\cfrac{\cfrac{\cfrac{Qz \Rightarrow Qz}{Ry \wedge Qz \Rightarrow Qz}}{\forall y(Ry \wedge Qz) \Rightarrow Qz}}{\forall x\forall y(Ry \wedge Qx) \Rightarrow Qz}}{\forall x\forall y(Ry \wedge Qx) \Rightarrow \forall zQz} \qquad \cfrac{Qy \Rightarrow Qy}{\forall zQz \Rightarrow Qy}}{\forall x\forall y(Ry \wedge Qx) \Rightarrow Qy}\,\text{Cut}$$

The impossibility of finding a cutfree proof is obviously connected with the fact that the variables can occur both free and bound in the same sequent. As will be shown later, if we permit renaming so as to keep bound and free variables disjoint, the addition of the Cut rule becomes conservative.

3.2.1. THEOREM. *(Closure under Cut) Any sequent* $\Gamma \Rightarrow \Delta$ *in which no variable occurs both free and bound, and which is provable in* **G**[**12**][**mic**] + Cut, *is also provable in* **G**[**12**][**mic**].

The proof will be postponed till section 4.1. We note that we can restrict attention to so-called pure-variable deductions, defined in 3.1.2.

3.2.1A. ♠* We form a calculus m-**G1i** from **G1c** by changing R→, R∀ to

$$\frac{\Gamma, A \Rightarrow B}{\Gamma \Rightarrow A \rightarrow B, \Delta} \qquad \frac{\Gamma \Rightarrow A[x/y]}{\Gamma \Rightarrow \forall xA, \Delta}$$

Equivalent to m-**G1i** is the variant m-**G1i**′ with for R→, R∀

$$\frac{\Gamma, A \Rightarrow B}{\Gamma \Rightarrow A \to B} \qquad \frac{\Gamma \Rightarrow A[x/y]}{\Gamma \Rightarrow \forall x A}$$

Show that the resulting calculi are equivalent to **G1i** in the sense that **G1i** $\vdash \Gamma \Rightarrow \bigvee \Delta$ iff m-**G1i** $\vdash \Gamma \Rightarrow \Delta$ iff m-**G1i**′ $\vdash \Gamma \Rightarrow \Delta$ (where $\bigvee$ denotes iterated disjunction; the empty disjunction is identified with $\bot$). What restriction on R→, R∀ in **G2c** produces equivalence to **G2i** and **G1i**? *Hint.* Use closure under Cut for **G1i**.

3.2.1B. ♠ If m-**G1i** $\vdash \Gamma \Rightarrow \Delta$, $|\Delta| \geq 1$, and Γ does not contain $\vee$, then m-**G1i** $\vdash \Gamma \Rightarrow A$ for some $A \in \Delta$. Prove this fact (Dragalin [1979]).

3.2.1C. ♠ As a generalization of the preceding exercise, prove the following. Let Γ not contain $\vee, \exists$. If m-**G1i** $\vdash \Gamma \Rightarrow \Delta\Delta'$, $|\Delta\Delta'| \geq 1$, Δ containing existential formulas, then m-**G1i** $\vdash \Gamma \Rightarrow A$ for some $A \in \Delta'$ or m-**G1i** $\vdash \Gamma \Rightarrow A[x/t]$ for some $\exists x A \in \Delta$ (Dragalin [1979]).

3.3 Equivalence of G- and N-systems

In this section we establish the equivalence between the N-systems and the corresponding G-systems. Let us write **N[mic]** $\vdash \Gamma \Rightarrow A$ iff there is a context $\Gamma^* \equiv u_1{:}\, A_1, \ldots, u_n{:}\, A_n$, such that $\Gamma \equiv A_1, \ldots, A_n$, and **N[mic]** $\vdash \Gamma^* \Rightarrow A$ in the sequential notation variant (cf. 2.1.8); in other words, $\vdash \Gamma \Rightarrow A$ in an N-system if there is a prooftree deriving A using the open assumptions in Γ; the multiplicity of a formula B in Γ is equal to the number of inhabited assumption classes with distinct labels containing occurrences of B.

Observe that the N-systems are closed under contraction and weakening; that is to say, if $\vdash \Gamma BB \Rightarrow C$ then $\vdash \Gamma B \Rightarrow C$, and if $\vdash \Gamma \Rightarrow C$ then $\vdash \Gamma B \Rightarrow C$. Contraction is achieved by identifying the labels for two distinct assumption classes containing the same formula B. Weakening may be achieved as follows. Let $\mathcal{D}$ derive A from assumptions Γ. Then the deduction (x a "fresh" label)

$$\cfrac{\cfrac{\begin{array}{c}\mathcal{D}\\ A\end{array} \quad B^x}{A \wedge B}}{A}$$

derives A from Γ, B.

3.3.1. THEOREM. **G[12][mi]** + Cut $\vdash \Gamma \Rightarrow A$ *iff* **N[mi]** $\vdash \Gamma \Rightarrow A$.

PROOF. We give the proof for **G2i** + Cut, **Ni**. The result for **G2m** is contained in that for **G2i**. Moreover, **G1[mi]** is equivalent to **G2[mi]**, and by the closure under Cut, also **G1[mi]** + Cut is equivalent to **G2[mi]** + Cut.

For the proof from left to right we use the fact that a sequent $\Gamma \Rightarrow A$ can always be proved by a deduction where all sequents have exactly one formula in the succedent. The proof proceeds by induction on the depth of a deduction in **G2i**; at each step in the proof we show how to construct from a G-deduction of $\Gamma \Rightarrow A$ an N-deduction of $\Gamma' \Rightarrow A$ for some Γ' with $\mathrm{Set}(\Gamma') \subset \Gamma$.

Basis. The base case starts from axioms $\Gamma, A \Rightarrow A$ or $\Gamma, \bot \Rightarrow A$, corresponding to deductions consisting of a single node A and deductions $\frac{\bot}{A}$ respectively.

Induction step. For the induction step, we have to review all the rules. As IH, we assume that to each deduction $\mathcal{D}$ of $\Gamma \Rightarrow A$ of depth at most k in **G2i** + Cut a deduction $\mathcal{D}^*$ of $\Gamma' \Rightarrow A$, $\Gamma' \subset \mathrm{Set}(\Gamma)$ has been found. The R-rules correspond to introduction rules in **Ni**, for example (on the left the sequent calculus deduction, on the right the corresponding deduction in **Ni**)

$$\frac{\begin{array}{cc}\mathcal{D}_0 & \mathcal{D}_1\\ \Gamma \Rightarrow A & \Gamma \Rightarrow B\end{array}}{\Gamma \Rightarrow A \wedge B} \qquad \text{goes to} \qquad \frac{\begin{array}{cc}\mathcal{D}_0^* & \mathcal{D}_1^*\\ A & B\end{array}}{A \wedge B}$$

$$\frac{\begin{array}{c}\mathcal{D}\\ \Gamma, A \Rightarrow B\end{array}}{\Gamma \Rightarrow A \to B} \qquad \text{goes to} \qquad \frac{\begin{array}{c}[A]\\ \mathcal{D}^*\\ B\end{array}}{A \to B}$$

etc. For the L-rules, we have to replace assumptions at a top node by an E-rule application deriving the assumption. Examples:

$$\frac{\begin{array}{c}\mathcal{D}\\ A, \Gamma \Rightarrow C\end{array}}{A \wedge B, \Gamma \Rightarrow C} \qquad \text{goes to} \qquad \begin{array}{c}\dfrac{A \wedge B}{[A]}\\ \mathcal{D}^*\\ C\end{array}$$

$$\frac{\begin{array}{cc}\mathcal{D}_0 & \mathcal{D}_1\\ \Gamma \Rightarrow A & \Gamma, B \Rightarrow C\end{array}}{A \to B, \Gamma \Rightarrow C} \qquad \text{goes to} \qquad \begin{array}{c}\dfrac{A \to B \quad \begin{array}{c}\mathcal{D}_0^*\\ A\end{array}}{[B]}\\ \mathcal{D}_1^*\\ C\end{array}$$

etc. Cut is treated by substitution:

$$\frac{\begin{array}{cc}\mathcal{D}_0 & \mathcal{D}_1\\ \Gamma \Rightarrow A & \Gamma' A \Rightarrow B\end{array}}{\Gamma\Gamma' \Rightarrow B} \qquad \text{goes to} \qquad \begin{array}{c}\mathcal{D}_0^*\\ [A]\\ \mathcal{D}_1^*\\ B\end{array}$$

For the direction from right to left, we use induction on the depth of prooftrees in **Ni**.

Basis. A corresponds to $A \Rightarrow A$.

Induction step. Let us assume that for deductions in **Ni** of depth at most k, corresponding deductions $\mathcal{D}^+$ in **G2i** + Cut have been constructed. Again the I-rules correspond to R-rules in the sequent calculus, for example

$$\dfrac{\begin{matrix}\mathcal{D}_0 & \mathcal{D}_1\\ A & B\end{matrix}}{A \wedge B} \quad \text{goes to} \quad \dfrac{\dfrac{\dfrac{\overset{\mathcal{D}_0^+}{\Gamma_0 \Rightarrow A}}{\Gamma_0, \Gamma_1 \Rightarrow A} \quad \dfrac{\overset{\mathcal{D}_1^+}{\Gamma_1 \Rightarrow B}}{\Gamma_0, \Gamma_1 \Rightarrow B}}{\Gamma_0, \Gamma_1 \Rightarrow A \wedge B}}{(\Gamma_0, \Gamma_1) \Rightarrow A \wedge B}$$

(Γ_0, Γ_1) is short for $\text{Set}(\Gamma_0, \Gamma_1)$. The upper double line indicates some (possibly zero) weakenings; the lower double line refers to some (possibly zero) contractions. The elimination rules are translated with help of the corresponding L-rule and the Cut rule:

$$\dfrac{\overset{\mathcal{D}}{\bot}}{A} \quad \text{goes to} \quad \dfrac{\overset{\mathcal{D}^+}{\Gamma \Rightarrow \bot} \quad \bot \Rightarrow A}{\Gamma \Rightarrow A}\,\text{Cut}$$

$$\dfrac{\overset{\mathcal{D}}{A \wedge B}}{A} \quad \text{goes to} \quad \dfrac{\overset{\mathcal{D}^+}{\Gamma \Rightarrow A \wedge B} \quad \dfrac{A \Rightarrow A}{A \wedge B \Rightarrow A}}{\Gamma \Rightarrow A}\,\text{Cut}$$

$$\dfrac{\begin{matrix}\mathcal{D}_0 & \mathcal{D}_1\\ A \to B & A\end{matrix}}{B} \quad \text{goes to} \quad \dfrac{\dfrac{\overset{\mathcal{D}_0^+}{\Gamma_0 \Rightarrow A \to B} \quad \dfrac{\overset{\mathcal{D}_1^+}{\Gamma_1 \Rightarrow A} \quad B \Rightarrow B}{A \to B, \Gamma_1 \Rightarrow B}}{\Gamma_0, \Gamma_1 \Rightarrow B}\,\text{Cut}}{(\Gamma_0, \Gamma_1) \Rightarrow B}$$

etc. ⊠

Remark. The step for translating L→ in the first half of the proof is not uniformly "economical" as to the size of the translated proof tree, since $\mathcal{D}_0^*$ may have to be copied a number of times. More economical is the following translation:

$$\dfrac{\dfrac{\begin{matrix}[B]^x\\ \mathcal{D}_1^*\\ C\end{matrix}}{B \to C}\,x \quad \dfrac{A \to B \quad \overset{\mathcal{D}_0^*}{A}}{B}}{C}$$

A similar remark holds for the translation of the Cut rule.

3.3.1A. ♠ Complete the proof of the theorem.

3.3.2. *The classical case*

DEFINITION. *c-equivalence* between sequents is the reflexive, symmetric and transitive closure of the relation R consisting of the pairs $((\Gamma \Rightarrow \Delta, A), (\Gamma, \neg A \Rightarrow \Delta))$ and $((\Gamma, A \Rightarrow \Delta), (\Gamma \Rightarrow \Delta, \neg A))$. Thus sequents $\Gamma, \Delta, \neg\Theta \Rightarrow \Gamma', \neg\Delta', \Theta'$ and $\Gamma, \Delta', \neg\Theta' \Rightarrow \Gamma', \neg\Delta, \Theta$ are c-equivalent.

Sequents of the form $\Rightarrow \forall\vec{x}(\neg\neg A \to A)$ are called *stability axioms*; "Stab" is the set of all stability axioms. ⊠

LEMMA. *(Shifting from left to right and vice versa)*

(i) If S, S' are c-equivalent sequents, then **G1c**+Cut $\vdash S$ *iff* **G1c**+Cut $\vdash S'$;

(ii) in **G1i** + Cut *we have* $\vdash \neg\neg A \to A, \Gamma \Rightarrow A$ *if* $\vdash \neg\neg A \to A, \Gamma, \neg A \Rightarrow$, *and* $\vdash \Gamma, \neg A \Rightarrow$, *if* $\vdash \Gamma \Rightarrow A$, *and also* $\vdash \Gamma \Rightarrow \neg A$ *iff* $\vdash \Gamma, A \Rightarrow$.

PROOF. The proof follows from the following deductions:

$$\dfrac{\Gamma \Rightarrow A \qquad \dfrac{\bot \Rightarrow}{\Gamma, \bot \Rightarrow}}{\Gamma, \neg A \Rightarrow} \qquad\qquad \dfrac{\dfrac{\dfrac{\Gamma, \neg A \Rightarrow}{\Gamma, \neg A \Rightarrow \bot}}{\Gamma \Rightarrow \neg\neg A} \qquad \dfrac{A \Rightarrow A}{\Gamma, A \Rightarrow A}}{\neg\neg A \to A, \Gamma \Rightarrow A}$$

$$\dfrac{\dfrac{\Gamma, A \Rightarrow}{\Gamma, A \Rightarrow \bot}}{\Gamma \Rightarrow \neg A} \qquad\qquad \dfrac{\Gamma \Rightarrow \neg A \qquad \dfrac{A \Rightarrow A \quad \bot \Rightarrow}{\neg A, A \Rightarrow}}{\Gamma, A \Rightarrow}$$

permitting shift of A from left to right, $\neg A$ from left to right, A from right to left, and $\neg A$ from right to left respectively. ⊠

PROPOSITION. **G1c** + Cut $\vdash \Gamma \Rightarrow \Delta$ *iff* **G1i** + Cut + Stab $\vdash \Gamma' \Rightarrow \Delta'$, *where* $\Gamma' \Rightarrow \Delta'$ *is c-equivalent to* $\Gamma \Rightarrow \Delta$, *and* Δ' *contains at most one formula. Hence* **G1c** + Cut $\vdash \Gamma \Rightarrow A$ *iff* **G1i** + Cut + Stab $\vdash \Gamma \Rightarrow A$.

PROOF. The direction from right to left is proved by a straightforward induction on the length of deductions. The direction from left to right is also proved by induction on the length of deductions.

Note that, whenever we can prove in **G1i** + Cut + Stab a sequent $\Gamma' \Rightarrow \Delta'$ which is c-equivalent to $\Gamma \Rightarrow \Delta$, then we can prove in **G1i** + Cut + Stab all sequents $\Gamma'' \Rightarrow \Delta''$ which are c-equivalent to $\Gamma \Rightarrow \Delta$, by the lemma.

We illustrate two cases of the induction step.

Case 1. The final inference in the proof of $\Gamma \Rightarrow \Delta$ is L→:

$$\dfrac{\Gamma' \Rightarrow A, \Delta' \qquad \Gamma', B \Rightarrow \Delta'}{\Gamma', A \to B \Rightarrow \Delta'}$$

By the induction hypothesis we have deductions in **G1i** + Cut + Stab of $\Gamma', \neg\Delta' \Rightarrow A$ and $\Gamma', B, \neg\Delta' \Rightarrow$; apply L$\rightarrow$ and find $\Gamma', \neg\Delta', A \rightarrow B \Rightarrow$.
Case 2. Let the final inference be R$\vee$:

$$\frac{\Gamma \Rightarrow A, \Delta}{\Gamma \Rightarrow A \vee B, \Delta}.$$

By the induction hypothesis we have in **G1i**+Cut+Stab a proof of $\Gamma, \neg\Delta \Rightarrow A$, from which we obtain a proof of $\Gamma, \neg\Delta \Rightarrow A \vee B$. ⊠

REMARK. Inspection of the proof shows that in transforming a deduction $\mathcal{D}$ in **G1c** + Cut into a deduction in **G1i** + Cut + Stab, we need only instances of $\Rightarrow \forall\vec{x}(\neg\neg A \rightarrow A)$ for formulas A occurring in $\mathcal{D}$.

3.3.3. THEOREM. **G[12]c** + Cut $\vdash \Gamma \Rightarrow A$ *iff* **Nc** $\vdash \Gamma \Rightarrow A$. ⊠

REMARK. The proof and the statement of this theorem also apply to all $\mathcal{X}$-fragments of **C** for which $\{\rightarrow, \bot\} \subset \mathcal{X} \subset \{\rightarrow, \bot, \wedge, \vee, \forall, \exists\}$. For fragments containing $\rightarrow$, but not $\bot$, we must proceed differently. A method which applies to these fragments is given in Curry [1950], cf. exercise 3.3.3B below.

3.3.3A. ♠ Complete the proof of the proposition and theorem.

3.3.3B. ♠* (a) For N-systems, let us write $\vdash \Gamma \Rightarrow A$ if A can be deduced from assumptions in Γ. Let **Nc**$'$ be **Nc** with $\bot_c$ replaced by the Peirce rule P, defined in 2.5.2. Show **Nc** $\vdash \Gamma \Rightarrow A$ iff **Nc**$'$ $\vdash \Gamma \Rightarrow A$.

In the following two parts of the exercise we extend the equivalence between N-systems and G-systems to fragments $\mathcal{X}$ such that $\{\rightarrow\} \subset \mathcal{X} \subset \{\rightarrow, \vee, \wedge, \forall, \exists\}$. Let **G** be $\mathcal{X}$-**G2c** + Cut.

(b) If $\mathcal{S} \equiv \Gamma \Rightarrow A, \Delta$, then a sequent $\mathcal{S}^* \equiv \Gamma, \Delta \rightarrow A \Rightarrow A$ is called a 1-equivalent of $\mathcal{S}$; $\Delta \rightarrow A$ abbreviates $B_1 \rightarrow A, B_2 \rightarrow A, \ldots, B_n \rightarrow A$ for $\Delta \equiv B_1, B_2, \ldots, B_n$. Show that any two 1-equivalents $\mathcal{S}^*, \mathcal{S}^{**}$ of a sequent $\mathcal{S}$ are provably equivalent in **G**.

(c) Show that **G** $\vdash \Gamma \Rightarrow A$ iff $\mathcal{X}$-**Nc**$'$ $\vdash \Gamma \Rightarrow A$. *Hint.* For the proof from left to right, show by induction on the depth of deductions in **G**, that whenever **G** $\vdash \mathcal{S}$, then for some 1-equivalent $\mathcal{S}^*$ of $\mathcal{S}$, **Nc**$'$ $\vdash \mathcal{S}^*$ (Curry [1950]).

3.3.3C. ♠ Prove equivalence of **G1i** with the Hilbert system **Hi** directly, that is to say, not via the equivalence of **G1i** with natural deduction.

3.3.4. *From Gentzen systems to term-labelled calculi*

In the discussion of term assignments for intuitionistic sequent calculi, the versions where empty succedents are possible are less convenient. Hence we consider slight modifications **G1i***, **G2i*** (these are ad hoc notations), where all sequents have exactly one formula in the succedent (cf. lemma 3.1.8). In fact, there are two options. In the case of **G2i**, we can replace $\Gamma, \bot \Rightarrow \Delta$, ($|\Delta| \leq 1$) by the more restricted $\Gamma, \bot \Rightarrow A$, or we may instead add a rule

$$\frac{\Gamma \Rightarrow \bot}{\Gamma \Rightarrow A}$$

Similarly, we may obtain a system **G1i*** from **G1i** by replacing $\bot \Rightarrow$ by axioms $\bot \Rightarrow A$, or by a rule as above. For definiteness, we keep to the first possibility, the modification in the axioms (instead of the addition of a new rule).

It is instructive to describe the assignment of natural deduction proofs to proofs in the sequent calculus in another way, namely by formulating the sequent calculus as a calculus with terms; the terms denote the corresponding natural deduction proofs.

A term-labelled calculus t-**G2i** corresponding to **G2i*** may be formulated as follows. Consider a term $t{:}\,B$ with $\mathrm{FV}(t) = \{u_1{:}\,A_1, \ldots, u_n{:}\,A_n\}$ as representing a deduction of $A_1, \ldots, A_n, \Gamma \Rightarrow B$ for arbitrary Γ. The assignment then becomes:

$$\Gamma, u{:}\,P \Rightarrow u{:}\,P \quad \text{(axiom)} \qquad\qquad \Gamma, u{:}\,\bot \Rightarrow \mathrm{E}^{\bot}_{C}(u){:}\,C \quad \text{(axiom)}$$

$$\frac{u_i{:}\,A_i, \Gamma \Rightarrow t{:}\,C}{w{:}\,A_0 \wedge A_1, \Gamma \Rightarrow t[u_i/\mathbf{p}_i w]{:}\,C} \qquad\qquad \frac{\Gamma \Rightarrow t{:}\,A \quad \Gamma \Rightarrow s{:}\,B}{\Gamma \Rightarrow \mathbf{p}(t,s){:}\,A \wedge B}$$

$$\frac{u : A, \Gamma \Rightarrow t_0{:}\,C \qquad v{:}\,B, \Gamma \Rightarrow t_1{:}\,C}{w{:}\,A \vee B, \Gamma \Rightarrow \mathrm{E}^{\vee}_{u,v}(w, t_0, t_1){:}\,C} \qquad\qquad \frac{\Gamma \Rightarrow t{:}\,A_i}{\Gamma \Rightarrow \mathbf{k}_i t{:}\,A_0 \vee A_1}$$

$$\frac{\Gamma \Rightarrow t{:}\,A \qquad u{:}\,B, \Gamma \Rightarrow s{:}\,C}{w{:}\,A \to B, \Gamma \Rightarrow s[u/wt]{:}\,C} \qquad\qquad \frac{u{:}\,A, \Gamma \Rightarrow t{:}\,B}{\Gamma \Rightarrow \lambda u.t{:}\,A \to B}$$

$$\frac{u{:}\,A[x/y], \Gamma \Rightarrow t{:}\,C}{w{:}\,\exists x A, \Gamma \Rightarrow \mathrm{E}^{\exists}_{u,y}(w,t){:}\,C} \qquad\qquad \frac{\Gamma \Rightarrow s{:}\,A[x/t]}{\Gamma \Rightarrow \mathbf{p}(t,s){:}\,\exists x A}$$

$$\frac{u{:}\,A[x/t], \Gamma \Rightarrow s{:}\,C}{w{:}\,\forall x A, \Gamma \Rightarrow s[u/wt]{:}\,C} \qquad\qquad \frac{\Gamma \Rightarrow t[x/y]{:}\,A[x/y]}{\Gamma \Rightarrow \lambda x.t{:}\,\forall x A}$$

$$\frac{u{:}\,A, v{:}\,A, \Gamma \Rightarrow t{:}\,B}{w{:}\,A, \Gamma \Rightarrow t[u,v/w,w]{:}\,B}$$

The w is always a fresh variable.

The same assignment works for **G1i***, where weakening does not change the term assignment:

$$\frac{\Gamma \Rightarrow t{:}\,A}{\Gamma, z{:}\,B \Rightarrow t{:}\,A}$$

N.B. The obvious map from t-**G2i**-derivations to derivations in **G2i*** with non-empty succedent is not one-to-one, cf. the following two t-**G2i**-derivations:

$$\frac{\dfrac{x{:}\,A, y{:}\,A \Rightarrow y{:}\,A}{y{:}\,A \Rightarrow \lambda x.y{:}\,A{\to}A}}{\Rightarrow \lambda yx.y{:}\,A{\to}(A{\to}A)} \qquad \frac{\dfrac{x{:}\,A, y{:}\,A \Rightarrow y{:}\,A}{x{:}\,A \Rightarrow \lambda y.y{:}\,A{\to}A}}{\Rightarrow \lambda xy.y{:}\,A{\to}(A{\to}A)}$$

which map to the same derivation in **G2i***.

Note that if the substitutions in the terms are conceived as a syntactical operation in the usual way, we cannot, from the variables and the term of the conclusion alone, read off the sequent calculus proof. Thus, for example, the deductions

$$\frac{\dfrac{\dfrac{B \Rightarrow B \quad C \Rightarrow C}{B, C \Rightarrow B \wedge C}}{B, C \wedge D \Rightarrow B \wedge C}}{A \wedge B, C \wedge D \Rightarrow B \wedge C} \qquad \frac{\dfrac{\dfrac{B \Rightarrow B \quad C \Rightarrow C}{B, C \Rightarrow B \wedge C}}{A \wedge B, C \Rightarrow B \wedge C}}{A \wedge B, C \wedge D \Rightarrow B \wedge C}$$

produce the same term assignment.

If we wish to design a term calculus which corresponds exactly to deductions in a Gentzen system, we must replace the substitution on the meta-level which takes place in the term-assignment for the left-rules into operations from which the rule used may be read off.

In particular, this will be needed in L$\wedge$, L$\to$, L$\forall$ and contraction. So write $\mathrm{let}_w(t{:}B, s{:}A)$ for the result of the operation of "taking s for w in t" (also called a "let-construct" and written as "let w be s in t"). $\mathrm{let}_w(t, s)$ denotes the same as $t[w/s]$, but is not syntactically equal to $t[w/s]$. Similarly, we need $\mathrm{contr}_{xyz}(t)$ for the result of replacing x and y by a single variable z in t. The rules L$\wedge$, L$\to$, L$\forall$ and contraction now read:

$$\mathrm{L}\wedge\ \frac{u_i{:}\,A_i, \Gamma \Rightarrow t{:}\,C}{w{:}\,A_0 \wedge A_1, \Gamma \Rightarrow \mathrm{let}_{u_i}(t, \mathbf{p}_i w){:}\,C} \qquad \mathrm{L}{\to}\ \frac{\Gamma \Rightarrow t{:}\,A \qquad \Gamma, u{:}\,B \Rightarrow s{:}\,C}{w{:}\,A{\to}B, \Gamma \Rightarrow \mathrm{let}_u(s, wt){:}\,C}$$

$$\mathrm{L}\forall\ \frac{u{:}\,A[x/t], \Gamma \Rightarrow s{:}\,C}{w{:}\,\forall x A, \Gamma \Rightarrow \mathrm{let}_u(s, wt){:}\,C} \qquad \mathrm{LC}\ \frac{u{:}\,A, v{:}\,A, \Gamma \Rightarrow s{:}\,B}{w{:}\,A, \Gamma \Rightarrow \mathrm{contr}_{uvw}(s){:}\,B}$$

REMARK. Instead of introducing "let" and "contr", we can also leave LC implicit, that is to say the effect of a contraction on a deduction represented by a term $t(x, y, \vec{z})$ is obtained by simply identifying the variables x, y instead of having an explicit operator; and instead of "let" we can treat substitution operations $[\vec{x}/\vec{t}]$ as an explicit operation of the calculus, instead of a meta-mathematical operator. If substitution is an operation of the term calculus, then, for example, t differs from $x[x/t]$.

3.3.4A. ♠ Check that the two proofs of $A \wedge B, C \wedge D \Rightarrow B \wedge C$ above are indeed represented by distinct terms if we use "let" and "contr".

3.3.4B. ♠* Show for the map N assigning natural deductions to derivations in **G2i** with inhabited succedent in 3.3.1 that $|\mathsf{N}(\mathcal{D})| < c2^{|\mathcal{D}|}$ (c positive integer). For the full system we can take $c = 2$, and $c = 1$ for the system without $\perp$.

3.3.4C. ♠ The following modification of the term assignment corresponds to the alternative mentioned in the remark of 3.3.1:

$$\frac{\Gamma \Rightarrow t\colon A \qquad u\colon B, \Gamma \Rightarrow s\colon C}{w\colon A \to B, \Gamma \Rightarrow (\lambda u.s)(wt)\colon C}$$

Adapt also the other clauses of the term assignment, where necessary, so as to achieve (for t-**G2i**, without Cut) $\mathrm{s}(\mathsf{N}(\mathcal{D})) \leq c(\mathrm{s}(\mathcal{D}))$, c a fixed natural number, and similarly with depth instead of size; here $\mathsf{N}(\mathcal{D})$ is the natural deduction proof assigned to the sequent calculus proof by the procedure.

3.4 Systems with local rules

The following section contains some quite general definitions, which however will be primarily used for G-systems.

3.4.1. DEFINITION. Deductions (of LR-systems to be defined below) are finite trees, with the nodes labelled by *deduction elements*. (Deduction elements may be formulas, sequents etc., depending on the type of formalism considered.)

An *n-premise rule* R is a set of sequences $\mathcal{S}_0, \ldots, \mathcal{S}_{n-1}, \mathcal{S}$ of length $n+1$, where $\mathcal{S}_i$, $\mathcal{S}$ are deduction elements. An element of R is said to be an *instance* or *application* of R. An instance is usually written

$$\frac{\mathcal{S}_0 \quad \mathcal{S}_1 \quad \ldots \quad \mathcal{S}_{n-1}}{\mathcal{S}}$$

$\mathcal{S}$ is the *conclusion*, and the $\mathcal{S}_i$ are the *premises* of the rule-application. Where no confusion is to be feared, we often talk loosely about a rule when an application of the rule is meant. An *axiom* is a zero-premise rule. Instances of axioms appear in prooftrees either simply as (labels of) top nodes, or equivalently as deduction elements with a line over them:

$$\overline{\mathcal{S}}$$

In principle, we shall assume that the premises are always exhibited in a standard order from left to right (cf. our convention for N-systems that the major premise is always the leftmost one), so that expressions like "rightmost branch of a prooftree" become unambiguous. (In exhibiting concrete prooftrees, it is sometimes convenient to deviate from this.) ⊠

3.4.2. DEFINITION. A formal system with *local rules*, or *LR-system*, is specified by a finite set of rules; a deduction tree or prooftree is a finite tree with deduction elements and (names of) rules assigned to the nodes, such that if $\mathcal{S}_0, \ldots, \mathcal{S}_{n-1}$ are the deduction elements assigned to the immediate successors of node ν, and $\mathcal{S}$ is assigned to the node ν, R is the rule assigned to ν, then $\mathcal{S}_0, \dots, \mathcal{S}_{n-1}, \mathcal{S}$ belong to rule R. Clearly, the rules assigned to top nodes must be axioms. The deduction element assigned to the root of the tree is said to be deduced by the tree. If we consider also deduction trees where some top nodes $\nu_0, \nu_1, \dots$ do not have names of axioms assigned to them, we say that the deduction tree derives $\mathcal{S}$ from $\mathcal{S}_0, \mathcal{S}_1, \dots$, where $\mathcal{S}$ is the deduction element assigned to the root, and $\mathcal{S}_0, \mathcal{S}_1, \dots$ are the deduction elements assigned to the top nodes which do not have an axiom assigned to them. ⊠

REMARKS. The rules of an LR-system are local in the sense that the correctness of a rule-application at a node ν can be decided locally, namely by looking at the name of the rule assigned to ν, and the proof-objects assigned to ν and its immediate successors (i.e., the nodes immediately above it). The G-systems described above are obviously local. The notion of a pure-variable proof is not local, but this is used at a meta-level only.

Not all systems commonly considered are LR-systems. For Hilbert systems the deduction elements are formulas, for G-systems sequents. If we want to bring the N-systems also under the preceding definition, we can take as deduction elements sequents $\Gamma \Rightarrow A$, where the Γ is of the form $u_1\colon A_1, \dots, u_n\colon A_n$, i.e. a set of formulas with deduction variables attached. The use of this format frees us from the reference to discharged assumptions occurring elsewhere in the prooftree.

3.4.3. NOTATION. We write $\mathcal{D} \vdash_n \mathcal{S}$ if a prooftree $\mathcal{D}$ derives $\mathcal{S}$ and has depth at most n, and $\mathcal{D} \vdash_{s \leq n} \mathcal{S}$ if $\mathcal{D}$ derives $\mathcal{S}$ and has size most n. We write $\vdash_n \mathcal{S}$, $\vdash_{s \leq n} \mathcal{S}$ if for some $\mathcal{D}$ we have $\mathcal{D} \vdash_n \mathcal{S}$, $\mathcal{D} \vdash_{s \leq n} \mathcal{S}$ respectively. If we want to stress the dependence on a system $\mathbf{T}$, we write $\vdash^{\mathbf{T}}_n$, $\vdash^{\mathbf{T}}_{s \leq n}$ etc. ⊠

3.4.4. DEFINITION. Let $\mathbf{T}$ be an LR-system, the rules specifying the system we call the *(primitive) rules* of the system. A rule R is said to be a *derivable* rule in $\mathbf{T}$, if for each instance $\mathcal{S}_0, \dots, \mathcal{S}_{n-1}, \mathcal{S}$ there is a deduction of $\mathcal{S}$ from the $\mathcal{S}_i$ be means of the rules of $\mathbf{T}$. That is to say, in this deduction the $\mathcal{S}_i$ are treated as additional axioms.

A rule R is said to be *admissible* for $\mathbf{T}$ (or $\mathbf{T}$ is *closed* under R), if for all instances $\mathcal{S}_0, \dots, \mathcal{S}_{n-1}, \mathcal{S}$ of R it is the case that

if for all $i < n$ $\vdash \mathcal{S}_i$, then $\vdash \mathcal{S}$.

R is said to be *depth-preserving admissible (dp-admissible)* for $\mathbf{T}$ (or $\mathbf{T}$ is *dp-closed* under R) if for all m

if for all $i < n$ $\vdash_m \mathcal{S}_i$, then $\vdash_m \mathcal{S}$.

An n-premise rule R of **T** is said to be *i-invertible* for **T** [*i-dp-invertible* for **T**] if the rule

$$\mathrm{R}_i \equiv \{(\mathcal{S}, \mathcal{S}_i) \ : \ (\mathcal{S}_0, \dots, \mathcal{S}_{n-1}, \mathcal{S}) \in \mathrm{R}\}$$

is admissible [dp-admissible]. R is *invertible* [*dp-invertible*] if R is i-invertible [i-dp-invertible] for all $0 \leq i < n$.

For two-premise rules, we may also use *left-invertible, right-invertible* for 0-invertible and 1-invertible respectively. ⊠

3.5 Absorbing the structural rules

We now consider Gentzen systems in which not only weakening but also contraction has been "absorbed" into the rules and axioms: the family of G3-systems. This has advantages in an upside down search procedure for proofs of a given sequent. (See also 4.2.7.)

A number of results in this section involve the notion of *depth* of a proof; but the proofs go through if we use the notion of *size* of a proof instead.

3.5.1. DEFINITION. (*The Gentzen systems* **G3c**, **G3m**, **G3i**) The system **G3c** is specified by the following axioms and rules:

$$\text{Ax} \ \ P, \Gamma \Rightarrow \Delta, P \ (P \text{ atomic}) \qquad \text{L}\bot \ \ \bot, \Gamma \Rightarrow \Delta$$

$$\text{L}\wedge \ \frac{A, B, \Gamma \Rightarrow \Delta}{A \wedge B, \Gamma \Rightarrow \Delta} \qquad \text{R}\wedge \ \frac{\Gamma \Rightarrow \Delta, A \qquad \Gamma \Rightarrow \Delta, B}{\Gamma \Rightarrow \Delta, A \wedge B}$$

$$\text{L}\vee \ \frac{A, \Gamma \Rightarrow \Delta \qquad B, \Gamma \Rightarrow \Delta}{A \vee B, \Gamma \Rightarrow \Delta} \qquad \text{R}\vee \ \frac{\Gamma \Rightarrow \Delta, A, B}{\Gamma \Rightarrow \Delta, A \vee B}$$

$$\text{L}{\to} \ \frac{\Gamma \Rightarrow \Delta, A \qquad B, \Gamma \Rightarrow \Delta}{A \to B, \Gamma \Rightarrow \Delta} \qquad \text{R}{\to} \ \frac{A, \Gamma \Rightarrow \Delta, B}{\Gamma \Rightarrow \Delta, A \to B}$$

$$\text{L}\forall \ \frac{\forall x A, A[x/t], \Gamma \Rightarrow \Delta}{\forall x A, \Gamma \Rightarrow \Delta} \qquad \text{R}\forall \ \frac{\Gamma \Rightarrow \Delta, A[x/y]}{\Gamma \Rightarrow \Delta, \forall x A}$$

$$\text{L}\exists \ \frac{A[x/y], \Gamma \Rightarrow \Delta}{\exists x A, \Gamma \Rightarrow \Delta} \qquad \text{R}\exists \ \frac{\Gamma \Rightarrow \Delta, A[x/t], \exists x A}{\Gamma \Rightarrow \Delta, \exists x A}$$

where in R∀, L∃ the y is not free in the conclusion.

The intuitionistic version **G3i** of **G3c** has the following form:

$$\text{Ax} \ \ P, \Gamma \Rightarrow P \ (P \text{ atomic}) \qquad \text{L}\bot \ \ \bot, \Gamma \Rightarrow A$$

$$\text{L}\wedge \frac{A, B, \Gamma \Rightarrow C}{A \wedge B, \Gamma \Rightarrow C} \qquad \text{R}\wedge \frac{\Gamma \Rightarrow A \qquad \Gamma \Rightarrow B}{\Gamma \Rightarrow A \wedge B}$$

$$\text{L}\vee \frac{A, \Gamma \Rightarrow C \qquad B, \Gamma \Rightarrow C}{A \vee B, \Gamma \Rightarrow C} \qquad \text{R}\vee \frac{\Gamma \Rightarrow A_i}{\Gamma \Rightarrow A_0 \vee A_1} \ (i = 0, 1)$$

$$\text{L}\!\to \frac{A \to B, \Gamma \Rightarrow A \qquad B, \Gamma \Rightarrow C}{A \to B, \Gamma \Rightarrow C} \qquad \text{R}\!\to \frac{A, \Gamma \Rightarrow B}{\Gamma \Rightarrow A \to B}$$

$$\text{L}\forall \frac{\forall x A, A[x/t], \Gamma \Rightarrow C}{\forall x A, \Gamma \Rightarrow C} \qquad \text{R}\forall \frac{\Gamma \Rightarrow A[x/y]}{\Gamma \Rightarrow \forall x A}$$

$$\text{L}\exists \frac{A[x/y], \Gamma \Rightarrow C}{\exists x A, \Gamma \Rightarrow C} \qquad \text{R}\exists \frac{\Gamma \Rightarrow A[x/t]}{\Gamma \Rightarrow \exists x A}$$

where in L∃ and R∀ the y is not free in the conclusion.

G3m is **G3i** with L⊥ left out, and $\bot, \Gamma \Rightarrow \bot$ added (to compensate for this missing instance of Ax). Alternatively, one can let the P in Ax range over prime, instead of atomic formulas; then **G3m** is simply a restriction of **G3i**. Sequents derivable in **G3m** always have a single formula in the succedent (just as for **G1m** and **G2m**).

The concepts of *principal* and *active* formula occurrence in an inference are copied from the systems **G1**[**mic**]; but note that in, for example, an application of L∀, only the occurrence of $\forall x A$ in the conclusion is *principal*. ⊠

3.5.1A. ♠ Show that $A \Rightarrow A$ is derivable in **G3**[**mic**] for arbitrary A. Show that in Ax, L⊥ in **G3cp** all formulas in $\Gamma\Delta$ may be taken to be atomic. What goes wrong for full **G3c**? And for **G3ip**?

3.5.1B. ♠ Give a proof of Peirce's law $((A \to B) \to A) \to A$ in the system **G3c**.

3.5.2. LEMMA. *(Substitution of terms) For the systems* **G[123][mic]**, *if* $\mathcal{D} \vdash \Gamma \Rightarrow \Delta$, *then* $\mathcal{D}[x/t] \vdash \Gamma[x/t] \Rightarrow \Delta[x/t]$, *provided t is free for x in* $\Gamma \Rightarrow \Delta$ *and does not contain variables used as proper parameters of* L∃, R∀. *The substitution does not change the size, depth or logical depth of the proof. Hence, by renaming proper parameters of* L∃, R∀*: if* $\vdash_n \Gamma \Rightarrow \Delta$ *then* $\vdash_n \Gamma[x/t] \Rightarrow \Delta[x/t]$ *provided t is free for x in* Γ, Δ.

PROOF. By induction on the depth of proofs. ⊠

3.5.3. LEMMA. *(dp-admissibility of weakening)* **G3[mic]** *is closed under weakening. That is to say, if* $\vdash$ *is deducibility in* **G3c**, *then*

$$\text{If } \vdash_n \Gamma \Rightarrow \Delta \text{ then } \vdash_n \Gamma, \Gamma' \Rightarrow \Delta, \Delta'$$

where $|\Delta\Delta'| \leq 1$ *for* **G3[mi]**. ⊠

N.B. This lemma is not true if we insist that in Ax, L$\bot$ all formulas of Γ, Δ are atomic.

3.5.4. PROPOSITION. *(Inversion lemma) Let* $\vdash$ *be deducibility in* **G3c**.

(i) If $\vdash_n A \wedge B, \Gamma \Rightarrow \Delta$ *then* $\vdash_n A, B, \Gamma \Rightarrow \Delta$.

(ii) If $\vdash_n \Gamma \Rightarrow \Delta, A \vee B$ *then* $\vdash_n \Gamma \Rightarrow \Delta, A, B$.

(iii) If $\vdash_n A_0 \vee A_1, \Gamma \Rightarrow \Delta$ *then* $\vdash_n A_i, \Gamma \Rightarrow \Delta$ $(i \in \{0, 1\})$.

(iv) If $\vdash_n \Gamma \Rightarrow \Delta, A_0 \wedge A_1$ *then* $\vdash_n \Gamma \Rightarrow \Delta, A_i$ $(i \in \{0, 1\})$.

(v) If $\vdash_n \Gamma \Rightarrow A \to B, \Delta$ *then* $\vdash_n \Gamma, A \Rightarrow B, \Delta$.

(vi) If $\vdash_n \Gamma, A \to B \Rightarrow \Delta$ *then* $\vdash_n \Gamma \Rightarrow \Delta, A$ *and* $\vdash_n \Gamma, B \Rightarrow \Delta$.

(vii) If $\vdash_n \Gamma \Rightarrow \Delta, \forall x A$ *then* $\vdash_n \Gamma \Rightarrow \Delta, A[x/y]$, *for any* y *such that* $y \notin \mathrm{FV}(\Gamma, \Delta, A)$.

(viii) If $\vdash_n \exists x A, \Gamma \Rightarrow \Delta$ *then* $\vdash_n A[x/y], \Gamma \Rightarrow \Delta$, *for any* y *such that* $y \notin \mathrm{FV}(\Gamma, \Delta, A)$.

The properties above, with the exception of (ii) and (vi), also hold for **G3[mi]**, *under the intuitionistic restriction on sequents. For* **G3[mi]** *one half of (vi) remains provable:*

(vi) If $\vdash_n \Gamma, A \to B \Rightarrow C$ *then* $\vdash_n \Gamma, B \Rightarrow C$.

PROOF. The proposition is proved by induction on n. As a typical example, we prove (vi) for **G3c**. Assume (vi) to have been proved for n, and all Γ, Δ. Let $\vdash_{n+1} A \to B, \Gamma \Rightarrow \Delta$ by a deduction $\mathcal{D}$. If $\mathcal{D}$ is an axiom, then $A \to B$ is not principal, and $\Gamma, B \Rightarrow \Delta$ as well as $\Gamma \Rightarrow \Delta, A$ are axioms. If $\mathcal{D}$ is not an axiom and $A \to B$ is not principal, we apply the IH to the premise(s) and then use the same rule to obtain deductions of $\Gamma \Rightarrow A, \Delta$ and $B, \Gamma \Rightarrow \Delta$.

If on the other hand $A \to B$ is principal, the deduction ends with

$$\frac{\Gamma \Rightarrow \Delta, A \qquad B, \Gamma \Rightarrow \Delta}{A \to B, \Gamma \Rightarrow \Delta}$$

and we can take the immediate subdeduction of premises. Similarly in the case of **G3i**, where only the second premise counts, if $A \to B$ is principal. ⊠

3.5.4A. ♠ Complete the proof of the inversion lemma.

3.5.5. PROPOSITION. *(dp-admissibility of contraction) Let* $\vdash$ *be deducibility in* **G3c**. *Then we have for all* A, Γ, Δ

(i) If $\vdash_n A, A, \Gamma \Rightarrow \Delta$, *then* $\vdash_n A, \Gamma \Rightarrow \Delta$.

(ii) If $\vdash_n \Gamma \Rightarrow \Delta, A, A$, *then* $\vdash_n \Gamma \Rightarrow \Delta, A$.

The first property, under the intuitionistic restriction, also holds for **G3[mi]**.

PROOF. By induction on n. We consider the first assertion; the second is treated symmetrically. Let $\mathcal{D}$ be a deduction of length $n+1$ of $A, A, \Gamma \Rightarrow \Delta$.

If A is not principal in the last rule applied in $\mathcal{D}$, apply IH to the premise. If A is principal in the last rule applied, we distinguish cases.

Case 1. The last rule applied is L$\wedge$:

$$\frac{\vdash_n A, B, A \wedge B, \Gamma \Rightarrow \Delta}{\vdash_{n+1} A \wedge B, A \wedge B, \Gamma \Rightarrow \Delta}$$

Apply the inversion lemma to the premise and find a proof of

$$\vdash_n A, B, A, B, \Gamma \Rightarrow \Delta$$

and use IH twice.

Case 2. The last rule applied is L$\exists$. Then

$$\frac{\vdash_n \Gamma, A[x/y], \exists x A \Rightarrow \Delta}{\vdash_{n+1} \Gamma, \exists x A, \exists x A \Rightarrow \Delta}$$

By the inversion lemma, there is a y' such that for some $\mathcal{D}'$

$$\mathcal{D}' \vdash_n \Gamma, A[x/y], A[x/y'] \Rightarrow \Delta,$$

and $y, y' \notin \mathrm{FV}(\Gamma\Delta)$, $y' \notin \mathrm{FV}(A[x/y])$, $y \notin \mathrm{FV}(A[x/y']$, $y \not\equiv y'$. Using the substitution lemma we may conclude that

$$\vdash_n \Gamma, A[x/z], A[x/z] \Rightarrow \Delta$$

where z is a fresh variable not occurring free in Γ, Δ. Then we apply the induction hypothesis w.r.t. $A[x/z]$ and find $\vdash_n \Gamma, A[x/z] \Rightarrow \Delta$.

Case 3. The last rule applied is L$\vee$:

$$\frac{\vdash_n A, A \vee B, \Gamma \Rightarrow \Delta \qquad \vdash_n B, A \vee B, \Gamma \Rightarrow \Delta}{\vdash_{n+1} A \vee B, A \vee B, \Gamma \Rightarrow \Delta}$$

We use the inversion lemma and apply the induction hypothesis.

Case 4. The last rule applied is L→:

$$\frac{\vdash_n A \to B, \Gamma \Rightarrow \Delta, A \qquad \vdash_n A \to B, B, \Gamma \Rightarrow \Delta}{\vdash_{n+1} A \to B, A \to B, \Gamma \Rightarrow \Delta}$$

By the inversion lemma applied to the first premise, $\vdash_n \Gamma \Rightarrow A, A, \Delta$, and applied to the second premise $\vdash_n \Gamma, B, B \Rightarrow \Delta$. We then use the IH and obtain $\vdash_n \Gamma \Rightarrow A, \Delta$ and $\vdash_n \Gamma, B \Rightarrow \Delta$, from which $\vdash_{n+1} \Gamma, A \to B \Rightarrow \Delta$.

In the case of **G3i** the treatment is slightly different, but we leave this to the reader (the occurrence of $A \to B$ in the left premise of L→ makes up for the missing half of the inversion lemma in this case).

Case 5. The last rule applied is L∀. Immediate. ⊠

3.5.6. Remarks. (i) If $A \to B$ is omitted in the left premise of L→ of **G3i**, the proof of the preceding proposition breaks down at Case 4. A counterexample in the implication fragment is provided by the sequent $(P, Q \in \mathcal{PV})$

$$(((P{\to}Q){\to}Q){\to}P){\to}Q, Q{\to}P \Rightarrow Q$$

We leave it to the reader to check this.

(ii) The inversion lemma may be stated as follows. In **G3c**, if $\vdash_n \Gamma, A \Rightarrow \Delta$ (respectively $\vdash_n \Gamma \Rightarrow A, \Delta$) then there is a proof of depth $\leq n+1$ with A as principal formula, if A is composite, but not of the form $\forall x A'$ (respectively composite, but not of the form $\exists x A'$); an appropriate adaptation holds for **G3i**.

It is possible to improve on the result for **G3c** as follows: we may take "depth $\leq n$" instead of "depth $\leq n+1$", *provided* we restrict attention to proofs where all axioms Ax, L⊥ are such that the formulas in Γ are atomic or ∀-formulas, and the formulas in Δ are atomic or ∃-formulas. But we have to pay a price for this: for this class of proofs, the weakening operation transforming a proof of $\Gamma \Rightarrow \Delta$ into a proof of $\Gamma, \Gamma' \Rightarrow \Delta, \Delta'$ cannot be done while preserving the depth (a corresponding observation for **G[12][mic]** was made in proposition 3.1.9).

(iii) If we drop the restriction on Ax, that is if we consider **G3[mic]** + GAx, where GAx is the axiom schema $\Gamma, A \Rightarrow A, \Delta$ without restrictions on the A, we can still formulate a version of the inversion lemma which is sometimes useful. For example, for **G3c** + GAx we have the following version of (vi) of the inversion lemma (3.5.4): if $\mathcal{D} \vdash_n \Gamma, A \to B \Rightarrow \Delta$ and $A \to B$ is not principal in an axiom in $\mathcal{D}$, then $\vdash_n \Gamma \Rightarrow A, \Delta$ and $\vdash_n \Gamma, B \Rightarrow \Delta$.

3.5.7. Proposition. *The dp-admissibility of weakening and contraction, dp-closure under substitution of terms and the dp-inversion lemma hold for* **G3[mic]** + $\mathrm{Cut_{cs}}$.

Proof. The proof for **G3[mic]** readily extends. ⊠

3.5.7A. ♠* Show that the example under (i) of the preceding remarks is indeed provable in **G3i**, but unprovable in **G3i** if $A \to B$ is omitted in the rule L→.

3.5.7B. ♠ Show that we can establish the stronger variant of the inversion lemma under the appropriate restriction on the axioms, as described above.

3.5.8. PROPOSITION. *In* **G3[ic]**, *if* $\vdash_n \Gamma \Rightarrow \bot, \Delta$, *then* $\vdash_n \Gamma \Rightarrow A, \Delta$.

PROOF. By induction on n. Let $\mathcal{D}$ be a proof of length n of $\Gamma \Rightarrow \bot, \Delta$. If $\mathcal{D}$ is an axiom, then either $\bot$ occurs in Γ, so then $\Gamma \Rightarrow A, \Delta$ is an axiom; or some P occurs in both Γ and Δ, and again $\Gamma \Rightarrow A, \Delta$ is an axiom. If $\mathcal{D}$ is not an axiom, we apply the IH to the premise(s) that the occurrence of $\bot$ derives from. ⊠

3.5.9. PROPOSITION. *(Equivalence)* **G1c** $\vdash \Gamma \Rightarrow \Delta$ *iff* **G3c** $\vdash \Gamma \Rightarrow \Delta$, *and* **G1[mi]** $\vdash \Gamma \Rightarrow A$ *iff* **G3[mi]** $\vdash \Gamma \Rightarrow A$.

PROOF. Straightforward by closure of **G3[ic]** under weakening and contraction. In both directions the proofs proceed by induction on the depth of deductions. ⊠

REMARK. As a corollary to the proof one obtains that

If **G1[mic]** $\vdash_n \Gamma \Rightarrow \Delta$ then **G3[mic]** $\vdash_n \Gamma \Rightarrow \Delta$.

But the converse does not hold: for $P \in \mathcal{PV}$, $P \vee \neg P$ has a proof of depth 2 in **G3c**, but not in **G1c**. Shortest proofs in **G1c**, **G3c** respectively are shown below:

$$\dfrac{\dfrac{\dfrac{\dfrac{\dfrac{P \Rightarrow P}{P \Rightarrow P, \bot}}{\Rightarrow P, P \to \bot}}{\Rightarrow P \vee \neg P, P \to \bot}}{\Rightarrow P \vee \neg P, P \vee \neg P}}{\Rightarrow P \vee \neg P} \qquad \dfrac{\dfrac{P \Rightarrow P, \bot}{\Rightarrow P, P \to \bot}}{\Rightarrow P \vee \neg P}$$

3.5.10. *Intuitionistic multi-succedent systems*

The systems in the following definition are used in 4.1.10 and some of the exercises only, so the definition may be skipped until needed.

DEFINITION. In 3.2.1A we already encountered a multi-succedent version of the system **G3i**. We may also define a multi-succedent version m-**G3i** of **G3i**, in which we keep as close as possible to **G3c**, permitting whenever possible a multiset in the succedent. The system m-**G3i** is obtained from **G3c** by restricting R→ and R∀ to

$$\text{R}\!\rightarrow \frac{\Gamma, A \Rightarrow B}{\Gamma \Rightarrow A \rightarrow B, \Delta} \qquad \text{R}\forall \frac{\Gamma \Rightarrow A[x/y]}{\Gamma \Rightarrow \forall x A, \Delta}$$

where in R∀ $x \notin \text{FV}(\Gamma)$, $y \equiv x$ or $y \notin \text{FV}(A, \Gamma)$, and L→ is modified into

$$\text{L}\!\rightarrow \frac{\Gamma, A \rightarrow B \Rightarrow A, \Delta \qquad \Gamma, B \Rightarrow \Delta}{\Gamma, A \rightarrow B \Rightarrow \Delta}$$

The system m-**G3m** is obtained from m-**G3i** by omitting the axiom L⊥, but then one has to add $\Gamma, \bot \Rightarrow \bot, \Delta$ as an instance of Ax.

A slight variant of m-**G3i**, m-**G3i′**, has a left premise in L→ of the form $\Gamma, A \rightarrow B \Rightarrow A$ (no Δ). ⊠

The substitution lemma (3.5.2), the lemma on dp-admissibility of weakening (3.5.3), a suitable version of inversion (3.5.4) and dp-admissibility of contraction (3.5.5) are valid also for m-**G3[mi]**.

Intuitionistic multi-succedent systems arise quite naturally in semantical investigations (cf. 4.9.1).

REMARK. We do not know of a designation of this type of system, that is completely satisfactory in the sense that it is mnemonically convenient, consistent, and not cumbersome. The classical systems are always "multi-succedent", so there we drop the prefix m-. Also, in a publication where only multi-succedent G-systems are discussed, the prefix m- is redundant.

3.5.10A. ♠ Check that the substitution lemma, dp-admissibility of weakening, and contraction and a suitable version of inversion are valid for m-**G3[mi]**.

3.5.11. *Kleene-style* G3-*systems*

The systems **G3[mic]** are inspired by Dragalin; the system closest to Dragalin's system for intuitionistic logic is m-**G3i**. Kleene's original systems of the G3-family differ in one important respect from **G3[mic]**: they are strictly cumulative, that is to say, if in the classical case $\Gamma \Rightarrow \Delta$ is the conclusion of an application of a rule of the system, then $\Gamma \Rightarrow \Delta$ appears as a subsequent of the premises of the application; and in the intuitionistic case, if $\Gamma \Rightarrow A$ is the conclusion of a rule-application, then Γ appears as a sub-multiset of the antecedents of the premises of the application. In other words, in the intuitionistic and minimal cases the antecedent can only increase when going from the conclusion to one of the premises, and in the classical case both the antecedent and the succedent can only increase.

Going downwards form premises to conclusion, any formula "introduced" on the left or on the right (classical case only) is already present in the premises; the active formula(s) in the antecedent and the succedent (in the classical case) are, so to speak, absorbed into the conclusion.

DEFINITION. (*The systems* **GK[mic]**) The rules for **GKi** are almost the same as for Kleene's system G3 in Kleene [1952a]. The subscript i appearing in some of the rules may be 0 or 1.

$$\text{Ax } P, \Gamma \Rightarrow P \ (P \text{ atomic}) \qquad \text{L}\bot \ \bot, \Gamma \Rightarrow A$$

$$\text{L}\wedge \ \frac{A_i, A_0 \wedge A_1, \Gamma \Rightarrow C}{A_0 \wedge A_1, \Gamma \Rightarrow C} \qquad \text{R}\wedge \ \frac{\Gamma \Rightarrow A \qquad \Gamma \Rightarrow B}{\Gamma \Rightarrow A \wedge B}$$

$$\text{L}\vee \ \frac{A_0, A_0 \vee A_1, \Gamma \Rightarrow C \qquad A_1, A_0 \vee A_1, \Gamma \Rightarrow C}{A_0 \vee A_1, \Gamma \Rightarrow C} \qquad \text{R}\vee \ \frac{\Gamma \Rightarrow A_i}{\Gamma \Rightarrow A_0 \vee A_1}$$

$$\text{L}\rightarrow \ \frac{A \rightarrow B, \Gamma \Rightarrow A \qquad A \rightarrow B, B, \Gamma \Rightarrow C}{A \rightarrow B, \Gamma \Rightarrow C} \qquad \text{R}\rightarrow \ \frac{A, \Gamma \Rightarrow B}{\Gamma \Rightarrow A \rightarrow B}$$

$$\text{L}\forall \ \frac{\forall x A, A[x/t], \Gamma \Rightarrow C}{\forall x A, \Gamma \Rightarrow C} \qquad \text{R}\forall \ \frac{\Gamma \Rightarrow A[x/y]}{\Gamma \Rightarrow \forall x A}$$

$$\text{L}\exists \ \frac{\exists x A, A[x/y], \Gamma \Rightarrow C}{\exists x A, \Gamma \Rightarrow C} \qquad \text{R}\exists \ \frac{\Gamma \Rightarrow A[x/t]}{\Gamma \Rightarrow \exists x A}$$

where in L$\exists$ and R$\forall$ the y is not free in the conclusion. The corresponding system **GKm** is obtained by dropping L$\bot$, and adding $\bot, \Gamma \Rightarrow \bot$.

The classical system **GKc** is obtained by extending the cumulativeness of the rules in a symmetric way to the succedent, and generalizing the axioms and rules to arbitrary contexts on the right. Thus we have

$$\text{R}\wedge \ \frac{\Gamma \Rightarrow \Delta, A_0, A_0 \wedge A_1 \qquad \Gamma \Rightarrow \Delta, A_1, A_0 \wedge A_1}{\Gamma \Rightarrow \Delta, A_0 \wedge A_1} \qquad \text{R}\vee \ \frac{\Gamma \Rightarrow \Delta, A_i, A_0 \vee A_1}{\Gamma \Rightarrow \Delta, A_0 \vee A_1}$$

$$\text{R}\rightarrow \ \frac{A, \Gamma \Rightarrow \Delta, B, A \rightarrow B}{\Gamma \Rightarrow \Delta, A \rightarrow B} \qquad \text{R}\forall \ \frac{\Gamma \Rightarrow \Delta, A[x/y], \forall x A}{\Gamma \Rightarrow \Delta, \forall x A}$$

etc. ⊠

The proof of dp-closure under contraction for these systems is virtually trivial; there is no need to appeal to an inversion lemma. But dp-inversion lemmas for the left rules become trivial (with the exception of left-inversion for L$\rightarrow$), since each premise is a left-weakening of the conclusion.

3.5.11A. ♠* Prove the following simple form of *Herbrand's theorem* for **G3[mic]**: if Γ, Δ and A are quantifier-free, and $\vdash_n \Gamma, \forall x A \Rightarrow \Delta$, then there are $t_1, \dots, t_m$ such that $\vdash_n \Gamma, A[x/t_1], \dots, A[x/t_m] \Rightarrow \Delta$. For **G3c** we also have: if $\vdash_n \Gamma \Rightarrow \Delta, \exists x A$ then for suitable $t_1, \dots, t_m \vdash_n \Gamma \Rightarrow \Delta, A[x/t_1], \dots, A[x/t_m]$.

3.5.11B. ♠ Describe an assignment N of natural deductions to **G3i**-deductions in such a way that, for a suitable constant $c \in \mathbb{N}$, $|\mathsf{N}(\mathcal{D})| \leq c|\mathcal{D}|$ (cf. exercise 3.3.4C). In fact, we can take $c = 8$.

3.5.11C. ♠ Check that an assignment G of proofs in **G3i** + Cut to proofs in **Ni** can be given such that $s(\mathsf{G}(\mathcal{D})) \leq c(s(\mathcal{D}))$ for some $c \in \mathbb{N}$ (in fact we can take $c = 5$), and similarly with depth instead of size.

3.5.11D. ♠ Prove that m-**G3i** $\vdash \Gamma \Rightarrow \Delta$ iff **G3i** $\vdash \Gamma \Rightarrow \bigvee \Delta$. (m-**G3i** was defined in 3.5.10.)

3.5.11E. ♠ Show that if m-**G3i** $\vdash \Gamma \Rightarrow \Delta$ for non-empty Δ, and Γ does not contain $\vee$, then for some $A \in \Delta$, m-**G3i** $\vdash \Gamma \Rightarrow A$ (m-**G3i** as in the preceding exercise).

3.5.11F. ♠ Formulate and prove lemmas on dp-invertibility of the rules of the systems **GK**[**mic**].

3.6 The one-sided systems for C

The symmetry present in classical logic permits the formulation of one-sided Gentzen systems, the *Gentzen–Schütte* systems; one may think of the sequents of such a calculus as obtained by replacing a two-sided sequent $\Gamma \Rightarrow \Delta$ by a one-sided sequent $\Rightarrow \neg\Gamma, \Delta$ (with intuitive interpretation the disjunction of the formulas in $\neg\Gamma, \Delta$), and if we restrict attention to one-sided sequents throughout, the symbol $\Rightarrow$ is redundant. Each of the systems **G**[**123**]**c** has its one-sided counterpart **GS**[**123**]. One may also think of "GS" as standing for "Gentzen-symmetric", since the symmetries of classical logic given by the De Morgan duality have been built in.

In order to achieve this, we need a different treatment of negation. We shall assume that formulas are constructed from *positive literals* P, P', P'', $R(t_0, \ldots, t_n)$, $R(s_0, \ldots, s_m)$ etc., as well as *negative literals* $\neg P$, $\neg P'$, $\neg P''$, $\neg R(t_0, \ldots, t_n)$, ... by means of $\vee, \wedge, \forall, \exists$. Both types of literals are treated as *primitives*.

3.6.1. DEFINITION. Negation $\neg$ satisfies $\neg\neg P \equiv P$ for literals P, and is defined for compound formulas by De Morgan duality:

(i) $\neg(A \wedge B) := (\neg A \vee \neg B)$;

(ii) $\neg(A \vee B) := (\neg A \wedge \neg B)$;

(iii) $\neg\forall x A := \exists x \neg A$;

(iv) $\neg\exists x A := \forall x \neg A$. ⊠

3.6.2. DEFINITION. The one-sided calculus **GS1** (corresponding to **G1c**) has the following rules and axioms:

$$\text{Ax} \quad P, \neg P$$

$$\text{RW}\,\frac{\Gamma}{\Gamma, A} \qquad \text{RC}\,\frac{\Gamma, A, A}{\Gamma, A}$$

$$\text{R}\vee_{\text{L}}\,\frac{\Gamma, A}{\Gamma, A \vee B} \quad \text{R}\vee_{\text{R}}\,\frac{\Gamma, B}{\Gamma, A \vee B} \qquad \text{R}\wedge\,\frac{\Gamma, A \quad \Gamma, B}{\Gamma, A \wedge B}$$

$$\text{R}\forall\,\frac{\Gamma, A[x/y]}{\Gamma, \forall x A} \qquad \text{R}\exists\,\frac{\Gamma, A[x/t]}{\Gamma, \exists x A}$$

under the obvious restrictions on y and t.

In the calculus **GS2** corresponding to **G2c** the axiom is generalized to

$$\Gamma, P, \neg P$$

and the rule W is dropped. Finally, in the calculus **GS3**, corresponding to **G3c**, the axioms are generalized to $\Gamma, P, \neg P$ (P atomic), the rules W and C are dropped and $\text{R}\vee, \text{R}\exists$ are replaced by

$$\text{R}\vee\,\frac{\Gamma, A, B}{\Gamma, A \vee B} \qquad \text{R}\exists\,\frac{\Gamma, A[x/t], \exists x A}{\Gamma, \exists x A}$$

The Cut rule takes the form

$$\text{Cut}\,\frac{\Gamma, A \qquad \Delta, \neg A}{\Gamma, \Delta}$$

⊠

The letter "R" in the designation of the rules may be omitted, but we have kept it since all the R-rules of the one-sided calculi are just the R-rules of the systems **G**[**123**]**c** for sequents of the form $\Rightarrow \Delta$.

3.6.2A. ♠ Prove an inversion lemma for **GS3**:

(i) If $\vdash_n \Gamma, A \vee B$ then $\vdash_n \Gamma, A, B$;

(ii) If $\vdash_n \Gamma, A \wedge B$ then $\vdash_n \Gamma, A$ and $\vdash_n \Gamma, B$;

(iii) If $\vdash_n \Gamma, \forall x A$ then $\vdash_n \Gamma, A[x/y]$ (y not free in Γ, and also $y \equiv x$ or $y \notin \text{FV}(A)$).

3.6.2B. ♠ Use the inversion lemma to prove closure of **GS3** under contraction.

3.7 Notes

3.7.1. *General.* Some papers covering to some extent the same ground as our chapters 1–6 are Gallier [1993], Bibel and Eder [1993]. For an introduction to Gentzen's work, see M. E. Szabo's introduction to Gentzen [1969].

3.7.2. *Gentzen systems; the calculi* **G1c**, **G1i**. Gentzen gave formulations for classical and intuitionistic logic; but, as already mentioned in the preceding chapter, Johansson [1937] was the first to give a Gentzen system for minimal logic.

Gentzen's original formulation LK differs from the subsystem **G1c** in the following respects:

(i) Instead of a primitive constant $\bot$, Gentzen uses a negation operator $\neg$ with rules

$$\text{L}\neg\ \frac{\Gamma \Rightarrow \Delta, A}{\neg A, \Gamma \Rightarrow \Delta} \qquad \text{R}\neg\ \frac{A, \Gamma \Rightarrow \Delta}{\Gamma \Rightarrow \Delta, \neg A}$$

(ii) In Gentzen's system, sequences instead of multisets were used; accordingly there were *exchange* or permutation rules (cf. 1.3.7):

$$\text{LE}\ \frac{\Gamma, A, B, \Gamma' \Rightarrow \Delta}{\Gamma, B, A, \Gamma' \Rightarrow \Delta} \qquad \text{RE}\ \frac{\Gamma \Rightarrow \Delta, A, B, \Delta'}{\Gamma \Rightarrow \Delta, B, A, \Delta'}$$

(iii) For L$\to$, Gentzen used the non-sharing version.

(iv) Gentzen defined his systems so as to include the Cut rule, whereas we have preferred to take the systems without Cut rule as basic.

LK is equivalent to **G1c** which may be seen as follows. If we define $\bot := A \wedge \neg A$ for some fixed A, we can derive $\bot \Rightarrow$ in Gentzen's system:

$$\dfrac{\dfrac{\dfrac{\dfrac{A \Rightarrow A}{A, \neg A \Rightarrow}\text{L}\neg}{A \wedge \neg A, \neg A \Rightarrow}\text{L}\wedge}{A \wedge \neg A, A \wedge \neg A \Rightarrow}\text{L}\wedge}{A \wedge \neg A \Rightarrow}\text{LC}$$

Conversely, defining $\neg A := A \to \bot$ as usual, we obtain in **G1c** the rules for L$\neg$ and R$\neg$ as a special case of L$\to$ and by an application of RW followed by R$\to$ respectively.

Kleene [1952b] gives the rules for **G1c** and **G1i** as in this text. **G1**[**ic**] + Multicut (where "Multicut" is a generalization of the Cut rule, defined in 4.1.9) is nearly identical with the G2-systems in Kleene [1952a].

3.7.3. *The calculi* **G[23][mic]**. As explained in the introduction to this chapter, the calculi **G2[mic]** serve only as a stepping stone to the more interesting systems **G3[mic]**.

The systems **G3[mic]** as presented here are inspired by Dragalin [1979], but are not quite identical with Dragalin's systems. The form of the rules L$\wedge$ and R$\wedge$ was first used in a sequent calculus for classical logic by Ketonen [1944].

Our **G3c** corresponds indeed to Dragalin's classical G-system, except that Dragalin has $\neg$ as an additional primitive. Dragalin's intuitionistic G-system corresponds most closely to m-**G3i**, with $\neg$ as an additional primitive. In a letter to H. A. J. M. Schellinx, dated 22-11-1990, Dragalin points out that instead of the form of L$\rightarrow$ in the intuitionistic system proposed in his book, the following form is preferable:

$$\frac{A \rightarrow B, \Gamma \Rightarrow A, \Delta \qquad B, \Gamma \Rightarrow \Delta}{A \rightarrow B, \Gamma \Rightarrow \Delta}$$

One of the advantages mentioned by Dragalin is dp-invertibility of this rule with respect to both premises.

The differences between our systems **GK[mic]** and the G3-systems formulated in Kleene [1952a] are the following.

(i) In Kleene's systems, $\neg$ is primitive, not $\bot$.

(ii) The rule L$\wedge$ has in Kleene's intuitionistic system the form

$$\text{L}\wedge\frac{\Gamma, A_0, A_1, A_0 \wedge A_1 \Rightarrow B}{\Gamma, A_0 \wedge A_1 \Rightarrow B}$$

and similarly for R$\vee$ and L$\wedge$ in the classical system.

(iii) Kleene wished to interpret these rules so that for every instance of a rule

$$\frac{\Gamma_1 \Rightarrow \Delta_1 \ \ldots \ \Gamma_n \Rightarrow \Delta_n}{\Gamma}$$

any instance

$$\frac{\Gamma_1' \Rightarrow \Delta_1' \ \ldots \ \Gamma_n' \Rightarrow \Delta_n'}{\Gamma'}$$

with $\text{Set}(\Gamma_i) = \text{Set}(\Gamma_i')$, $\text{Set}(\Gamma) = \text{Set}(\Gamma')$, $\text{Set}(\Delta_i) = \text{Set}(\Delta_i')$ and $\text{Set}(\Delta) = \text{Set}(\Delta')$, is also an application of the rule.

In other words, by the convention under (iii) the premises and conclusions of the rules may be read as finite sets, with $\Gamma\Delta A$ short for $\Gamma \cup \Delta \cup \{A\}$, etc.

It is perhaps worth pointing out that this means for, say, L$\wedge$ that if we write the premise as $A, B, A \wedge B, \Gamma$ with $A, B, A \wedge B \notin \Gamma$, then the conclusion may be any of the following:

$$A \wedge B, \Gamma \quad A, A \wedge B, \Gamma \quad B, A \wedge B, \Gamma \quad A, B, A \wedge B, \Gamma.$$

(The last possibility is actually redundant, since it results in repetition of the premise.) Our reason for deviating from Kleene as mentioned under (ii) above arises from the fact that the version presented here is in the case of **GKi** better suited for describing the correspondence between normal natural deductions and normal sequent calculus proofs, to be discussed later.

Pfenning [1994] has recently described a computer implementation of cut elimination for Kleene's G3-systems combined with context-sharing Cut_{cs}.

The semantic tableaux of Beth [1955], and the model sets of Hintikka [1955] are closely connected with G3-systems. See also 4.9.7.

Hudelmaier [1998, p. 25–50] introduces a generalized version of the notion of a multi-succedent G3-type system; for the systems falling under this definition, cut elimination is proved.

3.7.4. *Equivalence of G-systems and N-systems.* The construction in 3.3.1 of G-deductions with Cut from N-deductions is already found in Gentzen [1935]. The construction of a normal N-deduction from a cutfree G-deduction is outlined in Prawitz [1965, App. A, §2]. As to the assignment of cutfree G-deductions to normal N-deductions, see section 6.3.

3.7.5. *Gentzen systems with terms.* The assignment of typed terms to the sequents in a sequent calculus proof is something which might be said to be already present, for the case of implication logic, in Curry and Feys [1958, section 9F2], and follows from proofs showing how to construct a natural deduction from a sequent calculus proof, when combined with the formulas-as-types idea.

For a bijective correspondence between deductions in a suitable Gentzen system and a term calculus, less trivial than the one indicated at the end of 3.3.4, see, for example, Herbelin [1995].

Vestergaard [1998b,1998a] studies implicational G3-systems where the deductions are represented as terms, as in the paper by Herbelin mentioned above; the Cut rule is interpreted as an explicit substitution operator. The steps of the cut elimination procedure (the process of cut elimination is discussed in the first part of the next chapter) recursively evaluate the substitution operator. Vestergaard's results seem to indicate that $\to$**GKm** is computationally better behaved than $\to$**G3m**; for (his version of) the latter calculus Vestergaard presents an infinite sequence of deductions which (intuitively) represent distinct deductions but are mapped to the same deduction under cut elimination. Recent work by Grabmayer [1999] indicates that the results are highly sensitive to the precise formulation of the rules and the cut elimination strategy.

3.7.6. *Inversion lemmas.* Ketonen [1944] showed the invertibility of the propositional rules of his Gentzen system (using Cut, without preservation of depth). An inversion lemma of the type used in our text first appears in Schütte [1950b], for a calculus with one-sided sequents. Schütte does not explicitly state preservation of depth, but this is obvious from his proof, and in particular, he does not use Cut for showing invertibility. Curry [1963] contains inversion lemmas in practically the same form as considered here, with explicit reference to preservation of (logical) depth and (logical) size.

Related to the inversion lemmas is the so-called "inversion principle" for natural deduction. This principle is formulated by Prawitz [1965] as follows: the conclusion of an elimination does not state anything more than what must already have been obtained if the major premise had been obtained by an introduction. This goes back to Gentzen [1935, §5]: "The introductions [of **M**] represent, as it were, the 'definitions' of the symbols concerned, and the eliminations are no more, in the final analysis, than the consequences of these definitions." The term "inversion principle" was coined by Lorenzen [1950].

3.7.7. *The one-sided systems.* Gentzen systems with one-sided sequents for theories based on classical logic were first used by Schütte [1950b]. Schütte has negation for all formulas as a primitive and writes iterated disjunctions instead of multisets. The idea of taking negation for compound formulas as defined is found in Tait [1968]. Tait uses sets of formulas instead of multisets. Because of these further simplifications some authors call the GS-calculi "Tait calculi".

In the paper Rasiowa and Sikorski [1960] a system similar to **GS3** is found; however, negation is a primitive, and there are extra rules for negated statements. For example, $\neg(A \vee B)$ is inferred from $\neg A$ and $\neg B$. In addition there is a rule inferring $\neg\neg A$ from A. The inspiration for this calculus, which is halfway between the calculi of Schütte and Tait, derives from the ideas of Kanger and Beth, in other words, from semantic tableaux.

3.7.8. *Varying Gentzen systems.* In the literature there is a wide variety of "enrichments" of the usual Gentzen systems as described in this chapter. We give some examples.

(a) *Gentzen systems with head formulas.* Sequents take the form $\Gamma; \Gamma' \Rightarrow \Delta; \Delta'$; the formulas in $\Gamma\Delta$ are treated differently form the formulas in $\Gamma'\Delta'$. An example is given in 6.3.5, where sequents of the form $\Pi; \Gamma \Rightarrow A$ with $|\Pi| \leq 1$ are considered. In this example, if $\Pi; \Gamma \Rightarrow A$ has been obtained by a left rule, $|\Pi| = 1$ and the formula in Π is principal. See also 9.4 and Girard [1993].

(b) *Gentzen systems with labelled formulas.* Extra information may be added to the formulas in sequents; the term-labelled calculus described in 3.3.4 is an example.

(c) *Hypersequents* are finite sequences of ordinary sequents:

$$\Gamma_1 \Rightarrow \Delta_1 \mid \Gamma_2 \Rightarrow \Delta_2 \mid \cdots \mid \Gamma_n \Rightarrow \Delta_n.$$

Hypersequents were first introduced in Pottinger [1983] for the proof-theoretic treatment of certain modal logics, and have been extensively used and studied in a series of papers by Avron [1991,1996,1998]. They can be used to give cut-free formulations not only of certain modal logics, but also for substructural and intermediate logics. (In substructural logics the structural rules of weakening and contraction are not generally valid, intermediate logics are theories in the language of propositional logic or first-order predicate logic, which are contained in classical logic and contain intuitionistic logic.) An interesting example of an intermediate logic permitting a cutfree hypersequent formulation is the logic **LC**, introduced in Dummett [1959]; a Hilbert-type axiomatization for **LC** is obtained by adding axioms of the form $(A \to B) \vee (B \to A)$ to **Hip**.

The hypersequent formulation **CLC** of **LC** (Avron [1991]) uses sequents with a single formula in the succedent, and

$$\Gamma_1 \Rightarrow A_1 \mid \cdots \mid \Gamma_n \Rightarrow A_n$$

is interpreted as

$$(\bigwedge \Gamma_1 \to A_1) \vee \cdots \vee (\bigwedge \Gamma_n \to A_n).$$

A typical rule of **GLC** is the "commutation rule" showing interaction between various sequents in a hypersequent:

$$\frac{\Sigma_1 \mid \Gamma_1 \Rightarrow A_1 \mid \Sigma_1' \qquad \Sigma_2 \mid \Gamma_2 \Rightarrow A_2 \mid \Sigma_2'}{\Sigma_1 \mid \Sigma_2 \mid \Gamma_1 \Rightarrow A_2 \mid \Sigma_2 \Rightarrow A_1 \mid \Sigma_1' \mid \Sigma_2'}$$

where $\Sigma_1\Sigma_2, \Sigma_1', \Sigma_2'$ are hypersequents. (If $\Sigma_1\Sigma_2, \Sigma_1', \Sigma_2'$ are empty, $\Gamma_1 \equiv A_1$, $\Gamma_2 \equiv A_2$, this immediately yields $(A_1 \to A_2) \vee (A_2 \to A_1)$ on the interpretation.)

(d) *Labelled sequents.* Instead of labeling formula occurrences, we may also label the sequents themselves. An example is Mints [1997], where the sequents in hypersequents are labelled with finite sequences of natural numbers. Mints uses this device for a cutfree formulation of certain propositional modal logics. The indexing is directly related to the Kripke semantics for the logics considered.

(e) *Display Logic.* This is a very general scheme for Gentzen-like systems, introduced by Belnap [1982], where the comma in ordinary sequents has been replaced by a number of structural operations. Mints [1997] relates Display Logic to his hypersequents of indexed sequents, and Wansing [1998] shows how formulations based on hypersequents may be translated into formalisms based on display sequents. See furthermore Belnap [1990], Wansing [1994].

Chapter 4

Cut elimination with applications

The "applications of cut elimination" in the title of this chapter may perhaps be described more appropriately as "applications of cutfree systems", since the applications are obtained by analyzing the structure of cutfree proofs; and in order to prove that the various cutfree systems are adequate for our standard logics all we need to know is that these systems are closed under Cut (that is to say, Cut is a an admissible rule). Nevertheless there are good reasons to be interested in the process of cut elimination, as opposed to semantical proofs of closure under Cut. True, the usual semantical proofs establish not only closure under Cut, but also completeness for the semantics considered. On the other hand, the proof of cut elimination for **G3c** is at least as efficient as the semantical proof (although **G3cp** permits a very fast semantical proof of closure under Cut), and in the case of logics with a more complicated semantics (such as intuitionistic logic, and the modal logic **S4** in chapter 9) often more efficient. For linear logic in section 9.3, so far no semantical proof of closure under Cut has been published. Other reasons for being interested in the process of cut elimination will be found in certain results in sections 5.1 and 6.9, which describe bounds on the increase in size of deductions under cut elimination and normalization respectively.

4.1 Cut elimination

As mentioned before, "Cut" is the rule

$$\frac{\Gamma \Rightarrow \Delta, A \qquad A, \Gamma' \Rightarrow \Delta'}{\Gamma, \Gamma' \Rightarrow \Delta, \Delta'}$$

Closure under Cut just says that the Cut rule is admissible: if $\vdash \Gamma \Rightarrow \Delta A$ and $A\Gamma' \Rightarrow \Delta'$ in the system considered, then also $\vdash \Gamma\Gamma' \Rightarrow \Delta\Delta'$. This in itself does not give us an algorithm for constructing a deduction of $\Gamma\Gamma' \Rightarrow \Delta\Delta'$ from given deductions of $\Gamma \Rightarrow \Delta A$ and $A\Gamma' \Rightarrow \Delta'$. In the systems studied here the deductions are recursively enumerable. So, if we know that the system is closed under Cut, there exists, trivially, an uninteresting algorithm for finding a deduction of $\Gamma\Gamma' \Rightarrow \Delta\Delta'$ *given* the fact that $\Gamma \Rightarrow \Delta A$ and $A\Gamma' \Rightarrow \Delta'$ are deducible: just search through all deductions until one arrives at a deduction for $\Gamma\Gamma' \Rightarrow \Delta\Delta'$. For such a trivial algorithm we cannot find a bound on the depth of the cutfree proof in terms of the depth of the original proof.

We shall say that *cut elimination* holds, for system **S** + Cut, if there is a "non-trivial" algorithm for transforming a deduction in **S** + Cut into a deduction with the same conclusion in **S**. In our proofs such a non-trivial algorithm is based on certain local transformation steps, such as permuting rules upward over other rules, or replacing a cut on a compound formula A by some cuts on (A and) its immediate subformulas, and on certain simple global transformations on subdeductions, as for example the transformations implicit in the inversion lemma and closure under Contraction for **G3**[**mic**].

4.1.1. DEFINITION. The *level* of a cut is defined as the sum of the depths of the deductions of the premises; the *rank* of a cut on A is $|A|+1$. The *cutrank* of a deduction $\mathcal{D}$, $\mathrm{cr}(\mathcal{D})$, is the maximum of the ranks of the cutformulas occurring in $\mathcal{D}$. ⊠

4.1.2. NOTATION. For the deduction $\mathcal{D}^*$ with conclusion $\Gamma, \Gamma' \Rightarrow \Delta, \Delta'$ obtained by applying dp-admissible Weakening to $\mathcal{D}$ with conclusion $\Gamma \Rightarrow \Delta$ we write $\mathcal{D}[\Gamma' \Rightarrow \Delta']$. We say in this case that $\mathcal{D}$ has been *weakened with* $\Gamma' \Rightarrow \Delta'$; similarly for individual sequents.

In intuitionistic and minimal systems with at most one formula in the succedent this is usually weakening with sequents $\Gamma \Rightarrow$ (i.e., empty succedent); in this case $\mathcal{D}[\Gamma \Rightarrow]$ may be abbreviated as $\mathcal{D}[\Gamma]$. ⊠

4.1.3. LEMMA. *For all systems closed under dp-weakening, prooftrees with instances of Cut may be transformed in a prooftree with instances of contextsharing* $\mathrm{Cut_{cs}}$, *defined in 3.2; the transformation preserves depth. Hence eliminability of Cut is a consequence of eliminability of* $\mathrm{Cut_{cs}}$.

PROOF. If **T** is a system closed under dp-admissible Weakening, then so is $\mathbf{T} + \mathrm{Cut_{cs}}$: if we weaken premises and conclusion of an instance of $\mathrm{Cut_{cs}}$ with the same sequent, the result is again an instance of $\mathrm{Cut_{cs}}$.

Let $\mathcal{D}$ be a prooftree containing instances of Cut and $\mathrm{Cut_{cs}}$; take a topmost instance α of Cut, conclusion of a subdeduction $\mathcal{D}'$, and with premises $\Gamma \Rightarrow A, \Delta$ and $\Gamma', A \Rightarrow \Delta'$, derived by $\mathcal{D}_0, \mathcal{D}_1$ respectively. Then $\mathcal{D}_0[\Gamma' \Rightarrow \Delta']$, $\mathcal{D}_1[\Gamma \Rightarrow \Delta]$ have conclusions $\Gamma\Gamma' \Rightarrow A\Delta\Delta'$ and $\Gamma\Gamma' A \Rightarrow \Delta\Delta'$ respectively; now apply $\mathrm{Cut_{cs}}$ to to these deductions to obtain the original conclusion of $\mathcal{D}'$; replace $\mathcal{D}'$ by the transformed deduction. Thus we may successively replace all instances of Cut by instances of $\mathrm{Cut_{cs}}$. ⊠

4.1.4. NOTATION. In this section we adopt as a local convention, that the deduction(s) of the premise(s) of the conclusion of a prooftree $\mathcal{D}$ are denoted by $\mathcal{D}_0$ ($\mathcal{D}_0, \mathcal{D}_1$); the deductions of the premise(s) of the conclusion of $\mathcal{D}_i$ are $\mathcal{D}_{i0}, \mathcal{D}_{i1}$ etc. The depth of $\mathcal{D}$ is d, the depth of $\mathcal{D}_{i_1 \ldots i_n}$ is $d_{i_1 \ldots i_n}$. ⊠

4.1.5. THEOREM. *Cut elimination holds for* **G3[mic]** + Cut.

PROOF. We shall in fact establish cut elimination for **G3[mic]** + Cut_{cs}. Our strategy will be to successively remove cuts which are topmost among all cuts with rank equal to the rank of the whole deduction, i.e. topmost maximal-rank cuts. It suffices to show how to replace a subdeduction $\mathcal{D}$ of the form

$$\dfrac{\begin{matrix}\mathcal{D}_0 \\ \Gamma \Rightarrow \Delta, D\end{matrix} \qquad \begin{matrix}\mathcal{D}_1 \\ D, \Gamma \Rightarrow \Delta\end{matrix}}{\Gamma \Rightarrow \Delta}\,\text{Cut}_{\text{cs}}$$

where $\text{cr}(\mathcal{D}_i) \leq |D| = \text{cr}(\mathcal{D}) - 1$ for $i \in \{0, 1\}$, by a $\mathcal{D}^*$ with the same conclusion, such that $\text{cr}(\mathcal{D}^*) \leq |D|$. The proof proceeds by a main induction on the cutrank, with a subinduction on the level of the cut at the bottom of $\mathcal{D}$.

We treat the classical and the intuitionistic cases; the treatment for minimal logic is contained in the discussion of the intuitionistic case. For future use we shall also verify in the course of the proof the following property:

$$(*) \qquad d^* \leq d_0 + d_1 \text{ for } \mathbf{G3c}, \;\; d^* \leq 2(d_0 + d_1) \text{ for } \mathbf{G3[mi]}.$$

However, we recommend that initially the proof is read without paying attention to the verification of (*).

We use closure under Contraction and Weakening all the time. Recall that we consider only proofs with axioms where the principal formula is atomic. There are three possibilities we have to consider:

1. at least one of $\mathcal{D}_0, \mathcal{D}_1$ is an axiom;
2. $\mathcal{D}_0$ and $\mathcal{D}_1$ are not axioms, and in at least one of the premises the cutformula is not principal;
3. the cutformula is principal on both sides.

Case 1. At least one of $\mathcal{D}_0, \mathcal{D}_1$ is an axiom.
Subcase 1a. $\mathcal{D}_0$ is an instance of Ax, and D is not principal in $\mathcal{D}_0$. Then $\mathcal{D}$ is of the form

$$\dfrac{P, \Gamma \Rightarrow \Delta, P, D \qquad \begin{matrix}\mathcal{D}_1 \\ P, \Gamma, D \Rightarrow \Delta, P\end{matrix}}{P, \Gamma \Rightarrow \Delta, P}$$

The conclusion is an axiom, so we can take the conclusion for our $\mathcal{D}^*$. Similarly if $\mathcal{D}_0$ is an application of L$\bot$. The intuitionistic case is similar and the verification of $(*)$ trivial.

Subcase 1b. The premise on the left is an application of Ax and the succedent principal formula is also a cutformula:

$$\dfrac{P,\Gamma \Rightarrow \Delta, P \quad \begin{array}{c}\mathcal{D}_1\\ P,P,\Gamma \Rightarrow \Delta\end{array}}{P,\Gamma \Rightarrow \Delta}$$

$\mathcal{D}^*$ is obtained by applying closure under Contraction to $\mathcal{D}_1$. The intuitionistic case proceeds in the same way, and the verification of $(*)$ is trivial.

Subcase 1c. The premise on the right is an application of Ax or L$\bot$, and the antecedent principal formula is not a cutformula. This case is similar to subcase 1a.

Subcase 1d. The premise on the right is an application of the axiom Ax and the cutformula is also a principal formula of the axiom. This case is similar to subcase 1b.

Subcase 1e. The premise on the right is an application of the axiom L$\bot$ and the cutformula is also a principal formula of the axiom.

$$\dfrac{\begin{array}{c}\mathcal{D}_0\\ \Gamma \Rightarrow \Delta, \bot\end{array} \quad \bot,\Gamma \Rightarrow \Delta}{\Gamma \Rightarrow \Delta}$$

If $\mathcal{D}_0$ ends with a rule in which $\bot$ is principal, $\Gamma \Rightarrow \Delta, \bot$ is of the form $\Gamma', \bot \Rightarrow \Delta, \bot$ and hence is an instance of L$\bot$; then $\Gamma', \bot \Rightarrow \Delta$, which is the same as $\Gamma \Rightarrow \Delta$, is also an axiom. If $\mathcal{D}_0$ ends with a rule in which $\bot$ is not principal, say a two-premise rule, $\mathcal{D}$ is of the form

$$\dfrac{\dfrac{\begin{array}{c}\mathcal{D}_{00}\\ \Gamma' \Rightarrow \Delta', \bot\end{array} \quad \begin{array}{c}\mathcal{D}_{01}\\ \Gamma'' \Rightarrow \Delta'', \bot\end{array}}{\Gamma \Rightarrow \Delta, \bot}\text{R} \qquad \bot,\Gamma \Rightarrow \Delta}{\Gamma \Rightarrow \Delta}\text{Cut}_{\text{cs}}$$

This is transformed by permuting and duplicating the cut upwards over R on the left:

$$\dfrac{\text{Cut}_{\text{cs}}\dfrac{\begin{array}{c}\mathcal{D}_{00}\\ \Gamma' \Rightarrow \Delta'\bot\end{array} \quad \bot\Gamma' \Rightarrow \Delta'}{\Gamma' \Rightarrow \Delta'} \qquad \dfrac{\begin{array}{c}\mathcal{D}_{01}\\ \Gamma'' \Rightarrow \Delta''\bot\end{array} \quad \bot\Gamma'' \Rightarrow \Delta''}{\Gamma'' \Rightarrow \Delta''}\text{Cut}_{\text{cs}}}{\Gamma \Rightarrow \Delta}\text{R}$$

In this deduction $\mathcal{D}'$ we can apply the IH to $\mathcal{D}_0', \mathcal{D}_1'$. If $\mathcal{D}_0$ ends with a one-premise rule, the transformation is similar, but no duplication is needed. The intuitionistic case is also similar, but slightly simpler.

– *Verification of* $(*)$ *in the classical case*: if we replace, using the IH, in $\mathcal{D}'$ the immediate subdeductions $\mathcal{D}'_0, \mathcal{D}'_1$ by $(\mathcal{D}'_0)^*, (\mathcal{D}'_1)^*$ respectively, the depth of the resulting deduction $\mathcal{D}^*$ is

$$d^* \leq \max(d_{00} + 1, d_{01} + 1)$$

which is precisely what we have to prove for $(*)$. Similarly for the intuitionistic case.

Case 2. $\mathcal{D}_0$ and $\mathcal{D}_1$ are not axioms, and the cutformula is not principal in either the antecedent or the succedent. Let us assume that D is not principal in the succedent, and that $\mathcal{D}_0$ ends with a two-premise rule R:

$$\dfrac{\dfrac{\begin{matrix}\mathcal{D}_{00} \\ \Gamma' \Rightarrow \Delta' D\end{matrix} \quad \begin{matrix}\mathcal{D}_{01} \\ \Gamma'' \Rightarrow \Delta'' D\end{matrix}}{\Gamma \Rightarrow \Delta D}\,\mathrm{R} \qquad \begin{matrix}\mathcal{D}_1 \\ D\Gamma \Rightarrow \Delta\end{matrix}}{\Gamma \Rightarrow \Delta}$$

This is transformed by permuting the cut upwards over R on the left:

$$\dfrac{\mathrm{Cut_{cs}}\,\dfrac{\begin{matrix}\mathcal{D}_{00}[\Gamma\Rightarrow\Delta] \\ \Gamma\Gamma' \Rightarrow \Delta\Delta' D\end{matrix} \quad \begin{matrix}\mathcal{D}_1[\Gamma'\Rightarrow\Delta'] \\ D\Gamma\Gamma' \Rightarrow \Delta\Delta'\end{matrix}}{\Gamma\Gamma' \Rightarrow \Delta\Delta'} \qquad \dfrac{\begin{matrix}\mathcal{D}_{01}[\Gamma\Rightarrow\Delta] \\ \Gamma\Gamma'' \Rightarrow \Delta\Delta'' D\end{matrix} \quad \begin{matrix}\mathcal{D}_1[\Gamma''\Rightarrow\Delta''] \\ D\Gamma\Gamma'' \Rightarrow \Delta\Delta''\end{matrix}}{\Gamma\Gamma'' \Rightarrow \Delta\Delta''}\,\mathrm{Cut_{cs}}}{\Gamma\Gamma \Rightarrow \Delta\Delta}\,\mathrm{R}$$

Call the resulting deduction $\mathcal{D}'$; replacing $\mathcal{D}'_0, \mathcal{D}'_1$ by respectively $(\mathcal{D}'_0)^*$, $(\mathcal{D}'_1)^*$ given by the IH produces a deduction Δ'' which after use of closure under Contraction produces $\mathcal{D}^*$. Note that R may in particular be a cut with rank $\leq |D|$. The intuitionistic case is treated similarly, except where $\mathcal{D}_0$ ends with L→; then the cut is permuted upwards over one of the premises only.

– *Verification of* $(*)$ *in the classical case.* Note that $d'' = d^*$. We have to show $d^* \leq \max(d_{00}, d_{01}) + 1 + d_1$. By the IH,

$$(d'_0)^* \leq d_{00} + d_1, \quad (d'_1)^* \leq d_{01} + d_1,$$

and so $d^* \leq \max(d_{00} + d_1, d_{01} + d_1) + 1 = \max(d_{00}, d_{01}) + 1 + d_1$. The intuitionistic case is similar.

If the cutformula is not principal in the antecedent, the treatment is similar (symmetric in the classical case).

Case 3. The cutformula is principal on both sides. We distinguish cases according to the principal logical operator of D.

Subcase 3a. $D \equiv D_0 \wedge D_1$.

$$\dfrac{\dfrac{\begin{matrix}\mathcal{D}_{00} \\ \Gamma \Rightarrow \Delta, D_0\end{matrix} \quad \begin{matrix}\mathcal{D}_{01} \\ \Gamma \Rightarrow \Delta, D_1\end{matrix}}{\Gamma \Rightarrow \Delta, D_0 \wedge D_1} \qquad \dfrac{\begin{matrix}\mathcal{D}_{10} \\ D_0, D_1, \Gamma \Rightarrow \Delta\end{matrix}}{D_0 \wedge D_1, \Gamma \Rightarrow \Delta}}{\Gamma \Rightarrow \Delta}$$

becomes

$$\dfrac{\begin{array}{c}\mathcal{D}_{01}\\ \Gamma \Rightarrow \Delta, D_1\end{array} \quad \dfrac{\begin{array}{c}\mathcal{D}_{00}[D_1{\Rightarrow}]\\ D_1\Gamma \Rightarrow \Delta, D_0\end{array} \quad \begin{array}{c}\mathcal{D}_{10}\\ D_0, D_1, \Gamma \Rightarrow \Delta\end{array}}{D_1, \Gamma \Rightarrow \Delta}}{\Gamma \Rightarrow \Delta}$$

and similarly in the intuitionistic case.

– *Verification of* $(*)$. We have to show

$$d^* \leq \max(d_{00}, d_{01}) + 1 + d_{10} + 1 = \max(d_{00} + 1, d_{01} + 1) + d_{10} + 1.$$

In fact inspection of the constructed $\mathcal{D}^*$ shows

$$\begin{array}{rcl} d^* & = & \max(d_{01}, \max(d_{00}, d_{10}) + 1) + 1 \\ & = & \max(d_{01}, d_{00} + 1, d_{10} + 1) + 1 \\ & \leq & \max(d_{00} + 1, d_{01} + 1) + d_{10} + 1. \end{array}$$

Subcase 3b. $D \equiv D_0 \vee D_1$. The treatment in the classical case is symmetric to the preceding case; the intuitionistic case is somewhat simpler.

Subcase 3c. $D \equiv D_0 \to D_1$. The deduction in **G3c** has the form

$$\dfrac{\dfrac{\begin{array}{c}\mathcal{D}_{00}\\ \Gamma, D_0 \Rightarrow D_1, \Delta\end{array}}{\Gamma \Rightarrow D_0 \to D_1, \Delta} \quad \dfrac{\begin{array}{c}\mathcal{D}_{10}\\ \Gamma \Rightarrow D_0, \Delta\end{array} \quad \begin{array}{c}\mathcal{D}_{11}\\ \Gamma, D_1 \Rightarrow \Delta\end{array}}{\Gamma, D_0 \to D_1 \Rightarrow \Delta}}{\Gamma \Rightarrow \Delta}$$

This is transformed into a deduction $\mathcal{D}'$:

$$\dfrac{\begin{array}{c}\mathcal{D}_{10}\\ \Gamma, \Rightarrow D_0, \Delta\end{array} \quad \dfrac{\begin{array}{c}\mathcal{D}_{00}\\ \Gamma, D_0 \Rightarrow D_1, \Delta\end{array} \quad \begin{array}{c}\mathcal{D}_{11}[D_0{\Rightarrow}]\\ \Gamma, D_0, D_1 \Rightarrow \Delta\end{array}}{\Gamma, D_0 \Rightarrow \Delta}}{\Gamma \Rightarrow \Delta}$$

The new cut on $D_0 \to D_1$ is of a lower level than the original one, so by the subinduction hypothesis we can remove this cut. In the intuitionistic case we have

$$\dfrac{\dfrac{\begin{array}{c}\mathcal{D}_{00}\\ \Gamma, D_0 \Rightarrow D_1\end{array}}{\Gamma \Rightarrow D_0 \to D_1} \quad \dfrac{\begin{array}{c}\mathcal{D}_{10}\\ \Gamma, D_0 \to D_1 \Rightarrow D_0\end{array} \quad \begin{array}{c}\mathcal{D}_{11}\\ \Gamma, D_1 \Rightarrow A\end{array}}{\Gamma, D_0 \to D_1 \Rightarrow A}}{\Gamma \Rightarrow A}$$

which is replaced by

$$\dfrac{\dfrac{\dfrac{\dfrac{\begin{array}{c}\mathcal{D}_{00}\\ \Gamma D_0 \Rightarrow D_1\end{array}}{\Gamma \Rightarrow D_0{\to}D_1} \quad \begin{array}{c}\mathcal{D}_{10}\\ \Gamma, D_0{\to}D_1 \Rightarrow D_0\end{array}}{\Gamma \Rightarrow D_0}\,\text{Cut}_{\text{cs}} \quad \begin{array}{c}\mathcal{D}_{00}\\ \Gamma D_0 \Rightarrow D_1\end{array}}{\Gamma \Rightarrow D_1}\,\text{Cut}_{\text{cs}} \quad \begin{array}{c}\mathcal{D}_{11}\\ \Gamma D_1 \Rightarrow A\end{array}}{\Gamma \Rightarrow A}\,\text{Cut}_{\text{cs}}$$

The new cut on $D_0 \to D_1$ has a lower level, and can therefore be removed by the subinduction hypothesis; let $\mathcal{D}'_{00}$ be replaced by $(\mathcal{D}'_{00})^*$, the resulting deduction is $\mathcal{D}^*$ (and $(\mathcal{D}^*)_{00} = (\mathcal{D}'_{00})^*$).

– *Verification of* $(*)$. The classical case is similar to subcase 3a, and left to the reader. We verify the intuitionistic case. We have to show

$$
\begin{array}{lrcl}
(1) & d^* & \leq & 2(d_{00} + 1 + \max(d_{10}, d_{11}) + 1) \\
& & = & \max(2d_{00} + 2d_{10} + 4, 2d_{00} + 2d_{11} + 4).
\end{array}
$$

The deduction $(\mathcal{D}^*)_{00}$ satisfies $d^*_{00} \leq 2(d_{00} + 1 + d_{10})$ (using the IH), hence $\mathcal{D}^*$ satisfies

$$
\begin{array}{rcl}
d^* & \leq & \max(\max(2d_{00} + 2d_{10} + 2, d_{00}) + 1, d_{11}) + 1 \\
& = & \max(2d_{00} + 2d_{10} + 4, d_{00} + 2, d_{11} + 1),
\end{array}
$$

and this is obviously smaller than the right hand side of (1).

Subcase 3d. $D \equiv \forall x D_0$. We transform ($y \notin \mathrm{FV}(D_0, \Gamma, \Delta)$, $y \notin \mathrm{FV}(t)$)

$$
\dfrac{\dfrac{\begin{array}{c}\mathcal{D}_{00}\\ \Gamma \Rightarrow \Delta, D_0[x/y]\end{array}}{\Gamma \Rightarrow \Delta, \forall x D_0} \qquad \dfrac{\begin{array}{c}\mathcal{D}_{10}\\ \forall x D_0, D_0[x/t], \Gamma \Rightarrow \Delta\end{array}}{\forall x D_0, \Gamma \Rightarrow \Delta}}{\Gamma \Rightarrow \Delta}
$$

into

$$
\dfrac{\begin{array}{c}\mathcal{D}_{00}[y/t]\\ \Gamma \Rightarrow \Delta, D_0[x/t]\end{array} \qquad \dfrac{\dfrac{\begin{array}{c}\mathcal{D}_{00}[D_0[x/t] \Rightarrow]\\ D_0[x/t], \Gamma \Rightarrow \Delta, D_0[x/y]\end{array}}{D_0[x/t], \Gamma \Rightarrow \Delta, \forall x D_0} \qquad \begin{array}{c}\mathcal{D}_{10}\\ \forall x D_0, D_0[x/t], \Gamma \Rightarrow \Delta\end{array}}{D_0[x/t], \Gamma \Rightarrow \Delta}}{\Gamma \Rightarrow \Delta}
$$

The subdeduction $\mathcal{D}'$ ending in the cut on $\forall x D_0$ is of lower level, so may be replaced by the IH by a deduction $(\mathcal{D}')^*$. This produces the required $\mathcal{D}^*$.

– *Verification of* $(*)$ *in the classical case.* We have to show $d^* \leq d_{00} + d_{10} + 2$; for $\mathcal{D}^*$ we have

$$
\begin{array}{rcl}
d^* & \leq & \max(d_{00}, d_{00} + 1 + d_{10}) + 1 \\
& = & \max(d_{00} + d_{10} + 2, d_{00} + 1) \\
& = & d_{00} + d_{10} + 2.
\end{array}
$$

The treatment of the intuitionistic case is completely similar.

Subcase 3e. $D \equiv \exists x D_0$. The classical case is symmetric to the case of the universal quantifier; the intuitionistic case is simpler. We leave the verification to the reader. ⊠

4.1.5A. ♠ Supply the missing cases in the preceding proof.

From the proof we obtain the following lemma, which will be used in the next chapter to obtain an upper bound on the increase of depth of deductions as a result of cut elimination:

4.1.6. COROLLARY. *(Cut reduction lemma) Let $\mathcal{D}'$, $\mathcal{D}''$ be two deductions in* $\mathbf{G3c} + \mathrm{Cut}_{\mathrm{cs}}$, *with cutrank* $\leq |D|$, *and let* $\mathcal{D}$ *result by a cut:*

$$\dfrac{\begin{array}{cc}\mathcal{D}' & \mathcal{D}'' \\ \Gamma \Rightarrow D, \Delta & \Gamma, D \Rightarrow \Delta\end{array}}{\Gamma \Rightarrow \Delta}$$

Then we can transform $\mathcal{D}$ into a deduction $\mathcal{D}^*$ with lower cutrank and the same conclusion such that $|\mathcal{D}^*| \leq |\mathcal{D}'| + |\mathcal{D}''|$. A similar result holds for $\mathbf{G3i} + \mathrm{Cut}_{\mathrm{cs}}$, *with* $|\mathcal{D}^*| \leq 2(|\mathcal{D}'| + |\mathcal{D}''|)$.

4.1.7. REMARK. The cut elimination procedure as described above does not produce unique results: there is indeterminacy at certain steps. In particular when both cutformulas are non-principal, we can move a cut upwards on the right or on the left. By way of illustration, consider the following deduction:

$$\dfrac{\dfrac{P', P, Q \Rightarrow P, R, P' \vee Q'}{P', P \wedge Q \Rightarrow P, R, P' \vee Q'} \quad \dfrac{P \wedge Q, R, P' \Rightarrow P', Q', P}{P \wedge Q, R, P' \Rightarrow P' \vee Q', P}}{P', P \wedge Q \Rightarrow P, P' \vee Q'}$$

The cutformula R is not principal on either side. So we can permute the cut upwards on the right or on the left. Permuting upwards on the left yields

$$\dfrac{\dfrac{P \wedge Q, P, Q, P' \Rightarrow P, R, P' \vee Q' \quad \dfrac{P, Q, P \wedge Q, R, P' \Rightarrow P', Q', P}{P, Q, P \wedge Q, R, P' \Rightarrow P' \vee Q', P}}{P \wedge Q, P', P, Q \Rightarrow P, P' \vee Q'}}{P \wedge Q, P', P \wedge Q \Rightarrow P, P' \vee Q'}$$

Now we have to apply closure under Contraction to get the required conclusion; following the method of transformation of the deductions in the proof of dp-closure under Contraction (3.5.5) the result is

$$\dfrac{\dfrac{P, Q, P' \Rightarrow P, R, P' \vee Q' \quad \dfrac{P, Q, R, P' \Rightarrow P', Q', P}{P, Q, R, P' \Rightarrow P' \vee Q', P}}{P', P, Q \Rightarrow P, P' \vee Q'}}{P', P \wedge Q \Rightarrow P, P' \vee Q'}$$

The remaining cut is now simply removed by noting that the conclusion of the cut is an axiom (subcase 1a in the proof of 4.1.5):

$$\dfrac{P', P, Q \Rightarrow P, P' \vee Q'}{P', P \wedge Q \Rightarrow P, P' \vee Q'}$$

If we start, symmetrically, permuting the cut upwards on the right, we end up with

$$\dfrac{P', P \wedge Q \Rightarrow P, P', Q'}{P', P \wedge Q \Rightarrow P, P' \vee Q'}$$

These two results represent obviously different proofs.

Another source of indeterminacy in the cut elimination process appears in case 3, where in the subcases which reduce a cut of degree $n+1$ of level k into *two* cuts of degree n (and possibly a cut of degree $n+1$ and level less than k); one has to choose an order in which the cuts of degree n are applied.

4.1.8. *Variations*

The most commonly used strategy in proofs of cut elimination is the removal of topmost cuts; that is to say, we show how to replace a subdeduction $\mathcal{D}$ of the form

$$\dfrac{\begin{matrix}\mathcal{D}_0 \\ \Gamma \Rightarrow \Delta, D\end{matrix} \quad \begin{matrix}\mathcal{D}_1 \\ D, \Gamma \Rightarrow \Delta\end{matrix}}{\Gamma \Rightarrow \Delta}$$

where $\mathcal{D}_0, \mathcal{D}_1$ are cutfree, with a cutfree proof $\mathcal{D}^*$ with the same conclusion. The preceding proof can almost be copied for this strategy. We have to distinguish the same main cases, and the same subcases in case 3.

In permuting cuts upwards, as in subcase 1e and case 2, permuting over a cut of lower rank does not occur. In the subcases of case 3, we have to appeal not only to the subinduction hypothesis, but to the main IH as well. Take for example the prooftree obtained in subcase 3d after transformation. We first appeal to the subinduction hypothesis to remove the cut on $\forall x D_0$, and then to the IH to remove all cuts of lower degree.

In the proof above, we have removed instances of Cut_{cs}. Under the strategy of removing topmost cuts, we can also directly remove instances of Cut; the details are rather similar to those presented in the proof above, but the appeals to closure under Contraction appear at other places. For example, in the subcase 3d we now have:

$$\dfrac{\dfrac{\begin{matrix}\mathcal{D}_{00}[y] \\ \Gamma \Rightarrow \Delta, D_0[x/y]\end{matrix}}{\Gamma \Rightarrow \Delta, \forall x D_0} \quad \dfrac{\begin{matrix}\mathcal{D}_{10} \\ \forall x D_0, D_0[x/t], \Gamma' \Rightarrow \Delta'\end{matrix}}{\forall x D_0, \Gamma' \Rightarrow \Delta'}}{\Gamma, \Gamma' \Rightarrow \Delta, \Delta'}$$

into

$$\dfrac{\begin{matrix}\mathcal{D}_{00}[y/t] \\ \Gamma \Rightarrow \Delta, D_0[x/t]\end{matrix} \quad \dfrac{\dfrac{\begin{matrix}\mathcal{D}_{00}[y] \\ \Gamma \Rightarrow \Delta, D_0[x/y]\end{matrix}}{\Gamma \Rightarrow \Delta, \forall x D_0} \quad \begin{matrix}\mathcal{D}_{10} \\ \forall x D_0, D_0[x/t], \Gamma' \Rightarrow \Delta'\end{matrix}}{D_0[x/t], \Gamma, \Gamma' \Rightarrow \Delta, \Delta'}}{\Gamma\Gamma\Gamma' \Rightarrow \Delta\Delta\Delta'}$$

We remove the cuts by appeal to IH and sub-IH, and finally have to apply closure under Contraction. On the other hand, in the treatment of case 2 the appeal to dp-closure under Contraction is not any longer necessary.

The strategy of removal of topmost maximal-rank cuts does not work with Cut, since we cannot guarantee dp-closure under Contraction for the system with Cut.

4.1.9. *Gentzen's method of cut elimination*

There is another method, going back to Gentzen, which applies to **G[12][mic]**, not directly, but via a slight modification of these systems, and which works as follows.

If we try to prove cut elimination directly for **G2[mic]**, by (essentially) the same method as used above for **G3[mic]**, we encounter difficulties with the Contraction rule. We should like to transform a deduction

$$\dfrac{\begin{matrix}\mathcal{D}' \\ \Gamma \Rightarrow A\end{matrix} \quad \dfrac{\begin{matrix}\mathcal{D}'' \\ \Gamma', A, A \Rightarrow B\end{matrix}}{\Gamma', A \Rightarrow B}\,\mathrm{LC}}{\Gamma, \Gamma' \Rightarrow B}\,\mathrm{Cut}$$

into

$$\dfrac{\dfrac{\begin{matrix}\mathcal{D}' \\ \Gamma \Rightarrow A\end{matrix} \quad \dfrac{\begin{matrix}\mathcal{D}' \\ \Gamma \Rightarrow A\end{matrix} \quad \begin{matrix}\mathcal{D}'' \\ \Gamma', A, A \Rightarrow B\end{matrix}}{\Gamma, \Gamma', A \Rightarrow B}\,\mathrm{Cut}}{\Gamma, \Gamma, \Gamma' \Rightarrow B}\,\mathrm{Cut}}{\Gamma, \Gamma' \Rightarrow B}\,\mathrm{LC}$$

but this does not give a reduction in the height of the subtrees above the lowest new cut. The solution is to replace Cut by a *derivable* generalization of the Cut rule:

$$\text{Multicut} \quad \frac{\Gamma \Rightarrow \Delta, A^n \quad A^m, \Gamma' \Rightarrow \Delta'}{\Gamma, \Gamma' \Rightarrow B, \Delta, \Delta'} \quad (n, m > 0)$$

where A^k, $k \in \mathbb{N}$, stands for k copies of A. Multicut, also called "Mix", can then be eliminated from this modified calculus in the same way as Cut was eliminable from the **G3**-systems.

Rank and *level* of a Multicut application (a multicut) are defined as rank and level of a cut. We can apply either the strategy of removing topmost cuts, or the strategy of removing topmost maximal-rank cuts.

Under both strategies we use an induction on the rank of the multicut, with a subinduction on the level of the multicut, in showing how to get rid of a multicut of rank $k+1$ applied to two proofs with cutrank 0 (on the first strategy) or less than $k+1$ (on the second strategy). In the example above, the upper deduction is simply replaced by

$$\dfrac{\Gamma \Rightarrow A \quad \begin{matrix}\mathcal{D}'' \\ \Gamma', A, A \Rightarrow B\end{matrix}}{\Gamma, \Gamma' \Rightarrow B}\,\text{Multicut}$$

Instructive is the following case, the most complicated one: let $\mathcal{D}$ be obtained by a multicut on the following two cutfree deductions:

$$\dfrac{\begin{array}{c}\mathcal{D}_{00}\\ \Gamma A \Rightarrow B(A{\to}B)^m\Delta\end{array}}{\Gamma \Rightarrow (A{\to}B)^{m+1}\Delta}\,\mathrm{R}{\to} \qquad \dfrac{\begin{array}{c}\mathcal{D}_{10}\\ \Gamma'(A{\to}B)^n \Rightarrow A\Delta'\end{array} \quad \begin{array}{c}\mathcal{D}_{11}\\ \Gamma'(A{\to}B)^n B \Rightarrow \Delta'\end{array}}{\Gamma'(A{\to}B)^{n+1} \Rightarrow \Delta'}\,\mathrm{L}{\to}$$

In the case where $m, n > 0$ we construct $\mathcal{D}_a, \mathcal{D}_b, \mathcal{D}_c$:

$$\mathcal{D}_a \equiv \left\{ \dfrac{\begin{array}{c}\mathcal{D}_{00}\\ \Gamma A \Rightarrow B(A{\to}B)^m\Delta\end{array} \quad \dfrac{\begin{array}{c}\mathcal{D}_{10}\\ \Gamma'(A{\to}B)^n \Rightarrow A\Delta'\end{array} \quad \begin{array}{c}\mathcal{D}_{11}\\ \Gamma'(A{\to}B)^n B \Rightarrow \Delta'\end{array}}{\Gamma'(A{\to}B)^{n+1} \Rightarrow \Delta'}}{\Gamma\Gamma' A \Rightarrow B\Delta\Delta'} \right.$$

$$\mathcal{D}_b \equiv \left\{ \dfrac{\dfrac{\begin{array}{c}\mathcal{D}_{00}\\ \Gamma A \Rightarrow B(A{\to}B)^m\Delta\end{array}}{\Gamma \Rightarrow (A{\to}B)^{m+1}\Delta} \quad \begin{array}{c}\mathcal{D}_{10}\\ \Gamma'(A{\to}B)^n \Rightarrow A\Delta'\end{array}}{\Gamma\Gamma' \Rightarrow A\Delta\Delta'} \right.$$

$$\mathcal{D}_c \equiv \left\{ \dfrac{\dfrac{\begin{array}{c}\mathcal{D}_{00}\\ \Gamma A \Rightarrow B(A{\to}B)^m\Delta\end{array}}{\Gamma \Rightarrow (A{\to}B)^{m+1}\Delta} \quad \begin{array}{c}\mathcal{D}_{11}\\ \Gamma'(A{\to}B)^n B \Rightarrow \Delta'\end{array}}{\Gamma\Gamma' B \Rightarrow \Delta\Delta'} \right.$$

In each of these deductions the multicut on $A \to B$ has a lower level than in $\mathcal{D}$. Therefore we can construct by the IH their transforms $\mathcal{D}'_a, \mathcal{D}'_b, \mathcal{D}'_c$ of cutrank $\leq |A \to B|$ and combine these in

$$\dfrac{\dfrac{\dfrac{\begin{array}{c}\mathcal{D}'_a\\ \Gamma\Gamma' A \Rightarrow B\Delta\Delta'\end{array} \quad \begin{array}{c}\mathcal{D}'_b\\ \Gamma\Gamma' \Rightarrow A\Delta\Delta'\end{array}}{\Gamma\Gamma\Gamma'\Gamma' \Rightarrow B\Delta\Delta\Delta'\Delta'} \quad \begin{array}{c}\mathcal{D}'_c\\ \Gamma\Gamma' B \Rightarrow \Delta\Delta'\end{array}}{(\Gamma\Gamma')^3 \Rightarrow (\Delta\Delta')^3}}{\Gamma\Gamma' \Rightarrow \Delta\Delta'}\,\mathrm{C}$$

The multicuts are now all of lower rank.

4.1.9A. ♠ Show also for the other cases how to reduce the rank of an application of Multicut when the cutformula is principal in both premises.

4.1.9B. ♠ Argue that Gentzen's cut elimination procedure applies equally well to the system **G2i*** mentioned in 3.3.4. (One can save a few cases in the argument if the A in the axioms $\Gamma, \bot \Rightarrow A$ is restricted to be prime.) What happens to this argument if, instead of the axioms $\Gamma, \bot \Rightarrow A$, we adopt the rule "If $\Gamma \Rightarrow \bot$, then $\Gamma \Rightarrow A$"?

4.1.10. *Cut-elimination for* m-**G3i**

The proofs of cut elimination for **G3i** can be adapted to m-**G3i**. We shall not carry this out in detail, but instead provide a sketch.

LEMMA.

(i) In m-**G3i** *the following rule is depth-preserving admissible:*

$$\frac{\Gamma \Rightarrow \bot, \Delta}{\Gamma \Rightarrow \Delta}$$

(ii) L∨, L∧, L∃, R∨, R∧, R∃ *are invertible in* m-**G3i**.

(iii) m-**G3i** *is closed under depth-preserving left- and right-contraction.* ⊠

THEOREM. Cut *is eliminable from deductions in* m-**G3i** + Cut.

PROOF. We follow the standard strategy of removing topmost cuts; so we have to show how to remove a cut applied to two cutfree deductions of the premises. The main case distinctions are:

1. one of the premises of the cut is an axiom;
2. case 1 does not apply, and the left premise of the cut is obtained by a rule application for which the cutformula is not principal;
3. cases 1 and 2 do not apply, and the right premise of the cut is obtained by an application of rule R for which the cutformula is not principal;
4. the cutformula is principal in both premises.

The asymmetry between cases 2 and 3 is caused by the rules R→ and R∀ which deviate from the general pattern. The only new element, when compared with the proof for **G3i**, occurs under case 3, in particular where the rule R is R→ or R∀. For example, if R is R→, the proof ends with

$$\frac{\Gamma \Rightarrow \Delta, A \quad \dfrac{A, \Gamma', C \Rightarrow D}{A, \Gamma' \Rightarrow C \to D, \Delta'}}{\Gamma \Rightarrow C \to D, \Delta, \Delta'}$$

Since we are in case 3 and cases 1 and 2 do not apply, A is principal on the left. If the left premise is obtained by R→, or R∀, say R→, with $A \equiv A_1 \to A_2$, the proof ends

$$\frac{\Gamma, A_1 \Rightarrow A_2}{\Gamma \Rightarrow \Delta, A}$$

Then we transform the end of the proof simply into

$$\dfrac{\dfrac{\dfrac{\Gamma, A_1 \Rightarrow A_2}{\Gamma \Rightarrow A, \Delta} \quad A, \Gamma', C \Rightarrow D}{\Gamma, \Gamma', C \Rightarrow D}}{\Gamma, \Gamma' \Rightarrow C \to D, \Delta, \Delta'}$$

However, if the left premise has been obtained by one of the invertible rules R$\vee$, R$\wedge$, R$\exists$, we use inversion. For example, let R$\vee$ be the rule for the left premise, so the deduction ends with

$$\dfrac{\dfrac{\Gamma \Rightarrow \Delta, A, B}{\Gamma \Rightarrow \Delta, A \vee B} \quad \dfrac{A \vee B, \Gamma', C \Rightarrow D}{A \vee B, \Gamma' \Rightarrow C \to D, \Delta'}}{\Gamma, \Gamma' \Rightarrow \Delta, \Delta', C \to D}$$

we replace this by

$$\dfrac{\dfrac{\Gamma \Rightarrow \Delta, A, B \quad \dfrac{\dfrac{A \vee B, \Gamma', C \Rightarrow D}{A, \Gamma', C \Rightarrow D}\text{(Inv)}}{A, \Gamma' \Rightarrow C \to D, \Delta'}}{\Gamma, \Gamma' \Rightarrow \Delta, \Delta', B, C \to D} \quad \dfrac{\dfrac{A \vee B, \Gamma', C \Rightarrow D}{B, \Gamma', C \Rightarrow D}\text{(Inv)}}{B, \Gamma' \Rightarrow C \to D, \Delta'}}{\Gamma'\Gamma'\Gamma \Rightarrow \Delta\Delta'\Delta', C \to D}$$

Here the dotted line indicates an application of Inversion to transform a proof with conclusion as above the line into a proof of no greater depth with conclusion as below the line. ⊠

4.1.10A. ♠ Prove the lemma.

4.1.10B. ♠ Adapt the proof of cut elimination for m-**G3i** to elimination of Cut$_{\rm cs}$.

4.1.10C. ♠ Check that Cut or Cut$_{\rm cs}$ is also eliminable from m-**G3i**$'$ plus Cut or Cut$_{\rm cs}$.

4.1.11. *Semantic motivation for* **G3cp**

In the introduction we described for the case of implication logic a very natural way of arriving at a cutfree sequent calculus for **G3cp**. We extend this here to all of **G3cp**.

In testing the truth of a sequent $\Gamma \Rightarrow \Delta$, we try to give a valuation such that $\bigwedge \Gamma$ becomes true and $\bigvee \Delta$ becomes false, in other words, the valuation should make all formulas in Γ true and all formulas in Δ false.

In order to make $\Gamma \Rightarrow \Delta, A \to B$ false, we try to make Γ, A true and Δ, B false, i.e. we try to find a refuting valuation for the sequent $\Gamma, A \Rightarrow \Delta, B$. In order to make $\Gamma, A \to B \Rightarrow \Delta$ false, we try to make either Γ true and Δ, A false, or Γ, B true and Δ false. In other words, we try to find a refuting valuation either for $\Gamma \Rightarrow \Delta, A$ or for $\Gamma, B \Rightarrow \Delta$.

Thus at each step we reduce the problem of finding a refuting valuation to corresponding problems for less complex sequents. In the end we arrive at $\Gamma \Rightarrow \Delta$ with $\Gamma \Rightarrow \Delta$ consisting of atomic formulas only. These have refuting valuations if they are not axioms.

Our rules for reducing the problem of finding a refuting valuation for a sequent correspond to the following rules read upside down:

$$P, \Gamma \Rightarrow \Delta, P \qquad\qquad \bot, \Gamma \Rightarrow \Delta$$

$$\frac{\Gamma, A, B \Rightarrow \Delta}{\Gamma, A \land B \Rightarrow \Delta} \qquad \frac{\Gamma \Rightarrow \Delta, A \qquad \Gamma \Rightarrow \Delta, B}{\Gamma \Rightarrow \Delta, A \land B}$$

$$\frac{A, \Gamma \Rightarrow \Delta \qquad B, \Gamma \Rightarrow \Delta}{A \lor B, \Gamma \Rightarrow \Delta} \qquad \frac{\Gamma \Rightarrow \Delta, A, B}{\Gamma \Rightarrow \Delta, A \lor B}$$

$$\frac{\Gamma \Rightarrow \Delta, A \qquad B, \Gamma \Rightarrow \Delta}{\Gamma, A \to B \Rightarrow \Delta} \qquad \frac{A, \Gamma \Rightarrow \Delta, B}{\Gamma \Rightarrow \Delta, A \to B}$$

These rules are precisely the propositional part of **G3c**. If we start "bottom upwards", the different branches of the refutation search tree represent different possibilities for finding a refutation. This gives us the following:

THEOREM. *(Completeness for* **G3cp***) A sequent $\Gamma \Rightarrow \Delta$ is formally derivable by the rules listed above iff there is no refuting valuation.* ⊠

This idea may be extended to predicate logic **G3c**. Since Cut is obviously valid semantically, one thus finds a proof of closure under Cut by semantical means; but the proof does not provide a specific algorithm of cut elimination. See also 4.9.7.

4.2 Applications of cutfree systems

From the existence of cutfree formalizations of predicate logic one easily obtains a number of interesting properties.

Below, positive and negative occurrences of formulas in $\Gamma \Rightarrow \Delta$ are defined as positive and negative occurrences in the classical sense in $\bigwedge \Gamma \to \Delta$, where $\bigwedge$ is used for iterated conjunction.

4.2.1. PROPOSITION. *(Subformula property, preservation of signs in deductions) Let $\mathcal{D}$ be a cutfree deduction of a sequent $\Gamma \Rightarrow \Delta$ in* **G**[**mic**][**123**]. *Then for any sequent $\Gamma' \Rightarrow \Delta'$ in $\mathcal{D}$ we have*

(i) the formulas of Γ' occur positively in Γ, or negatively in Δ;

(ii) the formulas of Δ' occur either positively in Δ, or negatively in Γ; moreover, for **G1[mic]**,

(iii) if a formula A occurs only positively [negatively] in $\Gamma \Rightarrow \Delta$, then A is introduced in $\mathcal{D}$ by a right [left] rule.

PROOF. Immediate by inspection of the rules. ⊠

An appropriate formulation of the third property for systems **G[23][mic]** requires more care, since formulas may now also enter as context in an axiom.

COROLLARY. *(Separation property) Any provable sequent $\Gamma \Rightarrow A$ always has a proof using only the logical rules and/or axioms for the logical operators occurring in $\Gamma \Rightarrow A$.*

4.2.1A. ♠ Show that the separation property holds for **Ni**, and that it holds for **Hi** in the following form: let $\mathcal{X}$ be any subset of $\{\wedge, \vee, \to, \bot, \forall, \exists\}$ containing $\to$; show that the $\mathcal{X}$-fragment of **Hi** can be axiomatized by the axioms and rules involving operators from $\mathcal{X}$ only. *Hint.* Combine the equivalence proofs of 2.4.2 and 3.3 with the subformula property for **G3i**.

4.2.1B. ♠ Show that the separation property holds for **Nc** in the form: if $\Gamma \to A$ is provable in **Nc**, then it has a proof using only axioms/rules for logical operators occurring in $\Gamma \to A$, and possibly $\bot_c$. Extend the separation theorem to **Hc** for fragments containing at least $\to, \bot$. (For the case of $\to$**Hc** see 4.9.2, 6.2.7C, 2.1.8F.)

4.2.2. THEOREM. *(Relation between* **M** *and* **I***) Let P be a fixed proposition letter, not occurring in Γ, A. For arbitrary B not containing P let $B^* := B[\bot/P]$, and put $\Gamma^* := \{B^* : B \in \Gamma\}$. Then (with the notation of 1.1.6)*

$$\Gamma \vdash_{\mathrm{m}} A \ \ \textit{iff} \ \ \Gamma^* \vdash_{\mathrm{i}} A^*.$$

PROOF. If $\Gamma \vdash_{\mathrm{m}} A$, then $\Gamma^* \vdash_{\mathrm{m}} A^*$ (since $\bot$ behaves as an arbitrary proposition letter P in minimal logic), hence $\Gamma^* \vdash_{\mathrm{i}} A^*$. Conversely, if $\Gamma^* \vdash_{\mathrm{i}} A^*$, we can show $\Gamma^* \Rightarrow A^*$ by a cutfree proof in one of the intuitionistic systems; by the separation property for cutfree systems, the proof does not use the $\bot$-axiom, so $\Gamma^* \vdash_{\mathrm{m}} A^*$, hence $\Gamma \vdash_{\mathrm{m}} A$. ⊠

4.2.3. THEOREM. *(Disjunction property under hypotheses) In* **M** *and* **I***, if Γ does not contain a disjunction as s.p.p. (= strictly positive part, defined in 1.1.4), then, if $\Gamma \vdash A \vee B$, it follows that $\Gamma \vdash A$ or $\Gamma \vdash B$.*

PROOF. Suppose $\Gamma \vdash_{\mathrm{i}} A \vee B$, then we have a proof $\mathcal{D}$ in **G3i** of $\Gamma \Rightarrow A \vee B$, where Γ does not contain a disjunction as strictly positive part.

A sequent $\Gamma' \Rightarrow A \vee B$ in $\mathcal{D}$ such that Γ' does not contain a s.p.p. which is disjunctive, and where $A \vee B$ is not principal, has *exactly one* premise of the form $\Gamma'' \Rightarrow A \vee B$ where Γ'' has no disjunctive or existential s.p.p.'s. (Only L$\vee$ can cause two premises with $A \vee B$ in the succedent, but then Γ' would contain a disjunction, hence a disjunctive s.p.p.) Therefore there is a sequence

$$\Gamma_0 \Rightarrow A \vee B, \Gamma_1 \Rightarrow A \vee B, \ldots, \Gamma_n \Rightarrow A \vee B$$

in $\mathcal{D}$ such that $A \vee B$ is principal in the first sequent, $\Gamma_i \Rightarrow A \vee B$ is the premise of $\Gamma_{i+1} \Rightarrow A \vee B$ $(0 \leq i < n)$ and $\Gamma_n \Rightarrow A \vee B$ is the conclusion.

Note that $\Gamma_0 \Rightarrow A \vee B$ cannot be an axiom (except when $\bot \in \Gamma_0$, in which case the matter is trivial), because of the restriction to atomic principal formulas in axioms. Therefore $\Gamma_0 \Rightarrow A \vee B$ is preceded by $\Gamma_0 \Rightarrow A$ or $\Gamma_0 \Rightarrow B$, say the first; replacing in all $\Gamma_i \Rightarrow A \vee B$ the occurrence of $A \vee B$ by A and dropping the repetition of $\Gamma_0 \Rightarrow A$ results in a correct deduction. ⊠

There is a similar theorem for the existential quantifier:

4.2.4. THEOREM. *(Explicit definability under hypotheses) In* **M** *or* **I**

(i) if Γ does not contain an existential s.p.p., and $\Gamma \vdash \exists x A$, then there are terms $t_1, t_2, \ldots, t_n$ such that $\Gamma \vdash A(t_1) \vee \ldots \vee A(t_n)$,

(ii) if Γ contains neither a disjunctive s.p.p., nor an existential s.p.p., and $\Gamma \vdash \exists x A$, then there is a term t such that $\Gamma \vdash A(t)$. ⊠

REMARK. *Rasiowa–Harrop formulas* (in the literature also called *Harrop* formulas) are formulas for which no s.p.p. is a disjunction or an existential formula. For Γ consisting of Rasiowa–Harrop formulas 4.2.3 and 4.2.4 both hold.

4.2.4A. ♠ Prove theorem 4.2.4.

4.2.4B. ♠ Reformulate the arguments of theorem 4.2.3, 4.2.4 as proofs by induction on the depth of deductions.

4.2.4C. ♠ (Alternative method for proving the disjunction property) The method of the "Aczel slash" provides an alternative route to a proof of: if $\Gamma \vdash_i A \vee B$, then $\Gamma \vdash_i A$ or $\Gamma \vdash_i B$, for suitable Γ. We describe the method for propositional logic only. Let Γ be a set of sentences; $\Gamma | A$ is defined by induction on the depth of A by

(i) $\Gamma | P := \Gamma \vdash P$ for P atomic;
(ii) $\Gamma | A \wedge B := \Gamma | A$ and $\Gamma | B$;
(iii) $\Gamma | A \vee B := \Gamma | A$ or $\Gamma | B$;

(iv) $\Gamma|A \to B :=$ (If $\Gamma|A$ then $\Gamma|B$) and $\Gamma \vdash A \to B$.

By induction on the depth of A one can prove that, if $\Gamma|A$, then $\Gamma \vdash A$. By induction on the length of proofs in **Hi** one can show that, if one assumes $\Gamma|C$ for all $C \in \Gamma$ and $\Gamma \vdash A$, then $\Gamma|A$. Deduce from this that if $\Gamma|C$ for all $C \in \Gamma$, and $\Gamma \vdash A \vee B$, then $\Gamma \vdash A$ or $\Gamma \vdash B$. Show finally that if no formula in Γ contains a disjunction as a strictly positive subformula, then $\Gamma|C$ for all $C \in \Gamma$. Conclude from this: if $\vdash_{\rm i} \neg A \to B \vee C$, then $\vdash_{\rm i} \neg A \to B$ or $\vdash_{\rm i} \neg A \to C$.

4.2.5. THEOREM. *(Herbrand's theorem) A prenex formula B, say*

$$B \equiv \forall x \exists x' \forall y \exists y' \ldots A(x, x', y, y', \ldots),$$

A quantifier-free, is provable in **GS1**, *iff there is a disjunction of substitution instances of A of the form*

$$D \equiv \bigvee_{i=0}^{n} A(x_i, t_i, y_i, s_i, \ldots),$$

such that D is provable propositionally and B can be obtained from the sequent $A(x_0, t_0, y_0, s_0, \ldots), \ldots, A(x_n, t_n, y_n, s_n, \ldots)$ by structural and quantifier rules.

PROOF. For a detailed proof and a more precise statement see 5.3.7. The idea of the proof is as follows. Suppose B has a proof in **GS1**, then we can rearrange the proof in such a way that all quantifier-inferences come below all propositional inferences. For example, a succession of two rules as on the left may be rearranged as on the right ($x \notin \mathrm{FV}(\Gamma B)$):

$$\dfrac{\dfrac{A, B, \Gamma}{\forall x A, B, \Gamma}}{\forall x A, B \vee C, \Gamma} \qquad\qquad \dfrac{\dfrac{A, B, \Gamma}{A, B \vee C, \Gamma}}{\forall x A, B \vee C, \Gamma}$$

where, in case $x \in \mathrm{FV}(C)$, we have to rename the variable x in A in the inferences on the right. Ultimately we find a propositional sequent which is the last sequent of the propositional part and the first sequent of the quantifier part; this sequent must then consist of a multiset of formulas $A(x_i, t_i, y_i, s_i, \ldots)$. ⊠

4.2.5A. ♠ Give complete details of the permutation argument in the proof of Herbrand's theorem.

4.2.6. THEOREM. **Ip** *is decidable.*

PROOF. Let $\Gamma \Rightarrow A$ be a propositional sequent. We can construct a search tree, for "bottom-up" proof search in the system **G3ip**.

More generally, in order to describe the search tree we note that

(i) Each node of the search tree represents the problem of proving (simultaneously) a finite *set* of sequents $\Gamma_1 \Rightarrow A_1, \ldots, \Gamma_n \Rightarrow A_n$.

(ii) A predecessor of a problem is obtained by replacing a $\Gamma_i \Rightarrow A_i$ by $\Gamma' \Rightarrow A'$ or by a pair $\Gamma' \Rightarrow A'$, $\Gamma'' \Rightarrow A''$ such that

$$\frac{\Gamma' \Rightarrow A'}{\Gamma_i \Rightarrow A_i} \quad \text{or} \quad \frac{\Gamma' \Rightarrow A' \qquad \Gamma'' \Rightarrow A''}{\Gamma_i \Rightarrow A_i}$$

is a rule application of **G3i**.

We regard two problems $\{\Gamma_1 \Rightarrow A_1, \ldots, \Gamma_n \Rightarrow A_n\}$ and $\{\Delta_1 \Rightarrow B_1, \ldots, \Delta_n \Rightarrow B_m\}$ as *equivalent*, if to each $\Gamma_i \Rightarrow A_i$ there is a $\Delta_j \Rightarrow B_j$ such that $\mathrm{Set}(\Gamma_i) = \mathrm{Set}(\Delta_j)$ and $A_i \equiv B_j$. If along a branch of the search tree we meet with an axiom, the branch ends there; and if along a branch a repetition of a problem occurs, that is to say we encounter a problem equivalent to a problem occurring lower down the branch, the branch is cut off at the repeated problem.

(iii) Because of the subformula property, there are only finitely many problems. This puts a bound on the depth of the search tree. ⊠

REMARK. The proof that such a decision method works is still easier for Kleene's original calculus G3, or for **GKi**, defined in 3.5.11.

4.2.7. EXAMPLE. The following example illustrates the method. In order to shorten the verifications a bit, we note in advance that sequents of the forms

$$\Gamma, A_1 \to A_2, A_2 \to A_3, \ldots, A_{n-1} \to A_n \Rightarrow A_1 \to A_n$$

$$\Gamma, A_1, A_1 \to A_2, \ldots, A_{n-1} \to A_n \Rightarrow A_n$$

are derivable. Below, $P, Q, R \in \mathcal{PV}$. We drop $\to$ to keep formulas short. Let us now attempt in **G3i** a backward search for a proof of the sequent

$$(1) \qquad (QP)R, QR, RP \Rightarrow P$$

We add "$\surd, \dagger$" to indicate derivability and underivability, respectively. (N.B. We may conclude underivability if a sequent is obviously classically falsifiable.) "Indifferent" after a sequent indicates that the derivability for this sequent does not matter since the branch in the search tree breaks off already for other reasons.

(a) Apply L$\to$ with principal formula RP; this requires proofs of

$$(2) \qquad (QP)R, QR, RP \Rightarrow R,$$

$$(QP)R, QR, P \Rightarrow P \ \surd.$$

Since the second sequent is an axiom, the problem reduces to (2). We continue the search with (2) first.
(aa) Apply in (2) L→ with principal formula RP; we find

$$(QP)R, QR, RP \Rightarrow R \text{ and}$$

$$(QP)R, QR, P \Rightarrow R \text{ (indifferent)},$$

so this operation is useless, since we are back at (2).
(ab) Apply L→ with QR principal:

$$(3) \quad (QP)R, QR, RP \Rightarrow Q,$$

$$(QP)R, RP, R \Rightarrow R \checkmark.$$

We continue with (3).
(aba) Apply L→, with RP principal, to (3):

$$(QP)R, QR, RP \Rightarrow R \text{ (repetition)},$$

$$(QP)R, QR, P \Rightarrow Q \text{ (indifferent)},$$

so this track breaks down.
(abb) Apply L→, with QR principal, to (3):

$$(QP)R, QR, RP \Rightarrow Q \text{ (repetition)},$$

$$(QP)R, R, RP \Rightarrow Q \dagger,$$

so this track also breaks down.
(abc) Apply L→, with $(QP)R$ principal to (3):

$$(QP)R, QR, RP \Rightarrow QP \checkmark,$$

$$R, QR, RP \Rightarrow Q \dagger,$$

again breakdown. We return to (2).
(ac) Apply L→, with $(QP)R$ principal, to (2):

$$(QP)R, QR, RP \Rightarrow QP \checkmark,$$

$$QR, RP, R \Rightarrow R \checkmark,$$

so this track leads to a derivation. We have now investigated all possibilities for (2).

(b) Apply L$\to$, with QR principal, to (1):

$$(4) \qquad (QP)R, QR, RP \Rightarrow Q,$$

$$(QP)R, R, RP \Rightarrow P \surd.$$

(4) is identical with (3), so this track fails.
(c) Apply L$\to$, with $(QP)R$ principal, to (1):

$$(QP)R, QR, RP \Rightarrow QP \surd,$$

$$QR, RP, R \Rightarrow QP \surd,$$

so this leads to a derivation. All in all, we have found two roads leading to a deduction, all others failed.

4.2.7A. ♠* Show proof-theoretically that $\mathbf{I} \nvdash \forall x(P \vee Rx) \to P \vee \forall x Rx$ ($P \in \mathcal{PV}$, R a unary relation symbol). You may use classical unsatisfiability as a shortcut to see that a sequent cannot be derivable.

4.2.7B. ♠* Apply the decision procedure for **Ip** to the following sequents: $\Rightarrow (P \to Q) \vee (Q \to P)$, $\Rightarrow ((P \to Q) \to P) \to P$, $\Rightarrow \neg P \vee \neg\neg P$, $[((P \to Q) \to Q) \to P] \to Q, Q \to P \Rightarrow Q$ ($P, Q, R \in \mathcal{PV}$).

4.2.7C. ♠ Prove the following lemma for the calculi **G1**[**mic**]: a provable sequent always has a proof in which the multiplicity of any formula in antecedent or succedent is at most 2. Derive from this a decision method for **Ip** based on **G1i**.

Let **G1**[**mic**]° be the calculi obtained from **G1**[**mic**] by replacing L$\to$ by the original version of Gentzen:

$$\frac{\Gamma \Rightarrow \Delta, A \qquad B, \Gamma' \Rightarrow \Delta'}{A \to B, \Gamma, \Gamma' \Rightarrow \Delta, \Delta'}$$

Show that the statement above also holds for **G1**[**mic**]° if we read 'at most 3' for 'at most 2' (Gentzen [1935]).

4.2.7D. ♠* Let us call a proof in **G1**[**mic**]° (see the preceding exercise) *restricted* if in all applications of L$\to$

$$\frac{\Gamma \Rightarrow \Delta, A \qquad B, \Gamma' \Rightarrow \Delta'}{A \to B, \Gamma, \Gamma' \Rightarrow \Delta, \Delta'}$$

$A \to B$ does not occur in Γ'. Show that every [cutfree] proof of a sequent $\Gamma \Rightarrow \Delta$ can be transformed into a [cutfree] restricted proof of $\Gamma \Rightarrow \Delta$ (Došen [1987]).

4.2.7E. ♠ Use the preceding result to show that every sequent $\Gamma \Rightarrow \Delta$ provable in **G1[mic]**°, with multiplicity of formulas in Γ and in Δ at most 2, has also a deduction in which all sequents have multiplicity at most 2 for all formulas in antecedent and succedent (Došen [1987]).

4.2.7F. ♠* Let C be a formula of **I** not containing $\rightarrow$, and let $\Gamma = \{A_1 \rightarrow B_1, \ldots, A_n \rightarrow B_n\}$. Prove that if $\Gamma \vdash C$, then $\Gamma \vdash A_i$ for some $i \leq n$ (Prawitz [1965]).

4.2.7G. ♠ Show the decidability of prenex formulas in **I** for languages without function symbols and equality.

4.2.7H. ♠ Show that the following derived rule holds for intuitionistic logic: If $\mathbf{I} \vdash (A \rightarrow B) \rightarrow C \vee D$, then $\mathbf{I} \vdash (A \rightarrow B) \rightarrow C$ or $\mathbf{I} \vdash (A \rightarrow B) \rightarrow D$ or $\mathbf{I} \vdash (A \rightarrow B) \rightarrow A$.

Generalize the preceding rule to: If $E \equiv \bigwedge_i (A_i \rightarrow B_i)$ and $\mathbf{I} \vdash E \rightarrow C \vee D$, then $\mathbf{I} \vdash E \rightarrow C$ or $\mathbf{I} \vdash E \rightarrow D$ or $\mathbf{I} \vdash E \rightarrow A_i$ for some i.

4.3 A more efficient calculus for Ip

The fact that in the calculus **G3i** in the rule L→ the formula $A \rightarrow B$ introduced in the conclusion has to be present also in the left premise makes the bottom-up proof search inefficient; the same implication may have to be treated many times. Splitting L→ into four special cases, such that for a suitable measure the premises are strictly less complex than the conclusion, produces a much more efficient decision algorithm.

4.3.1. DEFINITION. The Gentzen system **G4ip** has the axioms and rules of **G3ip**, except that L→ is replaced by four special cases ($P \in \mathcal{PV}$):

$$\text{L0}\rightarrow \frac{P, B, \Gamma \Rightarrow E}{P \rightarrow B, P, \Gamma \Rightarrow E}$$

$$\text{L}\wedge\rightarrow \frac{C \rightarrow (D \rightarrow B), \Gamma \Rightarrow E}{C \wedge D \rightarrow B, \Gamma \Rightarrow E}$$

$$\text{L}\vee\rightarrow \frac{C \rightarrow B, D \rightarrow B, \Gamma \Rightarrow E}{C \vee D \rightarrow B, \Gamma \Rightarrow E}$$

$$\text{L}\rightarrow\rightarrow \frac{D \rightarrow B, C, \Gamma \Rightarrow D \qquad B, \Gamma \Rightarrow E}{(C \rightarrow D) \rightarrow B, \Gamma \Rightarrow E} \qquad \boxtimes$$

Note that all rules are invertible, except L→→, R∨.

4.3.1A. ♠ Observe that we do not have the subformula property in the strict sense for this new calculus; can you formulate a reasonable substitute?

It is not hard to obtain an upper bound on the length of branches in a bottom-up search for a deduction, once we define an appropriate measure.

4.3.2. DEFINITION. We assign to propositional formulas A a *weight* $\mathrm{w}(A)$ as follows:

1. $\mathrm{w}(P) = \mathrm{w}(\bot) := 2$ for $P \in \mathcal{PV}$,
2. $\mathrm{w}(A \wedge B) := \mathrm{w}(A)(1 + \mathrm{w}(B))$,
3. $\mathrm{w}(A \vee B) := 1 + \mathrm{w}(A) + \mathrm{w}(B)$,
4. $\mathrm{w}(A \to B) := 1 + \mathrm{w}(A)\mathrm{w}(B)$.

For sequents $\Gamma \Rightarrow \Delta$ we put

$$\mathrm{w}(\Gamma \Rightarrow \Delta) := \sum\{\mathrm{w}(B) : B \in \Gamma\Delta\}$$

where each $\mathrm{w}(B)$ occurs as a term in the sum with the multiplicity of B in $\Gamma\Delta$. ⊠

Now observe that for each rule of the calculus **G4ip**, the weight each of the premises is lower than the conclusion. So all branches in a bottom-up search tree for a proof of the sequent $\Gamma \Rightarrow A$ have length at most $\mathrm{w}(\Gamma \Rightarrow A)$. We now turn to the proof of equivalence between **G4ip** and **G3ip**.

The idea of the proof is to show, by induction on the weight of a sequent, that if **G3ip** $\vdash \Gamma \Rightarrow A$, then **G4ip** $\vdash \Gamma \Rightarrow A$. If **G3ip** $\vdash \Gamma \Rightarrow A$ and we can find one or two sequents ($\mathcal{S}, \mathcal{S}'$ say) from which $\Gamma \Rightarrow A$ would follow in **G4ip** by an invertible rule, the sequents $\mathcal{S}, \mathcal{S}'$ are also derivable in **G3ip** and have lower weight, so the IH applies.

If none of the invertible rules of **G4ip** is applicable, we must look at the last rule applied in the **G3ip**-proof. Except for one "awkward" case, we can then always show in a straightforward way that there are sequents of lower weight provable in **G3ip**, which by a rule of **G4ip** yield $\Gamma \Rightarrow A$. However, by lemma 4.3.4 we can show that we may restrict attention to proofs in **G3ip** in which the awkward case does not arise.

4.3.3. DEFINITION. (*Irreducible, awkward, easy*) A multiset Γ is called *irreducible* if Γ neither contains a pair $P, P \to B$ ($P \in \mathcal{PV}$), nor $\bot$, nor a formula $C \wedge D$, nor a formula $C \vee D$. A sequent $\Gamma \Rightarrow A$ is *irreducible* iff Γ is irreducible. A proof is *awkward* if the principal formula of the final step occurs on the left and is critical, that is to say of the form $P \to B$, otherwise it is *easy*. ⊠

4.3.4. LEMMA. *A provable irreducible sequent has an easy proof in* **G3ip**.

PROOF. We argue by contradiction. Assume that there are provable irreducible sequents without easy proofs. Among all the awkward proofs of such sequents, we select a proof $\mathcal{D}$ of such a sequent with a leftmost branch of minimal length. Let $\Gamma \Rightarrow C$ be the conclusion of that proof. $\Gamma \Rightarrow C \equiv P \to B, \Gamma' \Rightarrow C$ with $P \notin \Gamma'$, since Γ is irreducible. Hence $\mathcal{D}$ has the form

$$\frac{\begin{array}{c}\mathcal{D}'\\ P \to B, \Gamma' \Rightarrow P\end{array} \quad \begin{array}{c}\mathcal{D}''\\ B, \Gamma' \Rightarrow C\end{array}}{P \to B, \Gamma' \Rightarrow C}$$

$\mathcal{D}'$ cannot be an axiom since $P \notin \Gamma'$. $P \to B, \Gamma' \Rightarrow P$ is also irreducible, and not all possible deductions of this sequent can be awkward, for then $\mathcal{D}''$ would be an awkward proof with a leftmost branch shorter than the leftmost branch of $\mathcal{D}$, which is excluded by assumption. Hence $P \to B, \Gamma' \Rightarrow P$ must have an easy proof $\mathcal{D}'''$ and end with an application of a left rule; $\Gamma \equiv P \to B, \Gamma'$ is irreducible, so the last rule applied must have been L$\to$ with principal formula $D \to E$, D not atomic (since $\mathcal{D}'''$ was easy), i.e. if we replace $\mathcal{D}'$ by $\mathcal{D}'''$ we get a deduction of the form

$$\frac{\dfrac{\begin{array}{c}\mathcal{D}_0\\ P \to B, D \to E, \Gamma'' \Rightarrow D\end{array} \quad \begin{array}{c}\mathcal{D}_1\\ E, P \to B, \Gamma'' \Rightarrow P\end{array}}{P \to B, D \to E, \Gamma'' \Rightarrow P} \quad \begin{array}{c}\mathcal{D}''\\ B, D \to E, \Gamma'' \Rightarrow C\end{array}}{P \to B, D \to E, \Gamma'' \Rightarrow C}$$

where $\Gamma' \equiv D \to E, \Gamma''$. We permute the application of rules and obtain

$$\frac{\begin{array}{c}\mathcal{D}_0\\ P \to B, D \to E, \Gamma'' \Rightarrow D\end{array} \quad \dfrac{\begin{array}{c}\mathcal{D}_1\\ E, P \to B, \Gamma'' \Rightarrow P\end{array} \quad \begin{array}{c}\mathcal{D}''\\ B, D \to E, \Gamma'' \Rightarrow C\end{array}}{E, P \to B, \Gamma'' \Rightarrow C}}{P \to B, D \to E, \Gamma'' \Rightarrow C}$$

The new proof is easy. ⊠

4.3.5. THEOREM. **G3ip** *and* **G4ip** *are equivalent.*

PROOF. Let $\vdash$, $\vdash^*$ be derivability in **G3ip** and **G4ip** respectively. The rules of **G4ip** are derivable in **G3ip**, so **G4ip** $\subset$ **G3ip**.

For the converse, consider any **G3ip**-proof of a propositional sequent $\Gamma \Rightarrow E$; we show by induction on the weight of the sequent that we can find a **G4ip**-proof of the same sequent.

Case 1. If $\perp \in \Gamma$ we are done.

Case 2. Let $\Gamma \equiv \Gamma', A \wedge B \Rightarrow E$; then also $\vdash \Gamma', A, B \Rightarrow E$, and Γ', A, B has lower weight than Γ, so $\vdash^* \Gamma', A, B \Rightarrow E$.

Case 3. $\Gamma \equiv A \vee B, \Gamma'$: similarly.

Case 4. $\Gamma \equiv \Gamma', P, P \to B$. Then also $\vdash \Gamma', P, B \Rightarrow E$, so by the IH $\vdash^* \Gamma', P, B \Rightarrow E$; apply L0→.

Case 5. If none of the preceding cases applies, Γ is irreducible. By the preceding lemma, $\Gamma \Rightarrow E$ has a **G3ip**-proof, which is either an axiom, or has a last rule application with principal formula on the right, or has a principal formula on the left of the form $A \to B$, A not atomic.

Subcase 5.1. If the final step is an axiom, we are done.

Subcase 5.2. If the last rule is R∨, R∧ or R→, we are done since the premises are lower in weight than the conclusion.

Subcase 5.3. The last rule is L→ with $A \to B$ as principal formula, A not atomic.

• *5.3(i)* If $A \equiv C \wedge D$, then because of

$$\vdash (C \wedge D \to B) \leftrightarrow (C \to (D \to B))$$

also $C \to (D \to B), \Gamma' \Rightarrow E$ in **G3ip**, which by IH is provable in **G4ip**, hence with L∧→ $\vdash^* \Gamma \Rightarrow E$.

• *5.3(ii)* Similarly for $A \equiv C \vee D$, then $\vdash^* C \to B, D \to B, \Gamma' \Rightarrow E$; apply L∨→.

• *5.3(iii)* Let $A \equiv C \to D$; then the last rule application has the form

$$\frac{(C \to D) \to B, \Gamma' \Rightarrow C \to D \qquad B, \Gamma' \Rightarrow E}{(C \to D) \to B, \Gamma' \Rightarrow E}$$

In **G3m** we have generally $\vdash \Gamma, (A_0 \to A_1) \to A_2 \Rightarrow A_0 \to A_1$ iff $\vdash \Gamma, A_1 \to A_2 \Rightarrow A_0 \to A_1$ (exercise) iff $\Gamma, A_1 \to A_2, A_0 \Rightarrow A_1$ (by inversion); hence $\vdash (C \to D) \to B, \Gamma' \Rightarrow C \to D$ iff $\vdash D \to B, C, \Gamma' \Rightarrow D$, and this second sequent is lower in weight, so $\vdash^* D \to B, C, \Gamma' \Rightarrow D$; also $\vdash^* B, \Gamma' \Rightarrow E$; now apply L→→.

• *5.3(iv)* Let $A \equiv \bot$. Then $\vdash \bot \to B, \Gamma' \Rightarrow E$ iff $\vdash \Gamma' \Rightarrow E$ iff $\vdash^* \Gamma' \Rightarrow E$ (IH); use admissibility of Weakening to obtain $\vdash^* A \to B, \Gamma' \Rightarrow E$. ⊠

4.3.6. EXAMPLE. The following example illustrates what may be gained in reducing the possibilities for "backtracking". If we search for a proof of the purely implicational sequent (writing for brevity XY for $X \to Y$)

$$(QP)R, QR, RP \Rightarrow P,$$

we find in **G4ip** a single possibility:

$$\dfrac{\dfrac{\dfrac{PR, R, P, Q \Rightarrow P}{PR, RP, Q, R \Rightarrow P}\text{L0}\to}{PR, QR, RP, Q \Rightarrow P}\text{L0}\to \qquad \dfrac{R, P, QR \Rightarrow P}{R, QR, RP \Rightarrow P}\text{L0}\to}{(QP)R, QR, RP \Rightarrow P}\text{L}\to\to$$

Compare this with the proof search in 4.2.7 for the same sequent.

4.3.6A. ♠* Test the following formulas for derivability in **Ip**: $A \equiv (\neg\neg P \to P) \to \neg P \vee \neg\neg P$, $A \to ((\neg\neg P \to P) \vee (\neg P \vee \neg\neg P))$ and $[((P \to R) \to R) \to ((Q \to R) \to R)] \to (((P \to Q) \to R) \to R)$.

4.3.6B. ♠ Prove that in **G3m** $\vdash \Gamma, (A_0 \to A_1) \to A_2 \Rightarrow A_0 \to A_1$ iff $\vdash \Gamma, A_1 \to A_2 \Rightarrow A_0 \to A_1$.

4.4 Interpolation and definable functions

The interpolation theorem is a central result in first-order logic; therefore we have reserved a separate section for it. An important corollary of the interpolation theorem, historically preceding it, is Beth's definability theorem (4.4.2B).

4.4.1. NOTATION. In this section we adopt the following notation. $\mathrm{Rel}^+(\Gamma)$, $\mathrm{Rel}^-(\Gamma)$ are the sets of relation symbols occurring positively, respectively negatively in Γ. We put $\mathrm{Rel}(\Gamma) := \mathrm{Rel}^+(\Gamma) \cup \mathrm{Rel}^-(\Gamma)$. $\mathrm{Con}(\Gamma)$ is the set of individual constants occurring in Γ.

4.4.2. *Interpolation theorem for* **M**, **I**, **C**

An interpolant for a derivable implication $\vdash A \to B$ is a formula F such that $\vdash A \to F$, $\vdash F \to B$, and such that F satisfies certain additional conditions. For example, for propositional logic, one requires that F contains propositional variables occurring in both A and B only. For sequents $\Gamma \Rightarrow \Delta$, the obvious notion of interpolant would be a formula F satisfying additional conditions such that $\vdash \Gamma \Rightarrow F$, $\vdash F \Rightarrow \Delta$. However, in order to construct interpolants by induction on the depth of derivations of sequents, we need a more general notion of interpolant, as in the following theorem.

THEOREM. *Suppose* **G3[mic]** $\vdash \Gamma\Gamma' \Rightarrow \Delta\Delta'$ *(with* $|\Delta'| \leq 1$ *and* $\Delta = \emptyset$ *for* **G3[mi]**); *then there is an interpolation formula (interpolant) F such that*

(i) **G3[mic]** $\vdash \Gamma \Rightarrow \Delta F$, **G3[mic]** $\vdash \Gamma' F \Rightarrow \Delta'$;

(ii) $\mathrm{Rel}^i(F) \subset \mathrm{Rel}^i(\Gamma, \neg\Delta) \cap \mathrm{Rel}^i(\neg\Gamma', \Delta')$ *for* $i \in \{+, -\}$;

(iii) $\mathrm{Con}(F) \subset \mathrm{Con}(\Gamma, \neg\Delta) \cap \mathrm{Con}(\neg\Gamma', \Delta')$;

(iv) $\mathrm{FV}(F) \subset \mathrm{FV}(\Gamma, \neg\Delta) \cap \mathrm{FV}(\neg\Gamma', \Delta')$.

PROOF. By induction on the depth of cutfree deductions in **G3[mic]**. A *split sequent* is an expression $\Gamma; \Gamma' \Rightarrow \Delta; \Delta'$ such that $\Gamma\Gamma' \Rightarrow \Delta\Delta'$ is a sequent. A formula F is an *interpolant* of the split sequent $\Gamma; \Gamma' \Rightarrow \Delta; \Delta'$ if $\vdash \Gamma \Rightarrow \Delta, F$ and $\vdash \Gamma', F \Rightarrow \Delta'$. If F is an interpolant of $\Gamma; \Gamma' \Rightarrow \Delta; \Delta'$ we write

$$\Gamma; \Gamma' \xRightarrow{F} \Delta; \Delta'$$

Basis. We show below how the interpolants for an axiom $\Gamma, P, \Gamma' \Rightarrow \Delta, P, \Delta'$ are to be chosen, dependent on the splitting. The second line concerns cases which can arise in the classical system only.

$$\Gamma; P\Gamma' \xRightarrow{\bot \to \bot} \Delta; P\Delta' \qquad \Gamma P; \Gamma' \xRightarrow{P} \Delta; P\Delta'$$
$$\Gamma; P\Gamma' \xRightarrow{\neg P} \Delta P; \Delta' \qquad \Gamma P; \Gamma' \xRightarrow{\bot \to \bot} \Delta P; \Delta'$$

For axioms L$\bot$ the interpolants are given by

$$\Gamma\bot; \Gamma' \xRightarrow{\bot} \Delta; \Delta' \qquad \Gamma; \bot\Gamma' \xRightarrow{\bot \to \bot} \Delta; \Delta'$$

Induction step. We show for some cases of the induction step how to construct interpolants for a splitting of the conclusion from interpolants for suitable splittings of the premises. We first concentrate on the classical case; for **G3[mi]** slight adaptations are needed.
Case 1. The last rule is L$\to$. There are two subcases, according to the position of the principal formula in the splitting:

$$\frac{\Gamma'; \Gamma \xRightarrow{C} \Delta'; A\Delta \quad \Gamma B, \Gamma' \xRightarrow{D} \Delta; \Delta'}{\Gamma(A \to B); \Gamma' \xRightarrow{C \to D} \Delta; \Delta'} \qquad \frac{\Gamma; \Gamma' \xRightarrow{D} \Delta; A\Delta' \quad \Gamma; B\Gamma' \xRightarrow{C} \Delta; \Delta'}{\Gamma; (A \to B)\Gamma' \xRightarrow{C \land D} \Delta; \Delta'}$$

To see that, for example, the case on the left is indeed correct, note that by the IH for the premises we have (1) $\Gamma C \Rightarrow \Delta A$, (2) $\Gamma' D \Rightarrow \Delta'$, (3) $B\Gamma \Rightarrow D, \Delta$, (4) $\Gamma' \Rightarrow C\Delta'$. From (1) and (3), by closure under Weakening, $C\Gamma \Rightarrow \Delta AD$, $\Gamma BC \Rightarrow \Delta D$; from this with L$\to$, $\Gamma(A \to B)C \Rightarrow D\Delta$, and by R$\to$, $\Gamma(A \to B) \Rightarrow (C \to D)\Delta$. $\Gamma'(C \to D) \Rightarrow \Delta'$ is obtained from (2) and (4) by a single application of L$\to$.

$C \to D$ and $C \land D$ satisfy the requirements (i)–(iv) of the theorem, as is easily checked.
Case 2. The last rule applied is R$\to$. There are again two subcases:

$$\frac{A\Gamma; \Gamma' \xRightarrow{C} B\Delta; \Delta'}{\Gamma; \Gamma' \xRightarrow{C} A \to B, \Delta; \Delta'} \qquad \frac{\Gamma; \Gamma' A \xRightarrow{C} \Delta; \Delta' B}{\Gamma; \Gamma' \xRightarrow{C} \Delta; A \to B, \Delta'}$$

The first of these cases has no analogue in **G3[mi]**. To see for the first case that the interpolant for the premise is also an interpolant for the conclusion, note that $\Gamma A \Rightarrow \Delta C$ implies $\Gamma \Rightarrow \Delta, A \to B, C$. The checking of the other properties is left to the reader.

Case 3. The last rule is L$\forall$. The two subcases are

$$\frac{\Gamma, \forall x A, A[x/t]; \Gamma' \overset{C}{\Longrightarrow} \Delta; \Delta'}{\Gamma, \forall x A; \Gamma' \overset{\forall \vec{u}\vec{v} C[\vec{c}/\vec{v}]}{\Longrightarrow} \Delta; \Delta'} \qquad \frac{\Gamma; A[x/t], \forall x A, \Gamma' \overset{C}{\Longrightarrow} \Delta; \Delta'}{\Gamma; \forall x A, \Gamma' \overset{\exists \vec{z}\vec{y} C[\vec{d}/\vec{y}]}{\Longrightarrow} \Delta; \Delta'}$$

In the first subcase,

$$\begin{aligned}\vec{u} &= \mathrm{FV}(C) \setminus (\mathrm{FV}(\Gamma\Delta\forall x A) \cap \mathrm{FV}(\Gamma'\Delta')),\\ \vec{c} &= \mathrm{Con}(C) \setminus (\mathrm{Con}(\Gamma\Delta\forall x A) \cap \mathrm{Con}(\Gamma'\Delta')),\end{aligned}$$

$\vec{v}$ a sequence of fresh variables; in the second subcase,

$$\begin{aligned}\vec{z} &= \mathrm{FV}(C) \setminus (\mathrm{FV}(\Gamma\Delta) \cap \mathrm{FV}(\forall x A\, \Gamma'\Delta')),\\ \vec{d} &= \mathrm{Con}(C) \setminus (\mathrm{Con}(\Gamma\Delta) \cap \mathrm{Con}(\forall x A\, \Gamma'\Delta')),\end{aligned}$$

$\vec{y}$ a sequence of fresh variables. The case where the last rule is R$\exists$ is treated symmetrically.

Case 4. The last rule is L$\exists$.

$$\frac{\Gamma A; \Gamma' \overset{C}{\Longrightarrow} \Delta; \Delta'}{\Gamma, \exists x A; \Gamma' \overset{C}{\Longrightarrow} \Delta; \Delta'} \qquad \frac{\Gamma; A\Gamma' \overset{C}{\Longrightarrow} \Delta; \Delta'}{\Gamma; \exists x A, \Gamma' \overset{C}{\Longrightarrow} \Delta; \Delta'}$$

where $x \notin \mathrm{FV}(\Gamma\Gamma'\Delta\Delta')$. To see that the indicated interpolant is correct, note that C does not contain x free.

The case where the last rule is R$\forall$ can be treated symmetrically.

Let us now consider one of the least boring cases for **G3m**, namely where the last rule is L$\to$. The interesting subcase is

$$\frac{\Gamma'; A \to B, \Gamma \overset{C}{\Longrightarrow} A \qquad \Gamma, B; \Gamma' \overset{D}{\Longrightarrow} E}{\Gamma, A \to B; \Gamma' \overset{C \to D}{\Longrightarrow} E}.$$

By the IH we have $A \to B, C, \Gamma \Rightarrow A$ and $\Gamma, B \Rightarrow D$, and by weakening also $\Gamma, A \to B, C \Rightarrow D$; applying L$\to$ we find $\Gamma, A \to B, C \Rightarrow D$, hence with R$\to$ $\Gamma, A \to B \Rightarrow C \to D$.

By the IH we also have $\Gamma' \Rightarrow C$ and $\Gamma', D, \Rightarrow E$; weakening the first yields $\Gamma', C \to D \Rightarrow C$, and then by a single application of L$\to$ $\Gamma', C \to D \Rightarrow E$. ⊠

4.4.2A. ♠ Complete the proof of the interpolation theorem.

4.4.2B. ♠ (Beth's definability theorem) Let $A(X)$ be a formula with n-ary relation symbol X in language $\mathcal{L}$. Let R, R' be relation symbols not in $\mathcal{L}$, and assume $\vdash A(R) \wedge A(R') \Rightarrow \forall \vec{x}(R\vec{x} \leftrightarrow R'\vec{x})$. Show that there is a formula C in $\mathcal{L}$ such that $\vdash A(R) \Rightarrow \forall \vec{x}(C \leftrightarrow R\vec{x})$.

Interpolation with equality and function symbols

We shall now show how the interpolation theorem for logic with equality and function symbols may be obtained by a reduction to the case of pure predicate logic.

4.4.3. DEFINITION. We introduce the following notations:

Eq_0 $\quad \forall x(x = x) \wedge \forall xy(x = y \to y = x) \wedge \forall xyz(x = y \wedge y = z \to x = z)$,
$\mathsf{Eq}(f)$ $\quad \forall \vec{x}\vec{x}'\,(\vec{x} = \vec{x}' \to f\vec{x} = f\vec{x}')$,
$\mathsf{Eq}(R)$ $\quad \forall \vec{x}\vec{x}'\,(\vec{x} = \vec{x}' \wedge R\vec{x} \to R\vec{x}')$.

For a given language let Eq be the set consisting of Eq_0, $\mathsf{Eq}(f)$ for all function symbols f, and $\mathsf{Eq}(R)$ for all relation symbols R of the language. We put $\mathsf{Eq}(R_1, \dots, R_n) \equiv \mathsf{Eq}(R_1), \dots, \mathsf{Eq}(R_n)$, and $\mathsf{Eq}(f_1, \dots, f_n) \equiv \mathsf{Eq}(f_1), \dots, \mathsf{Eq}(f_n)$. For notational simplicity we shall assume below that our language has only finitely many relation symbols and function symbols. This permits us to regard Eq as a finite set, but is not essential to the argument.

We write $\mathsf{Eq}(A)$, $\mathsf{Eq}(\Gamma)$ for Eq with the function symbols and relation symbols restricted to those occurring in a formula A or a multiset Γ. Let

$$\exists! y\, Ay := \exists y(Ay \wedge \forall z(Az \to z = y)).$$

If F_i is a predicate variable of arity $p(i) + 1$, for $1 \le i \le n$, then

$$\mathsf{Fn}(F_i) := \forall \vec{x}\, \exists! y\, F_i(\vec{x}, y) \wedge \mathsf{Eq}(F_i),$$
$$\mathsf{Fn}(F_1, \dots, F_n) := \mathsf{Fn}(F_1), \dots, \mathsf{Fn}(F_n)$$

Let us assume that to each n-ary function symbol f_i in A there is associated the $n+1$-ary relation symbol F_i, F_i not occurring in A. Then $\mathsf{Fn}(A)$ consists of all $\mathsf{Fn}(F_i)$ for the f_i occurring in A. Similarly for $\mathsf{Fn}(\Gamma)$, Γ a multiset. ⊠

4.4.4. DEFINITION. Let $f_1, \dots, f_n$ be a fixed set of function symbols of the language, and $F_1, \dots, F_n$ be a corresponding set of predicate symbols, such that the arity of F_i is equal to the arity of f_i plus one. Relative to this set of function symbols we define for each term t of the language a predicate $t^*(x)$ ($x \notin \mathrm{FV}(t)$) by the clauses:

(i) $y^*(x) := (x = y)$;

(ii) $f_i(t_1, \dots, t_p)^*(x) := \forall x_1 \dots x_p(t_1^*(x_1) \wedge \dots \wedge t_p^*(x_p) \to F_i x_1 \dots x_p x)$;

(iii) $g(t_1, \dots, t_p)^*(x) := \forall x_1 \dots x_p(t_1^*(x_1) \wedge \dots \wedge t_p^*(x_p) \to g(x_1, \dots, x_p) = x)$, for all function symbols g distinct from the f_i.

We associate, relative to the same set of function symbols, to each formula A not containing $\vec{F}$ a formula A^*, by the clauses:

(iv) $(t_1 = t_2)^* := \forall x_1 x_2 (t_1^*(x_1) \wedge t_2^*(x_2) \to x_1 = x_2)$;

(v) $(Rt_1 \dots t_p)^* := \forall x_1 \dots x_p (t_1^*(x_1) \wedge \dots \wedge t_p^*(x_p) \to Rx_1 \dots x_p)$;

(vi) * is a homomorphism w.r.t. logical operators.

Finally, for any formula A we let A° be the formula obtained from A by replacing everywhere subformulas of the form $F_i t_1 \dots t_p t$ by $f_i t_1 \dots t_p = t$. ⊠

Note that $(\vec{x} = \vec{x}')^*$ is equivalent to $\vec{x} = \vec{x}'$, and that $(f_i(\vec{x}))^*(y)$ is equivalent to $F_i(\vec{x}, y)$.

Up till 4.4.12, we may take the discussion to refer to a fixed set $f_1, \dots, f_n$ of function symbols, with corresponding relation symbols $F_1, \dots, F_n$; we shall sometimes abbreviate $\mathsf{Fn}(\vec{F})$ as Fn.

4.4.5. LEMMA. *In* **G3[mic]** *we have*

(i) $\vdash \mathsf{Eq}, \Gamma \Rightarrow \Delta$ *iff* $\vdash \mathsf{Eq}(\Gamma\Delta), \Gamma \Rightarrow \Delta, \vdash \Gamma \Rightarrow \Delta$;

(ii) if $\vdash \mathsf{Eq}, \mathsf{Fn}(\vec{F}), \Gamma \Rightarrow \Delta$, *then* $\mathsf{Eq}, \Gamma^\circ \Rightarrow \Delta^\circ$.

Combining (i) and (ii) produces

(iii) if $\vdash \mathsf{Eq} \setminus \mathsf{Eq}(\vec{f}), \mathsf{Fn}(\vec{F}), \Gamma \Rightarrow \Delta$, *then* $\vdash \mathsf{Eq}, \Gamma^\circ \Rightarrow \Delta^\circ$.

PROOF. The first statement is proved by replacing in the proof of $\mathsf{Eq}, \Gamma \Rightarrow \Delta$ all occurrences $ft_1 \dots t_n$ of n-ary function symbols f not occurring in $\Gamma \Rightarrow \Delta$ by t_1, which amounts to interpreting f by the first projection function (from an n-tuple to the first component), and replacing all predicate symbols not occurring in $\Gamma \Rightarrow \Delta$ by $\top \equiv \bot \to \bot$. This makes the corresponding equality axioms trivially true; the result will be a derivation of a sequent $\mathsf{Eq}(\Gamma\Delta), \Sigma, \Gamma \Rightarrow \Delta$, where Σ is a set of derivable formulas which may be removed with Cut.

The proof of the second statement is easy: if we replace formulas $F_i t_1 \dots t_n t$ by $f_i t_1 \dots t_n = t$ throughout, $\mathsf{Fn}(\vec{F})$ becomes provable. ⊠

4.4.6. LEMMA. *The following sequents are provable in* **G3m**:

$$\mathsf{Eq} \Rightarrow (x = t) \leftrightarrow t^{*\circ}(x),$$
$$\mathsf{Eq} \Rightarrow (A \leftrightarrow A^{*\circ}),$$

where t is arbitrary, and does not contain x, and A is any formula not containing $F_1, \dots, F_n$.

PROOF. The first assertion is proved by induction on the complexity of t, the second statement by induction on the logical complexity of A, with the first assertion used in the basis case. ⊠

4.4.7. LEMMA. *For any term t,* $\mathbf{G3m} \vdash \mathsf{Eq}, \mathsf{Fn}(\vec{F}) \Rightarrow \exists! x(t^*(x))$.

PROOF. By induction on the construction of t. We consider the case $t \equiv f_i(t_1, \ldots, t_p)$; other cases are similar or simpler. To keep the notation simple, let $p = 1$. By the IH,

$$\mathsf{Eq}, \mathsf{Fn}, t_1^*(y), t_2^*(z) \Rightarrow y = z, \text{ so}$$
$$\mathsf{Eq}, \mathsf{Fn}, t_1^*(y), t_2^*(z), F(y, u) \Rightarrow F(z, u).$$

Now $t^*(z) \equiv \forall y(t_1^*(y) \to Fyz)$, hence

$$\mathsf{Eq}, \mathsf{Fn}, t_1^*(y), F(y, u) \Rightarrow t^*(u).$$

Fn contains $\forall x \exists! y Fxy$, hence

$$\mathsf{Eq}, \mathsf{Fn}, t_1^*(y) \Rightarrow \exists u t^*(u).$$

We can also prove

$$\mathsf{Eq}, t_1^*(y), t^*(z), t^*(u) \Rightarrow z = u,$$

which suffices for the statement to be proved. ⊠

4.4.8. LEMMA. *Let $t, s(y)$ be terms, $A(y)^*$ a formula not containing $F_1, \ldots, F_n$. Then we can prove in* **G3m***:*

$$\mathsf{Eq}, \mathsf{Fn}, t^*(x) \Rightarrow s(x)^*(z) \leftrightarrow s(t)^*(z),$$
$$\mathsf{Eq}, \mathsf{Fn}, t^*(x) \Rightarrow A(x)^* \leftrightarrow A(t)^*(z).$$

PROOF. The first statement is proved by induction on the complexity of s, the second statement by induction on the logical complexity of A, using the first statement in the basis case.

As a typical example of the inductive step in the proof of the first statement, let $s(x) \equiv g(s_1(x), \ldots, s_p(x))$ (g not one of the f_i). Then $s(t) \equiv g(s_1(t), \ldots, s_p(t))$. By the IH

$$\mathsf{Eq}, \mathsf{Fn}, t(x)^* \Rightarrow s_i(t)^*(y) \leftrightarrow s_i(x)^*(y) \quad (1 \leq i \leq p).$$

Assume now $t^*(x)$ and

$$g(s_1(t), \ldots, s_p(t))^*(y) \equiv \forall y_1 \ldots y_n(\bigwedge_i s_i(t)^*(y_i) \to y = g(y_1, \ldots, y_p)).$$

Using the IH, the displayed assumption is equivalent to

$$\forall y_1 \ldots y_n(\bigwedge_i s_i(x)^*(y_i) \to y = g(y_1, \ldots, y_p)),$$

i.e. to $g(s_1(x), \ldots, s_p(x))^*(y)$. ⊠

4.4.9. LEMMA. *Let $A(x)$ be a formula not containing a relation symbol from $\vec{F}$. Then we can prove in* **G3m***:*

$$\vdash \mathsf{Eq}, \mathsf{Fn}(\vec{F}), \forall x A(x)^* \Rightarrow A(t)^*; \quad \vdash \mathsf{Eq}, \mathsf{Fn}(\vec{F}), A(t)^* \Rightarrow \exists x A(x)^*.$$

PROOF. Immediate by lemmas 4.4.7, 4.4.8. ⊠

4.4.10. LEMMA. *Let $\Gamma \Rightarrow \Delta$ not contain relation symbols from $\vec{F}$. Then we have in* **G3[mic]***:*

$$\text{if } \vdash \mathsf{Eq}, \Gamma \Rightarrow \Delta \text{ then } \vdash \mathsf{Eq}, \mathsf{Fn}, \Gamma^* \Rightarrow \Delta^*.$$

Hence, by 4.4.5, if $\vdash \mathsf{Eq}, \Gamma \Rightarrow \Delta$, *then* $\vdash \mathsf{Eq} \setminus \mathsf{Fn}(\vec{f}), \mathsf{Fn}(\vec{F}), \Gamma^* \Rightarrow \Delta^*$.

PROOF. The easiest way to establish this result is to use the equivalence of sequent calculi with Hilbert-type systems; so it suffices to establish for **H[mic]** that

$$\text{If } \mathsf{Eq} \vdash A \text{ then } \mathsf{Eq}, \mathsf{Fn} \vdash A^*.$$

The proof is by induction on the depth of derivation of $\mathsf{Eq} \vdash A$ in the H-system.

Basis. If A is an element of Eq, other than $\mathsf{Eq}(f_i)$, the assertion of the theorem is trivial. If $A \equiv \mathsf{Eq}(f_i)$, A^* becomes equivalent to

$$\forall \vec{x}\vec{x}' (\vec{x} = \vec{x}' \to f_i\vec{x} = f_i\vec{x}')^*,$$

which is equivalent to

$$\forall \vec{x}\vec{x}' (\vec{x} = \vec{x}' \to \forall y y'(F_i\vec{x}y \wedge F_i\vec{x}'y' \to y = y')),$$

which follows from $\mathsf{Fn}(F_i)$.

If A is a propositional axiom, or a quantifier axiom of one of the following two types ($x \notin \mathrm{FV}(B)$, $y \equiv x$ or $y \notin \mathrm{FV}(A)$):

$$\forall x(B \to A) \to (B \to \forall y A[x/y]) \text{ or}$$
$$\forall x(A \to B) \to (\exists y A[x/y] \to B),$$

then A^* is an axiom of the same form. If A is one of the axioms

$$\forall x\, A \to A[x/t], \quad A[x/t] \to \exists x\, A,$$

the statement of the theorem follows from the lemma 4.4.9.

Induction step. It s readily seen that application of $\to$E or $\forall$I permutes with application of $*$. ⊠

4.4.11. THEOREM. *(Interpolation theorem for languages with functions and equality) For* **G[123][mic]**, *if* $\vdash \mathsf{Eq}, A \Rightarrow B$, *then there is an interpolating formula* C *such that*

(i) $\vdash \mathsf{Eq}, A \Rightarrow C, \vdash \mathsf{Eq}, C \Rightarrow B;$

(ii) all free variables, individual constants, function constants, predicate letters (not counting $=$*) in* C *occur both in* A *and in* B.

PROOF. Assume

$$\vdash \mathsf{Eq}, A \Rightarrow B$$

then by 4.4.5, (i)

$$\vdash \mathsf{Eq}(A), \mathsf{Eq}(B), A \Rightarrow B.$$

Applying 4.4.10 and 4.4.5(i) we find that

$$\vdash \mathsf{Eq}^-(A), \mathsf{Fn}(A), \mathsf{Eq}^-(B), \mathsf{Fn}(B), A^* \Rightarrow B^*,$$

where $\mathsf{Eq}^-(A) = \mathsf{Eq}(A) \setminus (\vec{f})$, $\mathsf{Eq}^-(B) = \mathsf{Eq}(B) \setminus (\vec{f})$. Hence

$$\vdash (\bigwedge \mathsf{Eq}^-(A) \wedge \bigwedge \mathsf{Fn}(A) \wedge A^*) \Rightarrow (\bigwedge \mathsf{Eq}^-(B) \wedge \bigwedge \mathsf{Fn}(B) \to B^*),$$

hence by the interpolation theorem for predicate logic without function symbols and equality, for some D

$$\vdash \bigwedge \mathsf{Eq}^-(A) \wedge \bigwedge \mathsf{Fn}(A) \wedge A^* \Rightarrow D,$$
$$\vdash D \Rightarrow \bigwedge \mathsf{Eq}^-(A) \wedge \bigwedge \mathsf{Fn}(A) \wedge A^* \to B^*,$$

that is to say

$$\vdash \mathsf{Eq}^-(A), \mathsf{Fn}(A), A^* \Rightarrow D \text{ and } \vdash \mathsf{Eq}^-(B), \mathsf{Fn}(B), D \Rightarrow B^*.$$

Then apply the mapping $^\circ$ and let $C \equiv D^\circ$, then by 4.4.5:

$$\vdash \mathsf{Eq}(A), A \Rightarrow C, \quad \vdash \mathsf{Eq}(B), C \Rightarrow B.$$

Condition (ii) for C readily follows from the corresponding condition satisfied by D. ⊠

As a by-product of the preceding arguments (in particular 4.4.10) we obtain the following theorem on definable functions.

4.4.12. THEOREM. *(Conservativity of definable functions) Let* $\mathbf{T}$ *be a first-order theory in a language* $\mathcal{L}$*, based on* $\mathbf{C}$, $\mathbf{I}$ *or* $\mathbf{M}$*. Let* $\mathbf{T} \vdash \forall \vec{x} \exists ! y A(\vec{x}, y)$ *(*$\vec{x}$ *of length* n*). Let* f *be a new function symbol not in* $\mathcal{L}$*. Then* $\mathbf{T} + \forall \vec{x} A(\vec{x}, fx)$ *is conservative over* $\mathbf{T}$ *w.r.t.* $\mathcal{L}$ *(i.e. no new formulas in* $\mathcal{L}$ *become provable when adding* $\forall \vec{x} A(\vec{x}, fx)$ *to* $\mathbf{T}$*).*

PROOF. Let Γ_0 be a finite set of non-logical axioms of $\mathbf{T}$, such that $\Gamma_0 \vdash \forall \vec{x} \exists ! y A(\vec{x}, y)$. Consider the mapping * of 4.4.4 with f for $f_1, \ldots, f_n$, and F for $F_1, \ldots, F_n$. If $\vdash \Gamma_1 \Rightarrow B$ in $\mathbf{T}$, with Γ_1 a finite subset of $\mathbf{T}$, Γ_1 in language $\mathcal{L} \cup \{f\}$, B in $\mathcal{L}$, then by 4.4.10 in the original $\mathbf{T}$ with F added to the language $\vdash \Gamma_1, \mathsf{Fn}(F) \Rightarrow B$ (since Γ_1 is not affected by *). But if we substitute $A(\vec{x}, y)$ for F, $\mathsf{Fn}(F)$ is provable relative to $\mathcal{L}$ from Γ_0, hence $\vdash \Gamma_0 \Gamma_1 \Rightarrow B$ in $\mathcal{L}$, i.e. $\mathbf{T} \vdash B$. ⊠

REMARK. The result extends to certain theories $\mathbf{T}$ axiomatized by axioms and axiom schemas, in which predicates appear as parameters. Let $\mathbf{T} \vdash \forall \vec{x} \exists ! y A(\vec{x}, y)$, let f be a new function symbol not in the language $\mathcal{L}$ of $\mathbf{T}$, and let $\mathbf{T}^*$ be $\mathbf{T}$ with the axiom schemas extended to predicates in the language $\mathcal{L} \cup \{f\}$. If now the new instances of the axiom schemas translate under * (substituting A for F) into theorems of $\mathbf{T}$, the result still holds. This generalization applies, for example, to first-order arithmetic, with induction as an axiom schema; the translation * transforms induction into other instances of induction.

Interpolation in many-sorted predicate logic

4.4.13. Many-sorted predicate logic is a straightforward extension of ordinary first-order predicate logic. Instead of a single sort of variables, there is now a collection of sorts J, and for each sort $j \in J$ there is a countable collection of variables of sort j; the logical operators are as before. We think of the collection of sorts as a collection of domains; variables of sort j range over a domain D_j.

Furthermore the language contains relation symbols, constants, and function symbols as before. In the standard version, the sorts of the arguments of an n-ary relation symbol of the language are specified, and the sorts of arguments and value of the function symbols is specified.

Only marginally different is a version where for the relation symbols the sort of the arguments is left open, so that $Rt_1 \ldots t_n$ is a well-formed formula regardless of the sorts of the terms $t_1, \ldots, t_n$.

Here we restrict attention to many-sorted languages with relation symbols, equality and constants, but no function symbols; the sort of the arguments of a relation symbol is left open.

4.4.14. DEFINITION. A quantifier occurrence α in A is *essentially universal* [*essentially existential*], if α is either a positive occurrence of $\forall$ of [$\exists$] or a negative occurrence of $\exists$ [of $\forall$]. A quantifier occurrence α in a sequent $\Gamma \Rightarrow \Delta$ is *essentially universal* [*essentially existential*] if the corresponding occurrence in $\bigwedge \Gamma \to \bigvee \Delta$ is essentially universal [essentially existential].

Let $\mathrm{Un}(A)$ be the collection of sorts such that A contains an essentially universal quantifier over that sort, and $\mathrm{Ex}(A)$ the collection of sorts such that A contains an essentially existential quantifier over that sort; and similarly for sequents. ⊠

Inspection of our proof of the interpolation theorem shows that essentially universal quantifiers in the interpolant derive from essentially universal quantifiers in $\Delta \Rightarrow \Gamma$ and that essentially existential quantifiers derive from essentially existential quantifiers in $\Gamma' \Rightarrow \Delta'$. So an interpolant C to $A \Rightarrow B$ contains an essentially universal [existential] quantifier if A [if B] contains an essentially universal [existential] quantifier.

This observation straightforwardly extends to many-sorted predicate logic, so that one obtains

4.4.15. THEOREM. *(Interpolation for many-sorted predicate logic without function symbols) If $\vdash A \to B$, there exists an interpolant C satisfying all the conditions of theorem 4.4.11 and in addition*

$$\mathrm{Sort}(C) \subset \mathrm{Sort}(A) \cap \mathrm{Sort}(B),$$
$$\mathrm{Un}(C) \subset \mathrm{Un}(A), \quad \mathrm{Ex}(C) \subset \mathrm{Ex}(B).$$

PROOF. For many-sorted languages without equality and without function symbols, we can check by looking at the induction steps of the argument for theorem 4.4.2 that an interpolant satisfying the conditions of the theorem may be found; this is then extended to the language with equality by taking from the argument for 4.4.11 what is needed for equality only. We leave the details as an exercise. ⊠

4.4.15A. ♠ Prove the interpolation theorem for many-sorted logic.

4.4.16. *Persistence.* We sketch an application of the preceding interpolation theorem to the model theory of classical logic.

A first-order sentence A is said to be *persistent*, if the truth of A in a model is preserved under model-extension. Let us consider, for simplicity, a language $\mathcal{L}$ with a single binary relation symbol R. For variables in $\mathcal{L}$ we use x, y, z. Persistence of a sentence A means that for any two models $\mathcal{M} \equiv (\mathrm{D}, \mathrm{R})$ and $\mathcal{M}' \equiv (\mathrm{D}', \mathrm{R}')$ such that

$$\mathrm{D} \subset \mathrm{D}', \quad \mathrm{R}' \cap (\mathrm{D} \times \mathrm{D}) = \mathrm{R},$$

we have

$$\text{If } \mathcal{M} \models \mathcal{A} \text{ then } \mathcal{M}' \models \mathcal{A}.$$

This may also be expressed by looking at two-sorted structures satisfying

$$(*) \qquad \mathcal{M}^* \equiv (\mathrm{D}, \mathrm{D}', \mathrm{R}, \mathrm{R}') \text{ with } \mathrm{D} \subset \mathrm{D}',\ \mathrm{R}' \cap (\mathrm{D} \times \mathrm{D}) = \mathrm{R}.$$

We extend $\mathcal{L}$ to a two-sorted language $\mathcal{L}'$ by adding a new sort of variables x', y', z' and a new predicate symbol R'. In this language the persistence of A in $\mathcal{L}$ may be expressed as

$$\vdash \mathrm{Ext} \wedge A \to A',$$

where Ext is the formula $\forall x \exists y'(x = y') \wedge \forall xy(R(x,y) \leftrightarrow R'(x,y))$, expressing the conditions on the two-sorted structure in $(*)$. A' is obtained from A by replacing quantifiers of $\mathcal{L}$ by quantifiers of the new sort, and replacing R by R'. By interpolation, we can find a formula F such that

$$\mathrm{Ext} \wedge A \to F', \quad F' \to A'.$$

Since Ext is not essentially universal w.r.t. the new sort of variables, F' contains only variables of the new sort and all quantifiers are essentially existential. From this we see that for F obtained from F' by replacing R' by R and new quantifiers by old quantifiers, that $\vdash F \leftrightarrow A$. So A is equivalent to a formula in which contains only essentially existential quantifiers.

It is not hard to prove the converse: if all quantifiers in A are essentially existential, then A is persistent. The obvious proof proceeds by formula induction, so we need an extension of the notion of persistence to formulas. A formula A with $\mathrm{FV}(A) = \vec{x}$ is *persistent* if for all sequences of elements $\vec{d}$ from D of the same length as $\vec{x}$ we have

$$\text{If } \mathcal{M} \models A[\vec{x}/\vec{d}] \text{ then } \mathcal{M}' \models A[\vec{x}/\vec{d}].$$

4.5 Extensions of G1-systems

This section is devoted to some (mild) generalizations of cut elimination, with some applications.

4.5.1. *Systems with axioms*

Extra (non-logical) axioms may be viewed as rules without premises; hence an application of an axiom A in a prooftree may be indicated by a top node labelled with A with a line over it.

It is possible to generalize the cut elimination theorem to systems with axioms in the form of extra sequents ("non-logical axioms"). In the case of the systems **G1[mic]** we again use Gentzen's method with the derived rule of Multicut (4.1.9). The statement is as follows:

PROPOSITION. *(Reduction to cuts on axioms) Let $\mathcal{D}$ be any deduction in* **G1**[**mic**] + Cut *from a set of axioms closed under substitution (i.e. if $\Gamma \Rightarrow \Delta$ is an axiom, then so is $\Gamma[\vec{x}/\vec{t}] \Rightarrow \Delta[\vec{x}/\vec{t}]$).*

Then there is also a deduction containing only cuts with one of the premises a non-logical axiom.

If the non-logical axioms consist of atomic formulas only, there is a deduction where all cuts occur in subdeductions built from non-logical axioms with Cut. Alternatively, if the non-logical axioms contain atomic formulas only and are closed under Cut, we can assert the existence of a cutfree proof.

PROOF. The cut elimination argument works as before, provided we count as zero the rank of a cut between axioms removing their principal formulas. ⊠

In particular, when the axioms contain only atoms, and possibly $\bot$, we conclude that a deduction of $\Gamma \Rightarrow A$ with cuts only in subdeductions constructed from non-logical axioms and Cut, contains only subformulas of $\Gamma \Rightarrow A$, atoms occurring in non-logical axioms, and $\bot$ (if it occurs in a non-logical axiom).

Below we present two examples; but in order to present the second example, we must first define primitive recursive arithmetic.

4.5.1A. ♠ Prove the proposition by carefully checking where Gentzen's proof of cut elimination (cf. 4.1.9) needs to be adapted.

4.5.2. *Primitive recursive arithmetic* PRA

This subsection is needed as background for the second example in the next subsection, and may be skipped by readers already familiar with one of the usual formalizations of primitive recursive arithmetic, a formalism first introduced by Skolem [1923]. For more information, see e.g. Troelstra and van Dalen [1988, 3.2, 3.10.2], where also further references may be found. **PRA** is based on **Cp**, with equality between natural numbers and function symbols for all primitive recursive functions. Specifically, there are the following equality axioms (in which $t, t', t'', \vec{t}, \vec{s}$ are arbitrary (sequences of) terms):

$$\begin{array}{l} t = t, \\ t = t'' \wedge t' = t'' \to t = t', \\ \vec{t} = \vec{s} \to f(\vec{t}) = f(\vec{s}), \end{array}$$

for all function symbols f of the language. For the functions we have as axioms

$$0 \neq \mathrm{S}0, \quad \mathrm{S}t = \mathrm{S}s \to t = s,$$

and defining axioms for all primitive recursive functions. For example, for addition f_+ we have axioms (t, s arbitrary terms)

$$f_+(t, 0) = t, \quad f_+(t, \mathrm{S}s) = \mathrm{S}(f_+(t, s)).$$

Finally, we have a quantifier-free rule of induction:

$$\text{If } \Gamma \vdash A(0) \text{ and } \Gamma \vdash A(x) \to A(\mathrm{S}x), \text{ then } \Gamma \vdash A(t),$$

for quantifier-free $A(x)$ and Γ not containing x free.

REMARKS. (i) If we define $\bot$ as $\mathrm{S}0 = 0$, the axiom $\mathrm{S}0 \neq 0$ becomes redundant: define by recursion a "definition-by-cases" function ϕ satisfying $\phi ts0 = t$, $\phi ts(\mathrm{S}r) = s$, then $0 = \mathrm{S}0 \to t = \phi ts0 = \phi ts(\mathrm{S}0) = s$, i.e. $0 = \mathrm{S}0 \to t = s$, and from this, by induction on $|A|$, $0 = \mathrm{S}0 \to A$, for all formulas A.

(ii) The axiom $\mathrm{S}t = \mathrm{S}s \to t = s$ is in fact a consequence of the presence of the primitive recursive predecessor function prd satisfying $\mathrm{prd}(\mathrm{S}t) = t$, $\mathrm{prd}(0) = 0$. For if $\mathrm{S}t = \mathrm{S}s$, then $t = \mathrm{prd}(\mathrm{S}t) = \mathrm{prd}(\mathrm{S}s) = s$.

(iii) The axioms may be formulated with variables, e.g. $x = z \wedge y = z \to x = y$, and the induction rule with conclusion $\Gamma \vdash A(x)$, provided we add a substitution rule:

$$\text{If } \Gamma(x) \vdash A(x), \text{ then } \Gamma(t) \vdash A(t).$$

(iv) The equality axioms for the primitive recursive functions are in fact provable from $t = s \to \mathrm{S}t = \mathrm{S}s$ and induction; the proof is long and tedious and left to the reader.

(v) If we base **PRA** on **Ip** instead of **Cp**, then decidability of equality, and hence by formula induction, $A \vee \neg A$ for all A, is provable (Troelstra and van Dalen [1988, 3.2]), hence this theory coincides with the theory based on classical logic.

(vi) **PRA** can be formulated as a calculus of term equations, without even propositional logic (the addition of propositional logic is conservative; references in Troelstra and van Dalen [1988, 3.10.2]).

4.5.3. EXAMPLES. *Example 1.* Logic with equality may be axiomatized by adding the following sequents as axioms:

$$\Rightarrow t = t$$
$$t = s, A[x/t] \Rightarrow A[x/s] \quad (A \text{ atomic})$$

Example 2. Primitive recursive arithmetic may be axiomatized by adding to the axiomatic sequents of example 1 the following:

$$\mathrm{S}t = \mathrm{S}s \Rightarrow t = s; \quad 0 = \mathrm{S}t \Rightarrow \ ;$$
$$\Rightarrow t = s \text{ for any defining equation } t = s$$
$$\text{of a primitive recursive function;}$$
$$t < 0 \Rightarrow , \quad s < t \Rightarrow s < \mathrm{S}t, \quad s = t \Rightarrow s < \mathrm{S}t;$$
$$s < \mathrm{S}t \Rightarrow s < t, s = t, \quad \Rightarrow s < t, u = t, t < s;$$
$$R(t_1, \ldots, t_n) \Rightarrow f_R(t_1, \ldots, t_n) = 0;$$
$$f_R(t_1, \ldots, t_n) = 0 \Rightarrow R(t_1, \ldots, t_n).$$

For any primitive recursive relation P there is among the primitive recursive functions a function t_P, such that $t_P(y, \vec{z}) := \min_u^y[u < y \wedge \neg P(\mathrm{S}u, \vec{z})]$, that is to say, $t_P(y, \vec{z})$ is the least $u < y$ such that $\neg P(\mathrm{S}u, \vec{z})$ if existing, and y otherwise. t_P is characterized by the equations (dropping the parameters $\vec{z}$ for notational simplicity):

$$t_P(0) := 0,$$

$$t_P(\mathrm{S}y) := \begin{cases} t_P(y) \text{ if } t_P(y) < y, \\ \mathrm{S}t_P(y) \text{ if } t_P(y) = y \wedge P(\mathrm{S}y), \\ t_P(y) \text{ if } t_P(y) = y \wedge \neg P(\mathrm{S}y). \end{cases}$$

These equations can easily be brought in the standard form of a primitive recursive definition. We can express induction w.r.t. P by the sequents:

$$P(0, \vec{z}) \Rightarrow P(t_P(u, \vec{z}), \vec{z}), P(u, \vec{z})$$
$$P(0, \vec{z}), P(\mathrm{S}t_P(u, \vec{z}), \vec{z}) \Rightarrow P(u, \vec{z}).$$

To see that the first sequent holds, observe that if $P(0, \vec{z})$, then either $t_P(u, \vec{z}) < u$, and then $P(t_P(u, \vec{z}), \vec{z})$, while $\neg P(\mathrm{S}t_P(u, \vec{z}), \vec{z})$, or $t_P(u, \vec{z}) = u$, and then $P(u, \vec{z})$. Also, by the preceding, if $P(0, \vec{z})$ and $P(\mathrm{S}t_P(u, \vec{z}), \vec{z})$, then $t_P(u, \vec{z}) < u$ is excluded, hence $t_P(u, \vec{z}) = u$ and then $P(u, \vec{z})$; this explains the second sequent.

We want to show that the following induction rule (x not free in Γ) is derivable: if $\vdash \Gamma \Rightarrow P(0)$, $\vdash \Gamma, Px \Rightarrow P(\mathrm{S}x)$, then $\vdash \Gamma \Rightarrow Px$. By Cut we get $\Gamma \Rightarrow P(t_P(x)), P(x)$ and $\Gamma, P(\mathrm{S}t_P(x)) \Rightarrow P(x)$. Then

$$\dfrac{\Gamma, P(\mathrm{S}t_P(x)) \Rightarrow P(x) \qquad \dfrac{\Gamma \Rightarrow P(t_P(x)), P(x) \quad \Gamma, P(t_P(x)) \Rightarrow P(\mathrm{S}t_P(x))}{\Gamma, \Gamma \Rightarrow P(x), P(\mathrm{S}t_P(x))}}{\Gamma, \Gamma, \Gamma \Rightarrow P(x), P(x)}$$

and then $\Gamma \Rightarrow P(x)$ by closure under Contraction. Application of the generalized cut elimination to the systems of these examples yields the following proposition (also easily proved model-theoretically).

4.5.4. Proposition.

(i) Predicate logic with equality is conservative over the propositional part of predicate logic with equality;

(ii) **PRA** *with full predicate logic is conservative over* **PRA** *with propositional logic only.* ⊠

REMARK. We also obtain an alternative proof of: if $\Gamma \Rightarrow \Delta$ is a sequent not containing $=$, which is provable in predicate logic with $=$, then in fact $\Gamma \Rightarrow \Delta$ is also provable without the axiom sequents for equality (cf. 4.7).

4.6 Extensions of G3-systems

Here we shall consider additional non-logical axioms and *rules* for G3-systems. We concentrate on the intuitionistic case, that is to say, extensions of **G3i**. The case of **G3c** can be dealt with similarly.

We consider additional rules of four types

4.6.1. DEFINITION. Let Γ be an arbitrary multiset, A an arbitrary formula, $\vec{P} \equiv P_1, \ldots, P_n$, P', $\vec{Q} \equiv Q_1, \ldots, Q_{k-1}$, Q' atomic formulas. (If $k = 0$, $\vec{Q}$ is taken to be empty.) We list four types of rules below, where the formulas in $\vec{Q}, Q'$ represent active formulas in the premises, and $\vec{P}, P'$ are principal in the conclusion.

$$\mathrm{Ru}_1(\vec{P}, P') \quad \Gamma, \vec{P} \Rightarrow P'$$

$$\mathrm{Ru}_2(\vec{P}) \quad \Gamma, \vec{P} \Rightarrow A$$

$$\mathrm{Ru}_3(\vec{P}; \vec{Q}) \ \frac{\Gamma, \vec{P}, Q_i \Rightarrow A \ (1 \leq i \leq k-1)}{\Gamma, \vec{P} \Rightarrow A}$$

$$\mathrm{Ru}_4(\vec{Q}, Q', P) \ \frac{\Gamma, \vec{Q} \Rightarrow Q'}{\Gamma \Rightarrow P}$$

If we want to deal with extensions of **G3c**, the treatment of antecedent and succedent must become symmetric.

REMARK. Addition of $\mathrm{Ru}_2(\vec{P})$ to **G3i** + $\mathrm{Cut_{cs}}$ is equivalent to addition of $\Gamma, \vec{P} \Rightarrow \bot$, as follows by cutting $\Gamma, \vec{P} \Rightarrow \bot$ with $\Gamma, \vec{P}, \bot \Rightarrow A$.

Addition of $\mathrm{Ru}_3(\vec{P}, \vec{Q})$ to **G3i** + $\mathrm{Cut_{cs}}$ is equivalent to the addition of

$$(1) \qquad \Gamma, \vec{P} \Rightarrow \bigvee \vec{Q}.$$

To see that (1) justifies $\mathrm{Ru}_3(\vec{P}; \vec{Q})$, let $\mathcal{D}$ derive $\Gamma, \vec{P}, \bigvee \vec{Q} \Rightarrow A$ from sequents $\Gamma, \vec{P}, Q_i \Rightarrow A$ by repeated use of L$\vee$. Then the rule is justified by

$$\frac{\Gamma, \vec{P} \Rightarrow \bigvee \vec{Q} \quad \overset{\mathcal{D}}{\Gamma, \vec{P}, \bigvee \vec{Q} \Rightarrow A}}{\Gamma, \vec{P} \Rightarrow A}\,\mathrm{Cut_{cs}}$$

Conversely, we obtain (1) by deriving from the axioms $\Gamma, \vec{P}, Q_i \Rightarrow Q_i$ by repeated R$\vee$ the sequent $\Gamma, \vec{P}, Q_i \Rightarrow \bigvee \vec{Q}$; then the rule yields $\Gamma, \vec{P} \Rightarrow \bigvee \vec{Q}$.

Ru_2 may be seen as a degenerate case of Ru_3 where the number of premises is zero.

Addition of $\mathrm{Ru}_4(\vec{Q}, Q', P)$ is equivalent to the addition of

$$(2) \qquad \Gamma, \bigwedge \vec{Q} \to Q' \Rightarrow P.$$

For if $\Gamma, \vec{Q} \Rightarrow Q'$ then by L$\wedge$ and R$\to$ we find $\Gamma \Rightarrow \bigwedge \vec{Q} \to Q'$, and $\mathrm{Cut_{cs}}$ with (2) yields $\Gamma \Rightarrow P$. Conversely,

$$\dfrac{\dfrac{\dfrac{\Gamma, \bigwedge \vec{Q} \to Q', \vec{Q} \Rightarrow Q_i \ (1 \le i < k)}{\Gamma, \bigwedge \vec{Q} \to Q', \vec{Q} \Rightarrow \bigwedge \vec{Q}}\mathrm{R}\wedge \qquad \Gamma, \vec{Q}, Q' \Rightarrow Q'}{\Gamma, \bigwedge \vec{Q} \to Q', \vec{Q} \Rightarrow Q'}\mathrm{L}\to}{\Gamma, \bigwedge \vec{Q} \to Q' \Rightarrow P}\mathrm{Ru}_4(\vec{Q}, Q', P)$$

Finally, the addition of $\mathrm{Ru}_1(\vec{P}, P')$ is equivalent to the addition of $\mathrm{Ru}_3(\vec{P}; P')$ by the following one-step deductions:

$$\mathrm{Cut_{cs}}\,\dfrac{\Gamma, \vec{P} \Rightarrow P' \qquad \Gamma, \vec{P}, P' \Rightarrow A}{\Gamma, \vec{P} \Rightarrow A} \qquad\qquad \mathrm{Ru}_3\,\dfrac{\Gamma, \vec{P}, P' \Rightarrow P'}{\Gamma, \vec{P} \Rightarrow P'}$$

4.6.1A. ♠ Assuming dp-closure under LC and LW, show that a generalization of Ru_3 to a rule with several active formulas in the premises, for example

$$\dfrac{\Gamma, P, Q_1, Q_2 \Rightarrow A \qquad \Gamma, P, Q_3, Q_4 \Rightarrow A}{\Gamma, P \Rightarrow A}$$

is equivalent to a finite set of rules of type Ru_3.

4.6.2. DEFINITION. A *basic cut* is a cut with atomic cutformula principal in the conclusions of instances of Ru_i $(1 \le i \le 4)$. ⊠

4.6.3. DEFINITION. A set of rules $\mathcal{X}$ is closed under substitution, if for every instance α of a rule from $\mathcal{X}$, the inference obtained by applying a substitution $[\vec{x}/\vec{t}]$ to each of the premises and the conclusion of α, is again an instance of a rule from $\mathcal{X}$.

A set of rules $\mathcal{X}$ is closed under (left-)contraction, if contraction on the conclusion of an instance of $\mathcal{X}$ yields the conclusion of another instance of $\mathcal{X}$.

A **G3i**-*system* is a system of sequents formulated in a first-order language, containing the axioms and rules of **G3i**, and in addition (1) a collection of rules of types Ru_i $(1 \le i \le 4)$ which is closed under substitution and Contraction plus (2) basic context-sharing cuts. ⊠

LEMMA. *Let* **S** *be any* **G3i***-system. Then*

(i) The inversion lemma of **G3i** *extends to* **S**.

(ii) **S** *and* **S** + $\mathrm{Cut_{cs}}$ *are dp-closed under left-weakening and left-contraction, and under the rule: if* $\vdash_n \Gamma \Rightarrow \bot$ *then* $\vdash_n \Gamma \Rightarrow A$ *(a variant of right-weakening).*

PROOF. (i) The proof for **G3i** also applies to **S**, since all the principal formulas in the additional rules are atomic.

(ii) Routine extension for the proofs for **G3i**. ⊠

4.6.4. THEOREM. *(Cut elimination for* **G3i***-systems) Context-sharing cuts, except possibly basic cuts, may be eliminated from proofs in* **S** + $\mathrm{Cut_{cs}}$.

PROOF. The proof for **G3i** extends to **G3i**-systems, since it remains true that whenever at least one of the cutformula occurrences (say α) in a cut is not a principal formula, then either the α belongs to the context of an axiom, and the whole cut is redundant, or α occurs in the conclusion of a rule, and the cut may be permuted upwards on that side where α occurs. The new extra axioms and rules, and the basic cuts, may be dealt with just as the other rules and axioms. Finally note that the new rules can never participate in a cut where both cutformula occurrences are principal unless the cut is basic, since the principal formulas of the logical rules are never atomic.

Let us check on the slight modifications in some of the cases.

Subcase 1a. $\mathcal{D}_0$ is a non-logical axiom, and D is not principal. So $\mathcal{D}$ must be of the form

$$\mathrm{Cut_{cs}}\ \frac{\Gamma, \vec{P} \Rightarrow D\,(\mathrm{Ru}_2) \qquad \begin{array}{c}\mathcal{D}_1\\ D, \Gamma, \vec{P} \Rightarrow A\end{array}}{\Gamma, \vec{P} \Rightarrow A}$$

Now the conclusion is an instance of the same non-logical axiom.

Subcase 1b. $\mathcal{D}_0$ is a non-logical axiom and D is principal. Then $\mathcal{D}$ is of the form

$$\frac{\Gamma, \vec{P} \Rightarrow P'\,(\mathrm{Ru}_1) \qquad \begin{array}{c}\mathcal{D}_1\\ P', \Gamma, \vec{P} \Rightarrow A\end{array}}{\Gamma, \vec{P} \Rightarrow A}\ \mathrm{Cut_{cs}}$$

If P' is context on the right, we can permute the cut upwards on the right. If P' is also principal on the right, then either $\mathcal{D}_1$ is an instance of Ax, so A is P', and the conclusion coincides with $\mathcal{D}_0$, or $\mathcal{D}_1$ is a non-logical rule, and then we have a basic cut.

Subcase 1c. $\mathcal{D}_1$ is a non-logical axiom, and D is not principal in $\mathcal{D}_1$. Then we have

$$\dfrac{\begin{array}{c}\mathcal{D}_0\\ \Gamma,\vec{P}\Rightarrow D\end{array} \quad D,\Gamma,\vec{P}\Rightarrow P'\ (\mathrm{Ru}_1)}{\Gamma,\vec{P}\Rightarrow P'}$$

and the conclusion is again an instance of Ru_1; similarly if $\mathcal{D}_1$ is an instance of Ru_2.

Subcase 1d. $\mathcal{D}$ is a non-logical axiom, say of type Ru_1, and D is principal in $\mathcal{D}_1$. Let $\vec{P}^* \equiv P_1,\ldots,P_{n-1}$. Then $\mathcal{D}$ is of the form

$$\dfrac{\begin{array}{c}\mathcal{D}_0\\ \Gamma,\vec{P}^*\Rightarrow P_0\end{array} \quad P_0,\Gamma,\vec{P}^*\Rightarrow P'\ (\mathrm{Ru}_1)}{\Gamma,\vec{P}^*\Rightarrow P'}$$

We may assume that $\mathcal{D}_0$ is not an axiom (otherwise one of the earlier subcases applies); now we may permute the cut upwards on the left. This is routine, except in the case where P_0 is principal; but then we have an instance of a Ru_4-rule, and the cut is a basic cut.

Case 2. Neither $\mathcal{D}_0$ nor $\mathcal{D}_1$ is an axiom, and D is not principal either on the left or on the right. Then we can permute upwards as before, for example

$$\dfrac{\dfrac{\begin{array}{c}\mathcal{D}_{0i}\\ \ldots\ \Gamma,\vec{P},Q_i\Rightarrow D\ \ldots (0\le i<k)\end{array}}{\Gamma,\vec{P}\Rightarrow D} \qquad \begin{array}{c}\mathcal{D}_1\\ \Gamma,\vec{P},D\Rightarrow A\end{array}}{\Gamma,\vec{P}\Rightarrow A}$$

becomes

$$\dfrac{\dfrac{\begin{array}{c}\mathcal{D}_{0i}\\ \Gamma,\vec{P},Q_i\Rightarrow D\end{array} \quad \begin{array}{c}\mathcal{D}_1[Q_i\Rightarrow]\\ \Gamma,\vec{P},Q_i,D\Rightarrow A\end{array}}{\Gamma,\vec{P},Q_i\Rightarrow A}\ (0\le i<k)}{\Gamma,\vec{P}\Rightarrow A}$$

Since we used "economical" thinning in this case, no appeal to closure under Contraction is needed.

Case 3. This can at most yield basic cuts in the new situations. ⊠

COROLLARY. *If the additional axioms and rules in* **S** *have their principal formulas on the left only (i.e., are of the types* $\mathrm{Ru}_2, \mathrm{Ru}_3$*), then cut elimination results in a cutfree proof (no basic cuts left).*

4.6.4A. ♠ Show that rules of the four types in 4.6.1 are sufficient to eliminate all logically compound principal and active quantifier-free formulas from extra axioms. *Hint.* A logically compound quantifier-free formula may be replaced by a set of atomic formulas linked together by suitable axioms; for example, if we have associated to B and C the atomic formulas P_B and P_C respectively, we can associate with $B \to C$ a new atomic formula $P_{B\to C}$ with additional axioms $P_{B\to C}, P_B \Rightarrow P_C$, and $P_B \to P_C \Rightarrow P_{B\to C}$.

4.6.4B. ♠ Formulate an adequate set of rule-types similar to 4.6.1 for the classical case.

4.6.4C. ♠ Give details of the cut elimination proof.

4.7 Logic with equality

4.7.1. DEFINITION. The theory of equality may be axiomatized by adding to **G3[mic]** the rules

$$\text{Ref}\,\frac{t = t, \Gamma \Rightarrow \Delta}{\Gamma \Rightarrow \Delta} \qquad \text{Rep}\,\frac{t = s, P[x/s], P[x/t], \Gamma \Rightarrow \Delta}{t = s, P[x/t], \Gamma \Rightarrow \Delta}$$

where P is atomic. Let us call these theories **G3[mic]**$^=$. By the remarks above, these theories are closed under Cut, Weakening and Contraction, provided the extra rules are themselves closed under Contraction. Duplication can happen in Rep if $P \equiv x = s$; in this case Rep concludes $t = s, t = s, \Gamma \Rightarrow \Delta$ from $t = s, t = s, s = s, \Gamma \Rightarrow \Delta$. But if we contract the instances of $t = s$ in premise and conclusion, the result is in fact an instance of Ref. ⊠

In order to prove equivalence with the extension of **G1[mic]** with axiomatic sequents for equality given before, we prove

4.7.2. LEMMA. *In* **G3[mic]**$^=$ *the following rules are admissible for all* A*:*

$$(i)\ \vdash t = s, A[x/t] \Rightarrow A[x/s] \quad (ii)\ \frac{t = s, A[x/s], A[x/t], \Gamma \Rightarrow \Delta}{t = s, A[x/t], \Gamma \Rightarrow \Delta}$$

PROOF. (i) is proved by induction on the depth of A. The most complicated induction step is where $A(x) \equiv B(x) \to C(x)$. In **G3[mi]**$^=$ this case is handled by the following deduction:

$$\dfrac{\dfrac{\dfrac{\dfrac{\dfrac{t = s, Bt \Rightarrow Bs\ (\text{IH})}{s = t, t = t, t = s, Bt \Rightarrow Bs}\,\text{W}}{s = t, t = t, Bt \Rightarrow Bs}\,\text{Rep}}{s = t, Bt \Rightarrow Bs}\,\text{Ref}}{s = t, Bs \to Cs, Bt \Rightarrow Bs}\,\text{W} \qquad \dfrac{s = t, Cs \Rightarrow Ct\ (\text{IH})}{s = t, Cs, Bt \Rightarrow Ct}\,\text{W}}{\dfrac{s = t, Bs \to Cs, Bt \Rightarrow Ct}{s = t, Bs \to Cs \Rightarrow Bt \to Ct}\,\text{R}\to}\,\text{L}\to$$

In **G3c**$^=$ the lower weakening on the left is left out. The other cases are left to the reader. (ii) is readily derivable from (i) and closure under Cut and Contraction. ⊠

4.7.2A. ♠ Supply the missing details in the preceding proof.

4.7.3. DEFINITION. Let us introduce a name for the following contracted instance of Rep:

$$\text{Rep}^* \; \frac{s = t, t = t, \Gamma \Rightarrow \Delta}{s = t, \Gamma \Rightarrow \Delta}$$

This is at the same time an instance of Ref, but in the next lemma we shall show how to remove Ref from deductions (except possibly Rep*). ⊠

LEMMA.

(i) If $\Gamma \Rightarrow \Delta$ is equality-free and derivable in **G3**[**mic**] + Ref + Rep* + Rep, *all sequents $\Gamma' \Rightarrow \Delta'$ in the proof have no equality in Δ'.*

(ii) If $\Gamma \Rightarrow \Delta$ is equality-free, and derivable in **G3**[**mic**] + Ref + Rep* + Rep, Ref *can be eliminated from the proof.*

PROOF. (i) If somewhere in the deduction of an equality-free $\Gamma \Rightarrow \Delta$ there appears a $\Gamma' \Rightarrow \Delta'$ containing $=$ in a formula A of Δ', A can only become active in a logical rule, but will then appear as a subformula of the conclusion.

(ii) We show how to eliminate a Ref-application appearing as the last inference in a deduction $\mathcal{D}$ containing no other applications of Ref. The proof is by induction on the depth of $\mathcal{D}$. Let

$$\frac{s = s, \Gamma' \Rightarrow \Delta'}{\Gamma' \Rightarrow \Delta'}$$

be the bottom inference of $\mathcal{D}$; if $s = s$ is not principal, we can permute the application of Ref upwards over the preceding rule and apply the IH. If $s = s$ is principal, it is principal in Rep or Rep*:

$$\frac{\dfrac{s = s, P[x/s], P[x/s], \Gamma'' \Rightarrow \Delta'}{s = s, P[x/s], \Gamma'' \Rightarrow \Delta'}\,\text{Rep}}{P[x/s] \Rightarrow \Delta'}$$

or

$$\frac{\dfrac{s = s, s = s, \Gamma' \Rightarrow \Delta'}{s = s, \Gamma' \Rightarrow \Delta'}\,\text{Rep}^*}{\Gamma' \Rightarrow \Delta'}$$

In both cases we can apply dp-closure under Contraction to the derivation of the topline, apply Ref and apply the IH to the resulting deduction. ⊠

4.7.4. THEOREM. **G3**[**mic**]$^=$ *is conservative over* **G3**[**mic**].

PROOF. Immediate from the preceding lemma: once an equality appears on the left, in a deduction without Ref, equalities remain present on the left, whereas the appearance of an equality on the right is excluded by the first half of the lemma. ⊠

4.8 The theory of apartness

4.8.1. Apartness is intended as a positive version of inequality. Thus, for example, in the intuitionistic theory of real numbers, two numbers are apart if the distance between them exceeds a positive rational number. The pure theory of apartness **AP** has besides equality a single primitive binary relation, $\#$; a simple description is to say that it can be formalized on the basis of **Ni** with the following axioms added:

REFL	$\forall x(x = x)$,
SYM	$\forall xy(x = y \to y = x)$,
TRA	$\forall xyz(x = y \wedge y = z \to x = z)$,
$\#$EQ	$\forall xyx(x \# y \wedge x = x' \wedge y = y' \to x' \# y')$,
AP1	$\forall xy(\neg x \# y \to x = y)$,
AP2	$\forall x(\neg x \# x)$,
AP3	$\forall xyz(x \# y \to x \# z \vee y \# z)$.

In this theory we can prove that $\neg t \# s \leftrightarrow t = s$. This permits us to consider instead of **AP** an equivalent theory **AP**′, formulated with a single binary relation $\#$, where equality is simply defined as

$$t = s := \neg t \# s.$$

The only extra axioms are now AP2 and AP3.

From AP3 one easily proves in **AP**′ the equality axiom $\#$EQ for $\#$. To see this, note that if $x \# y$, $x = x'$, then by the definition of equality $\neg(x \# x')$, and since by AP3 applied to $x \# y$ we find $x \# x' \vee x' \# y$, and the first disjunct is excluded, we have $x' \# y$, etc.

We can reformulate **AP**′ as a **G3i**-system **AP-G3i** as follows: there are an extra axiom of type Ru_2 and an extra rule of type Ru_3:

$$\#_1 \quad \Gamma, t \# t \Rightarrow A$$

$$\#_2 \; \frac{t \# s, t \# r, \Gamma \Rightarrow A \qquad t \# s, s \# r, \Gamma \Rightarrow A}{t \# s, \Gamma \Rightarrow A}$$

We define $t = s := \neg t \# s$. As an immediate consequence of the fact that the principal formulas appear on the left only, we have

4.8.2. THEOREM. $\mathrm{Cut}_{\mathrm{cs}}$ *is completely eliminable from* **AP-G3i** *(i.e. we need not even basic cuts).*

4.8.3. COROLLARY. *A cutfree deduction of* $\Gamma \Rightarrow A$ *in* **AP-G3i** *contains only subformulas of* Γ, A *and atomic formulas.* ⊠

We now want to characterize the equality fragment of the theory of apartness. We introduce "approximations" to apartness as a sequence of inequalities of ever increasing strength:

$$t \#^0 s := \neg t = s; \quad t \#^{n+1} s := \forall x(t \#^n x \vee s \#^n x).$$

DEFINITION. The theory **EQAP** consists of the pure theory of equality (REFL, SYM, TRA) with in addition the following axioms for all $n \in \mathbb{N}$:

$\mathrm{INEQ}_n \quad \neg t \#^{n+1} s \Rightarrow t = s.$ ⊠

It follows that

$$\text{For all } n, \ \neg t \#^n s \leftrightarrow t = s.$$

4.8.4. DEFINITION. We extend the notions of positive and negative context as follows. A *positive sequent-context* $\mathcal{S}^+[*]$ is of the form $\Gamma, \mathcal{N} \Rightarrow A$ or of the form $\Gamma \Rightarrow \mathcal{P}$, where $\mathcal{N}, \mathcal{P}$ are negative and positive formula contexts respectively. Similarly for a *negative sequent-context* $\mathcal{S}^-[*]$, with the roles of $\mathcal{N}, \mathcal{P}$ interchanged. ⊠

The proof of the following lemma is immediate.

LEMMA. *Let* $\mathcal{S}^+[*]$, $\mathcal{S}^-[*]$ *be a positive and a negative sequent-context respectively. Then*

$$\textit{if } \vdash \mathcal{S}^+[t \#^{n+1} s], \textit{ then } \vdash \mathcal{S}^+[t \#^n s],$$
$$\textit{if } \vdash \mathcal{S}^-[t \#^n s], \textit{ then } \vdash \mathcal{S}^-[t \#^{n+1} s].$$

4.8.5. DEFINITION. Let $\Gamma \Rightarrow A$ be a sequent in the language of **AP-G3i**. Then $(\Gamma \Rightarrow A)^n$ is obtained from $\Gamma \Rightarrow A$ by replacing all positive occurrences of $\#$ by $\neq$, and all negative occurrences by $\#^n$. For $(\Gamma \Rightarrow A)^n$ we also write $\Gamma^n \Rightarrow A^n$. ⊠

By repeated use of the lemma we have that if $\vdash (\Gamma \Rightarrow A)^n$ then $\vdash (\Gamma \Rightarrow A)^{n+1}$, for all $n \in \mathbb{N}$. Note that if all occurrences of $\#$ in $\Gamma \Rightarrow A$ occur under a negation (i.e., $\Gamma \Rightarrow A$ is a statement in equality theory), then $(\Gamma \Rightarrow A)^n$ holds in **EQAP** iff $\Gamma \Rightarrow A$ is provable in **EQAP**, since $\neg t \#^n s \leftrightarrow t = s \rightarrow \neg t \# s$.

4.8.6. THEOREM. *Let* $\Gamma \Rightarrow A$ *be a sequent in the language of pure equality which has been proved in* **AP-G3i** *by a deduction* $\mathcal{D}$. *Then we can effectively transform* $\mathcal{D}$ *into a deduction* $\mathcal{D}^*$ *in* **EQAP** *of* $\Gamma^m \Rightarrow A^m$ *for a suitable* m.

PROOF. We show by induction on the depth of deductions $\mathcal{D}$ in **AP-G3i** with conclusion $\Gamma \Rightarrow A$ that we can transform $\mathcal{D}$ into $\mathcal{D}^n$ in **EQAP** with conclusion $(\Gamma \Rightarrow A)^n$ for a suitable n. For definiteness we shall assume **EQAP** to be axiomatized with in a system based on **G3i** with Cut_{cs}, and logical axioms with principal formulas of arbitrary complexity. Furthermore below p will always be $\max(n, m)$.

Case 1. $\mathcal{D}$ is an axiom. For any axiom $\mathcal{D}$, $\mathcal{D}^0$ is provable in **EQAP**.

Case 2. $\mathcal{D}$ ends with an application of $\#_2$ and is of the form

$$\dfrac{\begin{matrix}\mathcal{D}_1 \\ t \# s, t \# r, \Gamma \Rightarrow A\end{matrix} \quad \begin{matrix}\mathcal{D}_2 \\ t \# s, s \# r, \Gamma \Rightarrow A\end{matrix}}{t \# s, \Gamma \Rightarrow A}$$

Remember that $t \#^{p+1} s = \forall z(t \#^p z \vee s \#^p = z)$. Then we can take for the transformed $\mathcal{D}$

$$\dfrac{\dfrac{\dfrac{\dfrac{\begin{matrix}\mathcal{D}_1^n \\ t \#^n s, t \#^n r, \Gamma^n \Rightarrow A^n\end{matrix}}{t \#^p s, t \#^p r, \Gamma^p \Rightarrow A^p} \quad \dfrac{\begin{matrix}\mathcal{D}_2^m \\ t \#^m s, s \#^m r, \Gamma^m \Rightarrow A^m\end{matrix}}{t \#^p s, s \#^p r, \Gamma^p \Rightarrow A^p}}{t \#^p s, t \#^p r \vee s \#^p r, \Gamma^p \Rightarrow A^p}\text{L}\vee}{t \#^{p+1} s, t \#^p r \vee s \#^p r, \Gamma^p \Rightarrow A^p}}{\dfrac{t \#^{p+1} s, \Gamma^p \Rightarrow A^p}{t \#^{p+1} s, \Gamma^{p+1} \Rightarrow A^{p+1}}}\text{L}\forall$$

The transitions marked by the dashed lines are justified by the preceding lemma, since $\Gamma^{n'} \Rightarrow A^{n'}$ $(n' > n)$ is obtained from $\Gamma^n \Rightarrow A^n$ by replacing a number of negative occurrences of $\#^n$ by $\#^{n'}$.

Case 3. $\mathcal{D}$ ends with a logical rule, for example a two-premise rule

$$\dfrac{\begin{matrix}\mathcal{D}_1 \\ \Gamma_1 \Rightarrow A_1\end{matrix} \quad \begin{matrix}\mathcal{D}_2 \\ \Gamma_2 \Rightarrow A_2\end{matrix}}{\Gamma \Rightarrow A}$$

Using the induction hypothesis for $\mathcal{D}_1, \mathcal{D}_2$, we see that we can take for the transform of $\mathcal{D}$

$$\dfrac{\dfrac{\begin{matrix}\mathcal{D}_1^n \\ \Gamma_1^n \Rightarrow A_1^n\end{matrix}}{\Gamma_1^p \Rightarrow A_1^p} \quad \dfrac{\begin{matrix}\mathcal{D}_2^m \\ \Gamma_2^m \Rightarrow A_2^m\end{matrix}}{\Gamma_2^p \Rightarrow A_2^p}}{\Gamma^p \Rightarrow A^p}$$

For single-premise rules the transformation is even simpler. ⊠

4.8.7. THEOREM. *The equality fragment of* **APP** *is* **EQAP**.

PROOF. A formula in the language of equality is expressed in the language with only $\#$ as a formula where all atomic formulas $t \# s$ occur negated. Let $\Gamma \Rightarrow A$ be such a statement, derived in **APP**. By the preceding proposition we then obtain a proof of $\Gamma^n \Rightarrow A^n$ for some n in **EQAP**. But since $\neg t \#^n s \leftrightarrow \neg t \# s$, this is in **EQAP** equivalent to $\Gamma \Rightarrow A$. ⊠

4.9 Notes

4.9.1. *The Cut rule.* The Cut rule is a special case of a rule considered in papers by Hertz, e.g. Hertz [1929]; Gentzen [1933b] introduces Cut. The proof of cut elimination in subsection 4.1.5 follows the pattern of Dragalin [1979]. The calculus in 4.1.11 is precisely the propositional part of the calculus of Ketonen [1944].

It was recently shown that the introduction of Multicut in establishing cut elimination for G1-systems (4.1.9) can be avoided; see von Plato [1999], Borisavljević [1999].

A detailed proof of cut elimination for m-**G3i** is found in Dyckhoff [1996].

In studying the computational behaviour of the cut elimination process, it makes sense to have both context-free and context-sharing logical rules within the same system (cf. refconrules), see Joinet et al. [1998], Danos et al. [1999].

Strong cut elimination means that Cut can be eliminated, regardless of the order in which the elementary steps of the process are carried out (the order may be subject to certain restrictions though). Strong cut elimination is treated in Dragalin [1979], Tahhan Bittar [1999], Grabmayer [1999].

The considerations in 4.1.11 may be extended to classical predicate logic, and used to give a perspicuous completeness proof relative to a cutfree system, thereby at one stroke establishing completeness and closure under Cut (not cut elimination). This was discovered in the fifties by several people independently: Beth [1955], Hintikka [1955], Schütte [1956], Kanger [1957]. For more recent expositions see, for example, Kleene [1967], Heindorf [1994], Socher-Ambrosius [1994]. The basic idea of all these proofs is to attempt, for any given sequent, to construct systematically cutfree proofs "bottom-up"; from the failure of the attempts to find a proof a counterexample to the sequent may be read off. It was Beth who introduced the name "semantic tableau" for such an upside-down Gentzen system; cf. 4.9.7.

This idea is also applicable to intuitionistic logic, using Beth models (Beth [1956,1959]) or Kripke models (Kripke [1965], Fitting [1969]), and to modal logics, using Kripke's semantics for modal logic (cf. Fitting [1983]). See 4.9.7 below. Beth was probably the first to use multi-succedent systems for intuitionistic logic, namely in his semantic researches just mentioned.

4.9.2. *Separation theorem.* Wajsberg [1938] proved the separation theorem for **Hip**; however, Wajsberg's proof is not completely correct; see Bezhanishvili [1987] and the references given there. In exercise 4.2.1B a proof of the separation property is sketched for all fragments of **Hc** except for the fragments not containing negation, such as the implicative fragment. In order to extend the separation theorem to this case, we consider the variant of **Hc** with $\neg\neg A \to A$ replaced by Peirce's law. Then the axioms **k**, **s** (1.3.9) and Peirce's law axiomatize $\to$**Hc**. This result is due to P. Bernays, and is not hard to prove by a semantical argument. For other axiomatizations of $\to$**Hc** see Curry [1963, p. 250].

Below we give a short semantical proof of Bernays' result, communicated to us by J. F. A. K. van Benthem. The argument consists in the construction of a Henkin set not containing a given unprovable formula A, i.e. a maximal consistent set of formulas not containing A; the required properties are proved with the help of the result of exercise 2.1.8F.

Suppose $\nvdash A$. Then, using a countable axiom of choice, we can find a maximal set $\mathcal{X}$ of formulas such that $\mathcal{X} \nvdash A$ (enumerate all formulas as $F_0, F_1, F_2, \ldots$, put $\mathcal{X}_0 = \emptyset$, $\mathcal{X}_{n+1} = \mathcal{X}_n$ if $\mathcal{X}_n \cup \{F_n\} \vdash A$, $\mathcal{X}_{n+1} = \mathcal{X}_n \cup \{F_n\}$ otherwise, and let $\mathcal{X} = \bigcup_{n\in\mathbb{N}} \mathcal{X}_n$). It is easy to see that $\mathcal{X}$ is deductively closed, that is to say, if $\mathcal{X} \vdash B$ then $B \in \mathcal{X}$. For all B, C

$$(*) \qquad B \to C \in \mathcal{X} \quad \text{iff} \quad B \notin \mathcal{X} \text{ or } C \in \mathcal{X}.$$

For the direction from left to right, let $B \to C \in \mathcal{X}$; if $B \in \mathcal{X}$ then by deductive closure $C \in \mathcal{X}$. Hence the right hand side of $(*)$ holds. For the direction from right to left, we argue by contraposition. *Case 1.* Assume $C \in \mathcal{X}$. Then $\mathcal{X} \vdash B \to C$, hence $\mathcal{X} \vdash A$; contradiction, hence $C \notin \mathcal{X}$. *Case 2.* Assume $B \notin \mathcal{X}$. Then $\mathcal{X}, B \vdash A$ (maximality), hence $\mathcal{X} \vdash B \to A$ (deduction theorem); also by the deduction theorem $\mathcal{X} \vdash (B \to C) \to A$. Hence with 2.1.8F $\mathcal{X} \vdash A$; again contradiction, so $B \in \mathcal{X}$.

By $(*)$ we can now define a valuation $v_{\mathcal{X}}$ (propositional model) from $\mathcal{X}$, for which we can prove for all formulas G: $G \in \mathcal{X}$ iff $v_{\mathcal{X}}(G) = \text{true}$.

We have not found a really *simple* syntactical proof of the separation theorem for $\to$**Hc**. A syntactical proof may be obtained from the result in Curry [1963, p. 227, corollary 2.3]; another syntactical proof is outlined in exercise 6.2.7C.

4.9.3. *Other applications.* 4.2.2 was proved in Malmnäs and Prawitz [1969], and 4.2.3, 4.2.4 were proved in Prawitz [1965] (in all three cases by using N-formalisms). Buss and Mints [1999] study the disjunction property and the explicit definability property from the viewpoint of complexity theory. The proof of 4.2.6 corresponds to Kleene's decision method for the system G3 in Kleene [1952a].

The notion of a Rasiowa–Harrop formula was discovered independently by H. Rasiowa and R. Harrop (Rasiowa [1954,1955], Harrop [1956,1960]). For more information on the "Aczel slash" of exercise 4.2.4C (Aczel [1968]) and similar relations, see the references in Troelstra [1992b].

Another type of application of cut elimination, not treated here, concerns certain axiomatization problems for intuitionistic theories, such as the axiomatization of the f-free fragment of the theory of a single Skolem function with axiom $\forall \vec{x} R(\vec{x}, f\vec{x})$, R a relation symbol, see Mints [1999]. For other examples, see Motohashi [1984a], Uesu [1984].

4.9.4. *The Gentzen system* **G4ip**. This calculus, with its splitting of L$\to$, was recently discovered independently by Dyckhoff [1992] and Hudelmaier [1989,1992]. Here we have combined the notion of weight as defined by Hudelmaier with Dyckhoff's clever proof of equivalence with the ordinary calculus. Long before these recent publications, a decision algorithm based on the same or very similar ideas appeared in a paper by Vorob'ev [1964]; but in that paper the present formalism is not immediately recognizable. Remarks on the history of this calculus may be found in Dyckhoff's paper.

In Dyckhoff and Negri [1999], a direct proof, without reference to **G3i**, of the closure of **G4ip** under Weakening, Contraction and Cut is presented. Such a proof is more suitable for generalization to predicate logic and to systems with extra rules, since induction on the weight of sequents may break down when rules are added. However, the proof is somewhat longer than the "quick and dirty" proof presented here.

A very interesting application of the system **G4ip** is found in a paper by Pitts [1992]. In particular, Pitts uses **G4ip** to show that in **Ip** there exist minimal and maximal interpolants (an interpolant M of $\vdash A \Rightarrow B$ is said to be *minimal* if for all other interpolants C we have $\vdash M \Rightarrow C$; similarly for *maximal* interpolants). The corresponding property of **Cp** is trivial. Since Pitts [1992] several semantical proofs of this result has been given; see for example Visser [1996].

4.9.5. *Interpolation theorem.* This was originally proved by Craig [1957a] for **C** without function symbols or equality, and in Craig [1957b] extended to **C** with equality. Craig's method is proof-theoretic, using a special modification of the sequent calculus. The theorem was inspired by the definability theorem of Beth [1953]. Craig's theorem turned out to have a model-theoretic counterpart, namely the consistency theorem of A. Robinson [1956].

The refinement taking into account positive and negative occurrences is due to Lyndon [1959]; Lyndon's method is algebraic–model-theoretic. The example $\exists x(x = c \wedge \neg Rx) \Rightarrow \neg Rc$ shows that we cannot extend the Lyndon refinement to constants or function symbols in the presence of equality (c occurs positively on the left, negatively on the right, but has to occur in every interpolant).

Schütte [1962] proved the interpolation theorem for **I**, using the method of "split sequents", which he apparently learned from Maehara and Takeuti [1961], but which according to Takeuti [1987] is originally due to Maehara [1960].

Nagashima [1966] extended the interpolation theorem to both **C** and **I** with function symbols, but without equality, using as an intermediate the theories with equality present. Inspection of his proof shows that with very slight adaptations it also works for languages with function symbols *and* equality. Kleene [1967] also proved interpolation for languages with equality and function symbols by essentially the same method. Our exposition of interpolation for languages with functions and equality combines features of both proofs. Felscher [1976] gives an essentially different proof for languages with function symbols but without equality (this paper also contains further historical references).

Oberschelp [1968] proved, by model-theoretic reasoning, for classical logic with equality, but without function symbols or constants, interpolation w.r.t. positive and negative occurrences of relation symbols plus the following condition on occurrences of $=$ in the interpolant: if $=$ has a positive [negative] occurrence in the interpolant C of $\vdash A \Rightarrow B$ (i.e., $\vdash A \Rightarrow C, \vdash C \Rightarrow B$) then $=$ has a positive occurrence in A [negative occurrence in B]. As noted by Oberschelp, we cannot expect a symmetric interpolation condition for $=$, as shown by the sequents $x = y \Rightarrow Rx \vee \neg Ry$ and $Rx \wedge \neg Ry \Rightarrow x \neq y$. Fujiwara [1978] extended this to interpolation w.r.t. functions, positive and negative occurrences of relations, and the extra condition for $=$. Motohashi [1984b] gave a syntactic proof of this result and moreover extended Oberschelp's result (i.e. without the interpolation condition for functions) to intuitionistic logic.

Schulte-Mönting [1976] gives a proof of the interpolation theorem which yields more fine-structure: one can say something about the complexity of terms and predicates in the interpolant.

For a model-theoretic proof of interpolation for **C** with functions and constants, but no $=$, see, for example, Kreisel and Krivine [1972]. There is an extensive literature on interpolation for extensions of first-order logic, most of it using model-theoretic methods.

The interpolation theorem for many-sorted languages, with the application to the characterization of persistent sentences (4.4.13–4.4.16), is due to Feferman [1968].

A syntactic proof of the eliminability of symbols for definable functions is already in Hilbert and Bernays [1934] (elimination of the ι-operator). Cf. also Schütte [1951], Kleene [1952a]. For our standard logics there are very easy semantical proofs of these results, using classical or Kripke models.

4.9.6. *Generalizations with applications.* Perhaps Ketonen [1944] may be said to be an early analysis of cutfree proofs in Gentzen calculi with axioms; but he considers the form of cutfree derivations in the pure calculus where axioms are present in the antecedent of the sequents derived. Schütte [1950b] considers for his one-sided sequent calculus derivations enriched with sequents of atomic formulas as axioms, and proves a generalized cut elimination theorem for such extensions. A proof for a context-free variant of the standard Gentzen formalism LK is given in Sanchis [1971]. Schütte and Sanchis require the axioms to be closed under Cut; whereas Girard [1987b]) does not require closure of the set of axioms under Cut.

The examples of this generalized cut elimination theorem are given as they appear in Girard [1987b].

The possibility of completely eliminating cuts for **G3[mic]** extended with suitable *rules* was first noticed in Negri [1999] for the intuitionistic theory of apartness, and is treated in greater generality in Negri and von Plato [1998]. Among the applications are predicate logic with equality, the intuitionistic theory of apartness just mentioned, and the intuitionistic theory of partial order.

The paper Nagashima [1966] axiomatizes intuitionistic and classical logic with equality by the addition of the two rules

$$\frac{t = t, \Gamma \Rightarrow \Delta}{\Gamma \Rightarrow \Delta} \qquad \frac{\Gamma \Rightarrow \Delta, A(s) \qquad A(t), \Gamma' \Rightarrow \Delta'}{s = t, \Gamma, \Gamma' \Rightarrow \Delta, \Delta'}$$

to Gentzen's systems LJ and LK. Nagashima states cut elimination for these extensions, and a corresponding subformula property: each formula in a cut-free proof in one of these extensions is either a prime formula or a subformula of a formula in the conclusion. From this he then derives the conservativity of the extensions over pure predicate logic without equality. This result is an essential ingredient of Nagashima's proof of the interpolation theorem for languages with functions but without equality.

The proof of this conservativity result presented in 4.7 is due to J. von Plato and S. Negri.

The characterization of the equality fragment of the predicate-logical theory of apartness in 4.8 is due to van Dalen and Statman [1979]. They obtained this result by normalization of an extension of natural deduction. The present simpler proof occurred to us after studying Negri [1999].

The latter paper also contains the interesting result that the propositional theory of apartness is conservative over the theory of pure equality plus the stability axiom (i.e., $\neg\neg t = s \to t = s$; Negri [1999] defines equality as the negation of apartness, which automatically ensures stability).

4.9.7. *Semantic tableaux.* Semantic tableaux may be described as a particular style of presentation of a certain type of Gentzen calculus – in fact, Kleene's G3-calculi mentioned in 3.5.11. In the literature, the details of the presentation vary.

The motivation of semantic tableaux, however, is in the semantics, not in the proof theory – in the spirit of the semantic motivation for **G3cp** in 4.1.11. They have been widely used in completeness proofs, especially for modal logics. For classical logic, there is a detailed treatment in Smullyan [1968]; for intuitionistic and modal logics, see Fitting [1969,1983,1988].

Let us describe a cumulative version of semantic tableaux for **Cp**, starting from **GK3c**. *Signs* are symbols **t**, **f**, and we use $s, s', s_1, \ldots$ for arbitrary signs. *Signed formulas* are expressions sA, A a formula. We encode a sequent $\Gamma \Rightarrow \Delta$, with Γ, Δ finite sets, as the set of signed formulas $\mathbf{t}\Gamma, \mathbf{f}\Delta$.

If we turn the rules of **GK3c** *upside down*, then, for example, L∧, R∧ take the form (Θ a set of signed formulas):

$$\mathbf{t}\wedge\ \frac{\Theta, \mathbf{t}(A \wedge B)}{\Theta, \mathbf{t}(A \wedge B), \mathbf{t}A, \mathbf{t}B} \qquad \mathbf{f}\wedge\ \frac{\Theta, \mathbf{f}(A \wedge B)}{\Theta, \mathbf{f}(A \wedge B), \mathbf{f}A \mid \Theta, \mathbf{f}(A \wedge B), \mathbf{f}B}$$

where we have used |, instead of spacing, to separate the possible "conclusions" (originally, in **GK3c**, premises). In this form the rules generate downward growing trees, called *semantic tableaux* (cumulative version). A tableau with $\mathbf{t}\Gamma, \mathbf{f}\Delta$ at the top node (root) is said to be a *tableau for* $\mathbf{t}\Gamma, \mathbf{f}\Delta$. The semantic reading of the rules is as follows (cf. the motivation for **G3cp** above): $\mathbf{t}\Gamma, \mathbf{f}\Delta$ represents the problem of finding a valuation making Γ true and Δ false; the problem represented by the premise of a rule application is solved if one can solve the problem represented by one of the conclusions. A node of the tableau is *closed* if it contains either $\mathbf{t}\bot$ or $\mathbf{t}A, \mathbf{f}A$ for some A. So a closed node represents a valuation (satisfiability) problem solved in the negative, and is in fact nothing but an axiom of the sequent calculus. We assume that a tableau is not continued beyond a closed node.

A *branch* of the tableau is a sequence of consecutive nodes starting at the root, which either is infinite, or ends in a closed node. A *subbranch* is an initial segment of a complete branch. A branch is said to be *closed* when ending in a closed node; a tableau is *closed* if all its branches are closed.

As an example we give a closed tableau for $\theta \equiv \mathbf{f}(A \vee B \to B \vee A)$:

$$\cfrac{\cfrac{\cfrac{\theta}{\theta, \mathbf{t}(A \vee B), \mathbf{f}(B \vee A)}}{\theta, \mathbf{t}(A \vee B), \mathbf{f}(B \vee A), \mathbf{f}B, \mathbf{f}A}}{\theta, \mathbf{t}(A \vee B), \mathbf{f}(B \vee A), \mathbf{f}B, \mathbf{f}A, \mathbf{t}A \ \mid\ \theta, \mathbf{t}(A \vee B), \mathbf{f}(B \vee A), \mathbf{f}B, \mathbf{f}A, \mathbf{t}B}$$

Obviously, a closed tableau is nothing but a **GK3c**-derivation, differently presented. All formulas obtained at a given node are repeated over and over again at lower nodes in this cumulative version of semantic tableaux; so, in attempting to construct a tableau by hand, it is more efficient to use a non-cumulative presentation, as in Beth's original presentation; see, for example, Smullyan [1968].

The use of signed formulas, due to R. M. Smullyan, is a convenient notational device which avoids the (in this setting) awkward distinction between formulas to the left and to the right of ⇒. If we do not want to use signed formulas, we can take the set $\Gamma, \neg\Delta$, corresponding to the sequent $\Gamma, \neg\Delta \Rightarrow$, instead of $\mathbf{t}\Gamma, \mathbf{f}\Delta$.

Another useful device of Smullyan is the notion of a formula type. In the tableaux, $\mathbf{t}(A \wedge B)$, $\mathbf{f}(A \vee B)$, $\mathbf{f}(A \to B)$ on the one hand, and $\mathbf{f}(A \wedge B)$, $\mathbf{t}(A \vee B)$, $\mathbf{t}(A \to B)$ on the other hand, show the same type of behaviour. Let us call formulas of the first group α-formulas, and of the second group β-formulas. To these formulas we assign components α_1, α_2 and β_1, β_2 according to the following tables:

α	α_1	α_2
$\mathbf{t}(A \wedge B)$	$\mathbf{t}A$	$\mathbf{t}B$
$\mathbf{f}(A \vee B)$	$\mathbf{f}A$	$\mathbf{f}B$
$\mathbf{f}(A \to B)$	$\mathbf{t}A$	$\mathbf{f}B$

β	β_1	β_2
$\mathbf{f}(A \wedge B)$	$\mathbf{f}A$	$\mathbf{f}B$
$\mathbf{t}(A \vee B)$	$\mathbf{t}A$	$\mathbf{t}B$
$\mathbf{t}(A \to B)$	$\mathbf{f}A$	$\mathbf{t}B$

The rules for the propositional operators may now be very concisely formulated as

$$\frac{\Gamma\alpha}{\Gamma\alpha\alpha_1\alpha_2} \qquad \frac{\Gamma\beta}{\Gamma\beta\beta_1|\Gamma\beta\beta_2}$$

The idea of formula types may be extended to predicate logic, by introducing two further types,

$$\text{type } \gamma: \ \mathbf{f}\forall xA, \mathbf{t}\exists xA \quad \text{and} \quad \text{type } \delta: \ \mathbf{t}\forall xA, \mathbf{f}\exists xA.$$

If we adopt the convention that for γ, δ as above, $\gamma(t), \delta(t)$ stand for $A[x/t]$, the tableau rules may be summarized by

$$\frac{\Gamma\gamma}{\Gamma\gamma\gamma(t)} \qquad \frac{\Gamma\delta}{\Gamma\delta\delta(y)} \quad (\ y \text{ a new variable}).$$

The simplification which is the result of the distinction of formula types may be compared to the simplification obtained by using one-sided sequents in the GS-calculi. Both devices use the symmetries of classical logic. However, the distinction of formula types is still useful for intuitionistic logic, whereas there is no intuitionistic counterpart to the GS-systems (cf. Fitting [1983, chapter 9]).

For intuitionistic logic, the semantic motivation underlying the tableaux is the construction of suitable valuations in a Kripke model. In the remainder of this subsection we assume familiarity with the notion of a Kripke model for **Ip**.

A top node $\mathbf{t}\Gamma, \mathbf{f}\Delta$ in a tableau now represents the problem of finding a Kripke model such that at the root $\bigwedge\Gamma$ is valid and $\bigvee\Delta$ invalid. Now cumulative rules such as $\mathbf{t}\wedge, \mathbf{f}\wedge$ mentioned above correspond to (in standard notation for Kripke models)

$$\begin{array}{lll} k \Vdash A \wedge B & \text{iff} & (k \Vdash A \ \text{ and } \ k \Vdash B), \ \text{and} \\ k \nVdash A \wedge B & \text{iff} & (k \nVdash A \ \text{ or } \ k \nVdash B) \end{array}$$

respectively (the classical truth conditions at each node). However, for $k \nVdash A \to B$ we must have a $k' \geq k$ with $k' \Vdash A, k' \nVdash B$. But if $k \nVdash C$, there is no guarantee that $k' \nVdash C$ for $k' \geq k$: only formulas forced at k are "carried over" to k'. Thus, to obtain a rule which reflects the semantic conditions for $k \nVdash A \to B$, we take

$$\mathbf{f}{\to}\ \frac{\Theta, \mathbf{f}(A \to B)}{\Theta^{\mathbf{t}}, \mathbf{t}B, \mathbf{f}A}$$

with $\Theta^{\mathbf{t}} := \{\mathbf{t}C : \mathbf{t}C \in \theta\}$. Putting $k \Vdash \mathbf{t}A$ iff $k \Vdash A$, $k \Vdash \mathbf{f}A$ iff $k \nVdash A$, this rule may be read as: in order to force $\Theta, \mathbf{f}(A \to B)$ at k, we must find another node $k' \geq k$, k' forcing $\Theta^{\mathbf{t}}, \mathbf{f}B, \mathbf{t}A$. So $\mathbf{f}\!\!\to$ is not strictly cumulative any more. Of course, if we interpret this rule as a rule of a Gentzen system, we find

$$\frac{\Gamma, A \Rightarrow B}{\Gamma \Rightarrow A \to B, \Delta}$$

exactly as in the calculi m-**G1i** in 3.2.1A and m-**G3i** in 3.5.11D, but for the fact that Γ and Γ, A, etc. are interpreted as sets, not as multisets. Thus we see that multiple-conclusion Gentzen systems for **I** appear naturally in the context of semantic tableaux.

Chapter 5

Bounds and permutations

This chapter is devoted to two topics: the rate of growth of deductions under the process of cut elimination, and permutation of rules.

It is not hard to show that there is a hyperexponential upper bound on the rate of growth of the depth of deductions under cut elimination. For propositional logic much better bounds are possible, using a clever strategy for cut elimination. This contrasts with the situation for normalization in the case of N-systems (chapter 6), where propositional logic is as bad as predicate logic in this respect.

In contrast to the case of normalization for N-systems, it is not easy to extract direct computational content from the process of cut elimination for G-systems, since as a rule the process is non-deterministic, that is to say the final result is not a uniquely defined "value". Recent proof-theoretical studies concerning linear logic (9.3) lead to a more or less satisfactory analysis of the computational content in cut elimination for **C** (and **I**); in these studies linear logic serves to impose a "fine structure" on sequent deductions in classical and linear logic (some references are in 9.6.5).

We also show that in a GS-system for **Cp** with Cut there are sequences of deduction with proofs linearly increasing in size, while the size of their cutfree proofs has exponentially increasing lower bounds.

These results indicate that the use of "indirect proof", i.e. deductions that involve some form of Cut play an essential role in formalized versions of proofs of theorems from mathematical practice, since otherwise the length of proofs would readily become unmanageable.

The second topic of this chapter is the permutation of rules. Permutation of rules permits further standardization of cutfree deductions, and in particular one can establish with their help a (version of) Herbrand's theorem. Permutation arguments also play a role in the theory of logic programming; see 7.6.3.

5.1 Numerical bounds on cut elimination

In this section we refine the analysis of cut elimination by providing numerical bounds. We analyze the cut elimination procedure according to the proof of 4.1.5. We recall the following corollary to that proof:

5.1.1. LEMMA. *(Cut reduction lemma) Let $\mathcal{D}'$, $\mathcal{D}''$ be two deductions in* $\mathbf{G3c} + \mathrm{Cut_{cs}}$, *with cutrank* $\leq |D|$, *and let $\mathcal{D}$ result by a cut:*

$$\dfrac{\begin{array}{c}\mathcal{D}'\\ \Gamma \Rightarrow D, \Delta\end{array} \quad \begin{array}{c}\mathcal{D}''\\ \Gamma, D \Rightarrow \Delta\end{array}}{\Gamma \Rightarrow \Delta}$$

Then we can transform $\mathcal{D}$ into a deduction $\mathcal{D}^*$ with lower cutrank such that $|\mathcal{D}^*| \leq |\mathcal{D}'| + |\mathcal{D}''|$. A similar result holds for $\mathbf{G3i} + \mathrm{Cut_{cs}}$, *with* $|\mathcal{D}^*| \leq 2(|\mathcal{D}'| + |\mathcal{D}''|)$.

REMARKS. (i) Taking **G1c**,**G1i** instead of **G3c**, **G3i**, and using Gentzen's method based on the Multicut rule, we can only give a cut reduction lemma with an estimate in terms of *logical depth* $||\ ||$. $||\mathcal{D}||$ is defined as $|\mathcal{D}|$ except that applications of W and C do not increase the logical depth. We then find for both **G1c** and **G1i** $||\mathcal{D}^*|| \leq 2(||\mathcal{D}'|| + ||\mathcal{D}''||)$, although the proof has to proceed by induction on $|\mathcal{D}'| + |\mathcal{D}''|$. Let us illustrate the proof by the case of a multicut on an implication which is principal in both premises. Let $\mathcal{D}'$ be the deduction

$$\dfrac{\begin{array}{c}\mathcal{D}_0\\ \vdash_{d-1} \Gamma A \Rightarrow B(A{\to}B)^m\Delta\end{array}}{\vdash_d \Gamma \Rightarrow (A{\to}B)^{m+1}\Delta}$$

and let $\mathcal{D}''$ be the deduction

$$\dfrac{\begin{array}{c}\mathcal{D}_1\\ \vdash_{d'-1} \Gamma'(A{\to}B)^n \Rightarrow A\Delta'\end{array} \quad \begin{array}{c}\mathcal{D}_2\\ \vdash_{d'-1}\colon \Gamma'(A{\to}B)^n B \Rightarrow \Delta'\end{array}}{\vdash_{d'}\colon \Gamma'(A{\to}B)^{n+1} \Rightarrow \Delta'}$$

In the case where $m, n > 0$ we construct $\mathcal{D}_3', \mathcal{D}_4', \mathcal{D}_5'$, as in section 4.1.9. An easy computation yields bounds on their logical depth of $2d + 2d' - 2$ in each case, and then the final deduction has depth $2d + 2d'$:

$$\dfrac{\dfrac{\begin{array}{c}\mathcal{D}_3'\\ \vdash_{2d+2d'-2} \Gamma\Gamma' A \Rightarrow B\Delta\Delta'\end{array} \quad \begin{array}{c}\mathcal{D}_4'\\ \vdash_{2d+2d'-2} \Gamma\Gamma' \Rightarrow A\Delta\Delta'\end{array}}{\vdash_{2d+2d'-1} (\Gamma\Gamma')^2 \Rightarrow B(\Delta\Delta')^2} \quad \begin{array}{c}\mathcal{D}_5'\\ \vdash_{2d+2d'-2} \Gamma\Gamma' B \Rightarrow \Delta\Delta'\end{array}}{\vdash_{2d+2d'} (\Gamma\Gamma')^3 \Rightarrow (\Delta\Delta')^3}$$

etc.

(ii) In the absence of the $\rightarrow$-rules, we can use the estimate $||\mathcal{D}^*|| \leq (||\mathcal{D}'|| + ||\mathcal{D}''||)$. We can also use this estimate in the case of **G1c**, if we split R$\rightarrow$ into

$$\text{R}{\rightarrow_0}\ \frac{\Gamma, A \Rightarrow \Delta}{\Gamma \Rightarrow A \rightarrow B, \Delta} \qquad \text{R}{\rightarrow_1}\ \frac{\Gamma \Rightarrow B, \Delta}{\Gamma \Rightarrow A \rightarrow B, \Delta}$$

5.1.2. NOTATION. Let "hyp" be the *hyperexponential function* defined by

$$\text{hyp}(x, 0, z) = z, \quad \text{hyp}(x, Sy, z) = x^{\text{hyp}(x,y,z)}$$

We abbreviate

$$2^i_k := \text{hyp}(2, k, i), \quad 2_k := \text{hyp}(2, k, 1),$$

and similarly for $4^i_k, 4_k$. ⊠

5.1.3. THEOREM. *(Hyperexponential bounds on cut elimination) To each* $\mathcal{D}$ *in* **G3**[**mic**] + Cut *of cutrank* k *there is a cutfree* $\mathcal{D}^*$ *with the same conclusion, obtained by eliminating cuts from* $\mathcal{D}$*, such that*

$$|\mathcal{D}^*| \leq 2^{|\mathcal{D}|}_k \ (\textit{for } \mathbf{G3c} + \text{Cut}), \quad |\mathcal{D}^*| \leq 4^{|\mathcal{D}|}_k \ (\textit{for } \mathbf{G3[mi]} + \text{Cut}).$$

PROOF. We show by induction on $|\mathcal{D}|$ that, whenever $\text{cr}(\mathcal{D}) > 0$, then there is a $\mathcal{D}^*$ with $\text{cr}(\mathcal{D}^*) < \text{cr}(\mathcal{D})$, $|\mathcal{D}^*| \leq 2^{|\mathcal{D}|}$ (for **G3c**) or $|\mathcal{D}^*| \leq 4^{|\mathcal{D}|}$ (for **G3i**).

If $\mathcal{D}$ does not end with a cut, or ends with a cut of rank less than $\text{cr}(\mathcal{D})$, we can apply the IH to the immediate subdeduction(s) of the premise(s), and find (in the case of **G3c**)

$$|\mathcal{D}^*| = |\mathcal{D}^*_0| + 1 \leq 2^{|\mathcal{D}_0|} + 1 \leq 2^{|\mathcal{D}|} \text{ (1-premise rule)},$$
$$|\mathcal{D}^*| = \max(|\mathcal{D}^*_0|, |\mathcal{D}^*_1|) + 1 \leq \max(2^{|\mathcal{D}_0|}, 2^{|\mathcal{D}_1|}) + 1 \leq$$
$$2^{|\mathcal{D}_0|} + 2^{|\mathcal{D}_1|} \leq 2^{\max(|\mathcal{D}_0|,|\mathcal{D}_1|)+1} = 2^{|\mathcal{D}|} \text{ (2-premise rule)}.$$

There remains the case where $\mathcal{D}$ ends with a cut on A, $|A| + 1 = \text{cr}(\mathcal{D})$. Then we can apply the reduction lemma. ⊠

5.1.4. *The inversion-rule strategy*

It follows from known results (see, for example, the remark at the end of 6.11.1) that hyperexponential bounds are unavoidable in the case of predicate logic, in the sense that no bounded iteration of exponentiation can provide bounds for cut elimination.

On the other hand, if we restrict attention to propositional logic, considerable improvements in the estimate are possible, by using a different strategy for eliminating cuts. This points to an essential difference between normalization for natural deduction and cut elimination by the inversion-rule strategy. More about this in 6.9. Before explaining this strategy for the case of classical implication logic, we first give a definition:

DEFINITION. The *cutlength* $\text{cl}(\mathcal{D})$ of a deduction $\mathcal{D}$ is the maximum value of

$$\sum\{\text{s}(A) : A \text{ occurrence of cutformula in } \sigma\}$$

taken over all branches σ of the prooftree. ⊠

The strategy works as follows. We first show how to replace a deduction ending with a single cut on A (i.e. the deductions of the premises of the cut are cutfree) by a deduction of the same sequent, such that the cutlength decreases, i.e the cutlength of the new deduction is less than $\text{s}(A)$.

If the deduction ends with a cut on a prime formula, we use essentially the same transformations as in Gentzen's procedure; and if the cutformula is not atomic, we use inversion lemmas which permit us to replace the cut by cuts of lower rank. The Cut rule we use is *context-sharing* (for the reason behind this choice, see the remarks in the next section).

Then we define a transformation "Red" on arbitrary deductions by recursion on the construction of the deduction; this operator removes all uppermost cuts, i.e. all cuts without cuts above them.

NOTATION. Let us write $\mathcal{D} \vdash_n^d \Gamma \Rightarrow \Delta$ if $\mathcal{D}$ proves $\Gamma \Rightarrow \Delta$ with depth $\leq n$, cutlength $\leq d$. $\vdash_n^d \Gamma \Rightarrow \Delta$ if such a $\mathcal{D}$ exists. ⊠

5.1.5. The rules of $\to$**G3c** plus context-sharing Cut_{cs} can be stated as

$$\vdash_m^d \Gamma, P \Rightarrow P, \Delta \quad \text{for } P \text{ atomic, all } d \text{ and } m$$

$$\frac{\vdash_m^d \Gamma, A \Rightarrow B, \Delta}{\vdash_{m+1}^d \Gamma \Rightarrow A \to B, \Delta} \qquad \frac{\vdash_m^d \Gamma \Rightarrow A, \Delta \qquad \vdash_m^d \Gamma, B \Rightarrow \Delta}{\vdash_{m+1}^d \Gamma, A \to B \Rightarrow \Delta}$$

In addition we have Cut_{cs}:

$$\frac{\vdash_m^d \Gamma \Rightarrow A, \Delta \qquad \vdash_m^d \Gamma, A \Rightarrow \Delta}{\vdash_{m+1}^{d+\text{deg}(A)} \Gamma \Rightarrow \Delta}$$

As already noted, the Cut rule is context-sharing. The reason for this is that with ordinary Cut, the transformations of deductions of $\Gamma \Rightarrow \Delta$ used yield deductions of sequents $\Gamma' \Rightarrow \Delta'$ where contraction would be needed to get $\Gamma \Rightarrow \Delta$ back; and we have depth-preserving closure under contraction for deductions without Cut, but not when ordinary context-free Cut is present. By the use of Cut_{cs} the need for contraction is avoided (moreover, depth-preserving contraction is derivable for the system with Cut_{cs}).
We need the following results:

5.1.6. Lemma. *(Weakening, Contraction, Inversion lemma for* $\rightarrow$**G3c***)*

(i) If $\vdash^0_n \Gamma \Rightarrow \Delta$ *then* $\vdash^0_n \Gamma, A \Rightarrow \Delta$*;*

(ii) if $\vdash^0_n \Gamma\Gamma'\Gamma' \Rightarrow \Delta\Delta'\Delta'$ *then* $\vdash^0_n \Gamma\Gamma' \Rightarrow \Delta\Delta'$*;*

(iii) if $\vdash^0_n \Gamma, A \to B \Rightarrow \Delta$ *then* $\vdash^0_n \Gamma \Rightarrow A, \Delta$, $\vdash^0_n \Gamma, B \Rightarrow \Delta$*;*

(iv) if $\vdash^0_n \Gamma \Rightarrow A \to B, \Delta$ *then* $\vdash^0_n \Gamma, A \Rightarrow B, \Delta$*.*

Proof. This has been established before. ⊠

5.1.7. Lemma. *(Cut elimination lemma for* $\rightarrow$**G3c***)*

(i) If $\vdash^0_m \Gamma \Rightarrow P, \Delta$ *and* $\vdash^0_n \Gamma, P \Rightarrow \Delta$ *then* $\vdash^0_{n+m} \Gamma \Rightarrow \Delta$*;*

(ii) if $\vdash^0_n \Gamma \Rightarrow A \to B, \Delta$ *and* $\vdash^0_n \Gamma, A \to B \Rightarrow \Delta$ *then* $\vdash^{\mathrm{s}(A)+\mathrm{s}(B)}_{n+2} \Gamma \Rightarrow \Delta$*.*

Proof. (i) Let two cutfree deductions

$$\begin{array}{c} \mathcal{D}' \\ \Gamma \Rightarrow P, \Delta \end{array} \qquad\qquad \begin{array}{c} \mathcal{D}'' \\ \Gamma, P \Rightarrow \Delta \end{array}$$

be given. We prove (i) by induction on $n + m$.
Basis. $n + m = 0$. $\mathcal{D}', \mathcal{D}''$ are axioms. If P is not principal in either $\mathcal{D}'$ or $\mathcal{D}''$, then also $\Gamma \Rightarrow \Delta$ is an axiom, so $\vdash^0_0 \Gamma \Rightarrow \Delta$. If P is principal in $\mathcal{D}'$ and $\mathcal{D}''$, then $\Gamma \equiv \Gamma', P$ and $\Gamma'P \Rightarrow P\Delta$, and $\mathcal{D}''$ becomes $\Gamma'PP \Rightarrow \Delta$. One of the occurrences of P in $\mathcal{D}''$ is not principal, so $\Gamma'P \Rightarrow \Delta$ is also an axiom, hence again $\vdash^0_0 \Gamma \Rightarrow \Delta$.
Induction step. If $n + m = k + 1$, then at least one of $\mathcal{D}', \mathcal{D}''$ is not an axiom, say $\mathcal{D}'$. If $\mathcal{D}'$ ends with L$\rightarrow$, then $\Gamma \equiv \Gamma', A{\to}B$, and we have deductions $\mathcal{D}_0 \vdash^0_{m-1} \Gamma' \Rightarrow AP\Delta$, $\mathcal{D}_1 \vdash^0_{m-1} \Gamma'B \Rightarrow P\Delta$; by the inversion lemma, we also have deductions $\mathcal{D}_2 \vdash^0_n \Gamma'P \Rightarrow A\Delta$, $\mathcal{D}_3 \vdash^0_n \Gamma'PB \Rightarrow \Delta$. Combining $\mathcal{D}_0$ with $\mathcal{D}_2$, $\mathcal{D}_1$ with $\mathcal{D}_3$, the IH yields $\vdash^0_{n+m-1} \Gamma' \Rightarrow A\Delta$, $\vdash^0_{n+m-1} \Gamma'B \Rightarrow \Delta$; with L$\rightarrow$ it follows that $\vdash^0_{n+m} \Gamma'A{\to}B \Rightarrow \Delta$. The case where $\mathcal{D}'$ ends with R$\rightarrow$ is even simpler.

As to (ii), let $\mathcal{D}' \vdash^0_n \Gamma \Rightarrow A{\to}B, \Delta$ and $\mathcal{D}'' \vdash^0_n \Gamma, A{\to}B \Rightarrow \Delta$. By the inversion lemma there are $\mathcal{D}_0 \vdash^0_n \Gamma, A \Rightarrow B, \Delta$, $\mathcal{D}_1 \vdash^0_n \Gamma \Rightarrow A, \Delta$, $\mathcal{D}_2 \vdash^0_n \Gamma, B \Rightarrow \Delta$; then

$$\dfrac{\text{Cut}_{\text{cs}}\dfrac{\begin{array}{c}\mathcal{D}_0\\ \vdash^0_n \Gamma A \Rightarrow B\Delta\end{array} \quad \begin{array}{c}\mathcal{D}_1^*\\ \vdash^0_n \Gamma \Rightarrow AB\Delta\end{array}}{\vdash^{\mathrm{s}(A)}_{n+1} \Gamma \Rightarrow B\Delta} \qquad \begin{array}{c}\mathcal{D}_2\\ \vdash^0_n \Gamma B \Rightarrow \Delta\end{array}}{\vdash^{\mathrm{s}(A)+\mathrm{s}(B)}_{n+2} \Gamma \Rightarrow \Delta}\text{Cut}_{\text{cs}}$$

where $\mathcal{D}_1^*$ is obtained by weakening all sequents in $\mathcal{D}_1$ on the right with B (weakening lemma). ⊠

5.1.8. THEOREM. *(Numerical bounds on cut elimination, classical case)*

(i) *In* $\rightarrow$**G3c** + $\mathrm{Cut_{cs}}$, *if* $\vdash^d_n \Gamma \Rightarrow \Delta$, *then* $\vdash^{d-1}_{2n+1} \Gamma \Rightarrow \Delta$;

(ii) *for each* $\mathcal{D}$ *in* $\rightarrow$**G3c** + $\mathrm{Cut_{cs}}$ *there is a cutfree* $\mathcal{D}^*$ *with the same conclusion, obtained by eliminating cuts from* $\mathcal{D}$, *such that*

$$|\mathcal{D}^*| \leq (|\mathcal{D}| + 1)2^{\mathrm{cl}(\mathcal{D})};$$

(iii) *for each* $\mathcal{D}$ *in* $\rightarrow$**G3c** + $\mathrm{Cut_{cs}}$ *there is a cutfree* $\mathcal{D}^*$ *with the same conclusion, obtained by eliminating cuts from* $\mathcal{D}$, *such that*

$$|\mathcal{D}^*| \leq (|\mathcal{D}| + 1)2^{(|\mathcal{D}|+1)2^{\mathrm{cr}(\mathcal{D})}}.$$

PROOF. (i) We define a transformation Red on deductions $\mathcal{D}$ as follows.

(1) $\mathcal{D}$ is an axiom: $\mathrm{Red}(\mathcal{D}) := \mathcal{D}$.

(2) $\mathcal{D}$ is obtained from cutfree $\mathcal{D}_0, \mathcal{D}_1$ by a cut. By the cut elimination lemma we construct a new deduction $\mathrm{Red}(\mathcal{D})$ with $\mathrm{cl}(\mathrm{Red}(\mathcal{D})) < \mathrm{cl}(\mathcal{D})$. Also $|\mathrm{Red}(\mathcal{D})| \leq 2|\mathcal{D}| + 1$ which may be seen by inspection ($n + 2 \leq n + (n + 1)$ since $n > 0$).

(3) $\mathcal{D}$ is obtained from other deductions $\mathcal{D}_0$, or from $\mathcal{D}_0, \mathcal{D}_1$, by a logical rule or a cut which is not a top cut. Then $\mathrm{Red}(\mathcal{D})$ is obtained by applying the same rule to $\mathrm{Red}(\mathcal{D}_0)$, or to $\mathrm{Red}(\mathcal{D}_0)$, $\mathrm{Red}(\mathcal{D}_1)$.

(ii) follows from (i) by iteration. As to (iii), let $\mathcal{D} \vdash^d_n \Gamma \Rightarrow \Delta$. Then $\mathrm{cl}(\mathcal{D}) \leq p(n+1)$ where p is the maximum size of cutformulas in $\mathcal{D}$; $p \leq 2^{\mathrm{cr}(\mathcal{D})}$, hence

$$\mathrm{cl}(\mathcal{D}) \leq 2^{\mathrm{cr}(\mathcal{D})}(|\mathcal{D}| + 1).$$ ⊠

5.1.9. *The inversion-rule strategy for intuitionistic implication logic*

In principle, the same strategy works for intuitionistic $\rightarrow$**GKi** + $\mathrm{Cut_{cs}}$, but we now have much more work to do since we do not have such strong inversion properties as in the classical case.

5.1.10. DEFINITION. Let the weight $\mathrm{w}(A)$ of a formula A be defined as in 4.3.2. We define the *cutweight* $\mathrm{cw}(\mathcal{D})$ of a deduction $\mathcal{D}$ as the maximum of

$$\sum\{\mathrm{w}(A) : A \text{ occurrence of cutformula in } \sigma\},$$

σ ranging over the branches of $\mathcal{D}$. ⊠

The notion of weight has the property

$$\mathrm{w}(A \rightarrow B) + \mathrm{w}(B \rightarrow C) < \mathrm{w}((A \rightarrow B) \rightarrow C)$$

since, if $a = \mathrm{w}(A)$, $b = \mathrm{w}(B)$, $c = \mathrm{w}(C)$, we have $1 + ab + 1 + bc = 2 + (a + c)b \leq 2 + abc$ (since $a, b, c \geq 2$) $< 1 + c + abc = 1 + (1 + ab)c$. Moreover we recall that

$$\mathrm{w}(A) \leq 2^{\mathrm{s}(A)} \leq 2^{2^{|A|+1}}.$$

5.1.11. DEFINITION. We define $\vdash^d_m \Gamma \Rightarrow A$, "$\Gamma \Rightarrow A$ is derivable with depth at most m and cutweight at most d" as follows:

$$\vdash^d_m \Gamma, P \Rightarrow P \text{ for } P \text{ atomic, all } d \text{ and } m$$

$$\frac{\vdash^d_m \Gamma, A \Rightarrow B}{\vdash^d_{m+1} \Gamma \Rightarrow A \to B} \qquad \frac{\vdash^d_m \Gamma, A \to B \Rightarrow A \qquad \vdash^d_m \Gamma, A \to B, B \Rightarrow C}{\vdash^d_{m+1} \Gamma, A \to B \Rightarrow C}$$

$$\frac{\vdash^d_m \Gamma \Rightarrow A \qquad \vdash^d_m \Gamma, A \Rightarrow B}{\vdash^{d+\mathrm{w}(A)}_{m+1} \Gamma \Rightarrow B}$$ ⊠

Note that the Cut rule is context-sharing (as in the classical case), and that moreover we have chosen a variant of L→ where the principal formula $A \to B$ occurs in the antecedent of *both* premises (cf. 3.5.11). The reason for the choice of context-sharing Cut is the same as for the classical case. On the other hand, we might have used the ordinary L→ for **G3i**, but then we must prove a slightly stronger form of depth-preserving contraction, namely that depth-preserving contraction is also derivable in the presence of *context-sharing* Cut.

5.1.12. LEMMA. *(Weakening, Contraction and Inversion for* →**GKi***)*

(i) If $\vdash^0_m \Gamma \Rightarrow A$, *then* $\vdash^0_m \Gamma, \Delta \Rightarrow A$;

(ii) If $\vdash^0_m \Gamma, A, A \Rightarrow B$, *then* $\vdash^0_m \Gamma, A \Rightarrow B$.

(iii) If $\vdash^0_m \Gamma \Rightarrow A \to B$, *then* $\vdash^0_m \Gamma, A \Rightarrow B$;

(iv) If $\vdash^0_m \Gamma, A \to B \Rightarrow C$, *then* $\vdash^0_m \Gamma, B \Rightarrow C$.

(v) If $\vdash^0_m \Gamma, (A \to B) \to C \Rightarrow D$, *then* $\vdash^0_m \Gamma, A, B \to C \Rightarrow D$.

PROOF. We use induction on m. Proofs of the first four statements have been given before (see the end of 3.5.11). As to the last statement, let us consider two typical cases. Assume $(A \to B) \to C$ to be principal in the final step of the deduction of $\vdash^0_m \Gamma, (A \to B) \to C \Rightarrow D$. Then

$$\dfrac{\dfrac{\dfrac{\dfrac{\vdash^0_{m-1} \Gamma, (A \to B) \to C \Rightarrow A \to B}{\vdash^0_{m-1} \Gamma, A, (A \to B) \to C \Rightarrow B}\,\mathrm{Inv}}{\vdash^0_{m-1} \Gamma, A, A, B \to C \Rightarrow B}\,\mathrm{IH}}{\vdash^0_{m-1} \Gamma, A, B \to C \Rightarrow B}\,\mathrm{LC} \qquad \dfrac{\vdash^0_{m-1} \Gamma, (A \to B) \to C, C \Rightarrow D}{\vdash^0_{m-1} \Gamma, A, B \to C, C \Rightarrow D}\,\mathrm{IH}}{\vdash^0_m \Gamma, A, B \to C \Rightarrow D}\,\mathrm{L}{\to}$$

In this "pseudo-prooftree" the dashed lines indicate transformations of deductions given by the lemmas on weakening ("LW"), contraction ("LC") and inversion ("Inv") as well as the IH.

If $(A \to B) \to C$ is not principal, $\vdash^0_m \Gamma, (A \to B) \to C \Rightarrow D$ follows from a one-premise rule or a two-premise rule with $(A \to B) \to C$ in the context. We consider the case of the one-premise rule; the case of the two-premise rule is similar. So let $\vdash^0_m \Gamma, (A \to B) \to C \Rightarrow D$ be obtained from $\vdash^0_{m-1} \Gamma', (A \to B) \Rightarrow D'$. Then

$$\dfrac{\dfrac{\vdash^0_{m-1} \Gamma', (A \to B) \to C \Rightarrow D'}{\vdash^0_{m-1} \Gamma', A, B \to C \Rightarrow D'}\,\text{IH}}{\vdash^0_m \Gamma, A, B \to C \Rightarrow D}$$

⊠

5.1.13. LEMMA. *(Cut elimination lemma for* $\to$**GKi**+ $\mathrm{Cut_{cs}}$*)*

(i) If $\vdash^0_n \Gamma \Rightarrow P$ *and* $\vdash^0_m \Gamma, P \Rightarrow B$, *then* $\vdash^0_{n+m} \Gamma \Rightarrow B$.

(ii) If $\vdash^0_n \Gamma \Rightarrow P \to B$ *and* $\vdash^0_m \Gamma, P \to B \Rightarrow C$, *and* $m \leq n$, *then* $\vdash^{\mathrm{w}(B)}_{n+m} \Gamma \Rightarrow C$.

(iii) If $\vdash^0_n \Gamma \Rightarrow P \to B$ *and* $\vdash^0_n \Gamma, B \Rightarrow C$ *and* $\vdash^0_m \Gamma, P \to B \Rightarrow P$ *and* $m \leq n$ *then* $\vdash^{\mathrm{w}(B)}_{n+m+1} \Gamma \Rightarrow C$.

(iv) If $\vdash^0_n \Gamma \Rightarrow (A \to B) \to C$ *and* $\vdash^0_m \Gamma, (A \to B) \to C \Rightarrow D$, *then* $\vdash^d_{n+m+2} \Gamma \Rightarrow D$ *with* $d = \mathrm{w}(A \to B) + \mathrm{w}(B \to C)$.

PROOF. (i) is proved as before.

(ii) and (iii) are proved simultaneously by induction on m. The case where $m = 0$ we leave to the reader.

Induction step for (ii): Assume $\vdash^0_n \Gamma \Rightarrow P \to B$, $\mathcal{D} \vdash^0_m \Gamma, P \to B \Rightarrow C$. Let us first assume that $P \to B$ is principal in the last rule of $\mathcal{D}$. Then we argue as represented schematically below (first step on the left by (iv) of the preceding lemma):

$$\dfrac{\dfrac{\vdash^0_m \Gamma, P \to B \Rightarrow C}{\vdash^0_n \Gamma, B \Rightarrow C}\,(m \leq n) \qquad \vdash^0_n \Gamma \Rightarrow P \to B \qquad \vdash^0_{m-1} \Gamma, P \to B \Rightarrow P}{\vdash^{\mathrm{w}(B)}_{n+m} \Gamma \Rightarrow C}\,\text{IH(iii)}$$

If $\mathcal{D}$ ends with an application of R$\to$, say

$$\dfrac{\vdash^0_{m-1} \Gamma, A, P \to B \Rightarrow A'}{\vdash^0_m \Gamma, P \to B \Rightarrow A \to A'}$$

where C in the statement of (ii) is $A \to A'$, then

$$\dfrac{\dfrac{\vdash^0_n \Gamma \Rightarrow P \to B}{\vdash^0_n \Gamma, A \Rightarrow P \to B}\,\text{LW} \qquad \vdash^0_{m-1} \Gamma, A, P \to B \Rightarrow A'}{\dfrac{\vdash^{w(B)}_{n+m-1} \Gamma, A \Rightarrow A'}{\vdash^{w(B)}_{n+m} \Gamma \Rightarrow A \to A'}\,\text{R}\to}\,\text{IH(ii)}$$

Finally, let $\vdash^0_m \Gamma, P \to B \Rightarrow C$ follow by L→, with principal formula $A \to A'$ distinct from $P \to B$, so $\Gamma \equiv \Gamma', A \to A'$. Then

$$\dfrac{\dfrac{\vdash^0_n \Gamma \Rightarrow P{\to}B}{\vdash^0_n \Gamma, A' \Rightarrow P{\to}B}\,\text{LW} \qquad \vdash^0_{m-1} \Gamma, A', P{\to}B \Rightarrow C}{\vdash^{w(B)}_{n+m-1} \Gamma, A' \Rightarrow C}\,\text{IH(ii)}$$

and

$$\dfrac{\vdash^0_n \Gamma \Rightarrow P{\to}B \qquad \vdash^0_{m-1} \Gamma, P{\to}B \Rightarrow A}{\vdash^{w(B)}_{n+m-1} \Gamma \Rightarrow A}\,\text{IH(ii)}$$

and the conclusions may be combined by

$$\dfrac{\vdash^{w(B)}_{n+m-1} \Gamma \Rightarrow A \qquad \vdash^{w(B)}_{n+m-1} \Gamma, A' \Rightarrow C}{\vdash^{w(B)}_{n+m} \Gamma \Rightarrow C}\,\text{L}\to$$

Induction step for (iii): Suppose $\vdash^0_n \Gamma \to P \Rightarrow B$, $\vdash^0_n \Gamma, B \Rightarrow C$ and $\mathcal{D} \vdash^0_m \Gamma, P \to B \Rightarrow P$, $m \leq n$. Suppose $P \to B$ to have been principal in the last rule of $\mathcal{D}$, then

$$\dfrac{\vdash^0_n \Gamma \Rightarrow P{\to}B \qquad \vdash^0_n \Gamma, B \Rightarrow C \qquad \vdash^0_{m-1} \Gamma, P \to B \Rightarrow P}{\vdash^{w(B)}_{n+m} \Gamma \Rightarrow C}\,\text{IH(iii)}$$

From this obviously $\vdash^{w(B)}_{n+m+1} \Gamma \Rightarrow C$. Now let the final rule of $\mathcal{D}$ be L→ with principal formula $A \to A'$ distinct from $P \to B$ and let $\Gamma \equiv \Gamma', A{\to}A'$. Then

$$\dfrac{\dfrac{\vdash^0_n \Gamma \Rightarrow P \to B}{\vdash^0_n \Gamma, A' \Rightarrow P \to B}\,\text{LW} \qquad \dfrac{\vdash^0_n \Gamma, B \Rightarrow C}{\vdash^0_n \Gamma, A', B \Rightarrow C}\,\text{LW} \qquad \vdash^0_{m-1} \Gamma, A', P{\to}B \Rightarrow P}{\vdash^{w(B)}_{n+m} \Gamma, A' \Rightarrow C}\,\text{IH(iii)}$$

and

$$\dfrac{\vdash^0_n \Gamma \Rightarrow P \to B \qquad \vdash^0_{m-1} \Gamma, P \to B \Rightarrow A}{\vdash^{w(B)}_{n+m-1} \Gamma \Rightarrow A}\,\text{IH(ii)}$$

from which $\vdash^{\mathrm{w}(B)}_{n+m} \Gamma \Rightarrow A$. Combining these with a final L→ yield

$$\vdash^{\mathrm{w}(B)}_{n+m+1} \Gamma \Rightarrow C.$$

Induction step for (iv). Assume $\vdash^0_n \Gamma \Rightarrow (A \to B) \to C$, $\mathcal{D} \vdash^0_m \Gamma, (A \to B) \to C \Rightarrow D$. The proof again proceeds by induction on m. We consider only the inductive subcase where $(A \to B) \to C$ is principal in the last rule of $\mathcal{D}$. Then

$$\dfrac{\dfrac{\dfrac{\dfrac{\vdash^0_n \Gamma \Rightarrow (A \to B) \to C}{\vdash^0_n \Gamma, A \to B \Rightarrow C}\,\text{Inv}}{\vdash^0_n \Gamma, B \Rightarrow C}\,\text{Inv}}{\vdash^0_{n+1} \Gamma \Rightarrow B \to C}\,\text{R}\!\to \qquad \dfrac{\dfrac{\dfrac{\dfrac{\vdash^0_{m-1} \Gamma, (A \to B) \to C \Rightarrow A \to B}{\vdash^0_{m-1} \Gamma, A, (A \to B) \to C \Rightarrow B}\,\text{Inv}}{\vdash^0_{m-1} \Gamma, A, A, B \to C \Rightarrow B}\,\text{Inv}}{\vdash^0_{m-1} \Gamma, A, B \to C \Rightarrow B}\,\text{LC}}{\vdash^0_m \Gamma, B \to C \Rightarrow A \to B}\,\text{R}\!\to}{\vdash^{\mathrm{w}(B \to C)}_{\max(n+2,m+1)} \Gamma \Rightarrow A \to B}\,\text{Cut}$$

and

$$\dfrac{\dfrac{\vdash^0_n \Gamma \Rightarrow (A \to B) \to C}{\vdash^0_n \Gamma, A \to B \Rightarrow C}\,\text{Inv} \qquad \dfrac{\dfrac{\vdash^0_{m-1} \Gamma, (A \to B) \to C \Rightarrow D}{\vdash^0_{m-1} \Gamma, C \Rightarrow D}\,\text{Inv}}{\vdash^0_{m-1} \Gamma, A \to B, C \Rightarrow D}\,\text{LW}}{\vdash^{\mathrm{w}(C)}_{\max(n+1,m)} \Gamma, A \to B \Rightarrow D}\,\text{Cut}$$

which then may be combined by a final application of Cut with result

$$\vdash^d_{n+m+2} \Gamma \Rightarrow D, \quad \text{where } d = \mathrm{w}(A{\to}B) + \mathrm{w}(B{\to}C). \qquad \boxtimes$$

5.1.13A. ♠ Supply proofs for the missing cases in proof of the preceding lemma.

5.1.13B. ♠ Check where changes in the argument are needed if we replace L→ of →**GKi** by standard L→ of **G3i**.

5.1.13C. ♠* Let us write $\vdash^0_n \Gamma \Rightarrow t{:}\,A$, where Γ is a sequence of typed variables $x_1{:}\,A_1, \ldots, x_n{:}\,A_n$, if there is a deduction $\mathcal{D}$ in **GKi** with $|\mathcal{D}| \leq n$, such that under the obvious standard assignment (cf. 3.3.4) of natural deduction terms to proofs in **GKi**, $\mathcal{D}$ gets assigned $t{:}\,A$. Show the following

(i) If $\vdash^0_n \Gamma, x{:}\,A, y{:}\,A \Rightarrow t{:}\,B$, then $\vdash^0_n \Gamma, x{:}\,A \Rightarrow t[x, y/x, x]{:}\,B$.

(ii) If $\vdash^0_n \Gamma \Rightarrow t{:}\,A \to B$, then $t \equiv \lambda x^A.t'$ and $\vdash^0_n \Gamma, x{:}\,A \Rightarrow t'{:}\,B$.

(iii) If $\vdash^0_n \Gamma, u{:}\,(A{\to}B){\to}C \Rightarrow t{:}\,D$, then there is a $t' =_\beta t[u/\lambda x^{A\to B}.z^{B\to C}(xy^A)]$ such that $\vdash^0_n \Gamma, y{:}\,A, z{:}\,B{\to}C \Rightarrow t'{:}\,D$.

5.1.14. THEOREM. *For deductions in* $\to$**G3i** + $\mathrm{Cut_{cs}}$ *as described, we have*

(i) *If* $\vdash_n^{d+1} \Gamma \Rightarrow A$ *then* $\vdash_{2n+1}^{d} \Gamma \Rightarrow A$;

(ii) *for all* $\mathcal{D}$ *there is a cutfree* $\mathcal{D}^*$ *with the same conclusion, obtained by eliminating cuts, such that*

$$|\mathcal{D}^*| \leq (|\mathcal{D}| + 1)2^{\mathrm{cw}(\mathcal{D})}.$$

(iii) *For all* $\mathcal{D}$ *there is a cutfree* $\mathcal{D}^*$ *with the same conclusion, obtained by eliminating cuts, such that*

$$|\mathcal{D}^*| \leq (|\mathcal{D}| + 1)2^{(|\mathcal{D}|+1)2^{2^{\mathrm{cr}(\mathcal{D})}}}.$$

PROOF. It suffices to prove the theorem for $\to$**GKi**. This is because deductions in $\to$**G3i** plus context-sharing Cut are easily transformed by dp-closure under weakening into deductions in **GKi** plus $\mathrm{Cut_{cs}}$, preserving depth and the measure for cuts. On cutfree deductions of the variant we can apply an inverse transformation with the help of dp-admissible weakening and contraction.

(i), (ii) are proved as in the classical case: we define an operator Red on deductions $\mathcal{D}$ which removes all top cuts from $\mathcal{D}$ in such a way that the cutweight (instead of the cutlength, as in the classical case) is reduced. The crucial case is the treatment of a top cut; for this we use the cut elimination lemma, (i), (ii) and (iv).

As to the proof of (iii),

$$|\mathcal{D}^*| \leq 2^{\mathrm{cw}(\mathcal{D})}(|\mathcal{D}| + 1) \leq 2^{m(|\mathcal{D})|+1)}(|\mathcal{D}| + 1),$$

where m is the maximum weight of cutformulas in $\mathcal{D}$, say $m = \mathrm{w}(A)$. We observe that $\mathrm{w}(A) \leq 2^{\mathrm{s}(A)} \leq 2^{2^{\mathrm{cr}|\mathcal{D}|}}$. ⊠

5.2 Size and cut elimination

In this section we present an example of an infinite sequence $\langle \mathcal{S}_n \rangle_n$ of sequents, with deductions in **GS3p** + Cut of size linear in n, while any cutfree proof of $\mathcal{S}_n$ in **GS3p** has size $\geq 2^n$. In fact, the result will be slightly stronger than just stated, since we shall consider a version of **GS3p** with the logical axiom generalized to principal formulas of arbitrary logical complexity:

$$\Gamma, A, \neg A$$

We call this version **GS3p***.

The result will entail a corresponding result for **G3p[mi]**, also with the axioms generalized. As an auxiliary system we shall use the following "strictly increasing" system **G** for classical propositional logic (which combines features of GK- and GS-formalisms):

$$\text{Ax}\quad \Gamma, A, \neg A$$

$$\text{R}\vee\ \frac{\Gamma, A, B, A \vee B}{\Gamma, A \vee B} \qquad \text{R}\wedge\ \frac{\Gamma, A \wedge B, A \qquad \Gamma, A \wedge B, B}{\Gamma, A \wedge B}$$

In what follows we use the notations $\vdash_{s \leq n}, \vdash^{\mathbf{S}}_{s \leq n}$ which have been introduced in 3.4.3. The following lemma is easily established, and its proof left as an exercise.

5.2.1. LEMMA. *If* **GS3p*** $\vdash_{s \leq n} \Gamma$ *then* **G** $\vdash_{s \leq n} \Gamma$.

5.2.2. DEFINITION. We define a sequent $\mathcal{S}_n$ for each positive n as follows:

$$\mathcal{S}_n := \mathcal{T}_n, P_{n+1}, Q_{n+1},$$

where

$$\begin{array}{l} \mathcal{T}_n := A_1 \wedge B_1, \ldots, A_{n+1} \wedge B_{n+1}; \\ F_0 := (P \vee \neg P), \quad F_{n+1} := (F_n \wedge (P_{n+1} \vee Q_{n+1})); \\ A_{n+1} := F_n \wedge \neg P_{n+1}, \quad B_{n+1} := F_n \wedge \neg Q_{n+1}. \end{array}$$

For use later on we also define the following abbreviations for $i \leq n$:

$$\begin{array}{l} \mathcal{T}_{n,i} := A_1 \wedge B_1, \ldots, A_{i-1} \wedge B_{i-1}, A_{i+1} \wedge B_{i+1}, \ldots, A_{n+1} \wedge B_{n+1}, \\ \mathcal{S}_{n,i} := \mathcal{T}_{n,i}, \neg P_i, P_{n+1}, Q_{n+1} \end{array}$$

⊠

It is not hard to see that $\mathcal{S}_n$ must be valid for all n: note that $\mathcal{S}_{n+1}$ is equivalent to $\neg(A_1 \wedge B_1) \wedge \ldots \wedge \neg(A_{n+1} \wedge B_{n+1}) \to P_{n+1} \vee Q_{n+1}$, and that $\neg(A_1 \wedge B_1)$ is equivalent to $P_1 \vee Q_1$, $\neg(A_i \wedge B_i)$ is equivalent to $\bigwedge_{1 \leq k \leq i}(P_i \vee Q_i) \to P_{i+1} \vee Q_{i+1}$. The validity follows more formally from the arguments below.

To simplify the proofs of the next few lemmas, we observe that if we can produce a derivation using instead of R$\wedge$ the more general rule

$$\frac{\Gamma, \Gamma', A \qquad \Gamma', \Gamma'', B}{\Gamma, \Gamma', \Gamma'', A \wedge B}$$

which generalizes both the context-sharing and the context-free versions of R$\wedge$, there is a derivation of the *same size* in **GS3p***: simply use the closure under weakening to transform a proof using the hybrid rule into a proof using only the context-sharing R$\wedge$.

5.2.3. LEMMA. *In* **GS3p*** *there are a cutfree deduction* $\mathcal{D}_k$ *of size* 7 *of*

$$\neg F_k, A_{k+1} \wedge B_{k+1}, P_{k+1}, Q_{k+1},$$

and a cutfree deduction $\mathcal{E}_k$ *of size* 10 *of*

$$\neg F_k, A_{k+1} \wedge B_{k+1}, F_{k+1}.$$

PROOF. We take for $\mathcal{D}_k$

$$\dfrac{\dfrac{\neg F_k, F_k \quad \neg F_k, P_{k+1}, \neg P_{k+1}}{\neg F_k, A_{k+1}, P_{k+1}} \quad \dfrac{\neg F_k, F_k \quad \neg F_k, Q_{k+1}, \neg Q_{k+1}}{\neg F_k, B_{k+1}, Q_{k+1}}}{\neg F_k, A_{k+1} \wedge B_{k+1}, P_{k+1}, Q_{k+1}}$$

and for $\mathcal{E}_k$ we take

$$\dfrac{F_k, \neg F_k \quad \dfrac{\dfrac{\mathcal{D}_k}{\neg F_k, A_{k+1} \wedge B_{k+1}, P_{k+1}, Q_{k+1}}}{\neg F_k, A_{k+1} \wedge B_{k+1}, P_{k+1} \vee Q_{k+1}}}{\neg F_k, A_{k+1} \wedge B_{k+1}, F_{k+1}}$$

5.2.4. PROPOSITION. $\mathcal{S}_n$ *has a deduction in* **GS3p*** + Cut *of size* $11n + 10$.

PROOF. We first construct a deduction $\mathcal{F}_n$ (with cuts) of size $11n + 2$ of

$$A_1 \wedge B_1, \ldots, A_n \wedge B_n, F_n.$$

We use induction on n. The basis case for $n = 0$

$$\dfrac{P, \neg P}{F_0}$$

has size 2, and once $\mathcal{F}_n$ of size $11n + 2$ has been obtained, we find $\mathcal{F}_{n+1}$ as

$$\dfrac{\begin{matrix}\mathcal{F}_n \\ A_1 \wedge B_1, \ldots, A_n \wedge B_n, F_n\end{matrix} \quad \begin{matrix}\mathcal{E}_n \\ \neg F_n, A_{n+1} \wedge B_{n+1}, F_{n+1}\end{matrix}}{A_1 \wedge B_1, \ldots, A_n \wedge B_n, A_{n+1} \wedge B_{n+1}, F_{n+1}} \text{Cut}$$

which then has size $11n + 2 + 10 + 1 = 11(n + 1) + 2$. Now take for the deduction of $\mathcal{S}_n$

$$\dfrac{\begin{matrix}\mathcal{F}_n \\ F_n, A_1 \wedge B_1, \ldots, A_n \wedge B_n\end{matrix} \quad \begin{matrix}\mathcal{D}_n \\ \neg F_n, A_{n+1} \wedge B_{n+1}, P_{n+1}, Q_{n+1}\end{matrix}}{\mathcal{S}_n}$$

with size $11n + 2 + 7 + 1 = 11n + 10$. ⊠

5.2.5. DEFINITION. A cutfree deduction $\mathcal{D}$ in **G** is called *strict*, if along any branch of $\mathcal{D}$ no formula is introduced more than once. ⊠

In the remainder of this section until the final theorem we consider only cutfree proofs in **G**.

5.2.6. LEMMA. *If $\mathcal{D}$ is a [strict] cutfree derivation of Γ, A, A with $\mathrm{s}(\mathcal{D}) \leq n$, there is also a [strict] cutfree derivation $\mathcal{D}'$ of Γ, A with $\mathrm{s}(\mathcal{D}') \leq n$.*

PROOF. Straightforward by induction on the size of $\mathcal{D}$; since the conclusion in the rules of **G** is always contained in the premise(s), we can delete an occurrence of A throughout the deduction; strictness is not affected by this operation. ⊠

5.2.7. LEMMA. *(Inversion in* **G***) Let $\mathcal{D} \vdash_{\mathrm{s} \leq n} \Gamma, A$, and A not be principal in an axiom, then*

(i) if $A \equiv B \wedge C$ then there are deductions $\mathcal{D}' \vdash_{\mathrm{s} \leq n} \Gamma, A$ and $\mathcal{D}'' \vdash_{\mathrm{s} \leq n} \Gamma, B$,

(ii) if $A \equiv B \vee C$ then there is a deduction $\mathcal{D}' \vdash_{\mathrm{s} \leq n} \Gamma, A, B$.

Strictness is preserved in the construction of $\mathcal{D}', \mathcal{D}''$ from $\mathcal{D}$.

PROOF. Assume $\mathcal{D} \vdash_{\mathrm{s} \leq n} \Gamma, A \wedge B$, where $A \wedge B$ is not principal in axioms of $\mathcal{D}$. We use induction on the size of $\mathcal{D}$. If $\mathcal{D}$ is an axiom, $A \wedge B$ is not principal, so Γ, A and Γ, B are again axioms. If the last rule in $\mathcal{D}$ introduces $A \wedge B$, $\mathcal{D}$ has immediate subdeductions $\mathcal{D}_A$, $\mathcal{D}_B$ deriving $\Gamma, A \wedge B, A$ and $\Gamma, A \wedge B, B$ respectively.

Apply the induction hypothesis to $\mathcal{D}_A$, this yields a proof of Γ, A, A; then appeal to closure under contraction, etc.

If $\mathcal{D}$ ends with another rule, not introducing $A \wedge B$, we apply the induction hypothesis to the immediate subderivation(s).

Case (ii) of the lemma is proved similarly. ⊠

5.2.8. LEMMA. *If $\mathcal{D} \vdash_n \Gamma$, there is a strict $\mathcal{D}'$ such that $\mathcal{D}' \vdash_{\mathrm{s} \leq n} \Gamma$.*

PROOF. We show how to reduce the number of violations of strictness in $\mathcal{D}$. If $\mathcal{D}$ is not strict, there are a formula A and a branch σ in $\mathcal{D}$ such that A is introduced at least twice along σ. Let us consider the two lowest introductions α and β of A in σ, and assume $A \equiv B \wedge C$ (the case $A \equiv B \vee C$ is similar but simpler). If σ passes through the left premise of the lowest introduction of α (the case of the right premise is symmetric), $\mathcal{D}$ has the form

$$\dfrac{\dfrac{\dfrac{\begin{matrix}\mathcal{D}_2 \\ \Gamma_2, B \wedge C, B, B\end{matrix} \quad \begin{matrix}\mathcal{E}_2 \\ \Gamma_2, B \wedge C, B, C\end{matrix}}{\Gamma_2, B \wedge C, B}\beta}{\begin{matrix}\mathcal{D}_1 \\ \Gamma_1, B \wedge C, B\end{matrix}} \qquad \begin{matrix}\mathcal{E}_1 \\ \Gamma_1, B \wedge C, C\end{matrix}}{\begin{matrix}\Gamma_1, B \wedge C \\ \mathcal{D}_0 \\ \Gamma\end{matrix}}\alpha$$

with σ passing through $\Gamma_2, B \wedge C, B$ and $\Gamma_1, B \wedge C, B$, and no introductions of $B \wedge C$ along σ in $\mathcal{D}_1$ and $\mathcal{D}_0$. By closure under contraction we find from

$\mathcal{D}_2$ a $\mathcal{D}'_2$ such that

$$\mathcal{D}'_2 \vdash_{s \leq s(\mathcal{D}_2)} \Gamma_2, B \wedge C, B.$$

Now replace the subdeduction with conclusion $\Gamma_2, B \wedge C, B$ in $\mathcal{D}$ by $\mathcal{D}'_2$. The resulting proof has fewer violations of strictness and is smaller than $\mathcal{D}$. Repeating this we finally arrive at a strict proof $\mathcal{D}'$ with $s(\mathcal{D}') \leq s(\mathcal{D})$. ⊠

5.2.9. DEFINITION. If a formula A is not principal in the axioms of a deduction $\mathcal{D}$, and A is nowhere introduced in $\mathcal{D}$, we say that A is *passive.* ⊠

The following lemma is immediate.

LEMMA. *If $\mathcal{D} \vdash_{s \leq n} \Gamma, A$ with A passive in $\mathcal{D}$, then $\mathcal{D}[-A]s \vdash_{s \leq n} \Gamma$, where $\mathcal{D}[-A]$ is the deduction obtained by deleting one occurrence of A from all sequents in $\mathcal{D}$.*

If a strict $\mathcal{D}$ ends with an introduction of A, and A is not principal in any axiom of $\mathcal{D}$, then A is passive in the immediate subdeductions of $\mathcal{D}$.

5.2.10. LEMMA. *In any deduction of $\mathcal{S}_n, \mathcal{S}_{n,i}$ the formulas $A_k \wedge B_k$, A_k, B_k, F_k, $P_k \vee Q_k$ cannot occur as principal formula in an axiom.*

PROOF. Each of the subformulas listed has only positive occurrences in the sequent $\mathcal{S}_n, \mathcal{S}_{n,i}$, which means that everywhere in the deduction they can only occur positively. ⊠

5.2.11. LEMMA. *Let $1 \leq i \leq n$. If $\vdash_{s \leq n} \mathcal{S}_{n+1,i}$ then $\vdash_{s \leq n} \mathcal{S}_n$.*

PROOF. The rules and axioms in a cutfree deduction of $\mathcal{S}_{n+1,i}$ can use only subformulas of $\mathcal{S}_{n+1,i}$. On multisets of subformulas of $\mathcal{S}_{n+1,i}$ we define, for a fixed i, $1 \leq i \leq n+1$, a mapping * which is essentially nothing else but the erasing of $P_i, \neg P_i, Q_i$ everywhere. For the empty multiset Λ we put

$$\Lambda^* := \Lambda,$$

and for the singleton-multisets, which we identify with formulas, we put

$$\begin{array}{l}
L^* := \Lambda \text{ if } L \in \{P_i, Q_i, \neg P_i\},\ L^* := L \text{ for other literals } L; \\
(P_i \vee Q_i)^* := \Lambda,\ ^* \text{ is the identity on other disjunctions,} \\
F_k^* := F_k \text{ for } k < i,\ F_i^* := F_{i-1},\ F_{k+1}^* := F_k^* \wedge (P_k \vee Q_k) \text{ for } k > i; \\
A_k^* := F_{k-1}^* \wedge \neg P_k,\ B_k^* := F_{k-1}^* \wedge \neg Q_k\ \ (k \neq i); \\
(A_k \wedge B_k)^* := A_k^* \wedge B_k^*\ \ (k \neq i).
\end{array}$$

Finally, for multisets of arbitrary size we put

$$(\Gamma, A)^* := \Gamma^*, A^*.$$

Assume Δ to be a strict cutfree deduction such that $\mathcal{D} \vdash_{s \leq n} \mathcal{S}_{n,i}$. We transform $\mathcal{D}$ into a deduction $\mathcal{D}^*$; we do this by defining $*$ on subdeductions of $\mathcal{D}$, using recursion on the depth of $\mathcal{D}$.

(i) Any subdeduction with one or more of $P_i, Q_i, \neg P_i, P_i \vee Q_i$ in the conclusion is mapped under $*$ to the empty deduction. Observe that introductions of $P_i \vee Q_i$, and axioms with $P_i, \neg P_i$ as principal formulas can occur in $\mathcal{D}$ only above introductions of F_i, hence they will be discarded.

(ii) Axioms Γ not having P_i or Q_i as principal formula are translated into axioms Γ^*.

(iii) Subdeductions of the form

$$\dfrac{\begin{matrix}\mathcal{D}_1 \\ \Delta, F_i, F_{i-1}\end{matrix} \qquad \begin{matrix}\mathcal{D}_2 \\ \Delta, F_i, P_i \vee Q_i\end{matrix}}{\Delta, F_i} \quad \text{are replaced by} \quad \begin{matrix}\mathcal{D}_1^*[-F_i^*] \\ \Delta^*, F_{i-1}^*\end{matrix}$$

Note that the whole subdeduction on the right is deleted by the process, in keeping with the fact that such subdeductions are mapped to the empty deduction. In $\mathcal{D}_1$ the formula F_i is passive, since $\mathcal{D}_1$ is part of a strict deduction and F_i cannot be principal in an axiom (5.2.9). This property is preserved under $*$.

(iv) In all other cases

$$\dfrac{\begin{matrix}\mathcal{D}_1 \\ \Delta'\end{matrix} \quad \begin{matrix}\mathcal{D}_2 \\ \Delta''\end{matrix}}{\Delta} \quad \text{is translated as} \quad \dfrac{\begin{matrix}\mathcal{D}_1^* \\ \Delta'^*\end{matrix} \quad \begin{matrix}\mathcal{D}_2^* \\ \Delta''^*\end{matrix}}{\Delta^*}$$

and

$$\dfrac{\begin{matrix}\mathcal{D}_1 \\ \Delta'\end{matrix}}{\Delta} \quad \text{is translated as} \quad \dfrac{\begin{matrix}\mathcal{D}_1^* \\ \Delta'^*\end{matrix}}{\Delta^*}$$

The resulting structure is a correct deduction, since axioms, when not discarded, are translated into axioms, and applications of rules are translated into correct applications of rules. We end up with a deduction of

$$A_1 \wedge B_1, \ldots, A_{i-1} \wedge B_{i-1}, A_{i+1}^* \wedge B_{i+1}^*, \ldots, A_{n+2}^* \wedge B_{n+2}^*, \neg P_{n+2}, \neg Q_{n+2}.$$

By renaming propositional variables $P_{k+1} \mapsto P_k$, $Q_{k+1} \mapsto Q_k$ $(k \geq i)$, leaving the rest invariant, we obtain a deduction of size $\leq n$ of $\mathcal{S}_{n+1}$. ⊠

5.2.12. PROPOSITION. *If $\mathcal{D}$ is a cutfree derivation of $\mathcal{S}_n$, then* $\mathrm{s}(\mathcal{D}) \geq 2^{n+1}$ *for all* $n > 0$.

PROOF. Since $\mathcal{S}_0 \equiv A_1 \wedge B_1, P_1, Q_1$ is not an axiom, any proof of this sequent must have size ≥ 2.

Assume now the statement of the theorem to have been proved for $\mathcal{S}_n$. Let $\mathcal{D}$ be a strict cutfree proof of $\mathcal{S}_{n+1}$, i.e. of

$$A_1 \wedge B_1, \ldots, A_{n+2} \wedge B_{n+2}, P_{n+2}, Q_{n+2}.$$

$\mathcal{D}$ must necessarily end with the introduction of $A_i \wedge B_i$ for some i. We distinguish two cases.

Case 1. $\mathcal{D}$ ends with the introduction of $A_{n+2} \wedge B_{n+2}$, so there are $\mathcal{D}_A, \mathcal{D}_B$ such that

$$\mathcal{D}_A \vdash_{\mathrm{s} \leq m} \mathcal{T}_n, A_{n+2}, P_{n+2}, Q_{n+2}, \quad \mathcal{D}_B \vdash_{\mathrm{s} \leq m'} \mathcal{T}_n, B_{n+2}, P_{n+2}, Q_{n+2},$$
$$\mathrm{s}(\mathcal{D}) = m + m' + 1.$$

In $\mathcal{D}_A$ the formula A_{n+2} cannot be principal in an axiom (lemma 5.2.10) hence by lemma 5.2.7

$$\vdash_m \mathcal{T}_n, F_{n+1}, P_{n+2}, Q_{n+2}.$$

Since P_{n+2}, Q_{n+2} do occur only once in this sequent, they cannot be principal in axioms, hence $\vdash_{\mathrm{s} \leq m} \mathcal{T}_n, F_{n+1}$, i.e. $\vdash_{\mathrm{s} \leq m} \mathcal{T}_n, F_n \wedge (P_{n+1} \vee Q_{n+1})$; since F_n and $P_{n+1} \vee Q_{n+1}$ cannot be principal in axioms, we find, again by lemma 5.2.7, that there must be a deduction showing $\vdash_{\mathrm{s} \leq m} \mathcal{T}_n, P_{n+1}, Q_{n+1}$. Since $\mathcal{T}_n, P_{n+1}, Q_{n+1}$ is $\mathcal{S}_n$, we find by induction hypothesis $2^{n+1} \leq m$, and similarly we find $2^{n+1} \leq m'$, so $\mathrm{s}(\mathcal{D}) > 2^{n+2}$.

Case 2. $\mathcal{D}$ ends with the introduction of $A_i \wedge B_i$, $i \leq n$. Then there are $\mathcal{D}_A, \mathcal{D}_B$ such that

$$\mathcal{D}_A \vdash_{\mathrm{s} \leq m} \mathcal{T}_{n+1,i}, A_i, P_{n+2}, Q_{n+2}, \quad \mathcal{D}_B \vdash_{\mathrm{s} \leq m'} \mathcal{T}_{n+1,i}, B_i, P_{n+2}, Q_{n+2},$$
$$\mathrm{s}(\mathcal{D}) = m + m' + 1.$$

Since A_i cannot be principal in an axiom, we find by lemma 5.2.7

$$\vdash_{\mathrm{s} \leq m} \mathcal{T}_{n+1,i}, \neg P_i, P_{n+2}, Q_{n+2}, \text{ i.e., } \vdash_{\mathrm{s} \leq m} \mathcal{S}_{n+1,i}.$$

With an appeal to lemma 5.2.11, we see that $\vdash_{\mathrm{s} \leq m} \mathcal{S}_n$, hence $2^{n+1} \leq m$. Similarly $2^{n+1} \leq m'$ by consideration of $\mathcal{D}_B$, hence again $\mathrm{s}(\mathcal{D}) > 2^{n+2}$. ⊠

5.2.13. THEOREM. *In* **GS3p*** + Cut *there is a sequence* $\langle \mathcal{S}_n \rangle_n$ *with deductions* $\mathcal{D}_n$ *of* $\mathrm{s}(\mathcal{D}_n)$ *linear in* n*, while any proof of* $\mathcal{S}_n$ *in* **GS3p*** *has size* $\geq 2^n$*; a corresponding result holds for* **G3[mi]***, *i.e.* **G3[mi]** *with the axiom generalized to* $A \Rightarrow A$ *for arbitrary* A.

PROOF. The statement for the classical system is immediate form what has been proved before. To prove the statement for the intuitionistic systems, we use a lemma. Let A^c be the standard translation of A into negation-normal form, that is to say, we first replace subformulas $B \to C$ by $\neg B \vee C$ and then push negations inward until they appear in front of prime formulas only; finally we delete occurrences of $\neg\neg$. For sequents we put $(\Gamma \Rightarrow A)^c := (\neg\Gamma)^c, A^c$. Then one easily checks by induction on size of deductions that if

$$(*) \qquad \mathbf{G3m} \vdash_{s \le n} \Gamma \Rightarrow A \text{ then } \mathbf{GS3p^*} \vdash_{s \le n} (\Gamma \Rightarrow A)^c.$$

Now we define

$$\mathcal{S}'_n := \mathcal{T}'_n \Rightarrow P_{n+1} \vee Q_{n+1},$$

where

$$\begin{array}{l} \mathcal{T}'_n := A'_1 \vee B'_1, \dots, A'_{n+1} \vee B'_{n+1}; \\ F_0 := (P \to P), \quad F_{n+1} := (F_n \wedge (P_{n+1} \vee Q_{n+1})); \\ A'_{n+1} := F_n \to P_{n+1}, \quad B'_{n+1} := F_n \to Q_{n+1}. \end{array}$$

Noting that $(\mathcal{S}'_n)^c \equiv \mathcal{S}_n$, we see that the result follows from $(*)$ and the result for the classical system. ⊠

5.2.13A. ♠ Prove lemma 5.2.1.

5.2.13B. ♠ The deductions in 5.2.4 are easily transformed into deductions in **GS3p** (with axioms with atomic principal formulas only). Determine the size of the deductions of $\mathcal{S}_n$ in **GS3p**.

5.2.13C. ♠ Describe cutfree proofs $\mathcal{D}_n$ for $\mathcal{S}'_n$ (as defined in 5.2.13) in **G3m***
such that the depth of $\mathcal{D}_n$ is linearly bounded in n.

5.3 Permutation of rules for classical logic

S. C. Kleene has analyzed in detail for the calculi **G1c** and **G1i** when two successively applied rules of the sequent calculus may be reversed in the order of their application. A similar analysis applies to **GS1**. In this section we want to give the analysis for **GS1** and **G1c**; the next section will deal with **G1i**.

But first of all we need a precise definition of the notion of permutability of rules.

5.3.1. DEFINITION. (*Permutability of rules in Gentzen systems*) Let us call two logical inferences *adjacent* if they are separated by applications of structural rules only.

Rule R is said to be *permutable below* (or permutable *down over*) rule R′, if the following holds: for all inferences $\alpha \in$ R, $\beta \in$ R′, α adjacent to β, and above β, such that

1. the (descendant of the) principal formula of α is not active in β,
2. α has as premises a set of sequents $\mathcal{A}$, the conclusion of α yields after some structural inferences a sequent S, β takes premises from $\{S\} \cup \mathcal{B}$ ($\mathcal{B}$ is possibly empty) and yields a conclusion S',

there is a deduction of S' from $\mathcal{A} \cup \mathcal{B}$ in which one or more applications of R$'$, preceded by (zero or more) structural inferences, occur above and are adjacent to an application of R, which is followed by (zero or more) structural inferences. ⊠

In the remainder of this section, "inference" will always mean "logical inference".

Let us consider an example in **G1c**. A deduction

$$\dfrac{\dfrac{\dfrac{(A \wedge B)^n, A, \Gamma \Rightarrow \Delta}{(A \wedge B)^n, \Gamma, A \wedge B \Rightarrow \Delta}\,\mathrm{L}\wedge}{\Gamma', A \wedge B, C \Rightarrow D, \Delta'}}{\Gamma', A \wedge B \Rightarrow C \to D, \Delta'}\,\mathrm{R}\!\to$$

(Structural rules transform Γ into Γ', C; Δ into Δ', D; and $(A \wedge B)^n$, $A \wedge B$ into $A \wedge B$.)

can be rearranged as

$$\dfrac{\dfrac{\dfrac{\dfrac{(A \wedge B)^n, A, \Gamma \Rightarrow \Delta}{(A \wedge B)^n, A, \Gamma', C \Rightarrow \Delta', D}}{(A \wedge B)^n, A, \Gamma' \Rightarrow C \to D, \Delta'}\,\mathrm{R}\!\to}{(A \wedge B)^n, A \wedge B, \Gamma' \Rightarrow C \to D, \Delta'}\,\mathrm{L}\wedge}{A \wedge B, \Gamma' \Rightarrow C \to D, \Delta'}$$

This shows that L$\wedge$ can always be permuted below R$\to$. On the other hand, the sequent $\forall x A(x), \exists x(A(x) \to \bot) \Rightarrow$ has a derivation with L$\exists$ below L$\forall$:

$$\dfrac{\dfrac{\dfrac{Ax \Rightarrow Ax \quad \bot \Rightarrow}{Ax, Ax \to \bot \Rightarrow}}{\forall x Ax, Ax \to \bot \Rightarrow}}{\forall x Ax, \exists x(Ax \to \bot) \Rightarrow}$$

but there is no deduction where the application(s) of L$\exists$ appear above L$\forall$.

We now turn to the study of permutability for **GS1**.

5.3.2. LEMMA. *(Permutation lemma for* **GS1***) R can always be permuted below R$'$ except when R = R$\exists$, R$'$ = R$\forall$.*

PROOF. We give the schemas for permutation of R over R$'$. D^n for a formula D means n copies of D. Γ is the set of active formulas for the rule application $\alpha \in$ R, and Δ is a set of formulas which after applying some structural rules become Δ', the active formulas of the inference $\beta \in \mathrm{R}'$. Θ is a set of passive formulas changed by structural rules into Θ'. A double line marks the application of zero or more structural rules.

One-premise rule over one-premise rule:

$$\beta\,\dfrac{\dfrac{\alpha\,\dfrac{\Gamma A^n \Delta\Theta}{AA^n\Delta\Theta}}{A\Delta'\Theta'}}{AB\Theta'} \quad \text{becomes} \quad \dfrac{\dfrac{\dfrac{\dfrac{\Gamma A^n\Delta\Theta}{\Gamma A^n\Delta'\Theta'}}{\Gamma A^n B\Theta'}\,\beta}{AA^n B\Theta'}\,\alpha}{AB\Theta'}$$

One-premise rule over two-premise rule:

$$\beta\,\dfrac{\dfrac{\alpha\,\dfrac{A^n\Gamma\Delta\Theta}{A^nA\Delta\Theta}}{AB\Theta'} \qquad AC\Theta'}{A(B\wedge C)\Theta'} \quad \text{becomes} \quad \dfrac{\dfrac{\dfrac{\dfrac{A^n\Gamma\Delta\Theta}{AA^n\Gamma B\Theta'} \qquad \dfrac{AC\Theta'}{A^nA\Gamma C\Theta'}}{A^{n+1}\Gamma(B\wedge C)\Theta'}\,\beta}{A^{n+2}(B\wedge C)\Theta'}\,\alpha}{A(B\wedge C)\Theta'}$$

Two-premise rule over one-premise rule:

$$\dfrac{\dfrac{\dfrac{A(A\wedge B)^n\Delta\Theta \qquad B(A\wedge B)^n\Delta\Theta}{(A\wedge B)^{n+1}\Delta\Theta}\,\alpha}{(A\wedge B)\Delta'\Theta'}}{(A\wedge B)C\Theta'}\,\beta \quad \text{becomes} \quad \dfrac{\dfrac{\beta\,\dfrac{\dfrac{A(A\wedge B)^n\Delta\Theta}{A(A\wedge B)^n\Delta'\Theta'}}{A(A\wedge B)^nC\Theta'} \qquad \dfrac{\dfrac{B(A\wedge B)^n\Delta\Theta}{B(A\wedge B)^n\Delta'\Theta'}}{B(A\wedge B)^nC\Theta'}\,\beta}{(A\wedge B)^{n+1}C\Theta'}\,\alpha}{(A\wedge B)C\Theta'}$$

Two-premise rule over two-premise rule:

$$\dfrac{\dfrac{\dfrac{(A\wedge B)^nA\Delta\Theta \qquad (A\wedge B)^nB\Delta\Theta}{(A\wedge B)^{n+1}\Delta\Theta}\,\alpha}{(A\wedge B)C\Theta'} \qquad (A\wedge B)D\Theta'}{(A\wedge B)(C\wedge D)\Theta'}\,\beta$$

becomes

$$\dfrac{\dfrac{\beta\,\dfrac{\dfrac{(A\wedge B)^nA\Delta\Theta}{(A\wedge B)^{n+1}AC\Theta'} \qquad \dfrac{(A\wedge B)D\Theta'}{(A\wedge B)^{n+1}AD\Theta'}}{(A\wedge B)^{n+1}A(C\wedge D)\Theta'} \qquad \dfrac{\dfrac{(A\wedge B)^nB\Delta\Theta}{(A\wedge B)^{n+1}BC\Theta'} \qquad \dfrac{(A\wedge B)D\Theta'}{(A\wedge B)^{n+1}BD\Theta'}}{(A\wedge B)^{n+1}B(C\wedge D)\Theta'}\,\beta}{(A\wedge B)^{n+2}(C\wedge D)^n\Theta'}\,\alpha}{(A\wedge B)(C\wedge D)\Theta'}$$

⊠

5.3.3. Definition. (*Partitions, order-restriction*) Let $\mathcal{D}$ be a deduction of a sequent S. Let $\mathcal{C}_1, \ldots, \mathcal{C}_n$ be a partition of all occurrences of logical symbols in S into classes; a class $\mathcal{C}_i$ is said to be *higher* than $\mathcal{C}_j$ iff $i < j$. The partition is *admissible* if it satisfies the following two conditions:

(a) If c is a symbol occurrence within the scope of another symbol occurrence c', then c and c' are either in the same class or c is in a higher class than c'.

(b) If $\alpha_c, \beta_{c'}$ are the rules corresponding to occurrences c, c' respectively, c' is not within the scope of c, c is an occurrence of $\exists$, and c' is an occurrence of $\forall$, then c is in the same or a higher class than c'.

We shall say that $\mathcal{D}$ satisfies the *order-restriction* (corresponding to the given admissible partition) if in any branch of $\mathcal{D}$ no inference α occurs above an inference β if α corresponds to a symbol occurrence in a lower class than the symbol occurrence corresponding to β. ⊠

5.3.4. Lemma. (*Bottom-violation*) *Let $\mathcal{D}$ be a deduction in which at most the final logical inference α has an inference β above it which belongs to a lower class than α. Then we can rearrange $\mathcal{D}$ to obtain $\mathcal{D}'$ with the same conclusion, without violation of the order-restriction, and with all logical inferences in $\mathcal{D}'$ corresponding to logical inferences in $\mathcal{D}$. If $\mathcal{D}$ appears as a subdeduction of a pure-variable deduction $\mathcal{D}_0$, it can be arranged that the replacement of $\mathcal{D}$ by $\mathcal{D}'$ results in a $\mathcal{D}'_0$ with the pure-variable property.*

Proof. Let the *grade* $g(\mathcal{D})$ of $\mathcal{D}$ be the number of inferences β' above α violating the order-restriction. Apply induction on the grade. In case $g(\mathcal{D}) = 0$ we are done. So let $g(\mathcal{D}) = n+1$, and let β be the lowest inference violating the restriction relative to α; then β must be adjacent to α, and α, β introduce distinct formulas which do not overlap (i.e. neither is a subformula of the other). Now we can permute β below α. The only situation where this is not possible is are precisely the ones not causing violation of the order-restriction (part (b) of the admissibility condition). ⊠

5.3.5. Theorem. (*Ordering theorem for* **GS1**) *Let $\mathcal{D}$ be a deduction in* **GS1** *and let an order-restriction relative to a partition of the occurrences of logical symbols in the conclusion of $\mathcal{D}$ be given. Then we can transform $\mathcal{D}$ into a deduction of the same sequent satisfying the order-restriction.*

Proof. Let the *degree* $d(\mathcal{D})$ of $\mathcal{D}$ be the number of logical inferences α_c such that there is a logical inference $\beta_{c'}$ violating the order-restriction. For degree 0 we are done; so let $d(\mathcal{D}) = n + 1$. We look for an uppermost α_c with a violating inference above it; we apply the bottom-violation lemma to the subdeduction ending with α_c. This reduces the degree. ⊠

5.3.6. THEOREM. *Let A be a prenex formula obtained as the conclusion of a deduction $\mathcal{D}$ in* **GS1**. *We can rearrange $\mathcal{D}$ to obtain a deduction $\mathcal{D}'$ such that all quantifier-inferences come below the propositional inferences in any branch of $\mathcal{D}'$.*

PROOF. We can distribute the symbols in A into two classes: $\mathcal{C}_1$ contains all propositional operators, $\mathcal{C}_2$ all quantifiers. This gives rise to an order-restriction meeting (a) and (b) of the definition. ⊠

5.3.7. THEOREM. *(Herbrand's theorem) Let us consider predicate logic without equality. A prenex formula B, say*

$$B \equiv \forall x \exists x' \forall y \exists y' \dots A(x, x', y, y', \dots),$$

with A quantifier-free, is provable in **GS1**, *iff there is a sequent Γ consisting of substitution instances of A of the form $A(x_i, t_i, y_i, s_i, \dots)$ such that Γ is provable propositionally and B can be obtained from Γ by structural and quantifier rules only.*

PROOF. A proof was already sketched in 4.2.5. We can now obtain a proof as a corollary to the ordering theorem. Applying the ordering theorem for **GS1**, we can achieve that all quantifier rules are applied below all propositional rules in the derivation of B. So the final part of the deduction derives the formula B from the sequent corresponding to the disjunction D using only R$\forall$, R$\exists$ and contractions (there is obviously no need for weakenings). ⊠

REMARK. If the prenex formula B is of the form $\exists x \forall y A(x, y)$, A quantifier-free, Γ may be assumed to have the following form:

$$A(t_0, y_0), \dots, A(t_i, y_i), \dots, A(t_n, y_n)$$

where the t_i are such that $y_i \notin \mathrm{FV}(t_j)$ for $i \geq j$. This is because the first logical inference must be R$\forall$, to be applied to an $A(t_i, y_i)$ such that y_i does not occur in t_j for any j. So let us assume $i = n$. By the ordering theorem, we may assume without loss of generality that this is followed (if necessary preceded by some contractions) by R$\exists$ on $\forall y A(t_n, y)$, producing $\exists x \forall y A$. Then the next step must be R$\forall$ again, applied to some y_j, say y_{n-1}, such that y_j does not occur in any t_p for $p < n$, etc.

A similar analysis applies to the more complicated case of $B \equiv \forall x \exists y \forall z$ $A(x, y, z)$, A quantifier-free; in this case Γ may be assumed to have the form

$$(*) \qquad A(x_0, t_0, z_0), \dots, A(x_i, t_i, z_i), \dots, A(x_n, t_n, z_n),$$

where $z_i \notin \mathrm{FV}(t_j)$ for $j \leq i$. Without loss of generality we may assume that in the t_n no variables occur except the x_i and the y_j.

5.3.7A. ♠ Prove that the conditions on Γ are necessary and sufficient to guarantee that $\forall x \exists y \forall z A$ can be derived from $(*)$.

5.3.7B. ♠ Let Γ be the sequent

$$A(t_0(x_1), x_0, s_0, y_0), A(t_1(x_2), x_1, s_1(y_0), y_1),$$
$$A(t_2(x_3), x_2, s_2(y_1), y_2), A(t_3, x_3, s_3(y_2), y_3),$$

with A quantifier-free. All x_i, y_j occurring in the t_k, s_k are shown. Derive from Γ the prenex formula $\exists u \forall x \exists v \forall y A(u, x, v, y)$. (This example shows that the dependence of the terms on the variables in the Herbrand disjunction can become complicated for the case of more than two quantifier alternations.)

We turn now to the formulation of the ordering theorem for **G1c**. The proof runs largely parallel to the proof for the one-sided calculus **GS1**. The definition of partitions and ordering condition are practically the same as for **GS1**, except that the second condition (b) in the definition of admissible partition has to be adapted:

5.3.8. DEFINITION. (*Partitions, order-restriction*) The definition runs as before except that the second restriction in the definition of admissible partition is modified as follows:

(b) If $\alpha_c, \beta_{c'}$ are the rules corresponding to occurrences c, c' respectively, c' is not within the scope of c, α_c is an occurrence of L$\forall$ or R$\exists$, and $\beta_{c'}$ is an occurrence of L$\exists$ or R$\forall$, then c is in the same or a higher class than c'. ⊠

5.3.9. NOTATION. We may think of a sequent $A_1, \ldots, A_n \Rightarrow B_1, \ldots, B_m$ as a multiset of signed formulas $\mathbf{t}A_1$, ..., $\mathbf{t}A_n$, $\mathbf{f}B_1$,..., $\mathbf{f}B_n$. We write for sequents simply S, S', T, T', U, U',.... If $S := \Gamma \Rightarrow \Delta$, $S' := \Gamma' \Rightarrow \Delta'$, then $SS' := \Gamma\Gamma' \Rightarrow \Delta\Delta'$. We write s, s' for arbitrary signs from $\{\mathbf{t}, \mathbf{f}\}$. So sA may stand for $\mathbf{t}A$ or $\mathbf{f}A$. ⊠

The use of signs can be avoided in the classical context (e.g. by using one-sided sequents), but is especially useful in the intuitionistic case.

REMARK. Since in $\bigwedge A_i \to \bigvee B_j$ the A_i occur negatively and the B_j positively, it might seem more natural to use $+, -$ for $\mathbf{f}, \mathbf{t}$. But $\mathbf{f}, \mathbf{t}$ are customary (and natural) in semantic tableau theory (4.9.7), and we have adopted the same notation here.

5.3.10. Lemma. *(Local permutation lemma for* **G1c***) In the calculus* **G1c** *rule* R *is always permutable below* R′ *in pure variable deductions except when*

$$\mathrm{R} \in \{\mathrm{L}\forall, \mathrm{R}\exists\},\ \mathrm{R}' \in \{\mathrm{L}\exists, \mathrm{R}\forall\}.$$

Proof. In the following general transformation schemas involving two-premise rules we adopt as notational conventions: sA, $s'B$ are principal in R, R′ respectively; the letter S is reserved for sequents of active formulas in R, the letter T for sequents of active formulas in R′; the letter U refers to passive sequents. Indices $_{1,2}$ refer to first and second premise; primes serve to indicate the effect of structural rules, e.g. U' follows from U by structural rules.

The schemas are quite similar to the permutation schemas for **GS1**. The principal reason for exhibiting them in full is that we want to be able to refer to them in the case of **G1i** in the next section.

One-premise rule over one-premise rule:

$$\dfrac{\dfrac{\dfrac{S(sA)^nTU}{(sA)(sA)^nTU}\,\mathrm{R}}{\overline{(sA)T'U'}}}{(sA)(s'B)U'}\,\mathrm{R}' \quad \text{becomes} \quad \dfrac{\dfrac{\dfrac{\dfrac{S(sA)^nTU}{\overline{S(sA)^nT'U'}}}{S(sA)^n(s'B)U'}\,\mathrm{R}'}{(sA)(sA)^n(s'B)U'}\,\mathrm{R}}{\overline{(sA)(s'B)U'}}$$

Two-premise rule over one-premise rule:

$$\dfrac{\dfrac{\dfrac{S_1(sA)^nTU \quad S_2(sA)^nTU}{(sA)(sA)^nTU}\,\mathrm{R}}{\overline{(sA)T'U'}}}{(sA)(s'B)U'}\,\mathrm{R}' \quad \text{becomes} \quad \dfrac{\dfrac{\dfrac{\dfrac{S_1(sA)^nTU}{\overline{S_1(sA)^nT'U'}}}{S_1(sA)^n(s'B)U'}\,\mathrm{R}' \quad \dfrac{\dfrac{S_2(sA)^nTU}{\overline{S_2(sA)^nT'U'}}}{S_2(sA)^n(s'B)U'}\,\mathrm{R}'}{(sA)(sA)^n(s'B)U'}\,\mathrm{R}}{\overline{(sA)(s'B)U'}}$$

One-premise rule over two-premise rule:

$$\dfrac{\dfrac{\dfrac{S(sA)^nT_1U}{(sA)(sA)^nT_1U}\,\mathrm{R}}{\overline{(sA)T_1'U'}} \qquad (sA)T_2U'}{(sA)(s'B)U'}\,\mathrm{R}' \quad \text{becomes} \quad \dfrac{\dfrac{\dfrac{S(sA)^nT_1U}{\overline{S(sA)(sA)^nT_1U'}} \quad \dfrac{(sA)T_2U'}{\overline{S(sA)(sA)^nT_2U'}}}{S(sA)(sA)^n(s'B)U'}\,\mathrm{R}'}{\dfrac{(sA)(sA)(sA)^n(s'B)U'}{\overline{(sA)(s'B)U'}}}\,\mathrm{R}$$

Two-premise rule over two-premise rule:

$$\dfrac{\dfrac{\dfrac{S_1(sA)^nT_1U \quad S_2(sA)^nT_1U}{(sA)(sA)^nT_1U}\,\mathrm{R}}{\overline{(sA)T_1'U'}} \qquad (sA)T_2U'}{(sA)(s'B)U'}\,\mathrm{R}'$$

becomes

$$\dfrac{\dfrac{\dfrac{S_1(sA)^nT_1U}{S_1(sA)(sA)^nT_1'U'} \quad \dfrac{(sA)T_2U'}{S_1(sA)(sA)^nT_2U'}}{S_1(sA)(sA)^n(s'B)U'}\,\mathrm{R}' \qquad \dfrac{\dfrac{S_2(sA)^nT_1U}{S_2(sA)(sA)^nT_1'U'} \quad \dfrac{(sA)T_2U'}{S_2(sA)(sA)^nT_2U'}}{S_2(sA)(sA)^n(s'B)U'}\,\mathrm{R}'}{\dfrac{(sA)(sA)(sA)^n(s'B)U'}{(sA)(s'B)U'}}\,\mathrm{R}$$

In the case of quantifier rules being involved, we have to check whether the conditions on variables remain satisfied after transformation. This turns out to be the case except in the cases listed above. For example, if $\mathrm{R} = \mathrm{R}\forall$, $\mathrm{R}' = \mathrm{L}\forall$, the variable condition on the given piece of deduction requires that TU does not contain the proper parameter y of the R-inference free; this remains correct since new variables introduced by weakening in U' must be distinct from y in a pure variable deduction. The counterexamples are:

(i) L$\forall$ over R$\forall$: $\forall x Ax \Rightarrow \forall x(Ax \vee B)$,

(ii) L$\forall$ over L$\exists$: $\forall x Ax, \exists x(Ax \to \bot) \Rightarrow$,

(iii) R$\exists$ over L$\exists$: $\exists x(Ax \wedge B) \Rightarrow \exists x Ax$,

(iv) R$\exists$ over R$\forall$: $\Rightarrow \exists x Ax, \forall x(Ax \to \bot)$.

The second example has been discussed before. ⊠

5.3.10A. ♠ Show that the counterexamples in the preceding proof are indeed counterexamples.

REMARK. All these examples are "absolute", that is to say in each case there is a deduction of the provable sequent with R over R$'$, but no deduction where all occurrences of R$'$ are over R, separated by structural rules only. Giving relative counterexamples, where the given sequent can be derived from some other given sequents with R over R$'$, but not with R$'$ over R, is easier. The following describes an application of the permutation theorem for **G1c**.

We can now prove a "bottom-violation lemma" and an "ordering theorem" exactly as for **GS1**.

5.3.11. THEOREM. *The ordering theorem holds for* **G1c** *(cf. 5.3.5).* ⊠

5.4 Permutability of rules for G1i

As in the preceding section, "inference" is short for "logical inference" = application of a logical rule. Permutability of rules has already been defined in 5.3.1.

5.4.1. Lemma. *(Local permutation lemma) In pure-variable deductions in* **G1i**, R *is permutable below* R′ *except in the following so-called* forbidden *cases:*

$$
\begin{array}{c||c|c|c|c}
\mathrm{R} & \mathrm{L}\forall & \mathrm{L}\forall,\mathrm{R}\exists & \mathrm{L}{\to} & \mathrm{R}{\to},\mathrm{L}{\to},\mathrm{R}\wedge,\mathrm{R}\vee,\mathrm{R}\forall,\mathrm{R}\exists \\
\mathrm{R}' & \mathrm{R}\forall & \mathrm{L}\exists & \mathrm{R}{\to} & \mathrm{L}\vee
\end{array}
$$

Proof. Observe first of all that certain combinations of rules need not be considered, due to the intuitionistic restriction on sequents. In particular, permutation of a right rule below another right rule is not possible, since the definition of permutability of rules will never apply: the principal formulas concerned are nested. For the rest, the proof in this case follows the general pattern of the proof in the classical case, but now we have to check the intuitionistic restriction on the succedent, and moreover must pay special attention to the L→ rule since the Δ appears only in the second premise.

Let us check the intuitionistic restriction for the case of a 2-premise rule over a 1-premise rule. Let us first observe that in the transformations indicated for the classical case, the structural rules leading from U to U' may be postponed, in the transformation, till after the application of R shown.

Difficulties can arise, if among the structural rules applied below R in the given deduction there is an application of RW serving to introduce a positive formula serving as an active formula for the next R′-application. (If the RW concerns U, the discussion is usually simpler.) If this is the case, $sA = -A$, $(T_1U)^+ = \Lambda$. This application may become incorrect after transformation if S_1 or S_2 contains a positive formula, which is only possible if R = L→, when S_1 consists of a positive active formula. The following R′-inference can be R∀, R∃, R∨ (for L→ over R→ nothing is claimed).

Let us consider the case of R′ = R∨. Then we transform as follows.

$$
\cfrac{\cfrac{\cfrac{\Gamma \Rightarrow A \quad \Gamma, B \Rightarrow}{\Gamma, A \to B \Rightarrow}}{\Gamma', A \to B \Rightarrow C}}{\Gamma', A \to B \Rightarrow C \vee D}
\qquad \text{becomes} \qquad
\cfrac{\cfrac{\Gamma \Rightarrow A \quad \cfrac{\cfrac{\Gamma, B \Rightarrow}{\Gamma, B \Rightarrow C}}{\Gamma, B \Rightarrow C \vee D}}{\Gamma, A \to B \Rightarrow C \vee D}}{\Gamma', A \to B \Rightarrow C \vee D}
$$

A similar analysis applies for a 2-premise rule over a 2-premise rule. Consider e.g. L→ over L→ with an essential RW in between.

$$
\cfrac{\cfrac{\cfrac{\Gamma \Rightarrow A \quad \Gamma, B \Rightarrow}{\Gamma, A \to B \Rightarrow}}{\Gamma', A \to B \Rightarrow C} \quad \Gamma', A \to B, D \Rightarrow E}{\Gamma', A \to B, C \to D \Rightarrow E}
$$

becomes

$$\dfrac{\dfrac{\Gamma \Rightarrow A}{\Gamma', A \to B, C \to D \Rightarrow A} \qquad \dfrac{\dfrac{\dfrac{\Gamma, B \Rightarrow}{\Gamma', A \to B, B \Rightarrow C} \quad \dfrac{\Gamma', A \to B, D \Rightarrow E}{\Gamma', A \to B, B, D \Rightarrow E}}{\Gamma', A \to B, C \to D, B \Rightarrow E}}{}}{\dfrac{\Gamma', (A \to B)^2, C \to D \Rightarrow E}{\Gamma', A \to B, C \to D \Rightarrow E}}$$

etc. Counterexamples establishing the exceptions: for (R,R$'$) = (L$\forall$,R$\forall$), (L$\forall$,L$\exists$),(R$\exists$,L$\exists$) we can use the same examples as in the classical case. For

(i) (L$\to$,R$\to$) take $A \to (A \to \bot) \Rightarrow A \to B$,

(ii) (L$\to$,L$\vee$) take $A \vee B, A \to B \Rightarrow B$,

(iii) (R$\vee$,L$\vee$) take $A \vee B \Rightarrow B \vee A$,

(iv) (R$\exists$,L$\vee$) take $Ax \vee Ay \Rightarrow \exists z Az$.

These examples are "absolute" in the sense that the provable sequents exhibited simply have no deductions where all the R$'$-inferences occur above the R-inferences. In the following cases we can only give local counterexamples, that is to say the deduction deduces a sequent from some other sequents, such that the permutation is impossible. The derivations

$$\dfrac{\dfrac{C, A \Rightarrow B}{C \Rightarrow A \to B} \quad D \Rightarrow A \to B}{C \vee D \Rightarrow A \to B} \qquad \dfrac{\dfrac{C \Rightarrow A \quad C \Rightarrow B}{C \Rightarrow A \wedge B} \quad D \Rightarrow A \wedge B}{C \vee D \Rightarrow A \wedge B}$$

$$\dfrac{\dfrac{C \Rightarrow Ax}{C \Rightarrow \forall x Ax} \quad D \Rightarrow \forall x Ax}{C \vee D \Rightarrow \forall x Ax}$$

give counterexamples for the pairs (R,R$'$) = (R$\to$,L$\vee$), (R$\wedge$,L$\vee$), (R$\forall$,L$\vee$) respectively. ⊠

5.4.1A. ♠ Show that the counterexamples for (L$\to$,R$\to$), (L$\to$,L$\vee$), (R$\vee$,L$\vee$), (R$\exists$,L$\vee$) are indeed counterexamples.

The argument concerning possibly awkward cases of RW can also be dealt with by considering only so-called W-normal deductions in **G1i**$'$, according to the next proposition.

5.4.2. PROPOSITION. *Let $\mathcal{D}$ be a deduction in* **G1i**. *$\mathcal{D}$ can be transformed into a deduction $\mathcal{D}'$ by moving applications of weakening downward such that the order of the logical inferences is retained (except that some of them become redundant and are removed), and such that weakenings occur only*

(i) immediately before the conclusion, or

(ii) to introduce one of the active formulas of an application of L$\to$, the other active formula not being introduced by weakening, or

(iii) for introducing a formula of the context of one the premises of a 2-premise rule, the occurrence of the same formula in the context of the other premise not being introduced by weakening (but not in the right hand context of intuitionistic L→).

As a result, any formula not just introduced by weakening can be traced to an axiom.

PROOF. By induction on the depth of $\mathcal{D}$. ⊠

5.4.2A. ♠ Give details of the proof.

DEFINITION. A deduction $\mathcal{D}$ satisfying (i)–(iii) of the preceding proposition we call *W-normal.* ⊠

5.4.3. DEFINITION. (*Partitions, order-restriction*) We copy the definition of an *admissible partition* from the definition for **G1c** and **GS1** in the preceding section, but with the following modification:

(b) Let R, R′ be the rules corresponding to occurrences c, c' respectively, and let c' not be within the scope of c. In the following cases c is in the same or a higher class than c':

R	L∀	L∀, R∃	L→	L→, R∨, R∃
R′	R∀	L∃	R→	L∨

⊠

5.4.3A. ♠ If we relax (ii) in 5.4.2 by replacing "L→" by "L→, L∨, L∃", we can achieve that the transformation from $\mathcal{D}$ to $\mathcal{D}'$ leaves the term denoting a corresponding assigned natural deduction unchanged. Show that this may fail if we do not permit L∨, L∃.

The following lemma is easily proved.

5.4.4. LEMMA. *Instances of a rule* R *adjacent to, and above, a rule* R′*, can be permuted if*

(i) R ∈ {R→, R∧, R∀} *and* R′ = L∨, *and*

(ii) over the other premise of the R′*-instance (i.e. not the one deriving from the* R*-application) either a weakening or another instance of* R *introduces the active formula.* ⊠

"Permuted" is here meant in a slightly more general sense than in the definition, since in the transformation the rule introducing the active formula in the second premise is also involved.

5.4.4A. ♠ Prove the lemma.

5.4.5. LEMMA. *(Bottom-violation lemma) Let $\mathcal{D}$ be a W-normal deduction in which only the lowest inference violates the order-restriction relative to a given partition. Then $\mathcal{D}$ can be rearranged so that the order-restriction is met, and the resulting deduction is W-normal.*

PROOF. Let $\mathcal{D}$ be given; we assume all instances of the axioms to be atomic, and $\mathcal{D}$ W-normal. Let α be the last inference in $\mathcal{D}$; α violates the order-restriction w.r.t. a partition $\mathcal{C}_1, \ldots, \mathcal{C}_k$. Then there is an inference β adjacent to α violating the order-restriction. As before we let the grade of $\mathcal{D}$ be the total number of inferences violating the order-restriction relative to the lowest inference. We show how to reduce the grade by considering cases.
Case 1. The grade can be reduced by permuting β below α according to the local permutation lemma, except when $\beta \in \{\mathrm{R}{\to}, \mathrm{R}\forall, \mathrm{R}\wedge\}$ and $\alpha \in \mathrm{L}\vee$. (Note that the other exceptions mentioned in the local permutation lemma cannot play a role since these cannot produce a violation by the conditions on admissible partitions.)
Case 2. $\beta \in \mathrm{R}\wedge, \mathrm{R}\forall, \mathrm{R}{\to}$, $\alpha \in \mathrm{L}\vee$, and the principal formula A of β in the other premise of α introduced by weakening; we can then reduce the grade by applying the special permutation lemma.
Case 3. $\beta \in \mathrm{R}\wedge, \mathrm{R}\forall, \mathrm{R}{\to}$, $\alpha \in \mathrm{L}\vee$, and the principal formula A of β in the second premise of α not introduced by weakening. We have two subcases.
Subcase 3.1. A is also a principal formula in the second premise of α, and has therefore been introduced by an inference β' of the same type as β, necessarily belonging to the same class of the partition; then by the special permutation lemma we can reduce the grade.
Subcase 3.2. A is not a principal formula in the second premise, and the inference β' above the second premise adjacent to α must in fact be a left-rule application. Since A was not introduced by weakening, it follows from W-normality that there is above β' an inference γ of the same rule as the rule of β, introducing A in the second premise of α; then γ and β belong to the same class and it follows that β' also violates the order-restriction, since α is the only inference with respect to which violations occur. But then we can apply the local permutation lemma w.r.t. β' and α. ⊠

We now get as before

5.4.6. THEOREM. *The ordering theorem with order-restriction as defined above holds for* **G1i** *(cf. 5.3.5).* ⊠

5.4.6A. ♠ *(Herbrand's theorem for negations of prenex formulas in the language without = and function symbols)* Let B be a prenex formula

$$\forall \vec{x}_0 \exists \vec{y}_0 \forall \vec{x}_1 \exists \vec{y}_1 \ldots A(\vec{x}_0, \vec{y}_0, \vec{x}_1, \vec{y}_1, \ldots),$$

A quantifier-free, and assume $\mathbf{I} \vdash \neg B$. Then we can find a finite conjunction C of substitution instances of A such that $\mathbf{I} \vdash \neg C$ and $\neg B$ is provable from $\neg C$. In particular, if $B \equiv \exists\vec{x}\forall\vec{y}\exists\vec{z}\,A(\vec{x},\vec{y},\vec{z})$ with A quantifier-free, the members of the conjunction are of the form $A(\vec{x}_i,\vec{t}_i,\vec{z}_i)$, where we may assume the $\vec{t}_i$ to consist of variables in $\{\vec{x}_0,\ldots\vec{x}_n,\vec{z}_0,\ldots,\vec{z}_{i-1}\}$ (Kreisel [1958]).

5.4.6B. ♠ For **G1i** we define two classes of formulas simultaneously:

$$\mathcal{G} \equiv \mathrm{At} \mid \mathcal{G}\wedge\mathcal{G} \mid \forall x\mathcal{G} \mid \exists x\mathcal{G} \mid \mathcal{D}\to\mathcal{G}$$
$$\mathcal{D} \equiv \mathrm{At} \mid \mathcal{G}\to\mathrm{At} \mid \forall x\mathcal{D} \mid \mathcal{D}\wedge\mathcal{D}$$

where At is the set of atomic formulas, excluding $\bot$. $\mathcal{D}$ is called the set of *hereditary Rasiowa–Harrop formulas.* Use the ordering theorem for **G1i** to show that if a sequent $\Gamma \Rightarrow G$ with $\Gamma \subset \mathcal{D}$, $G \in \mathcal{G}$ is derivable in **G1i**, there is a deduction such that (i) all sequents occurring in the deduction are of of the form $\Gamma' \Rightarrow G'$ with $\Gamma' \subset \mathcal{D}$, $G' \in \mathcal{G}$, and (ii) if the succedent formula G' is non-atomic, it is a principal formula.

Try to give a simple direct proof of this theorem, not relying on the ordering theorem. Note that (a) if we add $\top$ as a primitive with axioms $\Gamma \Rightarrow \top$, we can add a clause $\mathcal{G} \equiv \top \mid \ldots$, and that (b) actually the conditions mentioned in the local permutation lemma already suffice.

5.5 Notes

5.5.1. *Bounds on cut elimination.* Tait [1968] explicitly states the hyperexponential bound on the depth of deductions for a one-sided Gentzen system for classical logic. Girard [1987b] presents a detailed proof of the hyperexponential growth under cut elimination for Gentzen's system; the proof is also given, with a slight emendation, in Gallier [1993]. Here we have lifted the argument to **G3**-systems, thereby removing the need for distinguishing between "depth" and "logical depth" of deductions.

Curry's proof (Curry [1963]) of cut elimination for **G1i** plus Peirce's rule is analysed in Felscher [1975], who showed that Curry's procedure cannot be formalized in primitive recursive arithmetic; Gordeev [1987] showed that a different strategy, using appropriate inversion lemmas for a suitable system with Peirce's rule, produces the same hyperexponential estimates as in the case of the standard Gentzen systems. The better bounds for **Ip** obtained by the inversion-rule strategy have been found by Hudelmaier [1989,1992]; the exposition here is indebted to Schwichtenberg [1991], and hence indirectly to unpublished notes by Buchholz. See also Hudelmaier [1993].

The result in 5.2 is a slightly modified version of the presentation in Fitting [1996], which in its turn originated in a proof in Statman [1978], as simplified by S. Buss and G. Takeuti.

An area which we have not touched upon in this book is complexity theory for propositional and predicate logic. For an illuminating survey, see Urquhart [1995]. From this it will be seen that seemingly slight modifications in systems, which seem irrelevant from a general theoretical (metamathematical) point of view, may result in quite different behaviour from the viewpoint of complexity theory.

5.5.2. *Permutation of rules.* The results in the sections on permutation of rules are entirely based on Kleene [1952b]. In our exposition we have at one point simplified Kleene's argument for the intuitionistic case. Curry [1952b] discusses permutation of rules in classical Gentzen systems.

Already the proof of the "verschärfter Hauptsatz" (sharpened Hauptsatz) in Gentzen [1935] contains a permutation argument. The sharpened Hauptsatz states that a cutfree deduction in classical logic may always be arranged in such a way that the propositional inferences precede all quantifier inferences; the quantifier part of the deduction is linear, the last sequent of the propositional part is called the *midsequent*, and hence the sharpened Hauptsatz is also known as the *midsequent theorem.* Obviously the midsequent theorem contains a version of Herbrand's theorem; in other respects Herbrand stated a more general result, not only for formulas in prenex normal form (cf. the Introductory Note to Herbrand [1930] in van Heijenoort [1967], where also corrections to Herbrand [1930] are discussed).

Clearly, a Herbrand disjunction contains more information than the prenex formula derived from it. This suggests that it might be profitable to look for Herbrand disjunctions in mathematical proofs; if we are lucky, we can extract from the terms appearing in the disjunction explicit information, such as bounds on the size or number of realizations of existential quantifiers. An example of such a "Herbrand analysis" is given in Luckhardt [1989]. Such analyses are not applications of Herbrand's theorem as such, since (1) we want more precise information than is provided by Herbrand's theorem in its original form, and (2) interesting proofs for analysis will go beyond pure logic.

Chapter 6

Normalization for natural deduction

We now embark on a more thorough study of natural deduction, normalization and the structure of normal derivations. We describe a simple normalization strategy w.r.t. a specific set of conversions which transforms every deduction in **Ni** into a deduction in normal form; moreover, for $\rightarrow$**Nm** we prove that deductions are in fact *strongly* normalizable, i.e. every sequence of normalization steps terminates in a normal deduction, which is in fact *unique*.

As in the case of cut elimination, there is a hyperexponential upper bound on the rate of growth under normalization. From a suitable example we also easily obtain a hyperexponential lower bound. This still leaves open the possibility that each theorem might have at least some cutfree deduction of "modest" length; but this possibility is excluded by an example, due to Orevkov, of a sequence of statements C_n, $n \in \mathbb{N}$, with deductions linearly bounded by n, for which the minimum depth of *arbitrary normal* proofs has a hyperexponential lower bound.

This points to the very important role of indirect proof in mathematical reasoning: without indirect reasoning, exemplified by non-normal proofs, we cannot present proofs of manageable size for the C_n.

6.1 Conversions and normalization

In this and the next section we shall study the process of normalization for **Ni**, which corresponds to cut elimination for intuitionistic sequent calculi.

We shall assume, unless stated otherwise, that applications of $\bot_i$ have *atomic conclusions* in the deductions we consider.

As mentioned already in section 1.3.4, normalizations aim at removing local maxima of complexity, i.e. formula occurrences which are first introduced and immediately afterwards eliminated. However, an introduced formula may be used as a minor premise of an application of $\vee$E or $\exists$E, then stay the same throughout a sequence of applications of these rules, being eliminated at the end. This also constitutes a local maximum, which we should like to eliminate; for that we need the so-called permutation conversions. First we give a precise definition.

6.1.1. NOTATION. In order to be able to generalize conveniently later on, we introduce the term *del-rule* (from "disjunction-elimination-like"): the del-rules of **N**[**mic**] are $\exists$E, $\lor$E. ⊠

6.1.2. DEFINITION. A *segment* (of *length* n) in a deduction $\mathcal{D}$ of **Ni** is a sequence $A_1, \ldots, A_n$ of consecutive occurrences of a formula A in $\mathcal{D}$ such that

(i) for $1 < n$, $i < n$, A_i is a minor premise of a del-rule application in $\mathcal{D}$, with conclusion A_{i+1},

(ii) A_n is not a minor premise of a del-rule application,

(iii) A_1 is not the conclusion of a del-rule application.

(Note: An f.o. which is neither a minor premise nor the conclusion of an application of $\lor$E or $\exists$E always belongs to a segment of length 1.) A segment is *maximal*, or a *cut (segment)* if A_n is the major premise of an E-rule, and either $n > 1$, or $n = 1$ and $A_1 \equiv A_n$ is the conclusion of an I-rule. The *cutrank* $\mathrm{cr}(\sigma)$ of a maximal segment σ with formula A is $|A|$. The *cutrank* $\mathrm{cr}(\mathcal{D})$ of a deduction $\mathcal{D}$ is the maximum of the cutranks of cuts in $\mathcal{D}$. If there is no cut, the cutrank of $\mathcal{D}$ is zero. A *critical* cut of $\mathcal{D}$ is a cut of maximal cutrank among all cuts in $\mathcal{D}$. We shall use σ, σ' for segments.

We shall say that σ is a *subformula* of σ' if the formula A in σ is a subformula of B in σ'. A deduction without critical cuts is said to be *normal*. ⊠

REMARK. The obvious notion for a cut segment of length greater than 1 which comes to mind stipulates that the first formula occurrence of the segment must be the conclusion of an I-rule; but it turns out we can handle our more general notion of cut in our normalization process without extra effort. Note that a formula occurrence can belong to more than one segment of length greater than 1, due to the ramifications in $\lor$E-applications.

6.1.3. EXAMPLE.

$$\dfrac{A\lor(B\lor C) \qquad \dfrac{\dfrac{A^u}{A\lor B}}{\text{(b}')\ (A\lor B)\lor C} \qquad \dfrac{B\lor C^v \qquad \dfrac{\dfrac{B^w}{A\lor B}}{\text{(a)}\ (A\lor B)\lor C} \qquad \dfrac{C^{w'}}{\text{(a}')\ (A\lor B)\lor C}}{\text{(b)}\ (A\lor B)\lor C}\,w,w'}{\text{(c)}\ (A\lor B)\lor C}\,u,v$$

In this deduction (a),(b),(c) and (a′),(b),(c) mark segments of length 3, and (b′),(c) a segment of length 2. We are now going to define the various conversion steps we shall consider for the calculus **Ni**.

6.1.4. *Detour conversions*

We first show how to remove cuts of length 1. We write "conv" for "converts to".

$\wedge$*-conversion:*

$$\dfrac{\dfrac{\begin{matrix}\mathcal{D}_1 & \mathcal{D}_2\\ A_1 & A_2\end{matrix}}{A_1 \wedge A_2}}{A_i} \quad \text{conv} \quad \begin{matrix}\mathcal{D}_i\\ A_i\end{matrix} \quad \text{for } i \in \{1,2\}.$$

$\vee$*-conversion:*

$$\dfrac{\dfrac{\begin{matrix}\mathcal{D}\\ A_i\end{matrix}}{A_1 \vee A_2} \quad \begin{matrix}[A_1]^u\\ \mathcal{D}_1\\ C\end{matrix} \quad \begin{matrix}[A_2]^v\\ \mathcal{D}_2\\ C\end{matrix}}{C}\, u, v \quad \text{conv} \quad \begin{matrix}\mathcal{D}\\ [A_i]\\ \mathcal{D}_i\\ C\end{matrix} \quad \text{for } i \in \{1,2\}.$$

$\to$*-conversion:*

$$\dfrac{\dfrac{\begin{matrix}[A]^u\\ \mathcal{D}\\ B\end{matrix}}{A \to B}\, u \quad \begin{matrix}\mathcal{D}_1\\ A\end{matrix}}{B} \quad \text{conv} \quad \begin{matrix}\mathcal{D}_1\\ [A]\\ \mathcal{D}\\ B\end{matrix}$$

$\forall$*-conversion:*

$$\dfrac{\dfrac{\begin{matrix}\mathcal{D}\\ A\end{matrix}}{\forall y A[x/y]}}{A[x/t]} \quad \text{conv} \quad \begin{matrix}\mathcal{D}[x/t]\\ A[x/t]\end{matrix}$$

$\exists$*-conversion:*

$$\dfrac{\dfrac{\begin{matrix}\mathcal{D}\\ A[y/t]\end{matrix}}{\exists x A[y/x]} \quad \begin{matrix}[A]^u\\ \mathcal{D}'\\ C\end{matrix}}{C}\, u \quad \text{conv} \quad \begin{matrix}\mathcal{D}\\ [A[y/t]]\\ \mathcal{D}'[y/t]\\ C\end{matrix}$$

6.1.5. *Permutation conversions*

In order to remove cuts of length > 1, we permute E-rules upwards over minor premises of $\vee$E, $\exists$E.

$\vee$*-perm conversion:*

$$\dfrac{\dfrac{\begin{matrix}\mathcal{D}\\ A \vee B\end{matrix} \quad \begin{matrix}\mathcal{D}_1\\ C\end{matrix} \quad \begin{matrix}\mathcal{D}_2\\ C\end{matrix}}{C} \quad \mathcal{D}'}{D}\,\text{E-rule} \quad \text{conv} \quad \dfrac{\begin{matrix}\mathcal{D}\\ A \vee B\end{matrix} \quad \dfrac{\begin{matrix}\mathcal{D}_1\\ C\end{matrix} \quad \mathcal{D}'}{D} \quad \dfrac{\begin{matrix}\mathcal{D}_2\\ C\end{matrix} \quad \mathcal{D}'}{D}}{D}$$

∃-perm conversion:

$$\dfrac{\dfrac{\begin{array}{cc}\mathcal{D} & \mathcal{D}'\\ \exists xA & C\end{array}}{C} \quad \begin{array}{c}\mathcal{D}''\\ \end{array}}{D}\,\text{E-rule} \qquad \text{conv} \qquad \dfrac{\begin{array}{c}\mathcal{D}\\ \exists xA\end{array} \quad \dfrac{\begin{array}{cc}\mathcal{D}' & \\ C & \mathcal{D}''\end{array}}{D}}{D}$$

6.1.6. *Simplification conversions*

Applications of $\vee$E with major premise $A_1 \vee A_2$, where at least one of $[A_1]$, $[A_2]$ is empty in the deduction of the first or second minor premise, are redundant; we accordingly introduce simplifying conversions. Similarly, an application of $\exists$E with major premise $\exists xA$, where the assumption class $[A[x/y]]$ in the derivation of the minor premise is empty, is redundant. Redundant applications of $\vee$E or $\exists$E can be removed by the following conversions:

$$\dfrac{\begin{array}{ccc}\mathcal{D} & \mathcal{D}_1 & \mathcal{D}_2\\ A_1 \vee A_2 & C & C\end{array}}{C} \qquad \text{conv} \qquad \begin{array}{c}\mathcal{D}_i\\ C\end{array}$$

where *no* assumptions are discharged by $\vee$E in $\mathcal{D}_i$, and

$$\dfrac{\begin{array}{cc}\mathcal{D} & \mathcal{D}'\\ \exists xA & C\end{array}}{C} \qquad \text{conv} \qquad \begin{array}{c}\mathcal{D}'\\ C\end{array}$$

where *no* assumptions of $\mathcal{D}'$ are discharged at the final rule application. The simplification for $\vee$E introduces a non-deterministic element if both discharged assumption classes $[A_i]$ are empty.

6.1.7. *Term notation.* In term notation the conversions take the following compact form:

$$\mathbf{p}_i\mathbf{p}(t_0, t_1) \ \text{conv} \ t_i \ \ (i \in \{0, 1\}),$$
$$\mathrm{E}^{\vee}_{u_0u_1}(\mathbf{k}_it, t_0, t_1) \ \text{conv} \ t_i[u_i/t],$$
$$(\lambda u^A.t)s^A \ \text{conv} \ t[u/s],$$
$$(\lambda x.t)s \ \text{conv} \ t[x/s],$$
$$\mathrm{E}^{\exists}_{uy}(\mathbf{p}(t, s), t') \ \text{conv} \ t'[u, y/t, s];$$

$$f[\mathrm{E}^{\vee}_{uv}(t, t_0, t_1)] \ \text{conv} \ \mathrm{E}^{\vee}_{uv}(t, f[t_0], f[t_1]),$$
$$f[\mathrm{E}^{\exists}_{uy}(t, s)] \ \text{conv} \ \mathrm{E}^{\exists}_{uy}(t, f[s]);$$

$$\mathrm{E}^{\vee}_{u_0u_1}(t, t_0, t_1) \ \text{conv} \ t_i \ \ (u_i \text{ not free in } t_i),$$
$$\mathrm{E}^{\exists}_{uy}(t, t') \ \text{conv} \ t' \ \ (u, y \text{ not free in } t').$$

The first group expresses the detour conversions, the second group the permutation conversions and the third group the simplification conversions. In the second group, f is another elimination operator, with [ ] the argument corresponding to the main premise.

REMARK. The detour conversions are sometimes simply called β-conversions, after the typical case of $\rightarrow$-detour conversions.

Notationally, there is something to be said for reserving a special type (say "I") for individuals; $\forall$I then gives a term $\lambda x^I.t : \forall x A$.

The notion of a normal deduction can be defined very compactly, independently of the definition of cut segments as redexes, by stipulating that in a normal deduction each major premise of an E-rule is either an assumption or a conclusion of an application of an E-rule different from the del-rules.

6.1.8. THEOREM. *(Normalization) Each derivation $\mathcal{D}$ in* **Ni** *reduces to a normal derivation.*

PROOF. In applications of E-rules we always assume that the major premise is *to the left of* the minor premise(s), if there are any minor premises. We use a main induction on the cutrank n of $\mathcal{D}$, with a subinduction on m, the sum of lengths of all critical cuts (= cut segments) in $\mathcal{D}$.

By a suitable choice of the critical cut to which we apply a conversion we can achieve that either n decreases (and we can appeal to the main induction hypothesis), or that n remains constant but m decreases (and we can appeal to the subinduction hypothesis). Let us call σ a *t.c.c.* (top critical cut) in $\mathcal{D}$ if no critical cut occurs in a branch of $\mathcal{D}$ above σ. Now apply a conversion to the *rightmost* t.c.c. of $\mathcal{D}$; then the resulting $\mathcal{D}'$ has a lower cutrank (if the segment treated has length 1, and is the only maximal segment in $\mathcal{D}$), or has the same cutrank, but a lower value for m.

To see this in the case of an implication conversion, suppose we apply a conversion to the rightmost t.c.c. consisting of a formula occurrence $A \rightarrow B$

$$\dfrac{\dfrac{\begin{array}{c}[A]\\ \mathcal{D}'\\ B\end{array}}{A \rightarrow B} \quad \begin{array}{c}\mathcal{D}''\\ A\end{array}}{B} \qquad \text{conv} \qquad \begin{array}{c}\mathcal{D}''\\ [A]\\ \mathcal{D}'\\ B\end{array}$$

Then the repeated substitution of $\mathcal{D}''$ at each f.o. of $[A]$ cannot increase the value of m, since $\mathcal{D}$ does not contain a t.c.c. cut in $\mathcal{D}''$ above the minor premise A of $\rightarrow$E (such a cut would have to occur to the right of $A \rightarrow B$, contrary to our assumption). We leave it to the reader to verify the other cases. ⊠

REMARK. If we use the term notation for deductions, our strategy may be described by saying that we look for the *rightmost* redex of maximal degree not containing another redex of maximal degree.

It is worth noting that this strategy also produces normal deductions if we assume that our deductions obey the Complete Discharge Convention (cf. 2.1.9).

6.1.8A. ♠ Do the remaining cases of the proof of the theorem.

6.1.8B. ♠ If we permit in **Ni** non-atomic applications of $\bot_i$, a local maximum of complexity may arise if the conclusion of $\bot_i$ is the major premise of an elimination rule. Devise extra conversions to remove such maxima and extend the normalization theorem.

6.1.9. REMARK. The term notation also suggests the possibility of a more general concept of conversion, namely (restricting attention to $\rightarrow$**Nm**)

$$(\lambda\vec{x}y.t)\vec{s}r \ \ \mathrm{conv}_{g\beta} \ \ (\lambda\vec{x}t[y/r])\vec{s}.$$

We call this *generalized beta-conversion* (gβ-conversion). Consideration of this conversion has advantages when computing bounds on the number of reduction steps needed to reach normal form. For this more general notion of conversion, essentially the same strategy leads to normal forms (see 6.10).

It is worth noting that the following theorem holds:

6.1.10. THEOREM. *Deductions in* **Ni** *are strongly normalizing w.r.t. the conversions listed, that is all reduction sequences terminate (every strategy produces normal forms).*

We do not prove this for the full system here; for references, see 6.12.2. The important case of $\rightarrow$-logic will be treated in section 6.8, with some extensions indicated in the exercises. Strong normalization also holds w.r.t. gβ-conversion, by essentially the same method. See also section 8.3.

6.1.11. REMARK. The system **Nc** is not as well-behaved w.r.t. to normalization as **Ni**. In particular, no obvious "formulas-as-types" parallel is available. Nevertheless, as shown by Prawitz, a form of normalization for **Nc** w.r.t. the $\bot\wedge\rightarrow\forall$-language is possible, by observing that $\bot_c$ for this language may be restricted to instances with atomic conclusions. For example, the left tree below may be transformed into the tree on the right hand side:

$$\dfrac{\begin{array}{c}[\neg(B\rightarrow C)]^u\\ \mathcal{D}\\ \bot\end{array}}{B\rightarrow C}\,u \qquad\qquad \dfrac{\dfrac{\begin{array}{c}\dfrac{\dfrac{\dfrac{(B\rightarrow C)^u \quad B^v}{C} \quad \neg C^w}{\bot}}{[\neg(B\rightarrow C)]}\,u\\ \mathcal{D}\\ \bot\end{array}}{C}\,w}{B\rightarrow C}\,v$$

6.1.11A. ♠ Extend normalization to $\bot\wedge\to\forall$-**Nc**. *Hint.* Use the preceding remark and the result of exercise 2.3.6A (Prawitz [1965]).

6.2 The structure of normal derivations

6.2.1. *Normal deductions in implication logic*

By way of introduction, let us first consider the structure of normal derivations in $\to$**Nm**. Let $\mathcal{D}$ be a normal derivation in $\to$**Nm**. A sequence of f.o.'s $A_0, \ldots, A_n$ such that (1) A_0 is a top formula (leaf) of the prooftree, and (2) for $0 \leq i < n$, A_{i+1} is immediately below A_i, and (3) A_i is not the minor premise of an $\to$E-application, is called a *track* of the deduction tree $\mathcal{D}$. A track *of order* 0 ends in the conclusion of $\mathcal{D}$; a track of *order* $n+1$ ends in the minor premise of an $\to$E-application with major premise belonging to a track of order n.

Since by normality E-rules cannot have the conclusion of an $\to$I-application as their major premise, the E-rules have to precede the I-rules in a track, so the following is obvious: a track may be divided into an E-part, say $A_0, \ldots, A_{i-1}$, a minimal formula A_i, and an I-part $A_{i+1}, \ldots, A_n$. In the E-part all rules are E-rules; in the I-part all rules are I-rules; A_i is the conclusion of an E-rule and, if $i < n$, a premise of an I-rule. It is also easy to see that each f.o. of $\mathcal{D}$ belongs to some track. Tracks are pieces of branches of the tree with successive f.o.'s in the subformula relationship: either A_{i+1} is a subformula of A_i or vice versa. As a result, all formulas in a track $A_0, \ldots, A_n$ are subformulas of A_0 or of A_n; and from this, by induction on the order of tracks, we see that every formula in $\mathcal{D}$ is a subformula either of an open assumption or of the conclusion.

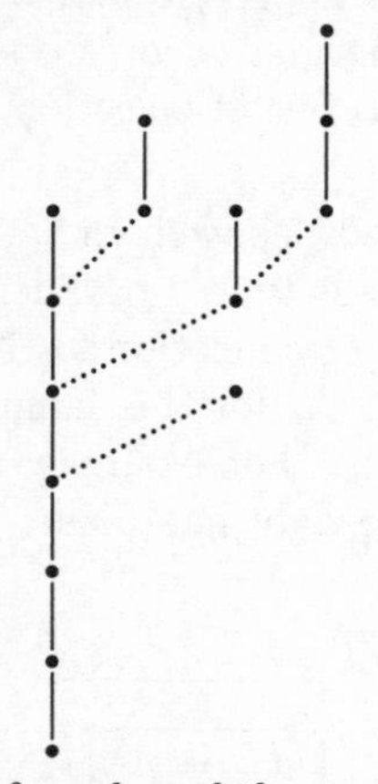

The tree to the left illustrates the structure of a normal derivation in $\to$**Nm**. A dotted line connects a minor premise of $\to$E with its conclusion; a solid line connects a (major) premise with the conclusion. The parts of branches made up of solid lines are the tracks; the unramified parts are always the I-part of a track. The tree shown has tracks of order 0–2. The only track of order 2 consists of three nodes.

The notion of track and the analysis given can readily be extended to $\to\wedge\forall\bot$-**Ni**. In the $\to\forall\bot$-fragment there is always a unique track of order 0, but as soon as $\wedge$ is added to the language, there may be several tracks of order 0 in a normal derivation.

However, if we want to generalize this type of analysis to the full system, we encounter a difficulty with the rules $\vee$E, $\exists$E. The conclusion of an application of $\vee$E or $\exists$E is not necessarily a subformula of the major premise. Hence restricting attention to pieces of branches of the prooftree does not lead to a satisfactory analysis of the form of normal deductions. The subformulas of a major premise $A \vee B$ or $\exists x A$ of an E-rule application do not appear in the conclusion, but among the assumptions being discharged by the application. This suggests the definition of track below.

The general notion of a track has been devised so as to retain the subformula property in case one passes through the major premise of an application of a del-rule. In a track, when arriving at an A_i which is the major premise of an application α of a del-rule, we take for A_{i+1} a hypothesis discharged by α.

6.2.2. Definition. A *track* of a derivation $\mathcal{D}$ is a sequence of f.o.'s $A_0, \ldots, A_n$ such that

(i) A_0 is a top f.o. in $\mathcal{D}$ not discharged by an application of del-rule;

(ii) A_i for $i < n$ is not the minor premise of an instance of $\to$E, and *either*

- (a) A_i is not the major premise of an instance of a del-rule and A_{i+1} is directly below A_i, *or*
- (b) A_i is the major premise of an instance α of a del-rule and A_{i+1} is an assumption discharged by α;

(iii) A_n is *either*

- (a) the minor premise of an instance of $\to$E, *or*
- (b) the conclusion of $\mathcal{D}$, *or*
- (c) the major premise of an instance α of a del-rule in case there are no assumptions discharged by α. ⊠

6.2.3. Example. Consider the following derivation:

$$
\dfrac{\dfrac{\dfrac{\dfrac{\forall x\exists y Pxy^{w}}{\exists y Puy}\,\forall\text{E} \qquad \dfrac{\dfrac{Puv^{w''} \qquad \dfrac{\dfrac{\dfrac{\dfrac{\forall xy(Pxy \to Pyx)^{w'}}{\forall y(Puy \to Pyu)}\,\forall\text{E}}{Puv \to Pvu}\,\forall\text{E} \qquad Puv^{w''}}{Pvu}\,\to\text{E}}{Puv \wedge Pvu}\,\wedge\text{I}}{\exists y(Puy \wedge Pyu)}\,\exists\text{I}}{\exists y(Puy \wedge Pyu)}\,\exists\text{E},w''}{\forall x\exists y(Pxy \wedge Pyx)}\,\forall\text{I}}{\forall xy(Pxy \to Pyx) \to \forall x\exists y(Pxy \wedge Pyx)}\,\to\text{I}\,w'}{\forall x\exists y Pxy \to [\forall xy(Pxy \to Pyx) \to \forall x\exists y(Pxy \wedge Pyx)]}\,\to\text{I}\,w
$$

The diagram below represents the tree structure of the derivation, with the rules and discharged assumption classes as labels. For easy reference we have also given a number to each node.

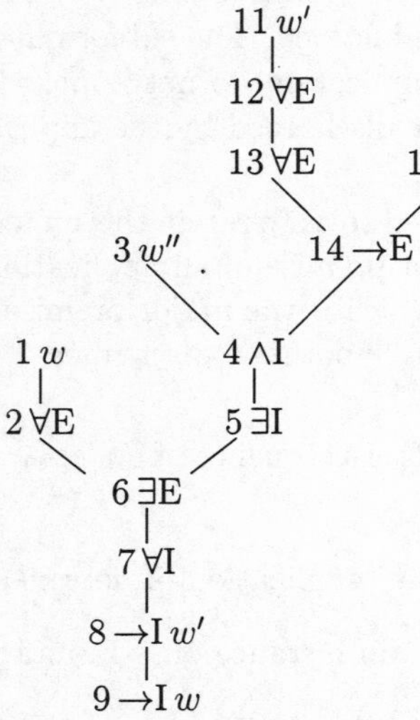

```
          11 w'
            |
          12 ∀E
            |
          13 ∀E      10 w''
              \       /
   3 w''       14 →E
        \       /
1 w       4 ∧I
|          |
2 ∀E      5 ∃I
    \     /
     6 ∃E
      |
     7 ∀I
      |
     8 →I w'
      |
     9 →I w
```

All nodes are numbered; top nodes have a variable as assumption label; below the top nodes the rule applied is indicated, plus the label of the discharged assumption class where applicable.

Tracks: 1–9;
11–14, 4–9;
1, 2, 10.

REMARK. To a deduction $\mathcal{D}$ we can associate a labelled tree $\langle\mathcal{D}\rangle$, by induction on $|\mathcal{D}|$, as follows. The labels are formulas, and each node in $\mathcal{D}$ labelled with a formula A corresponds to a set of nodes labelled with A in $\langle\mathcal{D}\rangle$. (It is sometimes convenient to think of the labels in $\langle\mathcal{D}\rangle$ as also containing the rule used to obtain the formula in the label, and if applicable, the discharged assumption classes.)

(a) If $\mathcal{D}$ ends with an f.o. A which is the conclusion of a rule $\mathrm{R} \notin \{\vee\mathrm{E}, \exists\mathrm{E}\}$, with deductions $\mathcal{D}_i$ of the premises A_i, then $\langle\mathcal{D}\rangle$ is obtained by putting A below the disjoint union of the partially ordered $\langle\mathcal{D}_i\rangle$;

(b) If $\mathcal{D}$ terminates with a $\vee$E, i.e. $\mathcal{D}$ ends with

$$\dfrac{\begin{array}{c}\mathcal{D}_0\\ A \vee B\end{array} \quad \begin{array}{c}[A]^u\\ \mathcal{D}_1\\ C\end{array} \quad \begin{array}{c}[B]^v\\ \mathcal{D}_2\\ C\end{array}}{C}\,u, v$$

then we insert a copy of $\langle\mathcal{D}_0\rangle$ above *each* formula occurrence in $\langle\mathcal{D}_1\rangle$ and $\langle\mathcal{D}_2\rangle$ corresponding to the occurrences in assumption classes labelled u, v. Below the resulting disjoint trees we place an occurrence of C. (The notion of corresponding occurrence is the obvious one.) Similarly if the final rule applied in $\mathcal{D}$ is $\exists$E. This second clause (b) is the reason that several occurrences of B in $\langle\mathcal{D}\rangle$ may correspond to a single occurrence B in $\mathcal{D}$. Thus the prooftree above yields a labelled tree with skeleton indicated below (the numbers are copied from the numbering of the corresponding occurrences in $\mathcal{D}$.)

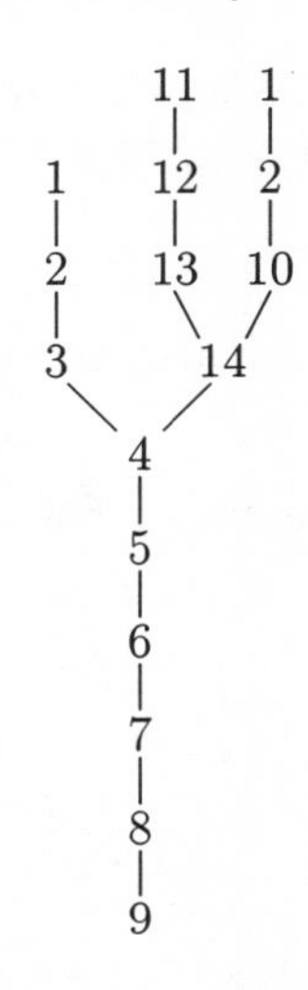

The tracks correspond in this tree to branches starting at a leaf and terminating either in a minor premise of an →E or in the conclusion. If we do not insist on having trees, we can also have partial orders associated to prooftrees such that the formula occurrences of the prooftree are in bijective correspondence with occurrences of the same formulas in the associated labelled partial order; in the second clause (b) of the description we then stipulate that a single copy of $\langle\mathcal{D}_0\rangle$ is above all occurrences in the classes labelled w and w' in the ordering of $\langle\mathcal{D}\rangle$. In the tree to the left this ordering would be obtained by identifying nodes with the same label (1 and 2 each occur twice).

6.2.4. PROPOSITION. *Let $\mathcal{D}$ be a normal derivation in* **I**, *and let $\pi \equiv \sigma_0, \ldots, \sigma_n$ be a track in $\mathcal{D}$. Then there is a segment σ_i in π, the* minimum segment *or* minimum part *of the track, which separates two (possibly empty) parts of π, called the* E-part *(*elimination part*) and the* I-part *(*introduction part*) of π such that:*

(i) for each σ_j in the E-part *one has $j < i$, σ_j is a major premise of an E-rule, and σ_{j+1} is a strictly positive part of σ_j, and therefore each σ_j is an s.p.p. of σ_0;*

(ii) for each σ_j in the I-part *one has $i < j$, and if $j \neq n$, then σ_j is a premise of an I-rule and an s.p.p. of σ_{j+1}, so each σ_j is an s.p.p. of σ_n;*

(iii) if $i \neq n$, σ_i is a premise of an I-rule *or a premise of $\perp_i$ (and then of the form $\perp$) and is an s.p.p. of σ_0.* ⊠

6.2.5. DEFINITION. A *track of order* 0, or *main track*, in a normal derivation is a track ending in a conclusion of $\mathcal{D}$. A *track of order* $n+1$ is a track ending in the minor premise of an →E-application, with major premise belonging to a track of order n.

A *main branch* of a derivation is a branch π in the prooftree such that π passes only through premises of I-rules and *major premises* of E-rules, and π begins at a top node and ends in the conclusion of the deduction.

If we do not include simplifications among our conversions, a track of order 0 ends either in the conclusion of the whole deduction or in the major premise of an application of a del-rule, provided the classes of assumptions discharged by the application are empty. ⊠

REMARK. If we search for a main branch going upwards from the conclusion, the branch to be followed is unique as long as we do not encounter an $\wedge$I-application.

6.2.6. PROPOSITION. *In a normal derivation each formula occurrence belongs to some track.*

PROOF. By induction on the height of normal deductions. For example, suppose $\mathcal{D}$ ends with an $\vee$E-application:

$$\dfrac{\begin{matrix} \\ \mathcal{D}_1 \\ A_1 \vee A_1 \end{matrix} \quad \begin{matrix} [A_1]^u \\ \mathcal{D}_2 \\ C \end{matrix} \quad \begin{matrix} [A_2]^v \\ \mathcal{D}_3 \\ C \end{matrix}}{C}\, u, v$$

C in $\mathcal{D}_2$ belongs to a track π (induction hypothesis); either this does not start in $[A_1]^u$, and then π, C is a track in $\mathcal{D}$ which ends in the conclusion; or π starts in $[A_1]^u$, and then there is a track π' in $\mathcal{D}_1$ (induction hypothesis) such that π', π, C is a track in $\mathcal{D}$ ending in the conclusion. The other cases are left to the reader. ⊠

REMARK. In the case discussed in the proof, we can explicitly describe all tracks of $\mathcal{D}$, they are of the following four types:

(a) tracks of order > 0 in $\mathcal{D}_1$, or tracks of order > 0 in $\mathcal{D}_{i+1}$ not beginning in $[A_i]$, or

(b) of the form π_1, π_2 with π_1 a track of order 0 in $\mathcal{D}_1$ and π_2 a track of order > 0 in $\mathcal{D}_{i+1}$ beginning in $[A_i]$, or

(c) of the form π_1, π_2, C with π_1 is a track of order 0 in $\mathcal{D}_1$, π_2 a track of order 0 in $\mathcal{D}_{i+1}$ beginning in $[A_i]$, or

(d) of the form π_i, C, with π_i is a track of order 0 in $\mathcal{D}_{i+1}$ not beginning in $[A_i]$.

6.2.7. THEOREM. *(Subformula property) Let $\mathcal{D}$ be a normal deduction in* **I** *for $\Gamma \vdash A$. Then each formula in $\mathcal{D}$ is a subformula of a formula in $\Gamma \cup \{A\}$.*

PROOF. We prove this for tracks of order n, by induction on n. ⊠

6.2.7A. ♠ Give full details of the proof of the subformula property.

6.2.7B. ♠ For $\bot\wedge\to\forall$-**Nc** (cf. 6.1.11) the following subformula property holds: if $\mathcal{D}$ derives A from Γ, and $\mathcal{D}$ is normal with atomic instances of $\bot_c$ only, then every formula in $\mathcal{D}$ is either a subformula of A, Γ or the negation of an atomic formula in A, Γ. Prove this fact (Prawitz [1965]).

6.2.7C. ♠ Prove the separation theorem for $\to$**Hc** (cf. 4.9.2) via the following steps.

(i) If A is an implication formula derivable in **Hc**, it is derivable, say by a deduction $\mathcal{D}$, in $\to\bot$**Nc** with $\bot_i$, $\bot_c$ restricted to atomic conclusions.

(ii) Replace in $\mathcal{D}$ the instances of $\bot_c$ by uses of $\mathrm{P}_{Q,\bot}$, where $\mathrm{P}_{X,Y}$ is Peirce's Law, i.e., $((X \to Y) \to X) \to X$, and where $Q \in \mathcal{PV}$. The result is a derivation $\mathcal{D}'$ in $\to\bot$**Ni** of A from $\mathrm{P}_{Q_i,\bot}$ $(1 \leq i \leq n)$ with $Q_i \in \mathcal{PV}$.

(iii) Let $P_1, \ldots, P_m$ be the conclusions of instances of $\bot_i$ occurring in $\mathcal{D}'$, and let $B \equiv \bigwedge_i P_i$. Replace $\bot$ everywhere by B; by interpolating some steps, the instances of $\bot_i$ are transformed into a sequence of $\wedge$E-applications. The result is a proof $\mathcal{D}''$ in $\to\wedge$**Nm** from assumptions $\mathrm{P}_{Q_i,B}$ $(1 \leq i \leq n)$.

(iv) Use exercise 2.1.8G to transform this into a derivation of A from P_{Q_i,P_j} $(1 \leq i \leq n,\ 1 \leq j \leq m)$. By the subformula property for **Nm**, this reduces to a normal deduction in $\to$**Nm**; from this we readily construct a proof in **Hm** plus assumptions P_{Q_i,P_j}, hence a proof in **Hc**.

6.2.7D. ♠ Prove the following proposition. Let $\mathcal{D}$ be a deduction of A in **Ni** without open assumptions, which is normal w.r.t. detour conversions. If A is not atomic, $\mathcal{D}$ ends with an I-rule. Hence if **Ni** $\vdash A \vee B$, it follows that either **Ni** $\vdash A$ or **Ni** $\vdash B$ (Prawitz [1965]). *Hint.* Consider a main branch in $\mathcal{D}$.

6.2.7E. ♠ Prove theorems 4.2.3, 4.2.4 using normalization instead of cut elimination (Prawitz [1965]).

6.3 Normality in G-systems and N-systems

This section is not needed in the remainder of this chapter and may be skipped. In this section we study the relationship between normal natural deductions and cutfree G-proofs. We first present a simple construction of G-proofs from normal N-proofs; this motivates the study of the class of so-called normal cutfree G-proofs.

As noted before (3.3.4), several cutfree G-deductions may correspond to a single normal N-deduction. However, by imposing the extra condition of normality plus some less crucial conditions, we can achieve, for the right choice of G-system, a one-to-one correspondence between normal **Ni**-deductions and normal G-deductions. For our G-system we choose a term-labelled version of **GKi**.

6.3.1. *Constructing normal cutfree G-proofs from normal N-proofs.* As a preliminary warming-up for the more precise results later in the section, we describe first a simple construction of cutfree G-proofs from normal proofs in **Ni**. The argument takes no account of assumption markers, and applies therefore also to **Ni** under CDC. On the other hand, the construction is largely insensitive to the precise G-system for which we want cutfree deductions. Below we present the argument for **G1i**; for other systems small adaptations are necessary.

Let us write $\vdash_G \Gamma \Rightarrow D$ if $\mathbf{G1i} \vdash \Gamma \Rightarrow D$, and $\Gamma \vdash_N D$ if there is a normal natural deduction proof of D from assumptions in Γ.

THEOREM. $\vdash_G \Gamma \Rightarrow D$ *iff* $\Gamma \vdash_N D$.

PROOF. We show by induction on the depth of normal N-deductions that if $\Gamma \vdash_N D$, then $\vdash_G \Gamma \Rightarrow D$.
Case 1. Suppose that $\mathcal{D}$ consists of the assumption A; this is translated as the axiom $A \Rightarrow A$.
Case 2. Let $\mathcal{D}$ be a normal derivation for $\Gamma \vdash_N D$, and suppose that the final step in $\mathcal{D}$ is an I-rule application. Let $\mathcal{E}$ be the deduction(s) in the sequent calculus corresponding by induction hypothesis to the immediate subdeduction(s) of $\mathcal{D}$; apply to $\mathcal{E}$ the corresponding R-rule. For example, if $\mathcal{D}$ ends with

$$\dfrac{\begin{matrix}[A]\\ \mathcal{D}'\\ B\end{matrix}}{A \to B}$$

we have by induction hypothesis a deduction showing $\vdash_G \Gamma, A \Rightarrow B$ (use weakening on the conclusion of $\mathcal{E}$ to introduce A on the left if necessary, hence by R$\to$ we have $\mathbf{G1i} \vdash \Gamma \Rightarrow A \to B$.
Case 3. The conclusion of $\mathcal{D}$ is the result of an E-rule application, or of $\bot_i$. We note beforehand, that if a main branch ending with an E-rule contains an application α of $\vee$E or $\exists$E, then α is the sole application of one of these rules and moreover it is the final rule applied in the main branch; for suppose not, then $\mathcal{D}$ is not normal, since in this case the uppermost occurrence β (which may, or may not coincide with α) of an application of $\vee$E or $\exists$E is followed by an E-rule, and it is possible to apply a permutation conversion.

Let τ be the main branch of $\mathcal{D}$. τ is unique, since the I-part of τ is empty (multiple main branches can only occur as a result of $\wedge$I). τ does not contain a minor premise, hence no assumption can be discharged along τ.

Thus the first f.o. C of τ belongs to Γ and is a major premise of an E-rule. Suppose e.g. $C \equiv C_1 \to C_2$. Then $\mathcal{D}$ has the form

$$\dfrac{C_1 \to C_2 \quad \begin{matrix}\mathcal{D}' \\ C_1\end{matrix}}{\begin{matrix}(C_2) \\ \mathcal{D}'' \\ A\end{matrix}}$$

where (C_2) refers to a single occurrence in $\mathcal{D}''$. The f.o. C_1 cannot depend on other assumptions besides the ones on which A depends, since no assumptions are discharged in τ, which passes through the f.o. C_2.

Thus, if $\mathcal{D}$ establishes $\Gamma, C_1 \to C_2 \vdash_{\mathrm{N}} A$, then $\mathcal{D}'$ establishes $\Gamma \vdash_{\mathrm{N}} C_1$, and $\mathcal{D}''$ shows $\Gamma, C_2 \vdash_{\mathrm{N}} A$. By the induction hypothesis,

$$\vdash_{\mathrm{G}} \Gamma \Rightarrow C_1, \qquad \vdash_{\mathrm{G}} \Gamma, C_2 \Rightarrow A.$$

and therefore by $\to$L:

$$\vdash_{\mathrm{G}} C_1 \to C_2, \Gamma \Rightarrow A.$$

To consider yet another subcase, suppose now that $C \equiv C_1 \vee C_2$. Then $\mathcal{D}$ has the form

$$\dfrac{C_1 \vee C_2 \quad \begin{matrix}[C_1]^u \\ \mathcal{D}_1 \\ A\end{matrix} \quad \begin{matrix}[C_2]^v \\ \mathcal{D}_2 \\ A\end{matrix}}{A}\,u, v$$

deriving A from $C_1 \vee C_2, \Gamma$. Then

$$\begin{matrix}[C_1] \\ \mathcal{D}_1 \\ A\end{matrix} \qquad \text{and} \qquad \begin{matrix}[C_2] \\ \mathcal{D}_2 \\ A\end{matrix}$$

are correct normal derivations of smaller depth and therefore by the induction hypothesis

$$\vdash_{\mathrm{G}} \Gamma, C_1 \Rightarrow A, \qquad \vdash_{\mathrm{G}} \Gamma, C_2 \Rightarrow A$$

Then the $\vee$L rule gives us a deduction showing $\vdash_{\mathrm{G}} \Gamma, C_1 \vee C_2 \Rightarrow A$. The other cases are left to the reader. ⊠

REMARK. If we adapt the argument to the construction of proofs in **G3i**, then either we have to generalize the axioms to $\Gamma, A \Rightarrow A$ for arbitrary A, or we must in the basis case use standard proofs of the axioms for compound A.

The argument also yields a quick proof of the subformula property for normal deductions, requiring a partial analysis of their structure only: the cutfree Gentzen proof constructed contains formulas from the original N-deduction only, and for Gentzen proofs the subformula property is immediate.

6.3.1A. ♠ Let $\mathcal{D}$ be a normal deduction in **Ni**, and let $\mathsf{G}(\mathcal{D})$ be the corresponding deduction in **GKi** constructed in the proof of the theorem above. Then $|\mathsf{G}(\mathcal{D})| \leq |\mathcal{D}|$.

6.3.1B. ♠ Show that for deductions in **G3i** + Cut or **GKi** + Cut (with atomic A in L$\bot$) we can find a normal proof $\mathsf{N}(\mathcal{D})$ in **Ni** such that $|\mathsf{N}(\mathcal{D})| < c2^{|\mathcal{D}|}$ for a positive integer c. For the full system we can take $c = 2$, for the system without $\bot$ it suffices to take $c = 1$ (cf. 3.3.4B).

6.3.2. If we analyse the construction in the proof of the preceding theorem, we discover that we do not obtain arbitrary cutfree proofs, but in fact proofs satisfying an extra property: whenever we encounter an application of L$\to$, L$\wedge$ or L$\forall$, the antecedent active formula in (one of) the premise(s) is itself principal. We call deductions obeying this condition *normal* deductions. The question rises, whether there is perhaps a one-to-one correspondence between normal natural deductions and normal cutfree G-proofs? Before we can answer this quewstion however, we first have to be more precise as to the systems we want to compare. For the N-systems the choice is canonical: standard **Ni**, which is isomorphic to a calculus of typed terms. On the side of the G-systems, we choose a term-labelled version of **GKi**.

6.3.3. NOTATION. Below we use u, v for deduction variables (with formula type), x, y for individual variables. If we wish to emphasize that a term is a deduction term, we use d, e. ⊠

6.3.4. DEFINITION. The system t-**GKi** with term labels is given by the following rules (i = 0,1).

$$\text{Ax}\quad \Gamma, u{:}\, P \Rightarrow u{:}\, P \qquad\qquad \text{L}\bot\quad \Gamma, u{:}\,\bot \Rightarrow \mathrm{E}_\bot(u){:}\, A$$

$$\text{L}\wedge\ \frac{\Gamma, u{:}\, A_0 \wedge A_1, v{:}\, A_i \Rightarrow t(u,v){:}\, B}{\Gamma, u{:}\, A_0 \wedge A_1 \Rightarrow t(u, \mathbf{p}_i u){:}\, B} \qquad \text{R}\wedge\ \frac{\Gamma \Rightarrow t_0{:}\, A_0 \qquad \Gamma \Rightarrow t_1{:}\, A_1}{\Gamma \Rightarrow \mathbf{p}(t_0, t_1){:}\, A_0 \wedge A_1}$$

$$\text{L}\to\ \frac{\Gamma, u{:}\, A_0 \to A_1 \Rightarrow s(u){:}\, A_0 \qquad \Gamma, u{:}\, A_0 \to A_1, v{:}\, A_1 \Rightarrow t(u,v){:}\, B}{\Gamma, u{:}\, A_0 \to A_1 \Rightarrow t(u, us(u)){:}\, B}$$

$$\text{R}\to\ \frac{\Gamma, u{:}\, A \Rightarrow t{:}\, B}{\Gamma \Rightarrow \lambda u.t{:}\, A \to B}$$

$$\text{L}\vee\ \frac{\Gamma, u{:}\, A_0 \vee A_1, v{:}\, A_0 \Rightarrow t_0(u,v){:}\, C \qquad \Gamma, u{:}\, A_0 \vee A_1, w{:}\, A_1 \Rightarrow t_1(u,w){:}\, C}{\Gamma, u{:}\, A_0 \vee A_1 \Rightarrow \mathrm{E}^\vee_{v,w}(u, t_0(u,v), t_1(u,w)){:}\, C}$$

$$\text{R}\vee\ \frac{\Gamma \Rightarrow t{:}\, A_i}{\Gamma \Rightarrow \mathbf{k}_i(t){:}\, A_0 \vee A_1}$$

$$\text{L}\forall\ \frac{\Gamma, u{:}\, \forall x A, v{:}\, A[x/t] \Rightarrow s(u,v){:}\, B}{\Gamma, u{:}\, \forall x A \Rightarrow s(u, ut){:}\, B} \qquad \text{R}\forall\ \frac{\Gamma \Rightarrow t{:}\, A}{\Gamma \Rightarrow \lambda x.t{:}\, \forall x A}$$

$$\text{L}\exists\ \frac{\Gamma, u{:}\, \exists x A(x), v{:}\, A(y) \Rightarrow s(u,v,y){:}\, B}{\Gamma, u{:}\, \exists x A(x) \Rightarrow \text{E}^{\exists}_{v,y}(u, s(u,v,y)){:}\, B} \qquad \text{R}\exists\ \frac{\Gamma \Rightarrow s{:}\, A[x/t]}{\Gamma \Rightarrow \mathbf{p}(t,s){:}\, \exists x A}$$

The variable restrictions on the rules are as usual: in R$\forall$ $x \notin \text{FV}(\Gamma, \forall x A)$, in L$\exists$ $x \notin \text{FV}(\Gamma, \exists x A, B)$.

By dropping all the term-labels from the formulas we obtain the system **GKi**. ⊠

6.3.5. DEFINITION. (*Normal and pruned deductions*) A deduction $\mathcal{D}$ in **GKi** or t-**GKi** is *normal*, if the active antecedent formula in any application of L$\forall$, L$\wedge$, L$\to$ is itself principal, and in the applications of L$\bot$ only atomic A are used.

A deduction $\mathcal{D}$ of $\Gamma \Rightarrow t{:}\, A$ in t-**GKi** is said to be *pruned*, if all deduction variables of Γ actually occur free in t. A deduction of $\Gamma \Rightarrow A$ in **GKi** is *pruned* if Γ is a set (a multiset where every formula has multiplicity 1) and $\mathcal{D}$ cannot be written as $\mathcal{D}'[\Gamma' \Rightarrow]$ for inhabited Γ'. (The notation $\mathcal{D}'[\Gamma' \Rightarrow]$ was explained in 4.1.2.) ⊠

The restriction to atomic A in applications of L$\bot$ in the definition of normality is needed because we had a corresponding restriction on $\bot_\text{i}$ in **Ni**.

One easily sees that every deduction in **GKi** of $\Gamma \Rightarrow A$ or in t-**GKi** of $\Gamma \Rightarrow t{:}\, A$ may be pruned to deductions of $\Gamma' \Rightarrow A$ or $\Gamma' \Rightarrow t{:}\, A$ respectively, with $\Gamma' \subset \Gamma$; the original deduction is obtained from the pruned deduction by a global weakening. The two notions of "pruned" do not fully correspond: if we have in the antecedent of the conclusion of a pruned deduction in t-**GKi** $x{:}\, B, y{:}\, B$, with x, y distinct, then stripping of the terms produces a deduction in **GKi** with in the conclusion multiple occurrences of B in the antecedent, so some further pruning is then needed.

DEFINITION. The system h-**GKi** with head formulas is specified as follows. Sequents are of the form $\Pi; \Gamma \Rightarrow A$ with $|\Pi| \leq 1$. To improve readability, we often write $-$ for an empty Π.

$$\text{Ax}\quad P; \Gamma \Rightarrow P \qquad\qquad \text{L}\bot\quad \bot; \Gamma \Rightarrow A$$

$$\text{L}\wedge\ \frac{A_i; A_0 \wedge A_1, \Gamma \Rightarrow B}{A_0 \wedge A_1; \Gamma \Rightarrow B} \qquad \text{R}\wedge\ \frac{\Pi; \Gamma \Rightarrow A_0 \qquad \Pi; \Gamma \Rightarrow A_1}{\Pi; \Gamma \Rightarrow A_0 \wedge A_1}$$

$$\text{L}{\to}\ \frac{-; \Gamma, A_0{\to}A_1 \Rightarrow A_0 \qquad A_1; \Gamma, A_0{\to}A_1 \Rightarrow B}{A_0{\to}A_1; \Gamma \Rightarrow B} \qquad \text{R}{\to}\ \frac{\Pi; \Gamma, A_0 \Rightarrow A_1}{\Pi; \Gamma \Rightarrow A_0{\to}A_1}$$

$$\text{L}\vee\ \frac{-; \Gamma, A_0{\vee}A_1, A_0 \Rightarrow C \qquad -; \Gamma, A_0{\vee}A_1, A_1 \Rightarrow C}{A_0{\vee}A_1; \Gamma \Rightarrow C} \qquad \text{R}\vee\ \frac{\Pi; \Gamma \Rightarrow A_i}{\Pi; \Gamma \Rightarrow A_0{\vee}A_1}$$

$$\text{L}\forall\ \frac{A[x/t]; \Gamma, \forall xA \Rightarrow B}{\forall xA; \Gamma \Rightarrow B} \qquad \text{R}\forall\ \frac{\Pi; \Gamma \Rightarrow A}{\Pi; \Gamma \Rightarrow \forall xA}$$

$$\text{L}\exists\ \frac{A[x/y]; \Gamma, \exists xA \Rightarrow B}{\exists xA; \Gamma \Rightarrow B} \qquad \text{R}\exists\ \frac{\Pi; \Gamma \Rightarrow A[x/t]}{\Pi; \Gamma \Rightarrow \exists xA}$$

$$\text{D}\ \frac{A; \Gamma \Rightarrow B}{-; A, \Gamma \Rightarrow B}$$

In R$\forall$ $x \notin \text{FV}(\Pi\Gamma)$, in L$\exists$ $y \notin \text{FV}(\Gamma, \exists xA, B)$; P atomic; and $i = 0$ or $i = 1$. The formula in front of the semicolon, if present, is called the *head formula.* The names of the rules are given as usual; "D" stands for *dereliction* (since A loses its status of head formula).

We may also formulate a combination ht-**GKi** in the obvious way. ⊠

Dropping in a deduction $\mathcal{D}$ in h-**GKi** the semicolons, and deleting the repetitions resulting from D, results in a normal deduction in **GKi**. Conversely, a normal deduction in **GKi** can straightforwardly be transformed into a deduction in h-**GKi**. The proof that every sequent derivable in **GKi** also has a normal proof may be established either by proving closure under Cut of h-**GKi**, or via the correspondence, to be established below, of normal derivations in **GKi** with normal derivations in **Ni**.

6.3.6. LEMMA. *(Contraction for* h-**GKi***)*

(i) If $\vdash_n A; \Gamma, A \Rightarrow B$ *then* $\vdash_n A; \Gamma \Rightarrow B$*, and*

(ii) if $\vdash_n \Pi; \Gamma, A, A \Rightarrow B$ *then* $\vdash_n \Pi; \Gamma, A \Rightarrow B$.

6.3.7. THEOREM. *(Cut elimination for* h-**GKi***) The system with head formulas is closed under the rules of Head-cut* (Cut_h) *and Mid-cut* (Cut_m)*:*

$$\text{Cut}_\text{h}\ \frac{\Pi; \Gamma \Rightarrow A \qquad A; \Gamma' \Rightarrow B}{\Pi; \Gamma\Gamma' \Rightarrow B} \qquad \text{Cut}_\text{m}\ \frac{-; \Gamma \Rightarrow A \quad \Pi; \Gamma', A \Rightarrow B}{\Pi; \Gamma\Gamma' \Rightarrow B}$$

PROOF. The proof follows the standard proof of Cut elimination for the system **GKi** without term labels. ⊠

6.3.8. THEOREM. *The systems* **GKi** *and* h-**GKi** *are equivalent:*

$$\mathbf{GKi} \vdash \Gamma \Rightarrow A \ \textit{ iff } \ \text{h-}\mathbf{GKi} \vdash -; \Gamma \Rightarrow A.$$

PROOF. The direction from right to left is trivial, as observed before. The other direction is proved by induction on the depth of deductions in **GKi**, using closure under Cut of h-**GKi**. We consider a typical case. Suppose the proof in **GKi** ends with

$$\frac{A{\to}B, \Gamma \Rightarrow A \qquad A{\to}B, B, \Gamma \Rightarrow C}{A{\to}B, \Gamma \Rightarrow C}$$

Then by the IH we have deductions in h-**GKi** of

$$-; A{\to}B, \Gamma \Rightarrow A \quad \text{and} \quad -; A{\to}B, B, \Gamma \Rightarrow C.$$

Then we obtain a proof of $-; A{\to}B, \Gamma \Rightarrow C$ with the help of Cut as follows:

$$\dfrac{-; A{\to}B, \Gamma \Rightarrow A \qquad \dfrac{\dfrac{\dfrac{\dfrac{\dfrac{A; A{\to}B \Rightarrow A}{-; A, A{\to}B \Rightarrow A} \quad B; A, A{\to}B \Rightarrow B}{A{\to}B; A \Rightarrow B}}{-; A{\to}B, A \Rightarrow B} \qquad -; A{\to}B, B, \Gamma \Rightarrow C}{-; (A{\to}B)^2, A, \Gamma \Rightarrow C}\,\text{Cut}}{-; (A{\to}B)^3, \Gamma^2 \Rightarrow C}\,\text{Cut}$$

where X^n is short for n copies of X. The desired conclusion follows by closure under contraction. If A or B are not atomic, we have to insert at the top standard deductions for $A; A{\to}B \Rightarrow A$ or $B, A{\to}B \Rightarrow B$ respectively. ⊠

6.3.9. *The correspondence between pruned normal proofs in* t-**GKi** *and normal proofs in* **Ni**. Instead of using the conventional notation $\mathbf{p}_0$ and $\mathbf{p}_1$ for the projections, we find it more convenient for the arguments below to switch notation and to introduce two constants $\mathbf{0}$ and $\mathbf{1}$ such that $\mathbf{p}_0 t \equiv t\mathbf{0}$, $\mathbf{p}_1 t \equiv t\mathbf{1}$. Then successive application of L→, L∧, L∀ results in a term of the form $ut_0 \dots t_{n-1}$ where each t_i is either a deduction term or an individual term or one of $\mathbf{0}$, $\mathbf{1}$, corresponding to L→, L∀ and L∧-applications respectively. The deduction variable u is called the head-variable.

6.3.10. Lemma. *The term t in a proof in* t-**GKi** *of $\Gamma \Rightarrow t{:}\,A$ represents a normal natural deduction.*

Proof. We only have to check that the applications of rules cannot introduce terms of the following form:

$$\mathbf{p}(-,-)-,\ (\lambda u.-)-,\ (\lambda x.-)-,\ \mathrm{E}^{\vee}_{v,w}(\mathbf{k}_i-,-,-),\ \mathrm{E}^{\exists}_{u,y}(\mathbf{p}(-,-),-),$$

$$\mathrm{E}^{\vee}_{u,v}((-)-,-,-),\ \mathrm{E}^{\exists}_{u,y}((-)-;-),\ \mathrm{E}^{\vee}_{u,v}(\mathrm{E},-,-),\ \mathrm{E}^{\exists}_{u,y}(\mathrm{E},-)$$

where the $-$ stand for arbitrary terms or $\mathbf{0}$ or $\mathbf{1}$ (at least when this makes sense syntactically). The different occurrences of $-$ do not necessarily represent the same expression. The E stands for a term of the form $\mathrm{E}^{\vee}_{u,v}(-,-,-)$ or $\mathrm{E}^{\exists}_{u,y}(-,-)$.

It is straightforward to carry out the check. ⊠

6.3.11. PROPOSITION. *To a normal term t, representing a normal deduction in* **Ni** *and deriving A from Γ, corresponds a unique pruned normal deduction $\mathcal{D}^t$ in* t-**GKi** *deriving $\Gamma \Rightarrow t\colon A$.*

PROOF. By induction on the *size* of t.

Case 1. If t begins with an I-operator, that is to say, if t has one of the following forms:

$$\mathbf{p}(d_0, d_1),\ \mathbf{p}(s, d),\ \lambda u.d,\ \lambda x.d,\ \mathbf{k}_i d$$

then the last rule applied in t is $\wedge$I, $\exists$I, $\to$I, $\forall$I, $\wedge$I, respectively, and $\mathcal{D}^t$ ends with a corresponding application of R$\wedge$, R$\exists$, R$\to$, R$\forall$, R$\wedge$. For example, if t ends with $\wedge$I, $\mathcal{D}^t$ has the form

$$\frac{\begin{array}{cc}\mathcal{D}^t_0 & \mathcal{D}^t_1 \\ \Gamma \Rightarrow d_0\colon A_0 & \Gamma \Rightarrow d_1\colon A_1\end{array}}{\Gamma \Rightarrow \mathbf{p}(d_0, d_1)\colon A_0 \wedge A_1}$$

and we can apply the IH to d_0 and d_1.

Case 2. Case 1 does not apply. Then t has one of the following forms:

$$t_0,\ \mathrm{E}^{\vee}_{v_1,v_2}(t_0, t_1, t_2),\ \ \mathrm{E}^{\exists}_{v,y}(t_0, t_1),\ \ \mathrm{E}_{\perp}(t_0)$$

where $t_0 \equiv u\varepsilon$, and ε is a (possibly empty) string $s_0, s_1 \ldots s_{n-1}$ such that each s_i is either a deduction term, or an individual term, or one of $\mathbf{0}$, $\mathbf{1}$. u is called the head-variable of t. Note that the fact that t is normal, and that case 1 does not apply, precludes that t_0 begins with an introduction operator (i.e., one of $\mathbf{p}$, λx, λv, $\mathbf{k}_i$), or that t is of the form $\mathrm{E}^{\vee}_{v_1,v_2}(t_0, t_1, t_2)$ or $\mathrm{E}^{\exists}_{v,y}(t_0, t_1)$, while t_0 begins with an elimination operator.

Subcase 2a. $t \equiv u$ or $t \equiv \mathrm{E}^{\vee}_{v_1,v_2}(u, t_1, t_2)$ or $t \equiv \mathrm{E}^{\exists}_{v,y}(u, t_1)$ or $t \equiv \mathrm{E}_{\perp}(u)$. Then $\mathcal{D}^t$ is an axiom Ax, or ends with L$\vee$, or ends with L$\exists$, or is an axiom L$\perp$ respectively, and the IH may be applied to t_1, t_2.

Subcase 2b. If $t \equiv t_0$ or $t \equiv \mathrm{E}^{\vee}_{u,v}(t_0, t_1, t_2)$ or $t \equiv \mathrm{E}^{\exists}_{v,y}(t_0, t_1)$ or $t \equiv \mathrm{E}_{\perp}(t_0)$, and t_0 is not a variable, then $t_0 \equiv ut'\varepsilon$. If $t \equiv t_0$, put $e(v) \equiv v\varepsilon$; if $t \equiv \mathrm{E}^{\vee}_{v_0,v_1}(t_0, t_1, t_2)$, let $e(v) \equiv \mathrm{E}^{\vee}_{v_0,v_1}(v\varepsilon, t_1, t_2)$, and if $t \equiv \mathrm{E}^{\exists}_{w,y}(t_0, t_1)$, put $e(v) \equiv \mathrm{E}^{\exists}_{w,y}(v\varepsilon, t_1)$, and if $t \equiv \mathrm{E}_{\perp}(t_0)$, put $e(v) \equiv \mathrm{E}_{\perp}(v\varepsilon)$. Then either

2b.1. t' is $\mathbf{0}$ or $\mathbf{1}$, $\mathcal{D}^t$ ends with an L$\wedge$-application

$$\frac{u\colon A_0 \wedge A_1, v\colon A_i, \Gamma \Rightarrow e(u, v\varepsilon)\colon B}{u\colon A_0 \wedge A_1, \Gamma \Rightarrow e(u, u\mathbf{i}\varepsilon)};$$

where $\mathbf{i}$ stands for $\mathbf{0}$ or $\mathbf{1}$, we may apply the IH to $e(u, v\varepsilon)$; or

2b.2. If t' is a deduction term, $\mathcal{D}^t$ ends with a L$\to$-application

$$\frac{u\colon A_0{\to}A_1, \Gamma \Rightarrow d'\colon A_0 \qquad u\colon A_0{\to}A_1,, v\colon A_1\Gamma \Rightarrow e(u, v\varepsilon)\colon B}{u\colon A_0{\to}A_1, \Gamma \Rightarrow e(u, ut'\varepsilon)\colon B},$$

and we can apply the IH to d', $e(u, v\varepsilon)$, or

2b.3. If t' is an individual term, $\mathcal{D}^t$ ends with L$\forall$:

$$\frac{u{:}\,\forall x A,\, v{:}\, A[x/t'] \Rightarrow e(u, v\varepsilon){:}\, B}{u{:}\,\forall x A \Rightarrow e(u, ut'\varepsilon){:}\, B},$$

and we can apply the IH to $e(u, v\varepsilon)$. ☒

6.3.12. THEOREM. *There is a bijective correspondence between pruned normal deductions in* t-**GKi** *and normal deductions in* **Ni**.

PROOF. Immediate from the preceding arguments. ☒

In fact, there is an effective procedure for transforming a non-normal deduction into a normal t-**GKi**-deduction by permuting rules; as shown by Schwichtenberg, even strong normalization holds for this process. See 3.7.4.

If we relax the restriction to atomic conclusions on $\bot_i$ in **Ni**, then we may also relax the restriction on L$\bot$ in normal proofs.

6.4 Extensions with simple rules

The extensions considered are extensions of **N**[**mi**] with rules involving only atomic premises, conclusions, and discharged assumptions. These extensions are the counterpart for N-systems of the extensions of G-systems considered in 4.5.

We recall that we shall assume restriction of $\bot_i$ to instances with *atomic conclusions* throughout.

6.4.1. DEFINITION. Let $\mathbf{N}^*$ be any system obtained from **N**[**mi**] by adding rules $\mathrm{Rule}_1(P_0, \ldots, P_{n-1}, Q)$ of *Type Ia* or $\mathrm{Rule}_1(P_0, \ldots, P_n, \bot)$ of *Type Ib* –

$$\frac{P_0 \quad \ldots \quad P_{n-1}}{Q} \quad \text{resp.} \quad \frac{P_0 \quad \ldots \quad P_n}{\bot}$$

– or $\mathrm{Rule}_1(R_0, \ldots, R_n; P_0, \ldots, P_n, Q)$ of *Type Ic* –

$$\frac{\begin{matrix} & [R_i]^{u_i} & \\ & \vdots & \\ \ldots & P_i & \ldots \end{matrix}}{Q}\, u_1, \ldots, u_n$$

– or rules $\mathrm{Rule}_2(Q_1, \ldots, Q_m; P_1, \ldots, P_n)$ of *type II* –

$$\frac{\begin{matrix} & & [P_i]^{u_i} & \\ & & \vdots & \\ Q_1 \ldots Q_m & \ldots & C & \ldots \end{matrix}}{C}\, u_1, \ldots, u_n$$

The Q_j are the major premises, the premises C are the minor premises. The P_i, Q_j, R_k are all atomic. The formula C in rules of type II is arbitrary.

We also require that the set of additional rules is closed under substitution of terms for individual variables, that is if we make in premises, assumptions and conclusion of one of the additional rules a substitution $[\vec{x}/\vec{t}]$, then the result is another rule of the same type.

In these extended systems the *del-rules* are the rules $\vee$E, $\exists$E and the type-II rules. ⊠

REMARKS. Note that addition of a rule of type II is equivalent to the addition of an axiom

$$Q_1 \wedge \ldots \wedge Q_m \to P_1 \vee \ldots \vee P_n$$

or a rule

$$(*) \qquad \frac{Q_1 \wedge \ldots \wedge Q_m}{P_1 \vee \ldots \vee P_m}$$

as may be easily checked by the reader.

Combination of two rules $\text{Rule}_1(P, Q)$ and $\text{Rule}_1(Q, R)$ produces a deduction containing

$$\begin{array}{c} \vdots \\ \underline{P} \\ \underline{Q} \\ R \\ \vdots \end{array}$$

This corresponds in G-systems to a basic Cut on Q. By our conventions the occurrence of Q is not counted as a cutformula in N-systems.

6.4.2. Conversions are now extended with permutative conversions and simplifications involving rules of type II. The notion of a maximal segment may be defined as before with our new notion of del-rule, and we can prove normalization.

LEMMA. *Each derivation in a system* **Ni*** *can be transformed into a normal derivation with the same conclusion.*

PROOF. The proof follows the same strategy as before, namely removing a cut of maximal complexity which is topmost in the rightmost branch of the deduction containing such cuts. ⊠

The definition of track is as before, but with our new notion of del-rule.

THEOREM. *Let $\mathcal{D}$ be a normal derivation in* $\mathbf{N}^*$, *and let π be a track in $\mathcal{D}$. Then π may be divided into three parts: an* E-part $\sigma_0, \ldots, \sigma_{i-1}$ *(possibly empty), a* minimal part $\sigma_i, \ldots, \sigma_k$, *and an* I-part $\sigma_{k+1}, \ldots, \sigma_n$ *(possibly empty) such that*

(i) for each σ_j in the E-part *one has $j < i$, σ_j is a major premise of an* E-*rule, and σ_{j+1} is a strictly positive part of σ_j, and therefore each σ_j is an s.p.p. of σ_0,*

(ii) for each σ_j in the I-*part one has $k < j$, and if $j \neq n$, then σ_j is a premise of an I-rule and an s.p.p. of σ_{j+1}, so each σ_j is an s.p.p. of σ_n,*

(iii) for $i \leq j < k$ the segment σ_j is a premise of a type-I rule or the major premise of a type-II rule, and for $i < j \leq k$ the segment σ_j is the conclusion of a type-I rule or an assumption discharged by a type-II rule,

(iv) if $k \neq n$, σ_k is also a premise of an I-rule or of $\bot_i$. ⊠

LEMMA. *In a normal deduction in a system* $\mathbf{Ni}^*$ *each f.o. belongs to a track.*

THEOREM. *Let $\mathcal{D}$ be a normal deduction of A from assumptions Γ in a system* $\mathbf{Ni}^*$. *Then each formula in $\mathcal{D}$ is either a subformula of Γ, A or an atomic formula occurring in one of the additional rules.*

6.4.2A. ♠ Check the proof of lemma and theorem.

6.5 E-logic and ordinary logic

E-logic is an adaptation of first-order predicate logic which accommodates possibly empty domains and possible non-denoting (undefined) terms. Such terms may arise, for example, in the theory of partial functions. This is achieved by adding a special unary predicate E, the *existence* predicate. Et means "t denotes" or "t is defined".

Two notions of equality play a role: *strict equality* $=$, where $t = s$ means that t and s are both defined and equal, and *weak equality* $\simeq$; $s \simeq t$ means that s is defined iff t is defined, and if one of these is defined, they are equal. In this section we want to compare a version of E-logic with modified quantifier rules with ordinary logic with some special axioms for E added.

This section makes use of the results of the preceding section on extensions of N-systems.

6.5.1. DEFINITION. Let **Nie** be the system obtained from **Ni** by adding a special unary predicate E, and modifying the quantifier rules of **Ni** as follows:

$$\forall\mathrm{I}\,\dfrac{\begin{array}{c}[Ey]^u\\ \mathcal{D}\\ A[x/y]\end{array}}{\forall x A}\,u \qquad \forall\mathrm{E}\,\dfrac{\begin{array}{cc}\mathcal{D}_0 & \mathcal{D}_1\\ \forall x A & Et\end{array}}{At}$$

$$\exists\mathrm{I}\,\dfrac{\begin{array}{cc}\mathcal{D}_0 & \mathcal{D}_1\\ A[x/t] & Et\end{array}}{\exists x A} \qquad \exists\mathrm{E}\,\dfrac{\begin{array}{cc} & [A[x/y]]^u\,[Ey]^v\\ \mathcal{D}_0 & \mathcal{D}_1\\ \exists x A & C\end{array}}{C}\,u,v$$

with the usual variable restrictions.

Nie$^{\simeq}$ is obtained from **Nie** by adding a predicate $\simeq$ for weak equality with axioms:

(a1) $t \simeq t$,
(a2) $s \simeq t \wedge A(s) \to A(t)$ (A atomic),
(a3) $(Es \vee Et \to s \simeq t) \to s \simeq t$,
(a4) $E(ft_1 \dots t_n) \to Et_i$ ($1 \le i \le n$),
(a5) $Rt_1 \dots t_n \to Et_i$ ($1 \le i \le n$)

(R a relation symbol, f a function symbol of the language). The corresponding theory is designated by **Ie**.

Ni^{E} is obtained from **Ni** by adding a unary predicate E and equality $=$, with extra axioms and rules

$$\frac{Et}{t = t} \qquad \frac{At \quad t = s}{As}\ (A \text{ atomic})$$

$$\frac{Rt_1 \dots t_n}{Et_i} \qquad \frac{t_0 = t_1}{Et_j} \qquad \frac{E(ft_1 \dots t_n)}{Et_i}$$

where $j \in \{0, 1\}$, $1 \le i \le n$, R is a predicate letter and f a function symbol of the language. The last three rules are called the *strictness* rules. ⊠

6.5.2. DEFINITION. We define the following map from formulas of **Nie**$^{\simeq}$ to formulas of **Ni**^{E}:

$$\begin{array}{ll}
(Et)^E & := Et,\\
(\bot)^E & := \bot,\\
(t_0 \simeq t_1)^E & := Et_0 \vee Et_1 \to t_0 = t_1,\\
(Rt_0 \dots t_{n-1})^E & := Rt_0 \dots t_{n-1},\\
(A \circ B)^E & := A^E \circ B^E \text{ for } \circ \in \{\wedge, \vee, \to\},\\
(\forall x A)^E & := \forall x[Ex \to A^E],\\
(\exists x A)^E & := \exists x[Ex \wedge A^E].
\end{array}$$

⊠

6.5.3. DEFINITION. Let us call a formula of $\mathbf{Ni}^{\mathrm{E}}$ *bounded* if all quantified subformulas of A are of the form $\forall x(Ex \to B)$ or $\exists x(Ex \wedge Bx)$. We define a mapping * from the bounded formulas of $\mathbf{Ni}^{\mathrm{E}}$ to the formulas of $\mathbf{Nie}^{\simeq}$ as follows:

$$\begin{array}{ll}(Et)^* & := Et, \\ (\bot)^* & := \bot, \\ (s = t)^* & := s \simeq t \wedge Es \wedge Et, \\ (Rt_0 \dots t_{n-1})^* & := Rt_0 \dots t_{n-1}, \\ (A \circ B)^* & := A^* \circ B^* \text{ for } \circ \in \{\wedge, \vee, \to\}, \\ (\forall x(Ex \to A))^* & := \forall x A^*, \\ (\exists x(Ex \wedge A))^* & := \exists x A^*. \end{array}$$ ⊠

6.5.4. LEMMA. $\mathbf{Nie}^{\simeq} \vdash A \leftrightarrow (A^E)^*$.

PROOF. By induction on the complexity of A. The most interesting case is that of prime formulas $t_0 \simeq t_1$:

$$\begin{array}{rcl}((t_0 \simeq t_1)^E)^* & \equiv & (Et_0 \vee Et_1 \to t_0 = t_1)^* \\ & \equiv & Et_0 \vee Et_1 \to (t_0 \simeq t_1 \wedge Et_0 \wedge Et_1) \\ & \leftrightarrow & (Et_0 \vee Et_1 \to t_0 \simeq t_1) \wedge (Et_0 \vee Et_1 \to Et_0 \wedge Et_1).\end{array}$$

The first half of this conjunction is by axiom (a3) equivalent to $t_0 \simeq t_1$, and the second part is derivable from $t_0 \simeq t_1$ with the help of (a2). ⊠

6.5.5. LEMMA. *If* $\mathbf{Nie}^{\simeq} \vdash \Gamma \Rightarrow A$*, then* $\mathbf{Ni}^{\mathrm{E}} \vdash \Gamma^E \Rightarrow A^E$.

PROOF. By induction on the depth of proofs in $\mathbf{Nie}^{\simeq}$. The more interesting cases concern the axioms. Let us consider the axiom

$$s \simeq t \wedge A[x/s] \to A[x/t],$$

for A atomic. For example, let $A \equiv r \simeq r'$. Assume

$$\begin{array}{l}A[x/s]^E \equiv Er[x/s] \vee Er'[x/s] \to r[x/s] = r'[x/s], \\ (s \simeq t)^E \equiv Es \vee Et \to s = t, \\ Er[x/t] \vee Er'[x/t].\end{array}$$

We have to show $r[x/t] = r'[x/t]$. Because of the strictness of function symbols, if x actually occurs in r, $Er[x/t] \vee Er'[x/t]$ implies Et. With the second assumption this yields $s = t$ and Es. By induction on the complexity of r'' one proves for all r'':

$$s = t \to r''[x/s] = r''[x/t],$$

hence in particular $r[x/s] = r[x/t]$, $r'[x/s] = r'[x/t]$, and so by strictness $Er[x/s] \vee Er'[x/s]$. Therefore $r[x/s] = r'[x/s]$ (first assumption), so $r[x/t] = r'[x/t]$. The rest of the proof is left to the reader. ⊠

6.5.6. LEMMA. *Let Γ, A consist of bounded formulas. If* $\mathbf{Ni}^{\mathrm{E}} \vdash \Gamma \Rightarrow A$ *then* $\mathbf{Nie}^{\simeq} \vdash \Gamma^* \Rightarrow A^*$.

PROOF. By induction on the $|\mathcal{D}|$, where $\mathcal{D}$ is a *normal* proof of $\Gamma \Rightarrow A$ in $\mathbf{Ni}^{\mathrm{E}}$, we construct a proof $\mathcal{D}^*$ of $\Gamma^* \Rightarrow A^*$. Let us consider the four cases where the last rule R applied in $\mathcal{D}$ is a quantifier rule. Observe that due to the subformula property for normal deductions, quantifiers appear only as the restricted quantifiers of bounded formulas.

Case 1. R = $\forall$I. The conclusion of $\mathcal{D}$ is of the form $\forall x(Ex \to Ax)$, and the conclusion and its premise belong to the I-part of the track to which they belong, so $\mathcal{D}$ is of the form shown on the left below. By the IH applied to $\mathcal{D}_0$ there exists a deduction $\mathcal{D}^*$ as shown on the right.

$$\dfrac{\dfrac{\begin{matrix}[Ex]^u\\ \mathcal{D}_0\end{matrix}}{Ex \to Ax}\,u}{\forall x(Ex \to Ax)} \qquad\qquad \dfrac{\begin{matrix}[Ex]^u\\ \mathcal{D}_0^*\\ A^*x\end{matrix}}{\forall x A^* x}\,u$$

Case 2. R = $\exists$I. Then for similar reasons $\mathcal{D}$ must be of the form on the left below, while by the IH we can then construct $\mathcal{D}^*$ as on the right below:

$$\dfrac{\dfrac{\begin{matrix}\mathcal{D}_0\\ Et\end{matrix} \quad \begin{matrix}\mathcal{D}_1\\ At\end{matrix}}{Et \wedge At}}{\exists x(Ex \wedge Ax)} \qquad\qquad \dfrac{\begin{matrix}\mathcal{D}_0^*\\ Et\end{matrix} \quad \begin{matrix}\mathcal{D}_1^*\\ A^*t\end{matrix}}{\exists x A^* x}$$

Case 3. R = $\forall$E. Then $\mathcal{D}$ is of the form on the left, and we construct, using the IH, a deduction as on the right:

$$\dfrac{\begin{matrix}\mathcal{D}_0\\ \forall x(Ex \to Ax)\end{matrix}}{Et \to At} \qquad\qquad \dfrac{\dfrac{\begin{matrix}\mathcal{D}_0^*\\ \forall x A^* x\end{matrix} \quad Et^u}{A^*t}}{Et \to A^*t}\,u$$

Case 4. R = $\exists$E. Then $\mathcal{D}$ is of the form on the left, and we construct, using the IH, a deduction as on the right:

$$\dfrac{\begin{matrix}\\ \mathcal{D}_0\\ \exists x(Ex \wedge Ax)\end{matrix} \qquad \begin{matrix}[Ex \wedge Ax]^u\\ \mathcal{D}_1\\ C\end{matrix}}{C}\,u \qquad\qquad \dfrac{\begin{matrix}\mathcal{D}_0^*\\ \exists x A^* x\end{matrix} \qquad \begin{matrix}\dfrac{Ex^u \quad A^*x^v}{[Ex \wedge A^*x]}\\ \mathcal{D}_1^*\\ C^*\end{matrix}}{C^*}\,u, v$$

All other cases are trivial. ⊠

The following is now immediate by combining the preceding lemmas.

6.5.7. THEOREM. $\mathbf{Nie}^{\simeq} \vdash \Gamma \Rightarrow A$ *iff* $\mathbf{Ni}^{\mathrm{E}} \vdash \Gamma^E \Rightarrow A^E$. ⊠

6.5.7A. ♠ Complete the proof of lemma 6.5.5.

6.6 Conservativity of predicative classes

6.6.1. We shall now give a proof-theoretic argument for the conservativity of the addition of predicative classes to intuitionistic first-order theories, as an application of normalization for natural deduction systems. This is another type of extension of the logical N-systems: we now add second-order quantifiers, and need a different argument to extend normalization.

Let **T** be a first-order theory formalized on the basis of intuitionistic predicate logic, plus some (individual) first-order axioms and a set of axiom schemas $\mathcal{F}_i$.

Let $\mathcal{L}$ be the language of **T**; we add n-argument predicate variables X_i^n, Y_i^n, Z_i^n ($i \in \mathbb{N}$) to $\mathcal{L}$, the extended language we call $\mathcal{L}'$.

An axiom schema $\mathcal{F}_i$ is a formula in the language $\mathcal{L}'$, say

$$\mathcal{F}_i(X_1^{n(i,1)}, \ldots, X_{r(i)}^{n(i,r(i))}),$$

where X_j is $n(i,j)$-ary. Each substitution of predicates of suitable arity definable in **T**, say

$$\lambda x_1 \ldots x_{n(i,j)}.A_j(x_1, \ldots, x_{n(i,j)}) \;\; (1 \le j \le r(i)),$$

for the $X_j^{n(i,j)}$ yields an axiom of **T**.

6.6.2. Definition. The *weak second-order extension* $\mathbf{T}^*$ (extension by predicative classes) of **T** is defined as follows. We add to the language of **T** relation variables $X^n, Y^n, Z^n, \ldots$ for n-ary relations, for each n, and the corresponding quantifiers $\forall X^n, \exists X^n$. If $t_1, \ldots, t_n$ are individual terms, then $X^n t_1 \ldots t_n$ is a prime formula. We add the *elementary comprehension schema*:

ECA $\quad \exists X^n \forall x_1 \ldots x_n [X^n(x_1, \ldots, x_n) \leftrightarrow A(x_1, \ldots, x_n)]$

for each A of $\mathbf{T}^*$ not containing bound relation variables. The axiom schemas $\mathcal{F}_i$ of **T** are replaced by corresponding axioms

$$\forall X_1^{n(i,1)} \ldots \forall X_{r(i)}^{n(i,r(i))} \mathcal{F}_i(X_1, \ldots, X_{r(i)}). \qquad \boxtimes$$

Example. Let **HA** be the system of intuitionistic first-order arithmetic, also called Heyting arithmetic, containing symbols for all primitive recursive functions, with as axioms the equality axioms as in 4.4.3, $\forall x(0 \ne \mathrm{S}x)$ (S successor function), $\forall xy(\mathrm{S}x = \mathrm{S}y \to x = y)$, defining axioms for all primitive recursive functions, and all instances of induction: $A[x/0] \land \forall x(A \to A[x/\mathrm{S}x]) \to \forall x A$. The weak second-order extension $\mathbf{HA}^*$ is defined as indicated above, with extra axioms $\forall Xxy(x = y \land Xx \to Xy)$, and the induction schema is replaced by the induction axiom $\forall X(X0 \land \forall n(Xn \to X(\mathrm{S}n)) \to \forall m Xm)$.

For weak second-order extensions we have the following theorem:

6.6.3. Theorem. $\mathbf{T}^*$ *is a conservative extension of* $\mathbf{T}$.

Proof. We make use of a mild extension of the normalization theorem for **Ni**. The axiom schema ECA can be dispensed with at the cost of introducing second-order quantifier rules of the following forms:

$$\frac{A[X^n/Y^n]}{\forall X^n A}\,\forall^2\mathrm{I} \qquad\qquad \frac{\forall X^n A}{A[X^n/\lambda\vec{x}.B]}\,\forall^2\mathrm{E}$$

$$\frac{A[X^n/\lambda\vec{x}.B]}{\exists X^n A}\,\exists^2\mathrm{I} \qquad\qquad \frac{\exists X^n A \qquad \begin{matrix}[A[X^n/Y^n]]^u\\ \vdots \\ C\end{matrix}}{C}\,u,\exists^2\mathrm{E}$$

Here $A[X^n/\lambda x_1 \ldots x_n.B]$ is the formula obtained from A by replacing each prime formula of the form $X^n t_1 t_2 \ldots t_n$ by $B[x_1, \ldots, x_n/t_1, \ldots, t_n]$.

In $\forall^2$I, Y^n does not occur in assumptions on which the formula occurrence $A(Y^n)$ depends; in $\exists^2$E Y^n does not occur free in assumptions on which C depends, except $A[X^n/Y^n]$, nor does Y^n occur free in C. B does not contain bound predicate variables.

To the conversions we add conversions for the second-order quantifiers, and permutative conversions for $\exists^2$E (and if desired simplifications as well).

The 2-*complexity* $\mathrm{c}^2(A)$ of A is the number of second-order quantifiers in A. The 2-*cutrank* of $\mathcal{D}$, $\mathrm{cr}^2(\mathcal{D})$, is the maximum of the 2-complexities of formulas in maximal segments. The 1-*cutrank* of $\mathcal{D}$, $\mathrm{cr}^1(\mathcal{D})$, is the maximum of $|A|$ for all cutformulas with $\mathrm{c}^2(A) = \mathrm{cr}^2(\mathcal{D})$. A *critical* cut is a maximal segment σ with $A \in \sigma$ and $|A| = \mathrm{cr}^1(\mathcal{D})$, $\mathrm{c}^2(A) = \mathrm{cr}^2(\mathcal{D})$. The *cutlength* of $\mathcal{D}$, $\mathrm{cl}(\mathcal{D})$ is the total number of cutformulas A in critical cuts.

A t.c.c. (topmost critical cut) is now defined as before. The notion of subformula is extended by stipulating that $A(\lambda x_1 \ldots x_n.B)$ is a subformula of $\forall X^n A(X^n)$, $\exists X^n A(X^n)$, for any B not containing bound relation variables. With this notion of subformula we prove normalization and give an analysis of the structure of tracks as before, and obtain the subformula property. The normalization proof uses a nested induction: a main induction on $\mathrm{cr}^2(\mathcal{D})$, with a subinduction on $\mathrm{cr}^1(\mathcal{D})$, and a sub-subinduction on $\mathrm{cl}(\mathcal{D})$. At each reduction step either $\mathrm{cr}^2(\mathcal{D})$ is lowered, or $\mathrm{cr}^1(\mathcal{D})$ is lowered and $\mathrm{cr}^2(\mathcal{D})$ stays the same, or $\mathrm{cl}(\mathcal{D})$ is lowered and $\mathrm{cr}^1(\mathcal{D})$ and $\mathrm{cr}^2(\mathcal{D})$ stay the same.

Alternatively, one may describe this as an induction on the lexicographically ordered triples $(\mathrm{cr}^2(\mathcal{D}), \mathrm{cr}^1(\mathcal{D}), \mathrm{cl}(\mathcal{D}))$. The induction hypothesis is then that for all $\mathcal{D}'$ with $(\mathrm{cr}^2(\mathcal{D}'), \mathrm{cr}^1(\mathcal{D}'), \mathrm{cl}(\mathcal{D}')) < (\mathrm{cr}^2(\mathcal{D}), \mathrm{cr}^1(\mathcal{D}), \mathrm{cl}(\mathcal{D}))$ the transformation into a normal deduction has already been achieved.

Now let A be a formula in the language of $\mathbf{T}$ such that $\mathbf{T}^* \vdash A$, and let $\mathcal{D}$ be a normal derivation in $\mathbf{T}^*$ with conclusion A. By the subformula property all formulas in $\mathcal{D}$ are subformulas either of A, or of axioms of $\mathbf{T}^*$. Each second-order axiom can only occur at a top node and it cannot appear as subformula of another formula occurring in $\mathcal{D}$; therefore it occurs as the first formula of a track followed by $\forall^2$E-applications, until a first-order formula has been reached; this first-order formula is then an instance of an axiom schema in $\mathbf{T}$. ⊠

REMARK. For a reader familiar with ordinals and transfinite induction, the argument may be described more simply: give formulas A a complexity $|A|^2 = \omega{\cdot}\mathrm{cr}^2(A) + |A|$, and give deductions $\mathcal{D}$ an induction value $\omega m + n$, with m the maximal complexity of formulas in cut segments of $\mathcal{D}$, n the total length of critical cut segments. Each reduction step then lowers the induction value, which is an ordinal below ω^3.

6.6.3A. ♠ Let $\forall^2\mathrm{E}'$, $\exists^2\mathrm{I}'$ be the versions of $\forall^2$E, $\exists^2$I where for B only a relation variable may appear. Show that $\forall^2\mathrm{E}'$, $\exists^2\mathrm{I}'$ plus ECA is equivalent to $\forall^2$E, $\exists^2$I relative to the other axioms and rules. What would go wrong in the proof of the theorem if B were completely unrestricted?

6.7 Conservativity for Horn clauses

The results in this section will be used in a proof of the completeness of a generalization of linear resolution. Throughout his section we restrict attention to the language without $\vee, \exists$.

6.7.1. DEFINITION. An *expansion* of a deduction $\mathcal{D}$ in **Nm** consists in the replacement of a subdeduction $\mathcal{D}'$ by another subdeduction according to one of the following three rules:

$$\text{(1)}\quad \begin{array}{c}\mathcal{D}'\\ A\to B\end{array} \quad\text{is replaced by}\quad \dfrac{\dfrac{\begin{array}{c}\mathcal{D}'\\ A\to B\end{array}\quad A^y}{B}}{A\to B}\,y \quad (y\text{ not free in }\mathcal{D}')$$

$$\text{(2)}\quad \begin{array}{c}\mathcal{D}'\\ \forall xA\end{array} \quad\text{is replaced by}\quad \dfrac{\dfrac{\begin{array}{c}\mathcal{D}'\\ \forall xA\end{array}}{A[x/y]}}{\forall xA} \quad (y\text{ not free in }\mathcal{D}')$$

$$\text{(3)}\quad \begin{array}{c}\mathcal{D}'\\ A\wedge B\end{array} \quad\text{is replaced by}\quad \dfrac{\dfrac{\begin{array}{c}\mathcal{D}'\\ A\wedge B\end{array}}{A}\quad \dfrac{\begin{array}{c}\mathcal{D}'\\ A\wedge B\end{array}}{B}}{A\wedge B}$$

In term notation the expansions correspond to replacing, respectively,

(1) $t^{A \to B}$ by $\lambda u^A.t^{A \to B}u$ ($u \notin \mathrm{FV}(t)$),

(2) $t^{\forall x A}$ by $\lambda y.t^{\forall x A}y$ ($y \notin \mathrm{FV}(t)$),

(3) $t^{A \wedge B}$ by $\mathbf{p}(\mathbf{p}_0 t, \mathbf{p}_1 t)$. ⊠

Cases (1) and (2) are often called η-expansions; this term is sometimes extended to case (3). The term "expansion" has been chosen since the inverse replacements (right hand side replaced by left hand side) are usually called in the terminology of the λ-calculus contractions; in particular the inverses of (1) and (2) are called η-contractions (in this book we have used "conversion" instead of "contraction" however).

Expansions may create new redexes. Therefore we want to allow them only in positions where no new redexes are created. We define:

6.7.2. Definition. (*Minimal position, long normal form*) A formula occurrence A is said to be in *end position* in a deduction $\mathcal{D}$, if A is either the conclusion of $\mathcal{D}$, or the minor premise of an application of $\to$E. A formula occurrence A is said to be in *minimal* position, if *either*

- A is the conclusion of an E-rule application and a premise of an I-rule application *or*

- A is in end position and the conclusion of an E-rule application.

A deduction is in *long normal form* if $\mathcal{D}$ is in normal form and no expansions at minimal positions are possible. ⊠

Remarks. (i) The expansion of an occurrence at a minimal position of a normal deduction does not create new redexes. Clearly, the minimal part of a path in a deduction in long normal form always consists of a single atomic formula. In order to construct, starting from a given deduction, a deduction in long normal form with the same conclusion, we first normalize, then apply expansions.

(ii) A deduction in long normal form is comparable to a sequent calculus deduction with the axioms $\Gamma, A \Rightarrow A, \Delta$ or $A \Rightarrow A$ restricted to atomic A. The construction of a deduction in $\to$**Nm** from a deduction in a Gentzen system, with atomic instances of the axioms only, as in 3.3 produces a deduction in long normal form. Conversely, the construction in 6.3 produces a deduction with atomic instances of the axioms from a deduction in long normal form.

6.7.3. Lemma. *Let $\mathcal{D}$ be a normal deduction. There is a terminating sequence of expansions transforming $\mathcal{D}$ into a deduction in long normal form.*

PROOF. Let ed($\mathcal{D}$), the *expansion degree*, of $\mathcal{D}$ be the sum of the sizes of formulas in minimal position. Assume $\mathcal{D}$ to be normal. Now search for an occurrence A of a compound formula in minimal position, such that no formula occurrence of this kind occurs above A. Then an expansion of $\mathcal{D}$ at A decreases the expansion degree of the deduction. ⊠

REMARKS. (i) The depth of the long normal form $\mathcal{D}'$ of $\mathcal{D}$ constructed according to the recipe above is at most $3|\mathcal{D}|$.

(ii) the transition to long normal form corresponds in Gentzen systems to the replacement of axioms with non-atomic active formulas by deductions of these axioms from axioms with atomic active formulas (cf. 3.1.3A).

6.7.4. DEFINITION. A *generalized Horn formula* is a formula of the form

$$\forall \vec{x}(A_0 \wedge \ldots \wedge A_{n-1} \to B)$$

where B is atomic and $A_0, \ldots, A_{n-1}$ are formulas without $\to$. A generalized Horn formula is called *definite* if B is atomic, not equal to $\bot$. If the A_i are atomic, we have *Horn formulas, definite Horn formulas* respectively. A *fact* is a Horn formula with $n = 0$, that is to say, a fact is of the form $\forall \vec{x} B$. ⊠

6.7.5. THEOREM.

(i) Let $\mathbf{Nc} \vdash \Gamma \Rightarrow \bot$, *where* Γ *is a set of generalized Horn formulas. Then* $\mathbf{Nm} \vdash \Gamma \Rightarrow \bot$, *by a deduction not involving* $\to$I.

(ii) Let $\mathbf{Nc} \vdash \Gamma \Rightarrow B$, *where* B *is atomic, and* Γ *is a set of definite generalized Horn formulas. Then* $\mathbf{Nm} \vdash \Gamma \Rightarrow B$, *by a deduction not involving* $\to$I.

(iii) If we drop the "generalized" from the preceding two statements, the deduction in $\mathbf{Nm}$ *may be assumed to contain applications of* $\wedge$I *and E-rules only.*

PROOF. Given $\mathbf{Nc} \vdash \Gamma \Rightarrow \bot$, there is by theorem 2.3.6 a deduction in $\mathbf{Nm}$ of $\Gamma, \Delta \Rightarrow \bot$, where Δ is a set of stability assumptions $\forall \vec{x}(\neg\neg R\vec{x} \to R\vec{x})$. By lemma 6.7.3 we may assume $\mathcal{D}$ to be in long normal form.

Closure of assumptions in an $\mathbf{Nm}$-deduction $\mathcal{D}$ in long normal form of $\Gamma, \Delta \Rightarrow A$, Γ a set of generalized Horn formulas, Δ a set of stability assumptions, A $\to$-free, can only occur in subdeductions of the following form (the double line stands for 0 or more $\forall$E-applications):

$$(1) \qquad \dfrac{\dfrac{\forall \vec{x}(\neg\neg R\vec{x} \to R\vec{x})}{\neg\neg R\vec{t} \to R\vec{t}}\,\forall\text{E} \qquad \dfrac{\begin{matrix}[\neg R\vec{t}\,]^u \\ \mathcal{D}' \\ \bot\end{matrix}}{\neg\neg R\vec{t}}\,{\to}\text{I},u}{R\vec{t}}\,{\to}\text{E}$$

This fact can be proved by induction on the size of deductions $\mathcal{D}$. Every formula occurrence in $\mathcal{D}$ belongs to a main branch or to a subdeduction ending in a minor premise of →E. A main branch must start in a generalized Horn formula or in a stability assumption. We note that along the main branch no →I can occur.

If the main branch started in a generalized Horn formula, we can apply the IH to the subdeductions of minor premises (which are implication-free) of →E along the main branch.

If the main branch started in a stability assumption, the minor premises of →E-applications along the main branch are of the form $\neg\neg R\vec{t}$ for R a relation symbol of the language. In this case, the subdeduction of the minor premise must end with an →I, since the whole deduction is in long normal form. Now note that the subdeduction $\mathcal{D}'$ of $\bot$ (as in the prooftree exhibited above) again may be seen as a deduction of an atomic formula from generalized Horn formulas, since $\neg R\vec{t}$ is itself a special case of a generalized Horn formula. So we can apply the IH to $\mathcal{D}'$.

Next we observe that in the subdeductions $\mathcal{D}'$ as above, the assumption $u : \neg R\vec{t}$ necessarily appears as the major premise of an instance of →E. Occurrence as the minor premise of an instance of →E is excluded since it would conflict with the long normal form. Occurrence as premise of an I-rule is excluded since this would lead to subformulas of a form not present in the conclusion and assumptions of the deduction. Hence the elements of $[\neg R\vec{t}]^u$ appear in a subdeduction of the form

$$(2) \qquad \dfrac{\neg R\vec{t} \qquad \begin{matrix}\mathcal{D}'' \\ R\vec{t}\end{matrix}}{\bot}$$

Case 1. Assume there are closed assumptions; then we may look for a subdeduction of type (2) in which a closed assumption $\neg R\vec{t}$ appears such that $\mathcal{D}''$ does not contain closed assumptions. Then we may replace the corresponding subdeduction of type (1) by $\mathcal{D}''$ and we have removed an application of a stability assumption. This may be continued till we arrive at

Case 2. There are no assumptions closed by →I. If there are no stability assumptions used, we are done. If there are still stability assumptions, we look for a subdeduction of type (1) such that $\mathcal{D}'$ does not contain stability assumptions. Then the whole deduction $\mathcal{D}$ may be replaced by $\mathcal{D}'$.

Part (ii) of the theorem for B distinct from $\bot$ is proved in a quite similar way, but now in case 2 the situation that a subdeduction without stability assumptions might derive $\bot$ is excluded (conflict with the subformula property of normal deductions). ⊠

REMARK. As follows from the preceding result, if Γ is a set of definite Horn formulas, the deduction $\mathcal{D}$ of $\bot$ or an atom from Γ is something like (double lines indicating possibly some ∀E-inferences)

$$\dfrac{\dfrac{H_1}{C_1 \land C_2 \to C} \qquad \dfrac{\dfrac{\dfrac{H_{11}}{C_{11} \land C_{12} \to C_1} \quad \dfrac{\begin{matrix}\mathcal{D}_{11}\\ C_{11}\end{matrix} \quad \begin{matrix}\mathcal{D}_{12}\\ C_{12}\end{matrix}}{C_{11} \land C_{12}}}{C_1} \qquad \dfrac{\dfrac{H_{12}}{C_{21} \land C_{22} \to C_2} \quad \dfrac{\begin{matrix}\mathcal{D}_{21}\\ C_{21}\end{matrix} \quad \begin{matrix}\mathcal{D}_{22}\\ C_{22}\end{matrix}}{C_{21} \land C_{22}}}{C_2}}{C_1 \land C_2}}{C}$$

Here H_1, H_{11}, H_{12} are definite Horn formulas which in the prooftree above have been assumed to have the form $\forall \vec{x}(A_1 \land A_2 \to B)$. The formulas $C_1 \land C_2 \to C$ etc. are substitution instances of the clauses obtained by repeatedly applying $\forall$E. $\mathcal{D}_{11}$, $\mathcal{D}_{12}$, $\mathcal{D}_{21}$, $\mathcal{D}_{22}$ are deductions of the same general shape as the whole $\mathcal{D}$. If some of the Horn formulas are facts, the structure of $\mathcal{D}$ is correspondingly simplified at those places.

A simplified presentation of such a deduction is an *implication tree*. The notion of an implication tree for a formula relative to a set of Horn formulas Γ is defined inductively by

- a substitution instance C' of a fact $\forall \vec{x} C$ from Γ is in itself a single-node implication tree for C';
- if $A_1 \land \ldots \land A_n \to B$ $(n > 0)$ is a substitution instance of a Horn formula of Γ, and the $\mathcal{D}_i$ are implication trees for the A_i, then

$$\frac{\mathcal{D}_1 \quad \ldots \quad \mathcal{D}_n}{B}$$

  is an implication tree for B.

The fact that the implication trees give a notion of derivation which is complete for derivations of atoms from Horn formulas Γ is also very easily proved by the following semantical argument, due to R. Stärk.

We construct a model $\mathcal{M}$ for Γ, such that (1) the domain of $\mathcal{M}$ is the set of all (open or closed) terms of the language; (2) function symbols f are interpreted by functions $f^{\mathcal{M}}$ given by $f^{\mathcal{M}}(\vec{t}) := f(\vec{t})$; and (3) relations R are interpreted by relations $R^{\mathcal{M}}$ such that $R^{\mathcal{M}}(\vec{t})$ holds in $\mathcal{M}$ iff $R(\vec{t})$ has an implication tree.

Then $\mathcal{M}$ is a model for Γ. If $\forall \vec{x} B \in \Gamma$ is a fact, then every substitution instance of B' of this fact is in itself an implication tree and hence valid in $\mathcal{M}$; hence the fact itself is valid in $\mathcal{M}$. If $A_1 \land \ldots \land A_n \to B$ is a substitution instance of an arbitrary $H \in \Gamma$, and $A_1, \ldots, A_n$ are true in $\mathcal{M}$, then they have implication trees, but then also B has an implication tree, and so is true in $\mathcal{M}$; therefore H is true in $\mathcal{M}$.

If now an atomic A semantically follows from Γ, then A holds in $\mathcal{M}$, and hence has an implication tree. Note that this gives us a semantical proof of the conservativeness of **Nc** over **Nm** for formulas of the form $\bigwedge \Gamma \to B$ (B atomic), since an implication tree obviously corresponds to a deduction in minimal logic.

6.8 Strong normalization for $\rightarrow$Nm and $\lambda_{\rightarrow}$

Strong normalization is a useful property to have: suppose we have a mapping ϕ from a term system **S** to a term system **S**$'$ such that a reduction step in **S** translates under ϕ into one or more reduction steps in **S**$'$. Then from strong normalization for **S**$'$ we may infer strong normalization for **S**. Normalization as a rule is not enough for such a transfer, unless we can indicate for **S** a strategy which translates under ϕ into a strategy for normalizing in **S**$'$. In preparation for strong normalization of intuitionistic second-order logic, we prove strong normalization for intuitionistic implication logic.

6.8.1. DEFINITION. $\mathrm{SN}(t) := t$ is strongly normalizing.

A term t is *non-introduced* if t is not of the form $\lambda x.s$. More generally, in term calculi a term t is *non-introduced* if the principal operator of t is not an operator corresponding to an introduction rule for the type of t.

So if conjunction is added to the type-forming operations, and pairing with inverses to the constant terms, non-introduced terms are the terms not of the form $\lambda x.s$ or $\mathbf{p}ts$. In the present case, we might also have used the term *non-abstract* for non-introduced. ⊠

6.8.2. DEFINITION. For each formula A we define by induction on the depth of A a "computability" predicate of type A, Comp_A, applicable to terms of type A, as follows:

$$\mathrm{Comp}_X(t) := \mathrm{SN}(t) \ \ (X \text{ a propositional variable}),$$

$$\mathrm{Comp}_{A\rightarrow B}(t) := \forall s(\mathrm{Comp}_A(s) \rightarrow \mathrm{Comp}_B(ts)). \qquad ⊠$$

6.8.3. LEMMA. *The following three properties hold for* Comp_A*:r.*

C1 *If* $\mathrm{Comp}_A(t)$, *then* $\mathrm{SN}(t)$.

C2 *If* $\mathrm{Comp}_A(t)$ *and* $t \succeq t'$, *then* $\mathrm{Comp}_A(t')$.

C3 *If* t *is non-introduced, then* $\forall t' {\prec_1} t\, \mathrm{Comp}_A(t')$ *implies* $Comp_A(t)$.

As a corollary of C3:

C4 *If* t *is non-introduced and normal, then* $\mathrm{Comp}_A(t)$.

PROOF. We establish C1–3 simultaneously by induction on $|A|$.
Basis. $A \equiv X$. C1, C2 are immediate. As to C3, for a non-introduced t such that $\forall t' {\prec_1} t\, \mathrm{Comp}_A(t')$, any reduction path starting from t passes through a $t' \prec_1 t$, $t' \in \mathrm{SN}$ by C1, so $t \in \mathrm{SN}$.
Induction step. $A \equiv B \rightarrow C$.

C1. Suppose $t \in \text{Comp}_{B\to C}$, and let x be a variable of type B. By C4, as a consequence of C3 for B, we have $x \in \text{Comp}_B$, hence $tx \in \text{Comp}_C$. Clearly the reduction tree of t is embedded in the reduction tree for tx, hence $\text{SN}(t)$, since $\text{SN}(tx)$ by C1 for C.

C2. Let $t \in \text{Comp}_{B\to C}$, $t' \preceq t$, $s \in \text{Comp}_B$. Then $ts \in \text{Comp}_C$, $ts \succeq t's$, so $\text{Comp}_C(t's)$ by C2 for C; s is arbitrary, so $t' \in \text{Comp}_{B\to C}$.

C3. Let t be non-introduced, and assume $\forall t' \prec_1 t(\text{Comp}_{B\to C}(t'))$. Let $s \in \text{Comp}_B$, then by induction hypothesis $\text{SN}(s)$; let h_s be the number of nodes in the reduction tree. We prove $ts \in \text{Comp}$ with a subinduction on h_s. If $ts \succ_1 t''$, then *either*

- $t'' \equiv t's$, $t \succ_1 t'$; by assumption for $B \to C$, $\text{Comp}_{B\to C}(t')$, hence $\text{Comp}_C(t's)$; *or*

- $t'' \equiv ts'$, $s \succ_1 s'$; by C2 $\text{Comp}_B(s')$, and $h_{s'} < h_s$, so by the subinduction hypothesis for s', $ts' \in \text{Comp}_C$.

There are no other possibilities, since t is non-introduced; therefore, using C3 for ts, we find that $ts \in \text{Comp}_C$. This holds for all $s \in \text{Comp}_B$ so $t \in \text{Comp}_{B\to C}$. ⊠

6.8.4. Lemma. $\forall s \in \text{Comp}_A(\text{Comp}_B(t[x/s]))$ *implies* $\text{Comp}_{A\to B}(\lambda x.t)$.

Proof. Assume $\forall s{\in}\text{Comp}_A(t[x/s] \in \text{Comp}_B)$; we have to show $(\lambda x.t)s \in \text{Comp}_B$ for all $s \in \text{Comp}_A$. We use induction on $h_s + h_t$, the sum of the sizes of the reduction trees of s and t. (Note that h_t is well-defined since our assumptions imply $\text{Comp}_B(t)$, using $\text{Comp}_A(x)$ for variable x, and by C1 of the preceding lemma $\text{SN}(t)$.) $(\lambda x.t)s$ is non-introduced; if $(\lambda x.t)s \succ_1 t''$, then *either*

- $t'' \equiv (\lambda x.t)s'$ with $s \succ_1 s'$, then by C2 $s' \in \text{Comp}$ and by induction hypothesis $t'' \in \text{Comp}_B$ follows; *or*

- $t'' \equiv (\lambda x.t')s$ with $t \succ_1 t'$, then by C2 $t' \in \text{Comp}$ and by induction hypothesis $t'' \in \text{Comp}_B$; *or*

- $t'' \equiv t[x/s]$, and $\text{Comp}_B(t'')$ holds by assumption.

Now apply C3. ⊠

6.8.5. Theorem. *All terms of* $\boldsymbol{\lambda}_{\to}$ *are strongly computable under substitution, that is to say if* $\text{FV}(t) \subset \{x_1{:}\, A_1, \ldots, x_n{:}\, A_n\}$, $s_i \in \text{Comp}_{A_i}$ $(1 \le i \le n)$, $t{:}\, B$ *then*

$$\text{Comp}_B(t[x_1, \ldots, x_n/s_1, \ldots, s_n]).$$

As a corollary, all terms are computable and therefore strongly normalizable.

PROOF. By induction on the construction of t. Let $r^* \equiv r[x_1, \dots, x_n/s_1, \dots, s_n]$ for all terms r.
Case 1. t is a variable: immediate.
Case 2. $t \equiv t_1 t_2$, Then $t^* \equiv t_1^* t_2^*$; by induction hypothesis $\mathrm{Comp}_{A\to B}(t_1^*)$, $\mathrm{Comp}_A(t_2^*)$. Then $\mathrm{Comp}_B(t^*)$ by definition of $\mathrm{Comp}_{A\to B}$.
Case 3. $t \equiv \lambda y^B.t_1C$. Let $\mathrm{FV}(t_1) \subset \{y, x_1{:}A_1, \dots, x_n{:}A_n\}$, $s \in \mathrm{Comp}_B$, $s_i \in \mathrm{Comp}_{A_i}$. Then by induction hypothesis, $t_1[y, x_1, \dots, x_n/s, s_1, \dots, s_n] \in \mathrm{Comp}_C$, i.e. $t_1^*[y/s] \in \mathrm{Comp}_C$. By the preceding lemma $\lambda y.t_1^* \in \mathrm{Comp}_{B\to C}$.⊠

We now immediately obtain

6.8.6. THEOREM. *All terms of* $\boldsymbol{\lambda}_\to$ *(deductions of* $\to$**Nm***) are strongly normalizable under* β*-reduction.* ⊠

Uniqueness of normal form is either proved directly, or readily follows from Newman's lemma (1.2.8). Strong normalizability for $\mathbf{CL}_\to$ may be proved by the same method as used above for $\boldsymbol{\lambda}_\to$, or can be reduced to strong normalization for $\boldsymbol{\lambda}_\to$ by the obvious embedding of terms of $\mathbf{CL}_\to$ into $\boldsymbol{\lambda}_\to$ (cf. the next subsection).

6.8.7. As a simple example of a *reduction* of strong normalization for a system of terms **S** to strong normalization for **S**′ via a mapping of terms which translates a one-step reduction in **S** into one or more reduction steps in **S**′, we take for **S** the term calculus $\boldsymbol{\lambda}_{\forall\to}$, and for **S**′ the calculus $\boldsymbol{\lambda}_\to$. The reduction map ψ is defined on formulas as follows:

$$
\begin{array}{ll}
\psi(Rt_1 \dots t_n) := R^* & (R^* \in \mathcal{PV}), \\
\psi(A \to B) := \psi A \to \psi B, & \\
\psi(\forall x A) := (Q \to Q) \to A & (Q \in \mathcal{PV} \text{ distinct from the } R^*).
\end{array}
$$

R^* is a propositional variable assigned to the relation letter R. We extend ψ to deductions by assigning to a singleton tree "A" the singleton tree "ψA", and extending the definition of ψ as a homomorphism relative to ∀I, ∀E, i.e.

$$
\dfrac{\begin{array}{c}[A]\\ \mathcal{D}\\ B\end{array}}{A \to B}\to\mathrm{I}
\quad\overset{\psi}{\mapsto}\quad
\dfrac{\begin{array}{c}[\psi A]\\ \psi\mathcal{D}\\ \psi B\end{array}}{\psi(A \to B)}\to\mathrm{I}
$$

etc. For ∀I, ∀E we translate

$$
\dfrac{\begin{array}{c}\mathcal{D}\\ A\end{array}}{\forall y A[x/y]}\forall\mathrm{I}
\quad\overset{\psi}{\mapsto}\quad
\dfrac{\begin{array}{c}\psi\mathcal{D}\\ \psi A\end{array}}{(Q \to Q) \to \psi A}\to\mathrm{I}
$$

$$\dfrac{\begin{matrix}\mathcal{D}\\ \forall x A\end{matrix}}{A[x/t]}\,\forall\text{E} \quad\overset{\psi}{\mapsto}\quad \dfrac{\begin{matrix}\psi\mathcal{D}\\ (Q\to Q)\to\psi A\end{matrix} \qquad \dfrac{Q}{Q\to Q}}{\psi A}\,\to\text{E}$$

Checking that this has the desired effect we leave as an exercise.

6.8.7A. ♠ Show that the embedding ψ just defined has the required properties for reducing strong normalization for $\boldsymbol{\lambda}_{\to\forall}$ to strong normalization for $\boldsymbol{\lambda}_{\to}$. The embedding ψ has the property that if $\Gamma \Rightarrow A$ is provable, then so is $\psi(\Gamma) \Rightarrow \psi(A)$. This property is not needed for the reduction; show that ψ may be somewhat simplified if this property is not required.

6.8.7B. ♠ Assuming strong normalization for $\boldsymbol{\lambda}_{\to\vee}$, show how to obtain strong normalization for $\boldsymbol{\lambda}_{\to\forall\vee\exists}$.

6.8.7C. ♠* Extend the proof of strong normalization via computability to $\boldsymbol{\lambda}_{\to\wedge}$, the term calculus for intuitionistic $\to\wedge$-logic, putting

$$\text{Comp}_{A\wedge B} := \text{Comp}_A(\mathbf{p}_0 t) \text{ and } \text{Comp}_B(\mathbf{p}_1 t).$$

The new detour-conversions are of course $\mathbf{p}_i(\mathbf{p}(t_0, t_1))$ cont t_i $(i = 0, 1)$. Show that lemma 6.8.3 extends to this case, and prove an extra lemma: if $t \in \text{Comp}_A$, $s \in \text{Comp}_B$, then $\mathbf{p}ts \in \text{Comp}_{A\wedge B}$; then prove the strong normalization theorem.

6.8.7D. ♠ Extend the uniqueness of normal form (modulo the renaming of bound variables) to the full calculus **Ni**, relative to detour- and permutative conversions. Including simplification conversions may spoil uniqueness of normal form; why?

6.8.8. *Failure of strong normalization under CDC*

The following example, due to R. Statman, shows that strong normalization fails for natural deduction under the CDC. Let $\mathcal{D}_0(P, Q)$ be the deduction

$$\dfrac{\dfrac{P}{Q\to P} \qquad \dfrac{\dfrac{Q}{P\to Q} \qquad \dfrac{\dfrac{P}{Q\to P} \quad Q}{P}}{Q}}{P}$$

and let $\mathcal{D}_{n+1}(Q, P)$ be

$$\dfrac{\dfrac{Q}{P\to Q} \qquad \mathcal{D}_n(P,Q)}{Q}$$

Note that $\mathcal{D}_n(P, Q)$ has conclusion P.

Let $P^{2n} \equiv P, Q^{2n} \equiv Q$, $P^{2n+1} \equiv Q, Q^{2n+1} \equiv P$. $\mathcal{D}_n(P, Q)$ has the form

$$\dfrac{\dfrac{P}{Q \to P} \qquad \dfrac{\dfrac{Q}{P \to Q} \qquad \dfrac{\dfrac{P}{Q \to P} \quad Q}{P}}{Q}}{\begin{array}{c} P \\ \vdots \\ P^n \end{array}}$$

We now start with a deduction $\mathcal{E}$:

$$\dfrac{\dfrac{\begin{array}{c}\mathcal{D}_0(P^0, Q^0) \\ P^0\end{array}}{Q^0 \to P^0} \qquad \mathcal{D}_0(Q^0, P^0)}{P^0}$$

After one reduction step at the cut shown we obtain a deduction containing as a subdeduction:

$$\dfrac{\dfrac{\begin{array}{c}\mathcal{D}_0(P^1, Q^1) \\ P^1\end{array}}{Q^1 \to P^1} \qquad \mathcal{D}_1(Q^1, P^1)}{P^1}$$

By induction on n we can prove that after the n-th reduction step we have obtained a deduction containing a subdeduction

$$\dfrac{\dfrac{\begin{array}{c}\mathcal{D}_n(P^n, Q^n) \\ P^n\end{array}}{Q^n \to P^n} \qquad \begin{array}{c}\mathcal{D}_{n+1}(Q^n, P^n) \\ Q^n\end{array}}{P^n}$$

where no assumption open in this subdeduction is cancelled in the remainder of the deduction. Applying a normalization step to the $Q^n \to P^n$ shown produces a deduction with as subdeduction

$$\dfrac{\dfrac{\mathcal{D}_{n+1}(Q^n, P^n)}{P^n \to Q^n} \qquad \dfrac{\dfrac{P^n}{Q^n \to P^n} \qquad \mathcal{D}_{n+1}(Q^n, P^n)}{P^n}}{Q^n}$$

with no assumption open in this subdeduction discharged in the remainder of the deduction. This subdeduction is equal to

$$\dfrac{\dfrac{\mathcal{D}_{n+1}(Q^n, P^n)}{P^n \to Q^n} \qquad \mathcal{D}_{n+2}(P^n, Q^n)}{Q^n}$$

which may be rewritten as

$$\dfrac{\dfrac{\mathcal{D}_{n+1}(P^{n+1}, Q^{n+1})}{Q^{n+1} \to P^{n+1}} \qquad \mathcal{D}_{n+2}(Q^{n+1}, P^{n+1})}{P^{n+1}}$$

As a result, the sequence of reduction steps indicated produces deductions forever increasing in size and depth.

6.9 Hyperexponential bounds

6.9.1. *Hyperexponential upper bounds on the growth of deductions.* It is not difficult to estimate the growth of the depth of a deduction on normalizing, by analogy with the result in 5.1. For deductions $\mathcal{D}$ in $\to$**Nm**, the cutrank $\mathrm{cr}(\mathcal{D})$ of $\mathcal{D}$ is simply the maximum of $|A|$ for all cutformulas A in $\mathcal{D}$.

LEMMA. *Let $\mathcal{D}$ be a deduction in $\to$**Nm** with cutrank $\leq k$. Then there is a deduction $\mathcal{D}^* \preceq \mathcal{D}$ with $\mathrm{cr}(\mathcal{D}) < k$ such that $|\mathcal{D}^*| \leq 2^{|\mathcal{D}|}$.*

PROOF. By induction on $|\mathcal{D}|$; the details are left as an exercise. ⊠

From this lemma we obtain immediately

THEOREM. *To each $\mathcal{D}$ in $\to$**Nm** there is a normal $\mathcal{D}^* \preceq \mathcal{D}$ with $|\mathcal{D}^*| \leq 2^{|\mathcal{D}|}_{\mathrm{cr}(\mathcal{D})}$ (which is equal to $\mathrm{hyp}(2, \mathrm{cr}(\mathcal{D}), |\mathcal{D}|)$).* ⊠

6.9.1A. ♠ Provide details of the proof of the lemma.

6.9.2. *Hyperexponential lower bounds on the growth of deductions.* We can easily show, by considering a particular example, that no elementary function (i.e. a primitive recursive function defined with recursions bounded by some finitely iterated exponentiation) can give a universal bound on the increase of the length of a deduction under normalization.

6.9.3. DEFINITION. Let X be a fixed proposition variable, and define the *iterated types* by

$$0X := X, \quad (k+1)X := kX \to kX.$$

The Church numerals of type kX are defined by (cf. 1.2.20)

$$\overline{n}_{kX} := \lambda y^{kX \to kX} x^{kX}.y^n(x).$$

Below we shall abbreviate $\bar{n}_{kX}$ as I^n_k. ⊠

Recall that, if we put

$$t \circ s := \lambda x.t(s(x)),$$

then

$$I_k^n(y) \circ I_k^m(y) =_\beta I_k^{n+m}(y),\ I_k^n \circ I_k^m =_\beta I_k^{nm},\ I_{k+1}^m I_k^n =_\beta I_k^{n^m}\ (m > 0).$$

The following deduction, logically trivial, represents the Church numeral $\overline{3}_A$:

$$\dfrac{\dfrac{\dfrac{f\colon A \to A \qquad \dfrac{f\colon A \to A \qquad \dfrac{f\colon A \to A \quad x\colon A}{fx\colon A}}{f^2x\colon A}}{f^3x\colon A}}{\lambda x.f^3x\colon A \to A}}{\overline{3}_A\colon (A \to A) \to (A \to A)}$$

Note that the deduction corresponding to $\overline{n}_A$ has depth $n+2$.

6.9.4. THEOREM. *We write $\mathcal{D}^{\mathrm{nf}}$ for the normal form of $\mathcal{D}$ in $\to$**Nm**. There is no fixed k such that we always have $|\mathcal{D}^{\mathrm{nf}}| \leq 2_k^{|\mathcal{D}|}$.*
PROOF. Consider the following special deduction term r_n:

$$r_n := I_{n-1}^2 I_{n-2}^2 \dots I_0^2 =_\beta I_0^{2_n}.$$

The depth of the left hand side is easily seen to be $n+3$, while on the right the depth is $2_n + 2$. ⊠

However, a still stronger result is possible; in 6.11.1 we shall exhibit a sequence of formulas (types) C_k with non-normal deductions of a size linear in k, such that *every* normal deduction of C_k contains at least 2_k nodes.

From the theorem above plus earlier results, it follows that the "inversion-rule strategy" of 5.1.9 cannot possibly correspond to normalization. To see this, we observe:

(a) A deduction $\mathcal{D}$ in $\to$**Nm** may be transformed into a deduction $\mathsf{G}(\mathcal{D})$ in **G3i*** + $\mathrm{Cut_{cs}}$ such that

$$|\mathsf{G}(\mathcal{D})| \leq c|\mathcal{D}|,$$

for a positive constant c, and the cutrank of $\mathsf{G}(\mathcal{D})$ is bounded by the maximum depth of formulas in $\mathcal{D}$. The proof is the same as for **G3i** + Cut (3.5.11C).

(b) For a derivation $\mathcal{D}$ in $\to$ **G3i*** + $\mathrm{Cut_{cs}}$ we can find a cutfree deduction $\mathcal{D}^*$ with bounds on $|\mathcal{D}^*|$ as in 5.1.14.

(c) For a deduction $\mathcal{D}$ in $\to$**G3i*** + $\mathrm{Cut_{cs}}$ we can construct a translation to a proof $\mathsf{N}(\mathcal{D})$ in $\to$**Nm** such that

$$|\mathsf{N}(\mathcal{D})| < 2^{|\mathcal{D}|}$$

(cf. 6.3.1B). Moreover, if $\mathcal{D}$ is cutfree, then $\mathsf{N}(\mathcal{D})$ is normal.

(d) Suppose that we apply (a), (b), (c) successively to the deductions r_n in the proof of the theorem. The maximal depth of formulas in r_n is easily seen to be $n+1$, hence $\mathsf{G}(r_n)$ has cutrank at most $n+2$. It is then readily seen that $|\mathsf{N}(\mathsf{G}(r_n)^*)|$ is bounded by 2_k^{n+3} for a fixed k.

(e) We finally observe that the mapping N under (a) is inverse to G under (c) in the sense that $\mathsf{NG}(\mathcal{D})$ and $\mathcal{D}$ have the same normal form. Therefore, for sufficiently large n, the normal form of $\mathsf{N}(\mathsf{G}(r_n)^*)$ cannot coincide with the normal form of r_n.

6.10 A digression: a stronger conversion

6.10.1. The following generalization of β-conversion, already mentioned in 6.1.9, is more readily suggested by the term notation than by deduction trees:

$$(\lambda \vec{u}v.t)\vec{s}r \ \text{ cont } \ (\lambda \vec{u}.t[v/r])\vec{s} \quad (\text{g}\beta\text{-conversion}).$$

The normal forms w.r.t. this notion of conversion however are the same as for β-conversion. For the purposes of illustration, let us also exhibit an instance of this conversion in tree form:

$$\dfrac{\dfrac{\dfrac{\dfrac{\begin{array}{c}[B]^v[A]^u\\ \mathcal{D}_0\\ t\colon C\end{array}}{\lambda v.t\colon B\to C}\,v}{\lambda uv.t\colon A\to(B\to C)}\,u \qquad \begin{array}{c}\mathcal{D}_1\\ s\colon A\end{array}}{(\lambda uv.t)s\colon B\to C} \qquad \begin{array}{c}\mathcal{D}_2\\ r\colon B\end{array}}{(\lambda uv.t)sr\colon C} \qquad \text{cont} \qquad \dfrac{\dfrac{\begin{array}{c}\mathcal{D}_2\\ [B]^v\,[A]^u\\ \mathcal{D}_0\\ t[v/r]\colon C\end{array}}{\lambda u.t[v/r]\colon A\to C}\,u \qquad \begin{array}{c}\mathcal{D}_1\\ s\colon A\end{array}}{(\lambda u.t[v/r])s\colon C}$$

This more general notion of reduction permits us to count the complexity of formulas in a more economical way, namely by the notion of (*implication-*) *level.*

Definition. Let $\preceq_{\text{g}\beta}$ be the reduction relation w.r.t. this more general conversion. A *g-cut* is simply a redex w.r.t. the generalized notion of conversion.

For implication formulas A we define the *level* $\text{lev}(A)$ by

$$\text{lev}(P) := 0 \ (P \text{ atomic}),\ \text{lev}(A\to B) := \max(\text{lev}(A)+1, \text{lev}(B)).$$

In a redex $(\lambda\vec{u}v^B.t)\vec{s}r^B$ we call B the *pre-cut formula.* The *l-rank* (*level-rank*) of a redex will be the level of its pre-cut formula plus 1. We write $\text{lcr}(\mathcal{D})$ for the maximum of the l-ranks of g-cuts (redexes) in $\mathcal{D}$.

A *critical g-cut* (*critical redex*) of a deduction $\mathcal{D}$ is a cut with pre-cut formula of maximal level among all pre-cut formulas of the deduction. ⊠

Suppose now we eliminate from a deduction the rightmost redex (in the term), or equivalently a critical g-cut which is topmost on the rightmost branch of the prooftree which contains critical g-cuts. Say the redex is $(\lambda u_1^{A_1} \dots \lambda u_n^{A_n} v^B.t)s_1^{A_1} \dots s_n^{A_n} r^B$. Then in the result of converting: $(\lambda u_1^{A_1} \dots \lambda u_n^{A_n}.t[v/r])s_1^{A_1} \dots s_n^{A_n}$ the only increase in critical g-cuts could arise from duplication of r when substituted in t for v; but this is excluded since r is free of critical g-cuts, if the original redex chosen was a rightmost redex.

Substitution of r^B in t might create new g-cuts, but necessarily of lower level, since they will be g-cuts with pre-cut formula B_1, $B \equiv B_1 \to B_2$, and $\mathrm{lev}(B_1) < \mathrm{lev}(B)$. (Observe that, if $n > 0$, the converted redex is again a redex with pre-cut formula A_1, but this redex was already present as a subredex $(\lambda u_1 \dots u_n(\lambda v.t)s_1 \dots s_n)$ in the original redex, so this is not a new critical redex.)

It is to be noted that, if we restrict attention to ordinary β-conversion, the notion of level as a measure for the complexity of (pre-)cuts fails: consider again our example above, and assume now that A is a pre-cut formula of maximal level. Reduction would produce

$$\cfrac{\cfrac{\begin{matrix}\mathcal{D}_1 \\ [B]^v\,[A]^u \\ \mathcal{D}_0 \\ C\end{matrix}}{B \to C}\,v \qquad \begin{matrix}\mathcal{D}_2 \\ B\end{matrix}}{C}$$

and we have obtained a new g-cut with pre-cut formula B, which may have the same level as A. Summing up, we have

6.10.2. THEOREM. *(Normalization for $\preceq_{g\beta}$) There is a standard strategy for obtaining a normal form w.r.t. $\preceq_{g\beta}$ (as described above).* ⊠

Virtually the same argument as for ordinary β-conversion yields:

6.10.3. THEOREM. *All terms of $\boldsymbol{\lambda}_\to$ are strongly normalizable under gβ-reduction and hence normal forms are unique (cf. 6.8.6).* ⊠

An upper bound for the number of reduction steps needed to normalize a term according to our standard strategy, for the extended notion of conversion, is easily given as a function of the leafsize of the prooftree.

THEOREM. *(Upper bound on the number of reduction steps) Let t be a deduction-term with $\mathrm{ls}(t) = p$. Put*

$$s_0(p) := 0, \;\; s_{k+1}(p) := s_k(p) + p^{2^{s_k(p)}}.$$

Then $s_k(p)$ is an upper bound on the number of steps needed to lower the g-cutrank of t by k.

PROOF. Observe that replacing $(\lambda\vec{u}u'.r)\vec{s}s'$ by $(\lambda\vec{u}.r[u'/s])\vec{s}$ in t can at most square the leafsize of t. Assume by IH that $s_k(p)$ is an upper bound on the number of steps needed to lower the g-cutrank by k. The leafsize of the term after this normalization is

$$p^{2^{s_k(p)}},$$

since each step at most squares the leafsize; hence we find that $s_{k+1}(p)$ as defined above is a bound on the number of steps needed to reduce the g-cutrank by one more. ⊠

REMARK. Replacing $(\lambda u.r)s$ by $r[u/s]$ at most squares the leafsize of t, a special case of the observation in the proof above. However, in the case of ordinary β-reduction, we work with a cutrank which is in general higher than the g-cutrank.

6.11 Orevkov's result

6.11.1. We present an example, due to V.P. Orevkov, of formulas C_k such that each C_k has a non-normal natural deduction of size linear in k, while on the other hand *every* normal derivation of C_k has at least $\text{hyp}(2,k,1) = 2_k$ nodes. So this is even worse than our example above, which demonstrated that *normalizing* a *given sequence* of deductions of A_k, which are linear in k, may produce deductions hyperexponential in k.

The example is analogous to Gentzen's proof of transfinite induction up to ω_k in arithmetic (cf. 10.2.2).

Let R be a ternary relation symbol for the graph of the function $\lambda yx.(y+2^x)$, i.e. $Ryxz$ is supposed to express $y + 2^x = z$. We introduce two axioms, which are in fact Horn formulas, fixing the meaning of R, in a language with a constant 0 for zero, and unary function symbol S for successor:

$$\begin{aligned}\text{Hyp}_1 &:= \forall y R(y, 0, Sy),\\ \text{Hyp}_2 &:= \forall yxzz_1(Ryxz \to Rzxz_1 \to R(y, Sx, z_1)).\end{aligned}$$

For C_k we take the formula expressing that $\text{hyp}(2,k,1)$ is defined:

$$C_k := \exists z_k \ldots z_0(R00z_k \wedge R0z_kz_{k-1} \wedge \ldots \wedge R0z_1z_0).$$

(Actually, our final choice will consist of variants C'_k of the C_k.) In the short deductions for C_k we use formulas A_i with parameter x:

$$\begin{aligned}A_0(x) &:= \forall y \exists z Ryxz,\\ A_{i+1}(x) &:= \forall y(A_i y \to \exists z(A_i z \wedge Ryxz)).\end{aligned}$$

To grasp the intuitive significance of A_i, put

$$\begin{array}{lcl} f_0(y_0, x) & := & y_0 + 2^x, \\ f_{n+1}(y_0, \ldots, y_n, y_{n+1}, x) & := & f_n(y_0, y_1, \ldots, y_n, y_{n+1} + 2^x). \end{array}$$

Using $\downarrow$ to express "is defined" we can say that $A_{i+1}(x)$ expresses

$$\forall y_{i+1}(f_i(y_0, \ldots, y_i, y_{i+1})\downarrow \to f_{i+1}(y_0, \ldots, y_{i+1}, x)\downarrow),$$

or

$$\forall y_{i+1}(f_i(y_0, \ldots, y_i, y_{i+1})\downarrow \to f_i(y_0, \ldots, y_{i+1} + 2^x)\downarrow).$$

6.11.2. Lemma. *In* **Nm** *for every* i $\mathrm{Hyp}_1 \to \mathrm{Hyp}_2 \to A_i 0$, *by a proof with size bounded by a constant (that is to say, not depending on* i*).*

Proof. We have to show how to construct formal proofs $\mathcal{E}_i$ of $A_i 0$. We leave the cases of $\mathcal{E}_0, \mathcal{E}_1$ to the reader. We define abbreviations:

$$\begin{array}{l} A_0(x, y) := \exists z Ryxz, \\ A_{i+1}(x, y) := A_i y \to \exists z(A_i z \wedge Ryxz). \end{array}$$

We construct deduction $\mathcal{E}'_{i+2}$:

$$\dfrac{\dfrac{A_i z_1 \wedge Rzxz_1{}^v}{Rzxz_1} \qquad \dfrac{\dfrac{A_i z \wedge Ryxz^{\,w}}{Ryxz} \qquad \dfrac{\mathrm{Hyp}_2}{Ryxz \to Rzxz_1 \to R(y, Sx, z_1)}\,\forall\mathrm{E}\,(4\times)}{Rzxz_1 \to R(y, Sx, z_1)}}{R(y, Sx, z_1)}$$

We construct deduction $\mathcal{E}_{i+2}$ from $\mathcal{E}'_{i+2}$:

$$\dfrac{\dfrac{\dfrac{\dfrac{\dfrac{\dfrac{A_{i+1}x^{\,u}}{A_{i+1}xy} \quad A_i y^{\,u'}}{\exists z(A_i z \wedge Ryxz)} \qquad \dfrac{\dfrac{\dfrac{A_{i+1}x^{\,u}}{A_{i+1}xz} \quad \dfrac{A_i z \wedge Ryxz^{\,w}}{A_i z}}{\exists z_1(A_i z_1 \wedge Rzxz_1)}\,\to\mathrm{E} \qquad \dfrac{\dfrac{\dfrac{A_i z_1 \wedge Rzxz_1{}^v}{A_i z_1} \quad \mathcal{E}'_{i+2}}{A_i z_1 \wedge R(y, Sx, z_1)}}{\exists z(A_i z \wedge R(y, Sx, z))}}{\exists z(A_i z \wedge R(y, Sx, z))}\,\exists\mathrm{E}, v}{\exists z(A_i z \wedge R(y, Sx, z))}\,\exists\mathrm{E}, w}{A_i y \to \exists z(A_i z \wedge R(y, Sx, z))}\,\to\mathrm{I}, u'}{A_{i+1}(Sx)}\,\forall\mathrm{I} \qquad \dfrac{\mathrm{Hyp}_1}{R(x, 0, Sx)}\,\forall\mathrm{E}}{\dfrac{\dfrac{A_{i+1}(Sx) \wedge R(x, 0, Sx)}{\exists z(A_{i+1}z \wedge Rx0z)}\,\exists\mathrm{I}}{\dfrac{A_{i+1}x \to \exists z(A_{i+1}z \wedge Rx0z)}{A_{i+2}0}\,\forall\mathrm{I}}\,\to\mathrm{I}, u}\,\wedge\mathrm{I}$$

⊠

6.11.3. PROPOSITION.

(i) In **Nm** $\mathrm{Hyp}_1 \to \mathrm{Hyp}_2 \to C_k$ *by a deduction linear in* k.

(ii) In $\to\wedge\forall\bot$-**Nm**, *for every* k, $\vdash \mathrm{Hyp}_1 \to \mathrm{Hyp}_2 \to C'_k$ *by a deduction linear in* k. *Here* C'_k *is a negative version of* C_k:

$$C'_k \equiv \neg\forall z_k \dots z_0(R00z_k \to R0z_kz_{k-1} \to R0z_1z_0 \to \bot).$$

PROOF. For the deductions $\mathcal{D}'_k$ of C_k from Hyp_1, Hyp_2 we introduce some further abbreviations.

$$z_{k+1} := 0,\ \ R_1 := R0z_1z_0,\ \ R_{i+1} := R0z_{i+1}z_i \wedge R_i\ \ (1 \le i \le k)$$
$$B_0z := R0z_1z,\ \ B_iz := A_{i-1}z \wedge R0z_{i+1}z.$$

For $1 < i \le k$:

$$\mathcal{D}_1 := \dfrac{\dfrac{B_1(z_1)}{A_0z}}{\exists zB_0z} \qquad \mathcal{D}_i := \dfrac{\dfrac{\dfrac{B_i(z_i)}{A_{i-1}z_i}}{A_{i-1}z_i0} \quad \begin{matrix}\mathcal{E}_{i-2}\\ A_{i-2}0\end{matrix}}{\exists zB_iz} \qquad \mathcal{D}_{k+1} := \dfrac{\dfrac{\begin{matrix}\mathcal{E}_k\\ A_k0\end{matrix}}{A_k00} \quad \begin{matrix}\mathcal{E}_{k-1}\\ A_{k-1}0\end{matrix}}{\exists zB_kz}$$

$$\mathcal{D}^*_1 := \dfrac{B_0(z_0)}{R0z_1z_0} \qquad \mathcal{D}^*_i := \dfrac{\dfrac{B_{i-1}(z_{i-1})}{R0z_iz_{i-1}} \quad \begin{matrix}\mathcal{D}_{i-1}\\ R_{i-1}\end{matrix}}{R_i} \quad (i > 1)$$

Finally we can construct the required deduction $\mathcal{D}'_k$ as follows:

$$\dfrac{\begin{matrix}\mathcal{D}_{k+1}\\ \exists zB_kz\end{matrix} \quad \dfrac{\begin{matrix}[B_k(z_k)]\\ \mathcal{D}_k\\ \exists zB_{k-1}z\end{matrix} \quad \dfrac{\begin{matrix}[B_{k-1}(z_{k-1})]\\ \mathcal{D}_{k-1}\\ \exists zB_{k-2}z\end{matrix} \quad \dfrac{\begin{matrix}[B_{k-2}(z_{k-2})]\\ \mathcal{D}_{k-2}\\ \exists zB_{k-3}z\end{matrix} \quad \begin{matrix}\dfrac{\begin{matrix}[B_1z_1]\\ \mathcal{D}_1\\ \exists zB_0z\end{matrix} \quad \dfrac{\begin{matrix}\mathcal{D}^*_k\\ R_k\end{matrix}}{C_k}\exists\mathrm{I}}{C_k}\\ \vdots\\ C_k\end{matrix}}{C_k}}{C_k}}{C_k}}{C_k}$$

If we now apply the Gödel–Gentzen negative translation, it is not hard to see that deductions $\mathcal{D}$ in **Nm** are translated into deductions $\mathcal{D}'$ in $\forall\wedge\to\bot$-**Nc** such that $|\mathcal{D}'| \le c|\mathcal{D}|$ for a constant c, provided that the instances of $\neg\neg A^{\mathrm{g}} \to A^{\mathrm{g}}$ have proofs of fixed depth. This is indeed the case, because the only critical cases in the deductions above are the applications of $\exists$E in but these are applied to existential formulas, which are translated as negations, and for which the property $\neg\neg A^{\mathrm{g}} \to A^{\mathrm{g}}$ is indeed provable with fixed depth, by specializing the standard proof of $\neg\neg\neg B \to \neg B$. As an example consider a deduction terminating with $\exists$E on the left, translated as on the right:

$$
\dfrac{\begin{matrix}\mathcal{D}_1\\ \exists x Ax\end{matrix} \qquad \begin{matrix}[Ax]\\ \mathcal{D}_2\\ \exists y By\end{matrix}}{\exists y By}
\qquad\qquad
\dfrac{\dfrac{\begin{matrix}\mathcal{D}_1^{\mathrm{g}}\\ \neg\forall x \neg A^{\mathrm{g}}x\end{matrix} \qquad \dfrac{\dfrac{\dfrac{\neg\neg\forall y \neg B^{\mathrm{g}}y \qquad \begin{matrix}[A^{\mathrm{g}}x]\\ \mathcal{D}_2^{\mathrm{g}}\\ \neg\forall y \neg B^{\mathrm{g}}y\end{matrix}}{\bot}}{\neg A^{\mathrm{g}}x}}{\forall x \neg A^{\mathrm{g}}x}}{\dfrac{\bot}{\neg\neg\neg\forall y \neg B^{\mathrm{g}}y}} \qquad \begin{matrix}\mathcal{F}\\ \neg\neg\neg\forall y \neg B^{\mathrm{g}}y \to \neg\forall y \neg B^{\mathrm{g}}y\end{matrix}}{\neg\forall y \neg B^{\mathrm{g}}y}
$$

Here $\mathcal{F}$ is a substitution instance of the standard proof of $\neg\neg\neg P \to \neg P$. From this we see that $\mathcal{D}'_k$ translates into a proof of

$$C_k^{\mathrm{g}} \equiv \neg\forall z_k \dots z_0(R00z_k \wedge R0z_kz_{k-1} \wedge \dots \wedge R0z_1z_0 \to \bot).$$

Finally there is a deduction of C'_k from C_k^{g} linear in k. (The sole reason for replacing C_k^{g} by C'_k is to get rid of $\wedge$, facilitating the comparison with the next proposition.)

6.11.4. PROPOSITION. *Any normal derivation in* $\to\wedge\forall\bot$-**Nm** *of* C'_k *from* Hyp_1, Hyp_2 *has at least* $\mathrm{hyp}(2, k, 1) = 2_k$ *nodes.*

PROOF. Let $\mathcal{D}$ be a normal derivation of $\bot$ from Hyp_1, Hyp_2 and the hypothesis

$$D := \forall z_k \dots z_0(R00z_k \to R0z_kz_{k-1} \to \dots \to R0z_1z_0 \to \bot).$$

Without loss of generality we may assume that there are no variables appearing free anywhere in the deduction (unless bound later by $\forall$I, a case which actually does not arise). If there are such variables, we can always replace them everywhere by 0.

The main branch of the derivation $\mathcal{D}$ must begin with D, since (1) Hyp_1 and Hyp_2 do not contain $\bot$, and (2) the main branch ends with an elimination, so $\bot$ is a subformula of the top formula, which cannot be discharged along the main branch.

The main branch starts with a series of $\forall$E-applications, followed by $\to$E-applications; all minor premises are of the form $R0\bar{n}\bar{m}$ ($\bar{k}$ abbreviates S^k0, as before).

Any normal deduction $\mathcal{D}'$ of $R\bar{m}\bar{n}\bar{k}$ from Hyp_1, Hyp_2 and D (1) actually does not use D, (2) has at least 2^n occurrences of Hyp_1, and (3) satisfies $k = m + 2^n$. (1) is readily proved by induction on the depth of the deduction, and is left to the reader. (2) and (3) are proved by induction on $\bar{n}$. For the induction step, assume that we have shown that any normal derivation of $R\bar{m}\bar{n}\bar{k}$ uses $\geq 2^n$ occurrences of Hyp_1, and satisfies $k = m + 2^n$. Consider a normal derivation $\mathcal{D}'$ of $R\bar{m}(S\bar{n})\bar{k}$. This must be of the form

$$\dfrac{\dfrac{\begin{matrix}\mathcal{D}_0'\\ R\bar{m}\bar{n}\bar{u} \to (R\bar{u}\bar{n}\bar{k} \to R\bar{m}(S\bar{n})\bar{k})\end{matrix} \quad \begin{matrix}\mathcal{D}_1'\\ R\bar{m}\bar{n}\bar{u}\end{matrix}}{R\bar{u}\bar{n}\bar{k} \to R\bar{m}(S\bar{n})\bar{k}} \quad \begin{matrix}\mathcal{D}_2'\\ R\bar{u}\bar{n}\bar{k}\end{matrix}}{R\bar{m}(S\bar{n})\bar{k}}$$

Application of the IH to $\mathcal{D}_1'$, $\mathcal{D}_2'$ produces (2), (3) for $\mathcal{D}'$.

Returning to the derivations of the minor premises along the main branch of the original $\mathcal{D}$, this observation tells us that they derive $R00\overline{2_0}$, $R0\overline{2_0}\,\overline{2_1}, \dots,$ $R0\overline{2_{k-1}}\,\overline{2_k}$. This uses at least $2^{2_{k-1}} = 2_k$ times Hyp_1. ⊠

Remark. The preceding result is transferable to Gentzen systems. From a cutfree proof $\mathcal{D}_k$ of C_k in **G3i** of depth $\leq 2_{k-2}$ we can construct an **Nm**-proof of depth $< 2_{k-1}$ (6.3.1B), hence of size $< 2_k$, contradicting the proposition, so any such $\mathcal{D}_k$ necessarily has depth $> 2_{k-2}$.

6.12 Notes

6.12.1. *Concepts concerning natural deduction prooftrees.* The notions of *segment, branch, track, track of order n* appear in Prawitz [1965], as *segment, thread, path, path of order n* respectively. We have replaced *thread* by *path*, as being the more usual terminology for trees in mathematics, and we have replaced *path* by *track*, in order to avoid confusion with the usual notion of a path in a tree. The notion of maximal segment used here slightly generalizes the notion in Prawitz [1965], as in Mints [1992a] (in Prawitz [1965] a maximal segment must be the conclusion of an I-rule). The concept of a *main branch* is taken from Martin-Löf [1971a].

Detour conversions and permutative conversions are from Prawitz [1965]; the *simplification conversions* from Prawitz [1971]. Prawitz has an extra simplification conversion in the case of **Ni**, simplifying

$$\dfrac{\begin{matrix}\dfrac{\begin{matrix}\mathcal{D}\\ A\end{matrix} \quad \neg A^u}{\bot}\\ \mathcal{D}'\\ \bot\end{matrix}}{A}\,u$$

to $\mathcal{D}$, provided no assumptions in $\mathcal{D}$ become bound in $\mathcal{D}'$. In Prawitz [1971, p. 254] it is also observed that a derivation in normal form may be expanded to a derivation in what is here called *long normal form.*

An example of a detour conversion is already present in Gentzen [1935] (end of section III.2).

6.12.2. *Normalization and its applications.* The proof of the normalization theorem follows Prawitz [1965]; this proof is in fact a straightforward extension of a very early (1942!) unpublished proof by A. M. Turing of normalization in simply typed lambda calculus (i.e. implication logic); see Gandy [1980]. The next proof known to us is in Curry and Feys [1958, theorem 9 in section 9F], where normalization is obtained via cut elimination. For normalization of **Nc** see Prawitz [1965], Smullyan [1965], Stålmarck [1991], Andou [1995] and the references given in these papers. In the case of Andou [1995] it is essential that $\neg$ is a primitive with rules $\neg$I, $\neg$E.

Extensions (section 6.4) with rules of type Ia are discussed in Prawitz [1971]. There is also a brief discussion of extensions of N-systems in Negri and von Plato [1998]; in that paper rules of type II are introduced.

The result in 6.5, relating E-logic to ordinary logic with some special axioms, is due to Scott [1979], who gave a semantic proof, but the idea of the present proof is due to G. R. Renardel de Lavalette (unpublished). There is a variant of E-logic where free variables stand for "existing" objects, the domains are always inhabited, but where terms need not be always defined: the logic of partial terms (LPT), called E^+-logic in Troelstra and van Dalen [1988].

As to the conservative extension of predicative classes (6.6.3), the corresponding result for classical theories is wellknown; a result of this type appears for example in Takeuti [1978]. A similar result for a theory based on intuitionistic logic, namely the conservativeness of $\mathbf{EM}_0\restriction$ over **HA**, is proved in Beeson [1985, p. 322] by means of Kripke models. The present proof is taken from Troelstra and van Dalen [1988, chapter 10].

Between 1965 and 1970 there appeared many proofs (cf. Troelstra [1973, 2.2.35]) of the fact that the terms of suitable term calculi for the primitive recursive functionals of Gödel [1958] could be brought into normal form, and hence the numerical terms evaluated. Usually these proofs implicitly establish normalization for $\boldsymbol{\lambda}_{\to}$. There is little or no attention given to strong normalization; exceptions are Sanchis [1967] (for a theory with combinators) and Howard [1970] (for lambda abstraction). But in most other cases the proofs might have been adapted to strong normalization without difficulty. For example, in Tait [1967], where the method using computability predicates is introduced, normalization for a system of terms with combinators and recursors is proved, but not strong normalization, although it is easy to adapt Tait's proof to strong normalization for a system of terms with lambda abstraction (see, for example, Troelstra [1973, section 2.2]). The proof of Diller [1970] can also be adapted so as to obtain a proof of strong normalization for simple type theory, etc. Strong normalization was firmly put on the map by Prawitz [1971], who proved strong normalization for a natural-deduction version of intuitionistic second-order logic, using Girard's extension of Tait's method. In this text we followed the Tait method, adapted to strong normalization.

For the proof of strong normalization for the full system, see Prawitz [1971], For an exposition, one may also look at Troelstra [1973, Chapter 4] (disregarding everything which concerns arithmetic).

Another method for proving strong normalization, by assigning suitable functionals to terms or derivations, is introduced in Gandy [1980]. Gandy did not treat permutative conversions; this step is taken in van de Pol and Schwichtenberg [1995].

A new elegant approach to proofs of normalization and strong normalization for systems of typed terms or typable terms, especially for $\boldsymbol{\lambda 2}$ and its extensions, is described in Matthes [1998]. In Joachimski and Matthes [1999] these methods are applied to a lambda calculus with sumtypes, where permutative conversions are also treated; these methods obviously also apply to strong normalization for **Ni**.

The conservative extension result in 6.7.5 is taken from Schwichtenberg [1992].

As to the failure of strong normalization under CDC, see also Leivant [1979]. The result in 6.9.4 and the generalization of β-conversion are taken from Schwichtenberg [1991]. The presentation of Orevkov's result in 6.11.1 is a slight modification of an exposition by Schwichtenberg, which in turn is an adaptation to N-systems of Orevkov [1979]. Orevkov's result is an adaptation of a result in Statman [1978] for languages containing function symbols. Other papers of Orevkov dealing with bounds are Orevkov [1984,1987].

6.12.3. *Comparing G-systems with N-systems.* (Continued from 3.7.4.) The natural map N from cutfree G-deductions to normal N-deductions, originally due to Prawitz [1965], is many-to-one, not one-to-one. Prawitz also described an inverse, G_{cf}, assigning a cutfree G-proof to a normal N-proof; this is the argument in 6.3.1 (Prawitz [1965, App.A §3]). The images under G_{cf} in fact not only are cutfree, but satisfy some extra conditions; they are so-called *normal* G-deductions, as in 6.3.5. This insight, with credit to Curry, is present in Howard [1980, section 5] which was written in 1969.

The precise notion of normality differs for the various G-systems, but in any case the antecedent active formulas in applications of L$\to$, L$\wedge$ and L$\forall$ have to be principal themselves. Zucker [1974] showed that in the negative fragment of LJ + Cut (that is to say, the fragment of $\to \wedge\forall\bot$) two deductions have the same image under N, if they are interreducible using permutations of rules and reductions of cuts. Pottinger [1977] extends Zucker's work. Mints [1996] proves normalization of cutfree proofs by permutation of rules, sharpening the notion of normality so as to obtain a one-to-one correspondence between normal natural deductions and normal proofs in a system which is practically identical with **G1i** (the treatment needs to be supplemented for contraction).

Dyckhoff and Pinto [1999] prove a result similar to the result of Zucker [1974] but for a cutfree calculus. Schwichtenberg [1999] proves strong normalization for the permutations involved. Troelstra [1999] describes a normalization procedure for cutfree G3-deduction in implication logic under CDC.

The treatment given in this book tries to avoid the complications arising from contraction and weakening in Mints [1996] and describes the correspondence between normal proofs in **GKi** and normal proofs in **Ni**, using the G-system with privileged "headformulas" as an intermediate. The use of headformulas is found in Herbelin [1995] and also crops up in the proof theory of linear logic (cf. 9.4).

6.12.4. *Generalized elimination rules.* In the papers von Plato [1998], Negri and von Plato [1999] a version of natural deduction is studied with generalized forms of $\wedge$E, $\to$E and $\forall$E:

$$\dfrac{A \wedge B \qquad \begin{matrix}[A]^u[B]^v \\ \vdots \\ C\end{matrix}}{C}\wedge\mathrm{E}^*,u,v \qquad \dfrac{A \to B \quad A \qquad \begin{matrix}[B]^u \\ \vdots \\ C\end{matrix}}{C}\to\mathrm{E}^*,u \qquad \dfrac{\forall x A \qquad \begin{matrix}A[x/t]^u \\ \vdots \\ C\end{matrix}}{C}\forall\mathrm{E}^*,u$$

The usual rules are readily seen to be special cases; for example, to obtain the usual $\to$E, take for the rightmost subdeduction simply the assumption B (with $B \equiv C$). The rule $\wedge$E* was already considered in Schroeder-Heister [1984].

Let us use **Ni*** as an ad hoc designation for this system. (N.B. In the papers just mentioned the assumption classes are not treated in quite the same way as for our **Ni**, but we shall disregard these differences here.)

Now all E-rules have the indirect form of $\vee$E, $\exists$E in **Ni**. Extra permutation conversions may be defined for the new extended rules.

If we define maximal segments just as before in **Ni**, namely that a segment is said to be *maximal* if it either is of length 1 and is the conclusion of an I-rule and major premise of an E-rule, *or* is of length greater than 1 and major premise of an E-rule, we can prove normalization as before. A *normal* deduction may now be defined as a deduction where major premises of E-rules are assumptions. For otherwise the deduction of some major premise either ends with an I-rule, and a detour conversion is possible, or ends with an E-rule and a permutation is possible.

Normal deductions $\mathcal{D}$ in **Ni*** can be translated in a straightforward way into cutfree proofs $\mathcal{D}^*$. Let us illustrate the idea for implication logic.

(i) A final application of $\to$I is translated as an application of R$\to$:

$$\dfrac{\begin{matrix}[A]^u \\ \mathcal{D}_0 \\ B\end{matrix}}{A \to B} \qquad \text{goes to} \qquad \dfrac{\begin{matrix}\mathcal{D}_0^* \\ \Gamma, A \Rightarrow B\end{matrix}}{\Gamma \Rightarrow A \to B}$$

(ii) A final application of $\rightarrow$E* with major premise an assumption is translated as an application of L$\rightarrow$:

$$\dfrac{A \rightarrow B \quad \begin{matrix} \mathcal{D}_1 \\ A \end{matrix} \quad \begin{matrix} [B]^u \\ \mathcal{D}_2 \\ C \end{matrix}}{C} \qquad \text{goes to} \qquad \dfrac{\begin{matrix} \mathcal{D}_1^* \\ \Gamma \Rightarrow A \end{matrix} \quad \begin{matrix} \mathcal{D}_2^* \\ B, \Gamma' \Rightarrow C \end{matrix}}{A \rightarrow B, \Gamma, \Gamma' \Rightarrow C}$$

(If the major premise had been derived by, say, $\mathcal{D}_0$, we would have needed a Cut to make the translation work.) Conversely, cutfree proofs in a G-system may readily be translated into normal deductions in **Ni***. The important difference with the correlation between **Ni** and, say, **G3i** is that now the order of the rules corresponds: the normal natural deduction is constructed from the cutfree proof by looking at each step at the last rule applied in order to find the last rule for the translated deduction.

By suitably choosing the N-system on the one hand and the G-system on the other hand one can easily obtain a one-to-one (bijective) correspondence.

In Negri and von Plato [1999] one considers for this purpose a G-system with context-free rules where multiple copies of the active formulas may occur (possibly zero); thus, for example, L$\wedge$ becomes

$$\frac{A^m, B^n, \Gamma \Rightarrow C}{A \wedge B, \Gamma \Rightarrow C}$$

For the corresponding N-system, written with sequents, one has

$$\frac{\Gamma \Rightarrow A \wedge B \quad A^m, B^n, \Gamma \Rightarrow C}{\Gamma \Rightarrow C}$$

This leads to a smooth correspondence, which may be extended to include Cut for the G-system and a rule of substitution for the N-system:

$$\text{Sub}\,\frac{\Gamma \Rightarrow A \quad A, \Delta \Rightarrow C}{\Gamma, \Delta \Rightarrow C}$$

For the classical propositional system one may add an atomic rule of the excluded middle,

$$\text{EM-At}\,\frac{\Gamma, P \Rightarrow C \quad \Delta, \neg P \Rightarrow C}{\Gamma, \Delta \Rightarrow C}\ (P \text{ atomic})$$

which corresponds on the natural deduction side to

$$\dfrac{\begin{matrix} [P] \\ \vdots \\ C \end{matrix} \quad \begin{matrix} [\neg P] \\ \vdots \\ C \end{matrix}}{C}$$

Summing up, the construction of the correspondence between G-deductions and a suitable variant of **Ni***-deductions makes us understand why in the correspondence for standard **Ni** (6.3) the rules L$\wedge$, L$\forall$, L$\rightarrow$ need to be treated differently from L$\vee$ and L$\exists$.

6.12.5. *Multiple-conclusion and sequence-conclusion natural deduction.* In Shoesmith and Smiley [1978] and Ungar [1992], systems are considered where the inferences produce finite sequences of assertions as conclusions; all formulas in such a conclusion may be used as premises for other inferences, simultaneously. That is to say a (fragment of a) proof may look like the following:

$$\begin{array}{cccc} A & & B & \\ \hline C & & D & \\ \cline{1-1}\cline{3-4} E & & F & G \end{array}$$

The deductions are therefore no longer trees.

Technically more manageable are systems with finite multisets or sequences of formulas as conclusions, one formula of which may be used as the active formula in a premise of the next inference. Such systems for **C** are considered in Boričić [1985] and Cellucci [1992]. For example, Boričić [1985] has the following rules for **Cp** (Γ, Δ multisets):

$$\dfrac{\begin{matrix}[A]^x\\ \vdots \\ \Gamma B\end{matrix}}{\Gamma(A \to B)}\, x, {\to}\mathrm{I} \qquad \dfrac{\Gamma(A \to B) \quad \Delta A}{\Gamma\Delta B}\, {\to}\mathrm{E} \qquad \dfrac{\begin{matrix}[A]^x\\ \vdots \\ \Gamma\end{matrix}}{\Gamma\neg A}\, x, \neg\mathrm{I}$$

$$\dfrac{\Gamma\neg A \quad \Delta A}{\Gamma\Delta}\, \neg\mathrm{E} \qquad \dfrac{\Gamma A \quad \Delta B}{\Gamma(A \wedge B)}\, \wedge\mathrm{I} \qquad \dfrac{\Gamma(A_0 \wedge A_1)}{\Gamma A_i}\, \wedge\mathrm{E}$$

$$\dfrac{A_i}{\Gamma(A_0 \vee A_1)}\, \vee\mathrm{I} \qquad \dfrac{\Gamma(A \vee B)}{\Gamma AB}\, \vee\mathrm{E} \qquad \dfrac{\Gamma}{\Gamma A}\, \mathrm{W} \qquad \dfrac{\Gamma AA}{\Gamma A}\, \mathrm{C}$$

(Actually, Boričić uses sequences instead of multisets, and hence also has a rule of exchange.) For predicate logic **C** one adds

$$\dfrac{\Gamma A}{\Gamma\forall xA}\, \forall\mathrm{I} \qquad \dfrac{\Gamma\forall xA}{\Gamma A[x/t]}\, \forall\mathrm{E} \qquad \dfrac{\Gamma A[x/t]}{\Gamma\exists xA}\, \exists\mathrm{I} \qquad \dfrac{\Gamma\exists xA \quad \begin{matrix}[A]^x\\ \vdots \\ B\end{matrix}}{B}\, \exists\mathrm{E}$$

with the obvious restrictions on x in $\forall\mathrm{I}, \exists\mathrm{E}$. $\exists\mathrm{E}$ rather spoils the regular pattern of the rules, so Cellucci [1992] considers a calculus in which $\exists\mathrm{E}$ has been replaced by

$$\dfrac{\Gamma\exists xA}{\Gamma A[x/\varepsilon_{\exists xA}]}\, \exists\mathrm{E}_\varepsilon$$

where $\varepsilon_{\exists xA}$ is an ε-term in the sense of Hilbert's ε-symbol (cf. Hilbert and Bernays [1939]), i.e. a term which satisfies $\exists xA \leftrightarrow A[x/\varepsilon_{\exists xA}]$. For these systems normalization with the usual consequences (subformula property etc.) is provable.

6.12.6. *Higher-order rules.* In Schroeder-Heister [1984] a generalization of natural deduction is considered, where not only formulas may appear as hypotheses, but also *rules*; a rule of order $n+1$ may contain rules of order n as hypotheses. Ordinary rules are rules of order 0.

Chapter 7

Resolution

In this chapter we study another form of inference, which forms the keystone of logic programming and certain theorem-proving systems. We do not aim at giving a complete introduction to the theory of logic programming; rather, we want to show how resolution is connected with other formalisms and to provide a proof-theoretic road to the completeness theorem for SLD-resolution.

The first three sections deal with propositional resolution, unification and resolution in predicate logic. The last two sections illustrate for **Cp** and **Ip** how deductions in a suitably chosen variant of the Gentzen systems can be directly translated into deductions based on resolution, which often permits us to lift strategies for proof search in Gentzen systems to resolution-based systems. The extension of these methods to predicate logic is more or less straightforward.

7.1 Introduction to resolution

Propositional linear resolution is a "baby example" of resolution methods, which is not of much interest in itself, but may serve as an introduction to the subject.

We consider *programs* consisting of finitely many sequents (*clauses*) of the form $\Gamma \Rightarrow P$, P a propositional variable and Γ a finite multiset of propositional variables ("definite clauses", "Horn clauses" or "Horn sequents"). A *goal* or *query* Γ is a finite (possibly empty) set of propositional variables, and may be identified with the sequent $\Gamma \Rightarrow$. [] is the empty goal. The so-called *(linear) resolution rule* is in the propositional case just an instance of Cut:

$$\frac{\Gamma, A \Rightarrow \qquad \Delta \Rightarrow A}{\Gamma, \Delta \Rightarrow}$$

A resolution derivation consists of a sequence of such instances of Cut, where the right premise is a rule from the given program. A *successful* derivation, starting from an initial goal Γ, is a finite derivation tree ending in the empty goal. Identifying a program clause $\mathcal{H} \equiv \Delta \Rightarrow A$ with the formula $(\bigwedge \Delta) \to A$, and a goal Γ with $\neg \bigwedge \Gamma$, we see that a successful derivation derives $\bot$ from the initial goal and thus provides in fact a refutation of the initial goal on the basis of the program clauses. In short, seen as a refutation of the initial goal, a resolution proof is nothing but a very special type of deduction in a Gentzen system.

7.1.1. Example. Consider atomic propositions S_m, S_n, W, H (for "Summer", "Sunny", "Warm", "Happy") with a program of four clauses:

$$\begin{array}{lll} (1) & S_m, W & \Rightarrow H \\ (2) & S_n & \Rightarrow W \\ (3) & S_m & \Rightarrow W \\ (4) & & \Rightarrow S_m \end{array}$$

The following are examples of, respectively, a successful and an unsuccessful derivation from this program:

$$\dfrac{\dfrac{\dfrac{\dfrac{H \quad (1)}{S_m, W} \quad (4)}{W} \quad (3)}{S_m} \quad (4)}{[\,]} \qquad\qquad \dfrac{\dfrac{\dfrac{H \quad (1)}{S_m, W} \quad (2)}{S_m, S_n} \quad (4)}{S_n}$$

The derivation on the right cannot be continued since there is no clause with S_n on the right. The left hand derivation in our example infers H from the assumptions embodied in the program clauses.

From the viewpoint of classical logic, *refuting* $\neg \bigwedge \Gamma$ is tantamount to proving $\bigwedge \Gamma$. This suggests that it is also possible to look at a resolution proof as an ordinary deduction of the initial goal constructed "backwards". Let a resolution proof be given,

$$\dfrac{\dfrac{\dfrac{\Gamma_0 \quad \mathcal{H}_0}{\Gamma_1} \qquad \mathcal{H}_1}{\Gamma_2}}{\begin{array}{c}\vdots \\ \dfrac{\Gamma_k \qquad\qquad \mathcal{H}_{k-1}}{\Gamma_k}\end{array}}$$

with n-th step

$$\dfrac{\Gamma_{n-1} \equiv \Gamma A \qquad \Delta \Rightarrow A}{\Gamma_n \equiv \Gamma\Delta}$$

and assume that we have already constructed a derivation $\mathcal{D}_n$ of $\mathcal{P}^* \Rightarrow \bigwedge \Gamma_n$, where $\mathcal{P}^*$ is the multiset consisting of formulas $(\bigwedge \Gamma') \to A'$, one formula occurrence for each clause $\Gamma' \Rightarrow A$ from the program $\mathcal{P}$. We construct a derivation $\mathcal{D}_{n-1}$ of $\mathcal{P}^* \Rightarrow \bigwedge \Gamma_{n-1}$ as follows.

Let $\mathcal{D}'_n$, $\mathcal{D}''_n$ be Gentzen-system deductions constructed in a standard way from $\mathcal{D}_n$, with conclusions $\mathcal{P}^* \Rightarrow \bigwedge \Delta$, $\mathcal{P}^* \Rightarrow \bigwedge \Gamma$ respectively, and let $\mathcal{D}'$ be

$$\dfrac{\mathcal{P}^* \Rightarrow \bigwedge\Delta \to A \qquad \dfrac{\bigwedge\Delta \Rightarrow \bigwedge\Delta \quad A \Rightarrow A}{\bigwedge\Delta \to A, \bigwedge\Delta \Rightarrow A}}{\mathcal{P}^*, \bigwedge\Delta \Rightarrow A}$$

From this we construct $\mathcal{D}_{n-1}$:

$$\dfrac{\begin{matrix}\mathcal{D}''_n \\ \mathcal{P}^* \Rightarrow \bigwedge\Gamma\end{matrix} \qquad \dfrac{\begin{matrix}\mathcal{D}'_n \\ \mathcal{P}^* \Rightarrow \bigwedge\Delta\end{matrix} \quad \begin{matrix}\mathcal{D}' \\ \mathcal{P}^*, \bigwedge\Delta \Rightarrow A\end{matrix}}{\mathcal{P}^* \Rightarrow A}}{\mathcal{P}^* \Rightarrow \bigwedge\Gamma \wedge A}$$

Another point worth noting is the following. Assume we have derived the empty sequent (goal) from an initial goal Γ_0 and a set of program clauses $\mathcal{P}$. We may regard the program clauses as axioms. Then the generalized form of the cut elimination theorem (cf. 4.5.1) tells us that an arbitrary classical Gentzen system deduction can be transformed into a deduction where all cut formulas occur in an axiom. Hence this may be read as a resolution proof; i.e. we have obtained a completeness theorem.

7.2 Unification

The present section contains some results on substitution operations needed in what follows.

7.2.1. NOTATION. A *substitution* is a mapping, say σ, of variables to terms such that the *domain* of σ, $\mathrm{dom}(\sigma) = \{x : \sigma x \neq x\}$, is finite. We may therefore represent a substitution by $[x_1/t_1, \dots, x_n/t_n]$, with all x_i distinct, and $x_i \neq t_i$ for $1 \leq i \leq n$. An equivalent notation is $[x_1, \dots, x_n/t_1, \dots, t_n]$. ϵ is the *identical substitution*, with empty domain. In many arguments we treat $[x_1/t_1, \dots, x_n/t_n]$ as a set of ordered pairs $\{(x_1, t_1), \dots, (x_n, t_n)\}$.

For a substitution $\sigma \equiv [x_1/t_1, \dots, x_n/t_n]$, $\mathrm{ranv}(\sigma) := \mathrm{FV}(\{t_1, \dots, t_n\})$. (We do not use "ran" as the abbreviation, since this suggests the range of a function in the usual sense.) σ is said to be a *variable substitution* if σx is a variable for all x.

For an arbitrary quantifier-free expression Θ and substitution σ, $\Theta\sigma$ is obtained by replacing every variable in Θ by its σ-image. $\Theta\sigma$ is called an *instance* (induced by σ) of Θ.

If σ, τ are substitutions, $\sigma\tau$, the *composition* of σ and τ, is the substitution defined by

$$x(\sigma\tau) = (x\sigma)\tau \text{ for all variables } x. \qquad \boxtimes$$

REMARK. Note that substitutions do not commute: $[x/y][y/z]$ is distinct from $[y/z][x/y]$.

Composition of substitutions may be defined in a more direct way by saying that if τ, σ are substitutions given by

$$\tau = [x_1/t_1, \ldots, x_n/t_n], \ \sigma = [y_1/s_1, \ldots, y_m/s_m],$$

then the substitution $\tau\sigma$ is the sequence ρ found by deleting from

$$[x_1/t_1\sigma, \ldots, x_n/t_n\sigma, y_1/s_1, \ldots, y_m/s_m]$$

the $x_i/t_i\sigma$ for which $t_i\sigma = x_i$, and the y_j/s_j for which $y_j \in \{x_1, \ldots, x_n\}$.

To see this, note that, for each variable x, $(x\sigma)\tau$ is the same as $x\rho$, where ρ is defined as above from σ and τ. (One considers three cases: $x = x_i$, $x \notin \{x_1, \ldots, x_n, y_1, \ldots, y_m\}$, and $x = y_j$ but $x \notin \{x_1, \ldots, x_n\}$.)

7.2.1A. ♠ Elaborate the preceding remark.

7.2.2. LEMMA. *Let θ, σ, τ be substitutions. Then:*

(i) $\sigma = \tau$ iff $t\sigma = t\tau$ for all terms t iff $x\sigma = x\tau$ for all variables x;

(ii) $\tau\epsilon = \epsilon\tau = \tau$;

(iii) $(t\tau)\sigma = t(\tau\sigma)$ for all terms t;

(iv) $\theta(\sigma\tau) = (\theta\sigma)\tau$.

PROOF. (ii) is obvious, (i) and (iii) are proved by a routine induction on terms, and (iv) is an immediate consequence of (i) and (iii). ⊠

REMARK. (iv) permits us to write $\theta\sigma\tau$ (without parentheses) for the composition of θ, σ, τ.

7.2.2A. ♠ Prove (i) and (iii) of the preceding lemma.

7.2.3. DEFINITION. A *variable-permutation* is a substitution σ with inverse σ^{-1} such that $\sigma\sigma^{-1} = \sigma^{-1}\sigma = \epsilon$. If Θ is a quantifier-free expression and σ is a variable-permutation, then $\Theta\sigma$ is called a *variant* of Θ.

$\sigma \leq \tau$ iff there is a θ such that $\sigma = \tau\theta$. σ and τ are said to be *equivalent* (notation $\sigma \sim \tau$) if $\sigma \leq \tau$ and $\tau \leq \sigma$. ⊠

7.2.4. LEMMA. *For equivalent σ, τ there is a permutation θ such that $\sigma\theta = \tau$, $\tau\theta^{-1} = \sigma$.*

PROOF. Let σ, τ be equivalent; then there are ρ, ρ' such that $\sigma = \tau\rho$, $\tau = \sigma\rho'$, and hence $\sigma\rho'\rho = \sigma$, $\tau\rho\rho' = \tau$. ρ' must be injective on $A = \bigcup\{\mathrm{FV}(x\sigma) : x \text{ variable}\}$, and map variables to variables, since $\rho'\rho$ is the identity on A. It is now easy to construct a variable-permutation θ which coincides with ρ' on A. ⊠

7.2.4A. ♠ Complete the proof of the lemma.

7.2.5. NOTATION. We call the expression $t \approx s$ an *equivalence*. We use $E, E', \ldots$ for finite multisets of equivalences $\{t_1 \approx s_1, \ldots, t_n \approx s_n\}$. The *inconsistent multiset of equivalences* is $\{\bot\}$. ⊠

7.2.6. DEFINITION. A substitution σ *unifies* E or σ is a *unifier* of E, if, for each $t \approx s$ in E, $t\sigma \equiv s\sigma$. No substitution unifies $\{\bot\}$.

σ is a *relevant* unifier of E, if $\mathrm{dom}(\sigma) \subset \mathrm{FV}(E)$, $\mathrm{ranv}(\sigma) \subset \mathrm{FV}(E)$.

σ is called a *most general unifier* (*m.g.u.* for short) if σ is a unifier of E, and for every other unifier τ of E we have $\tau \leq \sigma$.

σ is a *unifier* (*most general unifier*) of two atoms $P(t_1, \ldots, t_m)$,$Q(s_1, \ldots, s_n)$ if $P \equiv Q$ and $n = m$, and σ is a unifier (most general unifier) of $\{t_1 \approx s_1, \ldots, t_m \approx s_m\}$. ⊠

REMARK. If σ, σ' are m.g.u.'s of an expression Θ, and is t a term in Θ, then $t\sigma$ and $t\sigma'$ are variants (since $\sigma = \sigma'\theta$, $\sigma' = \sigma\theta'$, this follows by lemma 7.2.4).

EXAMPLE. $\{gx \approx fy\}$ has no unifier. $\{gx \approx gfy, fy \approx fgz\}$ has a unifier $[x/fgz, y/gz]$; this is an m.g.u., as we shall see.

7.2.7. DEFINITION. A substitution θ is *idempotent* if $\theta\theta = \theta$. ⊠

LEMMA. *θ idempotent iff $\mathrm{dom}(\theta) \cap \mathrm{ranv}(\theta) = \emptyset$.*

PROOF. Let $x \in \mathrm{dom}(\theta) \cap \mathrm{ranv}(\theta)$, and $y\theta = t$ with $x \in \mathrm{FV}(t)$, then $y(\theta\theta) = (y\theta)\theta = t\theta \neq t = y\theta$. Hence $\theta\theta \neq \theta$.

Conversely, if $\mathrm{dom}(\theta) \cap \mathrm{ranv}(\theta) = \emptyset$, then, for all variables x, $\mathrm{FV}(x\theta) \cap \mathrm{dom}(\theta) = \emptyset$. Hence $(x\theta)\theta = x\theta$; therefore by (iii) $\theta\theta = \theta$. ⊠

7.2.8. LEMMA. *Let θ be a unifier of Θ. Then θ is an idempotent m.g.u. iff $\sigma = \theta\sigma$ for all unifiers of Θ.*

PROOF. Let σ be unifier of Θ. Since θ is an m.g.u., we have $\sigma = \theta\tau$ for some substitution τ. Hence $\sigma = \theta\tau = \theta\theta\tau$ (idempotency) $= \theta\sigma$. The other direction is immediate, since θ itself is one of the σ. ⊠

7.2.9. DEFINITION. Let $E \rhd_\tau E'$ be defined by the following clauses:

(a) $\{t \approx x\} \cup E \rhd_\epsilon \{x \approx t\} \cup E$ if t is not a variable;

(b) $\{x \approx x\} \cup E \rhd_\epsilon E$;

(c) $\{f(s_1, \dots, s_n) \approx f(t_1, \dots, t_n)\} \cup E \rhd_\epsilon \{s_1 \approx t_1, \dots, s_n \approx t_n\} \cup E$, and $\{f(s_1, \dots, s_n) \approx g(t_1, \dots, t_m)\} \cup E \rhd_\epsilon \{\bot\}$, if $f \not\equiv g$;

(d) $\{x \approx t, s_1 \approx t_1, \dots, s_n \approx t_n\} \rhd_{[x/t]}$
$\{s_1[x/t] \approx t_1[x/s], \dots s_n[x/t] \approx t_n[x/t]\}$, if $x \notin \mathrm{FV}(t)$, and
$\{x \approx t\} \cup E \rhd_\epsilon \{\bot\}$, if $x \in \mathrm{FV}(t)$ and $t \not\equiv x$. ⊠

7.2.10. LEMMA. *Let $E \rhd_\rho E'$. Then*

(i) If σ unifies E', then $\rho\sigma$ unifies E.

(ii) If σ unifies E, then $\sigma = \rho\sigma$, and σ also unifies E'.

PROOF. By case distinction according to the definition of $E \rhd_\rho E'$. The only interesting case is the first part of (d) of the definition.

(i) If σ' is a unifier of E', then $[x/t]\sigma'$ is a unifier of E.

(ii) Let σ be a unifier of E. Then $x\sigma = t\sigma$, hence $[x/t]\sigma = \sigma$ (since both substitutions agree on all variables), and also $s_i[x/t]\sigma = s_i\sigma = t_i\sigma = t_i[x/t]\sigma$. Hence σ is also a unifier of E'. ⊠

7.2.11. THEOREM. *(Unification) Let E_1 be a finite multiset of equivalences. Then a sequence $E_1 \rhd_{\rho_1} E_2 \rhd_{\rho_2} E_3 \dots$ always terminates in an E_n which is either the empty set or the inconsistent set. In the first case, $\rho_1\rho_2\rho_3 \dots \rho_n$ is an idempotent and relevant most general unifier of E_1; in the second case, a most general unifier does not exist.*

PROOF. (i) The sequence $E_1, E_2, E_3, \dots$ terminates. To see this, we assign to a set of equivalences E a triple (n_1, n_2, n_3) where n_1 is the number of variables in E, n_2 the total number of occurrences of function symbols in E, and n_3 the total number of equations of the form $t = x$ in E, where t is not a variable. Case (a) of the definition of $\rhd_\rho$ lowers n_3, while n_1, n_2 remain the same; in cases (b),(c) n_2 is lowered and n_1 is not increased; in case (d) n_1 is lowered.

If the last $E_{n+1} = \emptyset$, then by (i) of the lemma $\rho_1 \dots \rho_n$ is a unifier of E_1. If θ is a unifier of E_1, then $\theta = \rho_1 \dots \rho_n\theta$. This is proved by induction on n, using (ii) of the preceding lemma.

If $E_{n+1} = \{\bot\}$, E is not unifiable. ⊠

EXAMPLE. If we apply the algorithm to $E_1 = \{gx \approx gfy, fy \approx fgz\}$, we obtain $E_1 \rhd_\epsilon \{x \approx fy, fy \approx fgz\} \rhd_\epsilon \{x \approx fy, y \approx gz\} \rhd_{[x/fy]} \{y \approx gz\}$ $\rhd_{[y/gz]} \emptyset$, producing an m.g.u. $[x, y/fgz, gz]$.

NOTATION. We write mgu(E) for a most general unifier according to the algorithm. ⊠

7.2.11A. ♠ Decide whether the following sets of equivalences are unifiable, and if this is the case, find an m.g.u.: $\{f(fx) \approx gyz, hx \approx fz\}$, $\{f(x, gx) \approx f(y, y)\}$, $\{f(hx, hv) \approx f(h(gu), h(fuw)\}$, $\{h(x, gx, y) \approx h(z, u, gu)\}$ (f, g, h function symbols).

7.3 Linear resolution

As in the preceding section, we have a fixed first-order language $\mathcal{L}$ which is kept constant throughout the section.

7.3.1. DEFINITION. A *Horn clause* is a sequent of the form $A_1, \ldots, A_n \Rightarrow B$ with the A_i and B atomic; a *definite* Horn clause is a Horn clause with $B \neq \bot$. We use $\mathcal{H}$, possibly sub- or superscripted, for Horn clauses. If $\mathcal{H} \equiv (\Gamma \Rightarrow B)$ is a Horn clause, let $\mathcal{H}^\forall$ be a corresponding formula $\forall \vec{x}(\bigwedge \Gamma \rightarrow B)$ where $\vec{x} = \mathrm{FV}(\Gamma \Rightarrow B)$.

A *definite program* consists of a finite set of definite Horn clauses. We use the letter $\mathcal{P}$ for programs. $\mathcal{P}^\forall = \{\mathcal{H}^\forall : \mathcal{H} \in \mathcal{P}\}$.

A *goal* or *query* Γ is a finite set of atomic formulas $A_1, \ldots, A_n$, $A_i \neq \bot$; a corresponding *goal formula* is $\bigwedge \Gamma = A_1 \wedge \ldots \wedge A_n$. We use Γ, possibly sub- or superscripted, for goals. The empty goal is denoted by $[\,]$. ⊠

REMARK. If we think of Γ and the antecedent of a Horn clause as multisets, the formulas $\bigwedge \Gamma$ and $\mathcal{H}^\forall$ are not determined uniquely, but only up to logical equivalence. This will not affect the discussion below.

Definite Horn clauses $A_1, \ldots, A_n \Rightarrow B$ are in the logic programming literature usually written as $B : - A_1, \ldots, A_n$.

7.3.2. DEFINITION. The *unrestricted resolution rule* $\mathrm{R_u}$ derives a goal Γ' from a goal Γ and program $\mathcal{P}$ via a substitution θ (Γ' is *derived unrestrictedly from* Γ *and* $\mathcal{P}$ *via* θ) if $\Gamma \equiv \Delta, A$, if there is a variant $\mathcal{H}' \equiv \Delta' \Rightarrow B$ of a clause $\mathcal{H}$ in $\mathcal{P}$; θ is a unifier of A and B, and $\Gamma' \equiv (\Delta, \Delta')\theta$. We can write

$$\frac{\Gamma \qquad \mathcal{H}'}{\Gamma'}\ \mathrm{R_u}, \theta$$

or, not specifying the variant $\mathcal{H}'$, we may also write

$$\Gamma' \xrightarrow{\theta}_{\mathrm{u}} \Gamma$$

In a still more precise notation, we may append the clause and/or the program. The implication

$$\bigwedge(\Gamma') \to \bigwedge(\Gamma\theta)$$

is called the *resultant* of the $\mathrm{R_u}$-inference; if Γ' is empty, we also identify the resultant with $\Gamma\theta$.

The *resolution rule* R is the special case of $\mathrm{R_u}$ where the resultant is most general among all possible resultants of unrestricted resolution with respect to the same program clause and the same atom occurrence in the goal Γ, that is to say, if $\bigwedge\Gamma' \to \bigwedge\Gamma\theta'$ is another resultant relative to the same rule and selected atom occurrence, then there is a substitution σ such that $(\bigwedge\Gamma' \to \bigwedge\Gamma\theta)\sigma = \bigwedge\Gamma' \to \bigwedge\Gamma\theta'$. We use

$$\Gamma' \xrightarrow{\theta} \Gamma$$

for a resolution step. ⊠

Remark. It readily follows that Γ' in a resolution step is unique modulo equivalence. For a resolution step it is not sufficient to require that we have an $\mathrm{R_u}$-application with θ an m.g.u. of A and B, as may be seen from the following example. Consider a program $\{Dy \Rightarrow Az\}$ and an instance of $\mathrm{R_u}$:

$$\frac{Cy, Ax \quad Dy \Rightarrow Az}{Cy, Dy}\ \mathrm{R_u},[z/x]$$

with resultant $Cy \wedge Dy \to Cy \wedge Ax$. A more general resultant is $Cy \wedge Du \to Cy \wedge Ax$, obtained from

$$\frac{Cy, Ax \quad Du \Rightarrow Az}{Cy, Du}\ \mathrm{R_u},[z/x]$$

Given the possibility of an unrestricted resolution step w.r.t. a partial goal, selection of atom in the goal, and rule from the program, a recipe for finding a resolution step for the same choice of rule and atom is given by the following:

7.3.3. Proposition. *Let Γ, A be a goal, $\mathcal{H}$ a rule from the program $\mathcal{P}$, and let there exist an unrestricted resolution step w.r.t. $\mathcal{H}$ and atom A in the goal. Then a resolution step for the same goal, and same choice of atom and rule, is obtained by taking a variable-permutation α such that $\mathrm{FV}(\mathcal{H}\alpha) \cap \mathrm{FV}(\Gamma) = \emptyset$, and then constructing an m.g.u. θ of A and B, where $\mathcal{H}\alpha = (\Delta \Rightarrow B)$.*

7.3.3A. ♠ Prove the proposition.

7.3.4. DEFINITION. An *unrestricted resolution derivation* from the program $\mathcal{P}$ is a "linear" finite or infinite tree of the form

$$\cfrac{\cfrac{\Gamma_0 \quad \mathcal{H}_0}{\Gamma_1}\,\theta_0 \qquad \mathcal{H}_1}{\Gamma_2}\,\theta_1$$

$$\vdots$$

$$\cfrac{\Gamma_{n-1} \qquad \mathcal{H}_{n-1}}{\Gamma_n}\,\theta_{n-1}$$

$$\vdots$$

with the Γ_i goals, $\mathcal{H}_i$ variants of program clauses, θ_i substitutions, and each rule an application of $\mathrm{R_u}$. Instead of the tree notation, one usually writes

$$\Gamma_0 \xrightarrow{\theta_0}_{\mathrm{u}} \Gamma_1 \xrightarrow{\theta_1}_{\mathrm{u}} \Gamma_2 \xrightarrow{\theta_2}_{\mathrm{u}} \cdots$$

The *resultant* of a finite unrestricted derivation

$$\Gamma_0 \xrightarrow{\theta_0}_{\mathrm{u}} \cdots \xrightarrow{\theta_{n-1}}_{\mathrm{u}} \Gamma_n$$

is the implication

$$\bigwedge \Gamma_n \to (\bigwedge \Gamma_0)\theta_0\theta_1 \ldots \theta_{n-1}. \qquad \boxtimes$$

7.3.5. It now seems natural to define a resolution derivation as an unrestricted resolution derivation where each unrestricted resolution step is in fact a resolution step. By this we expect to achieve that the resultant of any finite subderivation is always most general. The following example, however, shows that the requirement that every step is a resolution step is not quite enough to achieve this. Consider the program $\{Py \Rightarrow Rxy, \Rightarrow Py\}$, and the following derivation:

$$\cfrac{\cfrac{Rxy \quad Py \Rightarrow Rxy}{Py}\,\epsilon \qquad \Rightarrow Py}{[\,]}\,[y/x]$$

Each step is a resolution step, and the computed substitution is $[y/x]$ with resultant Rxx. This is not a most general resultant, since the derivation

$$\cfrac{\cfrac{Rxy \quad Py \Rightarrow Rxy}{Py}\,\epsilon \qquad \Rightarrow Pz}{[\,]}\,[y/z]$$

produces the more general resultant Rxz. The difficulty is caused by the fact that in the first derivation, the variable x, which disappeared at the first step (was *released*), was reintroduced at the next by the substitution $[y/x]$, and therefore occurred in the resultant. This motivates the following:

7.3.6. DEFINITION. A *resolution derivation* or *SLD-derivation* from $\mathcal{P}$ is an unrestricted resolution derivation with all applications of $\mathrm{R_u}$ in fact applications of R, and such that the variant of the clause chosen at each step has its free variables distinct from the free variables occurring in the resultant so far. ⊠

REMARK. The variable condition will guarantee that among "similar" derivations the resultant at every step will be most general. Two derivations are *similar* if they start from the same goal, and at each step the same clause is applied in both derivations, to the corresponding atom occurrences. In the set $\mathcal{X}$ of all derivations similar to a given derivation, a derivation $\mathcal{D}$ with resolution (instead of unrestricted resolution) will be most general in the sense that the resultants of any derivation in $\mathcal{X}$ may be obtained by substitution applied to the resultants of $\mathcal{D}$. This question is treated in greater detail in, for example, C. [1994]. The variable condition given is sufficient, but not necessary, for this result.

The abbreviation "SLD" stands for "Selection-driven Linear resolution for Definite clauses".

7.3.7. DEFINITION. An (unrestricted) resolution derivation is called *(unrestrictedly) successful* if terminating in the empty goal, and *(unrestrictedly) unsuccessful*, if terminating in a goal not permitting unification with the conclusion of a program clause.

A successful derivation via $\theta_0, \ldots, \theta_{n-1}$ starting from $\Gamma \equiv \Gamma_0$ using program $\mathcal{P}$ yields a *computed answer* on $\Gamma|\mathcal{P}$ of the form $\theta_0 \ldots \theta_{n-1} \restriction \mathrm{FV}(\Gamma_0)$. ⊠

7.3.8. DEFINITION. Let $\mathcal{P}$ be a program, Γ a goal; an *answer* to the query $\Gamma|\mathcal{P}$ ("can goal Γ be reached by program $\mathcal{P}$?") is a substitution σ with $\mathrm{dom}(\sigma) \subset \mathrm{FV}(\Gamma)$; a *correct answer* to $\Gamma|\mathcal{P}$ is a σ such that $\mathcal{P}^\forall \vdash_{\mathrm{c}} (\bigwedge \Gamma)\sigma$. ⊠

By the completeness theorem for classical logic $\mathcal{P}^\forall \vdash_{\mathrm{c}} (\bigwedge \Gamma)\sigma$ is equivalent to $\mathcal{P}^\forall \models (\bigwedge \Gamma)\sigma$ where $\models$ is the usual semantical consequence relation.

7.3.9. THEOREM. *(Soundness for linear resolution) Every computed answer is a correct answer.*

PROOF. For any resolution step relative to a program $\mathcal{P}$

$$\frac{\Gamma \qquad \Delta \Rightarrow A}{\Gamma'}$$

we can obviously prove in minimal logic

$$\mathcal{P}^\forall \vdash_{\mathrm{m}} \bigwedge \Gamma' \to (\bigwedge)\theta,$$

since $\Delta \Rightarrow A \in \mathcal{P}$, and since Γ and Γ' must be of the form $\Gamma''B$, $(\Gamma''\Delta)\theta$ respectively, with $B\theta = A\theta$. Hence, if we have an unrestricted resolution derivation of goal Γ_n from Γ_0:

$$\cfrac{\cfrac{\Gamma_0 \quad \mathcal{H}_0}{\Gamma_1} \theta_0 \qquad \mathcal{H}_1}{\Gamma_2} \theta_1$$

$$\vdots$$

$$\frac{\Gamma_{n-1} \qquad \mathcal{H}_{n-1}}{\Gamma_n} \theta_{n-1}$$

we have

$$\mathcal{P}^\forall \vdash_m \bigwedge \Gamma_n \to (\bigwedge_{n-1})\theta_{n-1},$$
$$\mathcal{P}^\forall \vdash_m \bigwedge \Gamma_{n-1} \to (\bigwedge_{n-2})\theta_{n-2},$$
$$\dots$$
$$\mathcal{P}^\forall \vdash_m \bigwedge \Gamma_1 \to (\bigwedge_0)\theta_0,$$

from which it readily follows that for all i, $0 \le i < n$,

$$\mathcal{P}^\forall \vdash_{\mathrm{m}} \bigwedge \Gamma_n \to (\bigwedge \Gamma_i)\theta_i\theta_{i+1}\dots\theta_{n-1}.$$

Hence, if Γ_n is an empty goal,

$$\mathcal{P}^\forall \vdash_{\mathrm{m}} (\bigwedge \Gamma_0)\theta_0\theta_1\dots\theta_{n-1}. \qquad \boxtimes$$

7.3.10. The interesting aspect of linear resolution is that more is achieved than just a refutation: there is also a computed substitution. This permits us to use linear resolution for computations. Consider e.g. the following program $\mathcal{P}_+$ in the language (sum, 0, S), where "sum" is a ternary relation, 0 a constant, and S a unary function:

$$\Rightarrow \mathrm{sum}(x, 0, x)$$
$$\mathrm{sum}(x, y, z) \Rightarrow \mathrm{sum}(x, \mathrm{S}y, \mathrm{S}z)$$

Suppose we take as goal $\mathrm{sum}(x, 3, y)$ (as usual, the numeral n abbreviates $\mathrm{S}(\mathrm{S}\dots(\mathrm{S}0))$ (n occurrences of S). We get

$$\begin{aligned} \mathrm{sum}(x,3,y) &\xrightarrow{[y/\mathrm{S}y]}_{\mathrm{u}} \mathrm{sum}(x,2,y) \\ &\xrightarrow{[y/\mathrm{S}y]}_{\mathrm{u}} \mathrm{sum}(x,1,y) \\ &\xrightarrow{[y/\mathrm{S}y]}_{\mathrm{u}} \mathrm{sum}(x,0,y) \\ &\xrightarrow{[y/x]}_{\mathrm{u}} [\,] \end{aligned}$$

with computed answer $[y/\mathrm{SSS}x]$. That is to say, refuting $\forall xy\, \neg\mathrm{sum}(x, 3, y)$ provides an instantiation of $\exists xy\, \mathrm{sum}(x, 3, y)$. A more complicated example is given by the program $\mathcal{P}_\times$ which is $\mathcal{P}_+$ plus

$$\Rightarrow \mathrm{mult}(x, 0, 0)$$
$$\mathrm{mult}(x, y, u), \mathrm{sum}(u, x, z) \Rightarrow \mathrm{mult}(x, \mathrm{S}y, z)$$

where "mult" is a new ternary relation added to the language. The following is a derivation from $\mathcal{P}_\times$:

$$\begin{array}{lll}
\mathrm{mult}(3,2,z) & \xrightarrow{\epsilon}_{\mathrm{u}} & \mathrm{mult}(3,1,u), \mathrm{sum}(u,3,z) \\
& \xrightarrow{\epsilon}_{\mathrm{u}} & \mathrm{mult}(3,0,u), \mathrm{sum}(u,3,z), \mathrm{sum}(v,3,u) \\
& \xrightarrow{[v/3]}_{\mathrm{u}} & \mathrm{sum}(u,3,z), \mathrm{sum}(3,3,u) \\
& \xrightarrow{[u/\mathrm{S}u]}_{\mathrm{u}} & \mathrm{sum}(\mathrm{S}u,3,z), \mathrm{sum}(3,2,\mathrm{S}u) \\
& \xrightarrow{[u/\mathrm{S}u]}_{\mathrm{u}} & \mathrm{sum}(\mathrm{S}^2u,3,z), \mathrm{sum}(3,1,\mathrm{S}^2u) \\
& \xrightarrow{[u/\mathrm{S}u]}_{\mathrm{u}} & \mathrm{sum}(\mathrm{S}^3u,3,z), \mathrm{sum}(3,0,\mathrm{S}^3u) \\
& \xrightarrow{[u/0]}_{\mathrm{u}} & \mathrm{sum}(3,3,z) \\
& \xrightarrow{[z/\mathrm{S}z]}_{\mathrm{u}} & \mathrm{sum}(3,2,z) \\
& \xrightarrow{[z/\mathrm{S}z]}_{\mathrm{u}} & \mathrm{sum}(3,1,z) \\
& \xrightarrow{[z/\mathrm{S}z]}_{\mathrm{u}} & \mathrm{sum}(3,0,z) \\
& \xrightarrow{[z/3]}_{\mathrm{u}} & \mathrm{sum}(3,0,3) \\
& \xrightarrow{\epsilon}_{\mathrm{u}} & [\,]
\end{array}$$

with computed answer substitution $[z/6]$. That is to say, we have derived $\mathrm{mult}(3,2,6)$.

7.3.10A. ♠ Is the unrestricted derivation above also a SLD-derivation?

7.3.11. THEOREM. *(Completeness of SLD-resolution) When $\mathcal{P}^\forall \vdash_c \bigwedge\Gamma\sigma$, then there exists a successful SLD-deduction of Γ from $\mathcal{P}$ with computed answer θ, such that there is a substitution γ such that $\Gamma\sigma = \Gamma\theta\gamma$.*

PROOF. The proof uses the conservative extension result of theorem 6.7.5. If $\mathcal{P}^\forall \vdash_c \bigwedge\Gamma\sigma$, we have $\mathbf{Nc} \vdash \mathcal{P}^\forall \Rightarrow (\bigwedge\Gamma)\sigma$; the deduction may be transformed into a deduction in $\mathbf{Nm}$ in long normal form. By recursion on k we construct SLD-derivations for $\mathcal{P}|\Gamma$, $\Gamma \equiv \Gamma_0$ of the form

$$\dfrac{\dfrac{\dfrac{\Gamma_0 \quad \mathcal{H}_0}{\Gamma_1}\,\theta_0 \qquad \mathcal{H}_1}{\Gamma_2}\,\theta_1}{\begin{array}{c}\vdots\\ \dfrac{\Gamma_{k-1} \qquad \mathcal{H}_{k-1}}{\Gamma_k}\,\theta_{k-1}\end{array}}$$

and substitutions γ_k^σ, deductions $\mathcal{D}_{k,i}^\sigma$ for the atoms in $\Gamma_k\gamma_k^\sigma$ such that

$$\Gamma_0\sigma = \Gamma_0\theta_0\ldots\theta_{k-1}\gamma_k^\sigma.$$

For $k=0$ we take $\Gamma \equiv \Gamma_0$, $\gamma_0^\sigma \equiv \sigma$. So by soundness $\bigwedge\Gamma_k \to (\bigwedge\Gamma_0)\theta_0\ldots\theta_{k-1}$.

Suppose we have carried out the construction up to k, and Γ_k is not empty, say $\Gamma_k \equiv \Gamma' A$. By the IH, there is an **Nm**-deduction $\mathcal{D}$ of $A\gamma_k^\sigma$ in long normal form from $\mathcal{P}^\forall$ in **Nm**, say with r nodes. The final step in $\mathcal{D}$ must be of the form

$$\dfrac{\dfrac{\mathcal{H}_k^\forall \equiv \forall\vec{x}(\bigwedge\Delta \to B)}{(\bigwedge\Delta \to B)\tau}\forall\text{E} \qquad \dfrac{\mathcal{D}'}{(\bigwedge\Delta)\tau}}{B\tau \equiv A\gamma_k^\sigma}\to\text{E}$$

where $\mathcal{D}'$ contains deductions of all atoms of $\Delta\tau$ with sum of sizes $\leq r-1$ nodes. Without loss of generality we may assume $\bigwedge\Delta \to B$ to be such that $\mathcal{H}_k \equiv (\Delta \Rightarrow B)$ is a variant of a clause in $\mathcal{P}$ with

$$(1) \qquad \mathrm{FV}(\mathcal{H}_k) \cap (\mathrm{FV}(\Gamma_k) \cup \mathrm{FV}(\Gamma\theta_0 \ldots \theta_{k-1})) = \emptyset.$$

If we take

$$\theta := \tau \restriction \mathrm{FV}(\mathcal{H}_k) \cup \gamma_k^\sigma \restriction (\mathrm{FV}(\Gamma_k) \cup \mathrm{FV}(\Gamma\theta_0 \ldots \theta_{k-1})),$$

it follows that $A\theta = B\theta$. So we may add a step

$$\dfrac{\begin{array}{c}\vdots\\ \overline{\Gamma_k}\end{array} \quad \mathcal{H}_k}{\Gamma_{k+1}}\theta_k$$

such that a new resolution derivation results. The condition (1) implies that the recipe of proposition 7.3.3 applies, i.e. that $\mathcal{H}_k$ can play the role of $\mathcal{H}\alpha$ in the proposition. θ_k is an m.g.u. of A and B; $\theta_k\gamma_{k+1}^\sigma = \theta$, for some γ_{k+1}^σ. Therefore

$$\begin{array}{rcl}\Gamma_{k+1}\gamma_{k+1}^\sigma &=& (\Gamma', \Delta)\theta_k\gamma_{k+1}^\sigma\\ &=& \Gamma'\theta_k\gamma_{k+1}^\sigma, \Delta\theta_k\gamma_{k+1}^\sigma = \Gamma'\theta, \Delta\theta = \Gamma'\gamma_k^\sigma, \Delta\tau.\end{array}$$

hence $\Gamma\theta_0 \ldots \theta_{k-1}\theta_k\gamma_{k+1}^\sigma = \Gamma\theta_0 \ldots \theta_{k-1}\theta = \Gamma\theta_0 \ldots \theta_{k-1}\gamma_k^\sigma = \Gamma\sigma$. We now have deductions in long normal form for the atoms in $\Gamma_{k+1}\gamma_{k+1}^\sigma$ with the total sum of sizes of these deductions less than the size for $\Gamma_k\gamma_k^\sigma$ (these deductions are encoded in $\mathcal{D}'$ above). ⊠

7.3.12. REMARKS. (i) Inspection of the proof shows that the choice of the atom A in Γ_k is irrelevant: the result will always be a successful SLD-resolution.

(ii) From the proof we also see that a deduction $\mathcal{D}$ in long normal form in **Nm** of $\bigwedge\Gamma\sigma$, for a goal Γ, in an obvious way encodes an unrestricted derivation of $\Gamma\sigma$. If $\bigwedge\Gamma\sigma$ is, say, $(A_1 \wedge A_2)\sigma$, the deduction must end with

$$\cfrac{\cfrac{\cfrac{\forall\vec{x}(\bigwedge\Delta\to B)}{(\bigwedge\Delta\to B)\sigma'}\forall\text{E}\qquad \begin{matrix}\mathcal{D}'\\(\bigwedge\Delta)\sigma'\end{matrix}}{B\sigma'}\qquad \cfrac{\cfrac{\forall\vec{y}(\bigwedge\Delta'\to B')}{(\bigwedge\Delta'\to B')\sigma''}\forall\text{E}\qquad \begin{matrix}\mathcal{D}''\\(\bigwedge\Delta')\sigma''\end{matrix}}{B'\sigma''}}{B\sigma'\wedge B'\sigma''\equiv(A_1\wedge A_2)\sigma}$$

Without loss of generality we can assume $\vec{x}\cap\vec{y}=\emptyset$, $A_1\sigma=B\sigma'$, $A_2\sigma=B'\sigma''$, which translates into the beginning of an unrestricted derivation

$$\cfrac{\cfrac{A_1,A_2\qquad \Delta\Rightarrow B}{B\sigma',\Delta\sigma'}\sigma'\qquad \Delta'\Rightarrow B'}{\Delta'\sigma'',\Delta\sigma'}\sigma''$$

which can be continued using $\mathcal{D}'$, $\mathcal{D}''$ in the same manner.

7.4 From Gentzen system to resolution

In this section P,Q,R will be positive literals.

7.4.1. At the end of section 7.1 we indicated how in propositional logic cutfree deductions of an empty goal could be read as resolution derivations, with "linear resolution", i.e. all program clauses were of the form $\Gamma\Rightarrow P$, Γ consisting of proposition letters only, or in one-sided notation $\Rightarrow\neg\Gamma,P$. So the program clauses in *linear* resolution contain a single *positive* literal.

A more general form of propositional resolution is just Cut on literals:

$$\frac{\Gamma,P\qquad \Delta,\neg P}{\Gamma\Delta}$$

We shall now describe a very simple method of transforming cutfree deductions of a Gentzen–Schütte system **GS5p** for **Cp** into a sytem based on general resolution. This method applies to program clauses of the form $\Gamma\Rightarrow\Delta$ (or in the form of a one-sided sequent, $\neg\Gamma,\Delta$).

The principal reason for discussing this very simple case is that it can serve as an introduction to the more complicated case of intuitionistic propositional logic, to be treated in the next section.

7.4.2. Definition. The axioms and rules of **GS5p** are

$$\text{Ax}\quad P,\neg P\qquad\qquad \text{R}\wedge\,\frac{\Gamma,A\qquad \Delta,B}{(\Gamma,\Delta,A\wedge B)}$$

$$\text{R}\vee_1\,\frac{\Gamma,A,B}{(\Gamma,A\vee B)}\qquad \text{R}\vee_2\,\frac{\Gamma,A}{(\Gamma,A\vee B)}\qquad \text{R}\vee_3\,\frac{\Gamma,B}{(\Gamma,A\vee B)}$$

Here (Γ) is the set corresponding to the multiset Γ. The system is not yet closed under weakening; if we add weakening

$$\text{RW}\ \frac{\Gamma}{\Gamma, \Delta}$$

we can always transform deductions such that weakening occurs only immediately before the final conclusion. ⊠

7.4.2A. ♠ Show that in **GS5p** plus RW the applications of RW can always be moved to the bottom of the deduction. Show that **GS5p** + RW is equivalent to, say, **GS1p**.

7.4.3. DEFINITION. A *clause formula* is a disjunction of literals $(\dots(L_1 \vee \dots) \vee L_n)$. We identify clause formulas which differ only in the order of the literals, and shall assume $L_1, \dots, L_n$ to be a *set*; this justifies the notation without parentheses: $L_1 \vee \dots \vee L_n$. A *clause* is a finite set $\{L_1, \dots, L_n\}$ of literals. The number n of elements of the disjunction in a clause formula [the number of elements in the clause] is called the *length* of the clause formula [clause]. ⊠

We might dispense with the separate syntactical category of clauses; but the distinction is convenient in discussing the connection between resolution and Gentzen system deductions in standard notation.

7.4.4. DEFINITION. (*The resolution system* **Rcp**) In this system clauses are derived from sets of clauses using the following axiom and rule:

$$\text{Axiom } P, \neg P \qquad \text{R}\ \frac{\neg P, \Gamma \qquad P, \Delta}{\Gamma, \Delta}$$

As before, "R" is called the *resolution rule* (but is in fact Cut on proposition letters). ⊠

7.4.5. PROPOSITION. *For each formula F constructed from literals using $\vee, \wedge$, there is a set of clause formulas* Clause(F) *and a proposition letter l_F such that in* **Cp** *F is derivable iff the sequent* Clause$(F) \Rightarrow l_F$ *is derivable in a Gentzen system for* **Cp**.

PROOF. Let F be an arbitrary formula constructed from positive and negative literals by means of $\vee, \wedge$. With each subformula A of F we associate a propositional variable l_A, the *label* of A. For atomic formulas P we assume $P \equiv l_P$. For $A \equiv B \wedge C$, $B \vee C$, $\neg P$ we let $A^* \equiv l_B \wedge l_C$, $l_B \vee l_C$, $\neg l_P$. Then

(1) $\qquad l_A \leftrightarrow A^*$

for all subformulas A of F guarantees that $l_A \leftrightarrow A$ for all subformulas A of F. The equivalences (1) may be expressed by a set of clause formulas Γ_F of length ≤ 3 as follows: for each subformula A let $\mathcal{C}_A \equiv \mathcal{C}_A^+ \cup \mathcal{C}_A^-$, where

$$\mathcal{C}_{B\wedge C}^+ := \{\neg l_B \vee \neg l_C \vee l_{B\wedge C}\}, \quad \mathcal{C}_{B\wedge C}^- := \{\neg l_{B\wedge C} \vee l_B, \neg l_{B\wedge C} \vee l_C\}$$

$$\mathcal{C}_{B\vee C}^+ := \{l_{B\vee C} \vee \neg l_B, l_{B\vee C} \vee \neg l_C\}, \quad \mathcal{C}_{B\vee C}^- := \{l_B \vee l_C \vee \neg l_{B\vee C}\}$$

$$\mathcal{C}_{\neg P}^+ \equiv \mathcal{C}_{\neg P}^- := \{l_P \vee l_{\neg P}\}$$

The formulas in $\mathcal{C}_A^+$ correspond to $A^* \to l_A$ (with positive occurrence of l_A), and the formulas in $\mathcal{C}_A^-$ to $l_A \to A^*$ (with negative occurrence of l_A). If we replace in $\Gamma_F \Rightarrow l_F$ all labels by the corresponding subformulas, all formulas in Γ become true, hence F holds if $\Gamma_F \Rightarrow l_F$ holds. The converse is obvious. ⊠

NOTATION. If $\Delta \equiv A_1, \ldots, A_n$, we write l_Δ for $l_{A_1}, \ldots, l_{A_n}$. ⊠

7.4.6. THEOREM. *There is a mapping* R *which transforms any deduction* $\mathcal{D}$ *of* $F_1, \ldots, F_n$ *in* **GS5p** *into a resolution proof* $\mathrm{R}(\mathcal{D})$ *of* $l_{F_1}, \ldots, l_{F_n}$ *from* $\mathcal{C}$ $:= \bigcup\{\mathrm{Cla}(F_i) : 1 \leq i \leq n\}$, *where* $\mathrm{Cla}(F)$ *is the set of clauses corresponding to* $\mathrm{Clause}(F)$.

PROOF. The deduction in **GS5p** has at its nodes (multi)sets of subformulas of $F_1, \ldots, F_n$. We show by induction on the depth of the deduction that if the multiset Δ has been derived, then in the resolution calculus we can derive l_Δ from the clauses in $\mathcal{C}$ as axioms.

Case 1. An axiom $P, \neg P$ translates into

$$\frac{l_{\neg P}, l_P \text{ (input)} \qquad l_P, \neg l_P}{l_P, l_{\neg P}}$$

where "(input)" means that the clause is an axiom clause from $\mathcal{C}$.

Case 2. If the final rule in $\mathcal{D}$ is $\mathrm{R}\vee_1$, as on the left, we add to the derivation in **Rp** existing by IH the part shown on the right:

$$\frac{\Gamma, A, B}{(\Gamma, A \vee B)} \qquad \frac{l_{A\vee B}, \neg l_B \text{ (input)} \qquad \dfrac{l_{A\vee B}, \neg l_A \text{ (input)} \quad (l_\Gamma, l_A, l_B)}{(l_{A\vee B}, l_B, l_\Gamma)}}{(l_{A\vee B}, l_{A\vee B}, l_\Gamma)}$$

The cases of $\mathrm{R}\vee_2$, $\mathrm{R}\vee_3$ are simpler and left to the reader.

Case 3. Similarly if the last rule is $\mathrm{R}\wedge$:

$$\frac{\Gamma, A \quad \Delta, B}{(\Gamma, \Delta, A \wedge B)} \qquad \frac{\dfrac{l_{A\wedge B}, \neg l_A, \neg l_B \text{ (input)} \quad l_\Gamma, l_A}{(l_{A\wedge B}, \neg l_B, l_\Gamma)} \qquad l_\Delta, l_B}{(l_{A\wedge B}, l_\Gamma, l_\Delta)}$$

⊠

7.4.6A. ♠ What goes wrong in the argument above if **GS5p** has $R\vee_1$, but not $R\vee_2, R\vee_3$? What is the reason behind the choice of a context-free version of $R\wedge$?

7.4.7. COROLLARY. *The calculus* **Rcp** *is complete for* **Cp** *(in the sense that there is a transformation of a formula F into a problem of deriving a letter l_F from a set of clauses, such that the the latter deduction is possible in* **Rcp** *iff F was derivable in* **Cp***).*

7.4.8. REMARK. In dealing with predicate logic, a standard procedure is to use "Skolemization": formulas are brought into prenex form, and existential quantifiers are eliminated by the introduction of new function symbols ("Skolem functions"). But there is also a method, due to N. Zamov, which avoids the process of prenexing and Skolemizing, and which is briefly sketched here.

In predicate logic, *literals* are atomic formulas or their negations, and *initial clauses* are disjunctions of literals $L_1 \vee \ldots \vee L_n$ $(n \geq 2)$, or existentially quantified disjunctions of literals $\exists y(L_1 \vee \ldots \vee L_n)$ $(n \geq 2)$. Clauses are simply disjunctions of literals (n arbitrary). (We make no distinction between clauses and clause formulas in this description.)

In the spirit of proposition 7.4.5 one can show that there is a finite set Cla_F of initial clauses such that a formula F holds if the universal closure of Cla_F is inconsistent.

Resolution appears in two versions, namely as ordinary resolution

$$\mathrm{R}\,\frac{L \vee C \qquad \neg L' \vee C'}{(C \vee C')\sigma}$$

with σ an m.g.u. of the literals L, L', and $\exists$-resolution

$$\mathrm{R}_\exists\,\frac{\exists y(L \vee C) \qquad \neg L' \vee C'}{(C \vee C')\sigma}$$

where σ is an m.g.u. of L, L' not containing the assignment y/t, and such that $(C \vee C')\sigma$ does not contain y.

7.5 Resolution for Ip

In this section $P, Q, R \in \mathcal{PV}$.

7.5.1. We now apply the ideas of the preceding section in the more complicated context of **Ip**. We start by introducing yet another variant of the intuitionistic Gentzen system, the system **G5ip**. The antecedents of sequents are *multisets*. There is no weakening rule, but weakening has been built into some of the logical rules. There is also no separate contraction rule, but most of the rules have contraction built explicitly into the conclusion. This is indicated by putting parentheses around the multiset of the antecedent; thus in the conclusions of the rules below, an expression like $(\Gamma, \Gamma', \ldots, A, A', \ldots)$ indicates the multiset obtained by taking the multiset union of $\Gamma, \Gamma', \ldots, \{A\}, \{A'\}, \ldots$ and contracting whenever possible, i.e. all formulas in the union are contracted to multiplicity one. In other words, (Γ) will be $\mathrm{Set}(\Gamma)$. Note that sets are representable as multisets in which every formula has multiplicity one. The rules below are such that in a deduction every multiset occurring in the antecedent of a sequent is in fact a set.

DEFINITION. (*The Gentzen systems* **G5ip**, **G5i**) The system **G5ip** is given by axioms and rules:

Axioms

$$\mathrm{Ax}\ A \Rightarrow A \qquad\qquad \mathrm{L}\bot\ \bot \Rightarrow A$$

Rules for the logical operators

$$\mathrm{L}\wedge_1 \frac{\Gamma, A_0 \Rightarrow C}{(\Gamma, A_0 \wedge A_1) \Rightarrow C} \qquad \mathrm{L}\wedge_2 \frac{\Gamma, A_1 \Rightarrow C}{(\Gamma, A_0 \wedge A_1) \Rightarrow C}$$

$$\mathrm{L}\wedge_3 \frac{\Gamma, A_0, A_1 \Rightarrow C}{(\Gamma, A_0 \wedge A_1) \Rightarrow C} \qquad \mathrm{R}\wedge \frac{\Gamma \Rightarrow A \qquad \Delta \Rightarrow B}{(\Gamma, \Delta) \Rightarrow A \wedge B}$$

$$\mathrm{L}\vee \frac{A, \Gamma \Rightarrow C \qquad B, \Delta \Rightarrow C}{(A \vee B, \Gamma, \Delta) \Rightarrow C} \qquad \mathrm{R}\vee_i \frac{\Gamma \Rightarrow A_i}{\Gamma \Rightarrow A_0 \vee A_1}\ (i = 0, 1)$$

$$\mathrm{L}{\to} \frac{\Gamma \Rightarrow A \qquad B, \Delta \Rightarrow C}{(\Gamma, \Delta, A \to B) \Rightarrow C} \qquad \mathrm{R}{\to}_1 \frac{A, \Gamma \Rightarrow B}{\Gamma \Rightarrow A {\to} B} \qquad \mathrm{R}{\to}_2 \frac{\Gamma \Rightarrow B}{\Gamma \Rightarrow A {\to} B}$$

The system can be extended to a system **G5i** for predicate logic by adding

$$\mathrm{L}\forall \frac{A[x/t], \Gamma \Rightarrow B}{(\forall x A, \Gamma) \Rightarrow B} \qquad \mathrm{R}\forall \frac{\Gamma \Rightarrow A[x/y]}{\Gamma \Rightarrow \forall x A}$$

$$\mathrm{L}\exists \frac{A[x/y], \Gamma \Rightarrow B}{(\exists x A, \Gamma) \Rightarrow B} \qquad \mathrm{R}\exists \frac{\Gamma \Rightarrow A[x/t]}{\Gamma \Rightarrow \exists x A}$$

where in L$\exists$, R$\forall$ $y \notin \mathrm{FV}(\Gamma, \Delta)$, and also $y \equiv x$ or $y \notin \mathrm{FV}(A)$. However, we shall not discuss predicate logic in the remainder of this section. ⊠

G5i has the following property.

7.5.2. PROPOSITION. *If we add the Weakening rule*

$$\text{LW}\ \frac{\Gamma \Rightarrow B}{A, \Gamma \Rightarrow B}$$

then any derivation in **G5i** + LW *can be transformed into a derivation where the only application of weakening appears just before the conclusion; the ordering of the subtree of the prooftree consisting only of nodes where a logical rule is applied, with the names of the rules attached to these nodes as labels, remains the same under this transformation.*

PROOF. The proof is straightforward, by "pushing down" applications of LW. The fact that there are several variants of certain rules, such as R→ and L∧, which incorporate an element of weakening, makes this possible. ⊠

7.5.3. PROPOSITION. *The system* **G5ip** *is equivalent to the system* **G1pi** *in the following sense: if* $\Gamma \Rightarrow A$ *is provable in* **G5ip**, *then it is also provable in* **G1pi**, *and if* $\Gamma \Rightarrow A$ *is provable in* **G1pi**, *then* $\Gamma' \Rightarrow A$ *is provable in* **G5ip** *for some* $\Gamma' \subset \Gamma$. ⊠

7.5.4. DEFINITION. An *intuitionistic clause* is a sequent of one of the following forms ($P_1, \ldots, P_n$ treated as a set):

$$(P \to Q) \Rightarrow R, \quad P \Rightarrow (Q \vee R), \quad P_1, \ldots, P_n \Rightarrow Q$$

($P, Q, R, P_1, \ldots, P_n$ atomic). A clause is *initial* if it is of the first, second or third type with $n \leq 2$. $n = 0$ is possible for clauses.

An *intuitionistic clause formula* is of the form

$$(P \to Q) \to R, \quad P \to (Q \vee R), \quad P_1 \to (P_2 \to \ldots (P_n \to Q) \ldots).$$

In the third case there corresponds more than one clause formula to a clause (because of permutations of the P_i). ⊠

7.5.5. THEOREM. *Let* F *be an arbitrary propositional formula, and suppose that we have associated with each subformula* A *of* F *a propositional variable* l_A *(the* label *of* A*). Then there is a set* Clause(F) *of initial clause formulas involving only labels of subformulas of* F *such that*

$$\Rightarrow F \quad \text{iff} \quad \text{Clause}(F) \Rightarrow l_F.$$

PROOF. For atomic formulas A we may take $l_A \equiv A$. For compound A of the forms $B \vee C$, $B \wedge C$, $B \to C$ we let A^* be $l_B \vee l_C$, $l_B \wedge l_C$, $l_B \to l_C$ respectively. The set of equivalences of the form

$$l_A \leftrightarrow A^*$$

fixes the relationship between the formulas and their labels, and entails

$$l_A \leftrightarrow A$$

for all subformulas A of F. We express $A^* \rightarrow l_A$ (where the label of A appears positively) by a set of clause formulas $\mathcal{C}_A^+$, and $l_A \rightarrow A^*$ (where l_A appears negatively) by a set of clause formulas $\mathcal{C}_A^-$.

$$\begin{array}{ll}
\mathcal{C}^+_{B\wedge D} = \{l_B, l_D \rightarrow l_{B\wedge D}\} & \mathcal{C}^-_{B\wedge D} = \{l_{B\wedge D} \rightarrow l_B, l_{B\wedge D} \rightarrow l_D\} \\
\mathcal{C}^+_{B\vee D} = \{l_B \rightarrow l_{B\vee D}, l_D \rightarrow l_{B\vee D}\} & \mathcal{C}^-_{B\vee D} = \{l_{B\vee D} \rightarrow l_B \vee l_D\} \\
\mathcal{C}^+_{B\rightarrow D} = \{(l_B \rightarrow l_D) \rightarrow l_{B\rightarrow D}\} & \mathcal{C}^-_{B\rightarrow D} = \{l_{B\rightarrow D} \rightarrow (l_B \rightarrow l_D)\}.
\end{array}$$

Now take for Clause(F) the collection

$$\bigcup \{\mathcal{C}_A^+ \cup \mathcal{C}_A^- \;:\; A \text{ non-atomic subformula of } F\}.$$

Then we have: if Clause(F) $\Rightarrow l_F$, then $\Rightarrow F$, for if we substitute A for the l_A everywhere, all the formulas of Clause(F) become true and l_F becomes F. Conversely, if $\Rightarrow F$, the Clause(F) entails $F \leftrightarrow l_F$, so Clause(F) $\Rightarrow l_F$ follows.

REMARK. The theorem can be extended to sequents $\Delta \Rightarrow F$ in a straightforward way.

7.5.6. DEFINITION. (*The resolution calculus* **Rip**)
Axioms

$$P \Rightarrow P, \qquad \bot \Rightarrow P \quad (A \text{ atomic}).$$

Inference rules

$$\rightarrow^r \; \frac{(P \rightarrow Q) \Rightarrow R \qquad \Delta, [P] \Rightarrow Q}{\Delta \Rightarrow R}$$

([P] indicates that P may be present or not).

$$\vee^r \; \frac{P \Rightarrow Q \vee R \qquad \Delta \Rightarrow P \qquad \Delta', Q \Rightarrow S \qquad \Delta'', R \Rightarrow S}{(\Delta\Delta'\Delta'') \Rightarrow S}$$

$$\text{Res} \; \frac{\Delta \Rightarrow P \qquad P, \Delta' \Rightarrow Q}{(\Delta\Delta') \Rightarrow Q}$$

with P, Q, R, S propositional variables. ⊠

7.5.7. THEOREM. *For any $\mathcal{D} : \Gamma \Rightarrow F$ in* **G5ip** *we may construct a derivation* $\mathrm{R}(\mathcal{D}) : \Delta \vdash_{\mathbf{Rpi}} l_\Gamma \Rightarrow l_F$ *where* $l_\Gamma \equiv l_{B_1}, \dots, l_{B_n}$ *if* $\Gamma \equiv B_1, \dots, B_n$, *and* Δ *consists of all initial clauses corresponding to the initial clause formulas constructed in 7.5.5.*

PROOF. By induction on the length of $\mathcal{D}$.
Basis. $\mathcal{D}$ is an axiom $A \Rightarrow A$ or $\bot \Rightarrow A$. Then $\mathrm{R}(\mathcal{D})$ is the corresponding axiom $l_A \Rightarrow l_A$, $\bot \Rightarrow l_A$.
Induction step. We show below on the left the last rule application in $\mathcal{D}$, and on the right the corresponding final steps in $\mathrm{R}(\mathcal{D})$; if we have on the left premises $\Gamma \Rightarrow A$, $\Delta \Rightarrow B$ etc. Then $l_\Gamma \Rightarrow l_A$, $l_\Delta \Rightarrow l_B$ are clauses derived by the induction hypothesis.

$$\frac{A, B, \Gamma \Rightarrow D}{(A \wedge B, \Gamma) \Rightarrow D} \qquad \frac{l_{A\wedge B} \Rightarrow l_B \qquad \dfrac{l_{A\wedge B} \Rightarrow l_A \qquad l_A, l_B, l_\Gamma \Rightarrow l_D}{(l_{A\wedge B}, l_B, l_\Gamma) \Rightarrow l_D}\,\mathrm{Res}}{(l_{A\wedge B}, l_\Gamma) \Rightarrow l_D}\,\mathrm{Res}$$

$$\frac{\Gamma \Rightarrow A \qquad B, \Delta \Rightarrow D}{(A \to B, \Gamma, \Delta) \Rightarrow D} \qquad \frac{l_\Gamma \Rightarrow l_A \qquad \dfrac{l_{A\to B}, l_A \Rightarrow l_B \qquad l_B, l_\Delta \Rightarrow l_D}{(l_{A\to B}, l_A, l_\Delta) \Rightarrow l_D}\,\mathrm{Res}}{(l_\Gamma, l_\Delta, l_{A\to B}) \Rightarrow l_D}\,\mathrm{Res}$$

$$\frac{A, \Gamma \Rightarrow D \qquad B, \Delta \Rightarrow D}{(A \vee B, \Gamma, \Delta) \Rightarrow D} \quad \frac{l_{A\vee B} \Rightarrow l_A \vee l_B \quad l_{A\vee B} \Rightarrow l_{A\vee B} \qquad l_A, l_\Gamma \Rightarrow l_D \quad l_B, l_\Delta \Rightarrow l_D}{(l_{A\vee B}, l_\Gamma, l_\Delta) \Rightarrow l_D}$$

etc. The other cases are left to the reader. ⊠

7.5.8. THEOREM. $\mathrm{Clause}(F) \vdash_{\mathbf{Rip}} l_F$ *iff* **G5ip** $\Rightarrow F$. ⊠

7.5.9. LEMMA. *Let $\mathcal{D} : \Gamma \Rightarrow F$ in* **G1pi** *or* **G5ip**. *Then for any sequent $A_1, \dots, A_n \Rightarrow B$ in $\mathcal{D}$, the A_i are negative subformulas of F or positive subformulas of Γ, and B is a positive subformula of F or a negative subformula of Γ.*

PROOF. By inspection of the rules. ⊠

7.5.10. COROLLARY. *(Refinement) For any $\Delta \Rightarrow F$, let* $\mathrm{Clause}^*(\Delta \Rightarrow F)$ *be the set of initial clauses containing only the $\mathcal{C}_A^-$ for A occurring positively in Γ or negatively in F, and the $\mathcal{C}_A^+$ for A occurring negatively in Γ and positively in F. If* **G5ip** $\vdash \Delta \Rightarrow F$, *then* $\mathrm{Clause}^*(\Delta \Rightarrow F) \vdash_{\mathbf{Rip}} l_\Delta \Rightarrow l_F$.

PROOF. By inspection of the proof of 7.5.7 and the preceding lemma. ⊠

7.5.11. EXAMPLE. We demonstrate the transformation of theorem 7.5.7 by an example. Since in this example only $\to$ is involved as operator, we write simply AB for $A \to B$. The purely implicational sequent $(QP)R, QR, RP \Rightarrow P$ has the following deduction in **G5ip**:

$$\dfrac{\dfrac{\dfrac{\dfrac{Q \Rightarrow Q \quad R \Rightarrow R}{Q, QR \Rightarrow R} \qquad P \Rightarrow P}{Q, QR, RP \Rightarrow P}}{QR, RP \Rightarrow QP} \qquad \dfrac{R \Rightarrow R \quad P \Rightarrow P}{R, RP \Rightarrow P}}{(QP)R, QR, RP \Rightarrow P}$$

By theorem 7.5.7, we can construct a deduction of

$$l_{(QP)R}, l_{QR}, l_{RP} \Rightarrow P$$

from clauses

$$\begin{array}{l} l_{QP}, l_{(QP)R} \Rightarrow R \\ l_{QR}, Q \Rightarrow R \\ l_{RP}, R \Rightarrow P \\ Q \to P \Rightarrow l_{QP} \end{array}$$

The proof gives the following deduction in **Rpi**:

$$\dfrac{\dfrac{Q \to P \Rightarrow l_{QP} \qquad \dfrac{\dfrac{Q \Rightarrow Q \qquad \dfrac{l_{QR}, Q \Rightarrow R \quad R \Rightarrow R}{l_{QR}, Q \Rightarrow R}}{l_{QR}, Q \Rightarrow R} \qquad \dfrac{l_{RP}, R \Rightarrow P \quad P \Rightarrow P}{l_{RP}, R \Rightarrow P}}{l_{QR}, l_{RP}, Q \Rightarrow P}}{l_{QR}, l_{RP} \Rightarrow l_{QP}} \qquad \mathcal{D}}{l_{(QP)R}, l_{RP}, l_{QR} \Rightarrow P}$$

where $\mathcal{D}$ is the deduction

$$\dfrac{l_{(QP)R}, l_{QP} \Rightarrow R \qquad \dfrac{P \Rightarrow P \qquad \dfrac{R, l_{RP} \Rightarrow P \quad P \Rightarrow P}{R, l_{RP} \Rightarrow P}}{R, l_{RP} \Rightarrow P}}{l_{(QP)R}, l_{QP}, l_{RP} \Rightarrow P}$$

This can be simplified to

$$\dfrac{\dfrac{Q \Rightarrow P \Rightarrow l_{QP} \qquad \dfrac{l_{QR}, Q \Rightarrow R \quad l_{RP}, R \Rightarrow P}{l_{QR}, l_{RP}, Q \Rightarrow P}}{l_{QR}, l_{RP} \Rightarrow l_{QP}} \qquad \dfrac{l_{(QP)R}, l_{QP} \Rightarrow R \quad R, l_{RP} \Rightarrow P}{l_{(QP)R}, l_{QP}, l_{RP} \Rightarrow P}}{l_{(QP)R}, l_{RP}, l_{QR} \Rightarrow P}$$

7.5.12. *Permutability of rules*

We recall the definition 5.3.1 of the permutability of rules in Gentzen systems.

LEMMA. *In* **G5ip** *the following permutation properties hold:*

(i) R$\to$ *permutes over* L$\lor$, L$\to$*;*

(ii) L$\lor$ *permutes over* L$\to$, R$\to$, L$\lor$*;*

(iii) L$\to$ *permutes over* L$\to$, R$\to$*.*

PROOF. By checking the various cases. ⊠

From now on, we consider deductions in **G5ip** with *atomic* instances of the axioms only.

7.5.13. LEMMA. *Let $\mathcal{D}$ be a (cutfree) deduction in* **G5ip** *of a sequent of the form $\Gamma \Rightarrow P$, P atomic, Γ a set of initial clause formulas. Then $\mathcal{D}$ can be transformed into a deduction $\mathcal{D}'$ such that the following three conditions are met.*

(i) An f.o. $P \to Q$ is principal in the succedent if it is at the same time an active formula of an L$\to$ *application.*

(ii) An f.o. $P \to Q \lor R$ is principal in the antecedent, and the active occurrence of $Q \lor R$ in one of the premises is itself principal.

(iii) An f.o. $P \to (Q \to R)$ is principal in the antecedent, and the active occurrence of $Q \to R$ is itself principal.

PROOF. We note that because of the subformula property, every succedent formula must be of the form Q or $Q \to R$, and an antecedent formula has one of the following forms: Q, $Q \to R$, $Q \lor R$, $Q \to (R \to S)$, $(Q \to R) \to S$, $Q \to R \lor S$; formulas of the last three forms belong to Γ.

We now extensively use permutability of rules, as follows.

Step 1. We permute any application of R$\to$ introducing a formula $Q \to R$ downward over L$\lor$, L$\to$ until we reach an L$\to$ application with $Q \to R$ as active formula. It is easy to see that each permutation of this kind decreases the number of occurrences of implications in the succedents of the deduction. Note that premises of applications of L$\to$ with $Q \to R$ or $Q \lor R$ as active formula on the left will have an atomic formula as succedent.

Step 2. If the deduction does not yet meet the second or third condition, there will be applications of L$\to$ with an active formula occurrence of the form $Q \to R$ or $Q \lor R$ on the left such that this active occurrence is not the principal formula of an application of L$\to$ or L$\lor$. Such an occurrence and its ancestors we call "offending occurrences". We apply induction w.r.t. the number of offending occurrences of $Q \to R$ or $Q \vee R$. We search for a topmost application of this kind, and permute it upwards over the preceding two rules (L→, L→ or L∨, L→ or R→, L→) corresponding to an application of L→ meeting condition (i), (ii) or (iii). The number of offending occurrences of the offending active formula $Q \vee R$ or $Q \to R$ will diminish by this procedure. The final result will be a deduction meeting all requirements. ⊠

7.5.13A. ♠ Check the cases of Step 2 in detail.

7.5.14. THEOREM. $\mathbf{G5ip} \vdash \Gamma^* \Rightarrow P$ *iff* $\Gamma \vdash_{\mathbf{Rpi}} P$.

PROOF. The direction from right to left is straightforward. Conversely, let us assume $\mathbf{G5ip} \vdash \Gamma^* \Rightarrow P$. We may assume this to have been proved by a cutfree deduction $\mathcal{D}$ in standard form according to the preceding lemma. Each sequent occurring in $\mathcal{D}$ will be of the form

$$\Sigma, \Delta \Rightarrow A$$

where $\Sigma \subset \Gamma^*$. We prove by induction on the depth of $\mathcal{D}$ that $l_\Delta \Rightarrow l_A$ is derivable in **Rip** from Γ.

Case 1. $\mathcal{D}$ is an axiom; trivial.

Case 2. Suppose the deduction ends with L→ introducing $(P \to Q) \to R \in \Gamma^*$. Then the final part of the deduction has the form

$$\dfrac{\dfrac{\Sigma, \Delta, [P] \Rightarrow Q}{\Sigma, \Delta \Rightarrow P \to Q} \quad R, \Delta', \Sigma' \Rightarrow P'}{((P \to Q) \to R, \Sigma, \Sigma'\Delta, \Delta') \Rightarrow P'}$$

with $(P \to Q) \to R \in \Gamma^*$. By IH we have resolution proofs of

$$l_\Gamma, [P] \Rightarrow Q, \qquad R, l_{\Delta'} \Rightarrow P'.$$

We construct a resolution proof terminating in

$$\dfrac{\dfrac{\dfrac{P \to Q \Rightarrow l_{P \to Q} \quad l_\Delta, [P] \Rightarrow Q}{l_\Delta \Rightarrow l_{P \to Q}} \quad l_{P \to Q} \Rightarrow R}{l_\Delta \Rightarrow R} \quad R, l_{\Delta'} \Rightarrow P'}{(l_\Delta, l_{\Delta'}) \Rightarrow P'}$$

Case 3. The deduction ends with L→, introducing $P \to (Q \to R) \in \Gamma^*$; then the final part of the deduction has the form

$$\dfrac{\Sigma, \Delta \Rightarrow P \quad \dfrac{\Sigma', \Delta' \Rightarrow Q \quad \Sigma'', \Delta'', R \Rightarrow P'}{(\Sigma', \Sigma'', \Delta', \Delta'', Q) \to R \Rightarrow P'}}{(P \to (Q \to R), \Sigma, \Sigma', \Sigma'', \Delta, \Delta', \Delta'') \Rightarrow P'}$$

By IH we have resolution proofs of

$$l_\Delta \Rightarrow P, \qquad l_{\Delta'} \Rightarrow Q, \qquad l_{\Delta''}, R \Rightarrow P'.$$

We construct from these a resolution proof terminating in:

$$\dfrac{\dfrac{\dfrac{P, Q \Rightarrow R \quad l_\Delta \Rightarrow P}{(Q, l_\Delta) \Rightarrow R} \qquad l_{\Delta'} \Rightarrow Q}{(l_\Delta, l_{\Delta'}) \Rightarrow R} \qquad l_{\Delta''}, R \Rightarrow P'}{(l_\Delta, l_{\Delta'}, l_{\Delta''}) \Rightarrow P'}$$

Case 4. The deduction ends with L$\vee$ introducing $P \to (Q \vee R) \in \Gamma^*$. The final part of the deduction has the form

$$\dfrac{\Delta, \Sigma \Rightarrow P \qquad \dfrac{\Delta', \Sigma', Q \Rightarrow P' \quad \Delta'', \Sigma'', R \Rightarrow P'}{(\Delta', \Delta'', \Sigma', \Sigma'', Q \vee R) \Rightarrow P'}}{(\Delta, \Delta', \Delta'', \Sigma, \Sigma', \Sigma'', P \to (Q \to R)) \Rightarrow P'}$$

By IH we have resolution proofs of

$$l_\Delta \Rightarrow P, \qquad l_\Delta, Q \Rightarrow P', \qquad l_{\Delta''}, R \Rightarrow P'.$$

We construct from these a resolution proof by application of the $\vee$-rule:

$$\dfrac{P \Rightarrow Q \vee R \qquad l_\Delta \Rightarrow P \qquad l_{\Delta'}, Q \Rightarrow P' l_{\Delta''}, R \Rightarrow P'}{(l_\Delta, l_{\Delta'}, l_{\Delta''}) \Rightarrow P'}$$

Case 5. The deduction ends with introduction of $P \to Q \in \Gamma^*$ on the left:

$$\dfrac{\Delta, \Sigma \Rightarrow P \qquad \Delta', \Sigma', Q \Rightarrow R}{(\Delta, \Delta', \Sigma, \Sigma', P \to Q) \Rightarrow R}$$

By IH we have resolution proofs of

$$l_\Sigma \Rightarrow P, \qquad l_{\Sigma'}, Q \Rightarrow R.$$

We construct a resolution proof terminating in

$$\dfrac{\dfrac{l_\Sigma \Rightarrow P \quad P \Rightarrow Q}{l_\Sigma \Rightarrow Q} \qquad l_{\Sigma'}, Q \Rightarrow R}{(l_\Sigma, l_{\Sigma'}) \Rightarrow R}$$

etc. ⊠

7.5.14A. ♠ Show by a direct argument that for the resolution calculus **Rpi** there is a decision procedure for sequents.

7.5.14B. ♠ Give deductions in **Rpi** of $(P \lor Q \to R) \to (P \to R) \land (Q \to R)$ and $(P \to (P \to Q)) \to (P \to Q)$.

7.6 Notes

There is a very extensive literature on resolution methods, theorem proving and logic programming. Some good general sources are Lloyd [1987], Apt [1990], Eisinger and Ohlbach [1993], Hodges [1993], C. [1994] and Jäger and Stärk [1994]. For the older literature see Chang and Lee [1973]. For the connection with Maslov's so-called "inverse method", see Lifschitz [1989].

There are many important aspects of logic programming which have not even been mentioned in this chapter. For example, there is an extensive literature on the correct interpretation and handling of negation in logic programming. (For a first orientation, see C. [1994].) Another example of a neglected topic is the peculiarities of the search mechanism of languages such as PROLOG, which are not reflected in the deduction systems considered in this chapter. As to the possibility of expressing the search mechanism of PROLOG and related languages in a deduction system, see, for example, Kalsbeek [1994,1995].

7.6.1. *Unification.* The first to give an algorithm for unification together with an explicit proof of its completeness was J. A. Robinson [1965]. The idea for the unification algorithm presented here goes back to Herbrand [1930, section 2.4], where such an algorithm is sketched in a few lines. A full description together with a proof of the termination and correctness of this algorithm is first given in Martelli and Montanari [1982]; this proof has been followed here. Cf. also C. [1994, 3.10]. More on unification may be found in Baader and Siekmann [1994].

7.6.2. *Resolution.* The first to study resolution as a proof method in predicate logic was J. A. Robinson [1965]. One of the earliest references for SLD-resolution is Kowalski [1974]; around the same time PROLOG was developed by a group around A. Colmerauer. The concise proof of completeness for linear resolution given here is essentially due to Stärk [1990]; combination of Stärk's proof with the proof of the conservative extension result in Schwichtenberg [1992] transforms it into a proof-theoretic reduction of completeness for resolution to completeness for other systems.

The material in the last two sections is based on papers by Mints, in particular Mints [1990,1994b].

7.6.3. *Languages suitable for logic programming.* In the literature a good deal of attention has been given to discovering languages suitable for logic programming, and of greater expressive power than is provided by programs consisting of definite clauses and goals which are conjunctions of atoms.

Generalization is possible in several ways. For example, one may enlarge the classes of formulas which are used as goals and clauses of programs, while keeping the notion of derivability standard, that is to say, classical. A further generalization consists in considering also intuitionistic or minimal derivability. A more radical move is the consideration of logical languages with more or different logical operators, such as the languages of higher-order logic and of linear logic.

A good example of such an investigation is Miller et al. [1991]. The idea in this paper is to look for logical languages for which the "operational semantics" (which provides the computational meaning and which is similar to the BHK-interpretation of intuitionistic logic (2.5.1)) coincides with provability.

Let G be a goal formula, Γ be a program, specified as a finite set of formulas, and let $\Gamma \vdash_o G$ express that our search mechanism succeeds with program Γ for goal G. Then one requires for $\vdash_o$: $\Gamma \vdash_o \top$, $\Gamma \vdash_o G_1 \wedge G_2$ iff $\Gamma \vdash_o G_1$ and $\Gamma \vdash_o G_2$, $\Gamma \vdash_o G_1 \vee G_2$ iff $\Gamma \vdash_o G_1$ or $\Gamma \vdash_o G_2$, $\Gamma \vdash_o D \to G$ iff $\Gamma \cup \{D\} \vdash_o G$, $\Gamma \vdash_o \exists x A$ iff $\Gamma \vdash_o A[x/t]$ for some term t, $\Gamma \vdash_o \forall x A$ iff $\Gamma \vdash_o A[x/y]$ for some variable y, not free in Γ or $\forall x A$, and free for x in A. As a formalization of $\vdash_o$ one can take the notion of *uniform provability* in a Gentzen system. A deduction is said to be *uniform* if, whenever in the course of the deduction $\Gamma \Rightarrow G$ is proved for a non-atomic goal G, the last step of the deduction consists in the application of the right-introduction rule for the principal operator in G.

One now looks for triples $\langle \mathcal{F}, \mathcal{G}, \vdash \rangle$ with $\mathcal{D}, \mathcal{F}$ formula classes, and $\vdash$ a notion of derivability in a Gentzen system such that for $\Gamma \subset \mathcal{F}$, $G \in \mathcal{G}$ one has $\Gamma \vdash_o G$ iff there is a uniform deduction of $\Gamma \Rightarrow G$ iff $\vdash \Gamma \Rightarrow G$. Such triples are called *abstract programming languages* in Miller et al. [1991].

A simple example is obtained by taking for $\vdash$ classical provability (say in **G1c**), for $\mathcal{G}$ the class inductively characterized by

$$\mathcal{G} \equiv \top \mid \mathrm{At} \mid \mathcal{G} \wedge \mathcal{G} \mid \mathcal{G} \vee \mathcal{G} \mid \exists x \mathcal{G},$$

where At is the class of atomic formulas, and for $\mathcal{F}$ the class characterized by

$$\mathcal{D} \equiv \mathrm{At} \mid \mathcal{G} \to \mathrm{At} \mid \mathcal{F} \wedge \mathcal{F} \mid \forall x \mathcal{F}.$$

The proof that this triple is an abstract programming language involves showing that whenever a sequent of the right form is provable, it has a uniform deduction; the argument uses the permutability of certain rules in the Gentzen system (cf. 5.4.6B). Permutation arguments are also an important ingredient of the papers mentioned below.

In Harland [1994], Hodas and Miller [1994], Miller [1994], and Pym and Harland [1994] suitable (fragments of) linear logic are used as formalisms for logic programming (references to related work in Hodas and Miller [1994]). For examples of programming in such languages see Andreoli and Pareschi [1991] and Hodas and Miller [1994].

Andreoli [1992] shows how by choosing an appropriate formalization of linear logic one can greatly restrict the search space for deductions.

Chapter 8

Categorical logic

For this chapter preliminary knowledge of some basic notions of category theory (as may be found, for example, in Mac Lane [1971], Blyth [1986], McLarty [1992], Poigné [1992]) will facilitate understanding, but is not necessary, since our treatment is self-contained. Familiarity with chapter 6 is assumed.

In this chapter we introduce another type of formal system, inspired by notions from category theory. The proofs in formalisms of this type may be denoted by terms; the introduction of a suitable equivalence relation between these terms makes it possible to interpret them as arrows in a suitable category.

In particular, we shall consider a system for minimal $\to\wedge\top$-logic connected with a special cartesian closed category, namely the free cartesian closed category over a countable discrete graph, to be denoted by $\mathrm{CCC}(\mathcal{PV})$. In this category we have a decision problem: when are two arrows from A to B the same?

This problem will be solved by establishing a correspondence between the arrows of $\mathrm{CCC}(\mathcal{PV})$ and the terms of the extensional typed lambda calculus. For this calculus we can prove strong normalization, and the decision problem is thereby reduced to computing and comparing normal forms of terms of the lambda calculus.

Another interesting application of this correspondence will be a proof of a certain "coherence theorem" for $\mathrm{CCC}(\mathcal{PV})$. (A coherence theorem is a theorem of the form: "between two objects satisfying certain conditions there is at most one arrow".) The correspondence between lambda terms modulo $\beta\eta$-equality and arrows in $\mathrm{CCC}(\mathcal{PV})$ will enable us to use proof-theoretic methods.

In this chapter $P, P', P'', Q, R \in \mathcal{PV}$.

8.1 Deduction graphs

Deduction systems ("deduction graphs") inspired by category theory manipulate 1-sequents. A 1-sequent is a sequent of the form $A \Rightarrow B$. A sequent $\Rightarrow B$ is interpreted as a sequent $\top \Rightarrow B$. Below we shall describe deduction graphs for intuitionistic $\rightarrow\wedge\top$-logic.

8.1.1. DEFINITION. A *deduction graph* consists of a directed graph $\mathcal{A} \equiv (\mathcal{A}_0, \mathcal{A}_1)$ consisting of a set of *objects* (vertices, nodes) $\mathcal{A}_0$ and a set of *arrows* (directed edges) $\mathcal{A}_1$ such that (writing $f: A \Rightarrow B$ or $A \overset{f}{\Rightarrow} B$ for an arrow from A to B)

(i) If $f: A \Rightarrow B$, $g: B \Rightarrow C$, then there is an arrow $g \circ f: A \Rightarrow C$, the *composition* of f and g;

(ii) For each $A \in \mathcal{A}_0$ there is an *identity arrow* $\mathrm{id}^A: A \Rightarrow A$.

A is the *domain* of $f: A \Rightarrow B$, and B is the *codomain* of $f: A \Rightarrow B$. For the set of arrows from A to B we also write $\mathcal{A}_1(A, B)$.

A *tci-deduction graph* or *positive deduction graph* ("tci" from "truth, conjunction, implication") is a deduction graph $\mathcal{A} \equiv (\mathcal{A}_0, \mathcal{A}_1)$ with a special object $\top \in \mathcal{A}_0$ (*truth* or the *terminal object*) and with the objects closed under binary operations $\wedge, \rightarrow$, that is to say, if $A, B \in \mathcal{A}_0$, then also $A \wedge B \in \mathcal{A}_0$ (the *product* or *conjunction* of A and B), and $A \rightarrow B \in \mathcal{A}_0$ (the *implication*, *exponent* or *function object* from A to B). There are some extra arrows and arrow-constructors:

(iii) $\mathrm{tr}^A: A \Rightarrow \top$ (the *truth arrow*);

(iv) $\pi_0^{A,B}: A \wedge B \Rightarrow A$, $\pi_1^{A,B}: A \wedge B \Rightarrow B$, and if $f: C \Rightarrow A$, $g: C \Rightarrow B$, then $\langle f, g \rangle: C \Rightarrow A \wedge B$. We shall sometimes use $\mathbf{p}(f, g)$ for $\langle f, g \rangle$;

(v) $\mathrm{ev}^{A,B}: (A \rightarrow B) \wedge A \Rightarrow B$, and if $h: C \wedge B \Rightarrow A$ then $\mathrm{cur}(h): C \Rightarrow B \rightarrow A$. ("cur" is called the *currying* operator, "ev" is called *evaluation*.)

We shall frequently drop type superscripts, whenever we can do so without danger of confusion. Instead of $x: A$, $t: A$ we also write x^A, t^A. ⊠

8.1.2. EXAMPLES. (a) We obtain a deductive system for minimal $\top\wedge\rightarrow$-logic by taking as objects all formulas built from proposition variables and $\top$ by means of $\wedge, \rightarrow$, and as arrows all arrows constructed from id^A, tr^A, $\pi_0^{A,B}$, $\pi_1^{A,B}$, $\mathrm{ev}^{A,B}$ by closing under $\circ$, $\langle\, , \rangle$ and cur. The construction tree of an arrow corresponds to a deduction tree, or alternatively, the expressions for the arrows form a term system for deductions with axioms and rules:

$$\mathrm{id}^A: A \Rightarrow A; \quad \mathrm{tr}^A: A \Rightarrow \top$$

$$\frac{t\colon A \Rightarrow B \qquad s\colon B \Rightarrow C}{s \circ t\colon A \Rightarrow C}$$

$$\pi_0^{A,B}\colon A \wedge B \Rightarrow A; \quad \pi_1^{A,B}\colon A \wedge B \Rightarrow B$$

$$\frac{t\colon C \Rightarrow A \qquad s\colon C \Rightarrow B}{\langle t, s\rangle\colon C \Rightarrow A \wedge B}$$

$$\mathrm{ev}^{A,B}\colon (A \to B) \wedge A \Rightarrow B$$

$$\frac{t\colon A \wedge B \Rightarrow C}{\mathrm{cur}(t)\colon A \Rightarrow B \to C}$$

A formula A is said to be *derivable* in this system if we can deduce a sequent $\top \Rightarrow A$.

An example of a deduction, establishing associativity of conjunction (abbreviating "$A \wedge B$" as "AB", and leaving out the terms π_0 and and π_1 for the axioms to save space):

$$\frac{\dfrac{(DE)F \Rightarrow DE \qquad DE \Rightarrow D}{\pi_0 \circ \pi_0\colon (DE)F \Rightarrow D} \qquad \dfrac{\dfrac{(DE)F \Rightarrow DE \qquad DE \Rightarrow E}{\pi_1 \circ \pi_0\colon (DE)F \Rightarrow E} \qquad (DE)F \Rightarrow F}{\langle \pi_1 \circ \pi_0, \pi_1\rangle\colon (DE)F \Rightarrow EF}}{\alpha^{D,E,F}\colon (DE)F \Rightarrow D(EF)}$$

where

$$\alpha^{D,E,F} := \langle \pi_0 \circ \pi_0, \langle \pi_1 \circ \pi_0, \pi_1\rangle\rangle.$$

Adding constants $c : A \Rightarrow B$ amounts to the addition of extra axioms. Note that there may be different c for the same 1-sequent. Adding more than one constant for the same 1-sequent may arise quite naturally: a particular sequent may be assumed as axioms for different reasons, so to speak. Thus we obtain $\top \Rightarrow \top$ both as $\mathrm{tr}^{\top}$ and as $\mathrm{id}^{\top}$. Adding "variable arrows" $x\colon A \Rightarrow B$ corresponds to reasoning from assumptions $A \Rightarrow B$. See also example (c) below.

(b) Given a directed graph $\mathcal{A} \equiv (\mathcal{A}_0, \mathcal{A}_1)$ we may construct a free deductive graph $\mathrm{FG}(\mathcal{A})$ over $\mathcal{A}$ by taking as objects $\mathcal{A}_0$, and as arrows the expressions generated from elements of $\mathcal{A}_1$ and id^A by means of composition; the elements of $\mathcal{A}_1$ are treated as constants. Two arrows given by expressions are the same iff they are literally identical as expressions.

(c) The free tci-deduction graph $\mathrm{D}(\mathcal{A})$ over a directed graph $\mathcal{A}$ is obtained similarly; objects are obtained from $\top$ and elements of $\mathcal{A}_0$ by closing under $\wedge, \to$. The arrows are obtained by adding to the arrows of $\mathcal{A}_1$ $\mathrm{tr}^A, \mathrm{id}^A, \pi_0^{A,B}, \pi_1^{A,B}, \mathrm{ev}^{A,B}$ and closing under $\circ, \langle\, ,\, \rangle$ and cur. If we take $\mathcal{PV}$ for $\mathcal{A}$, we are back at our first example.

8.1.2A. ♠ Show that the free tci-deduction graph over the discrete graph of all propositional variables is equivalent, as a deduction system, to one of the usual formalisms for the $\top\wedge\to$-fragment of minimal logic.

8.1.3. In systems of natural deduction, the reduction relation generated by the usual conversions suggests an equivalence relation on the terms denoting deductions. Similarly we can impose identifications on the arrows of deduction graphs and tci-deduction graphs; this time the identifications are suggested by the notion of *category* and *cartesian closed category* respectively.

DEFINITION. A *category* $\mathcal{A}$ is a deduction graph such that the arrows satisfy

iden	$f \circ \mathrm{id}_A = f,\ \mathrm{id}_B \circ f = f$ for every arrow $f : A \Rightarrow B$,
ass	$f \circ (g \circ h) = (f \circ g) \circ h$
	for all arrows $f\colon C \Rightarrow D, g\colon B \Rightarrow C, h\colon A \Rightarrow B$. ⊠

DEFINITION. A *cartesian closed category* (a *"CCC"* for short) is a tci-deduction graph which is a category and in which the arrows satisfy for all $f\colon A \Rightarrow \top$:

true $\quad f = \mathrm{tr}^A$,

and for all $f\colon C \Rightarrow A$, $g\colon C \Rightarrow B$, $h\colon C \Rightarrow A \wedge B$:

proj	$\pi_0^{A,B} \circ \langle f, g\rangle = f,\ \pi_1^{A,B} \circ \langle f, g\rangle = g,$
surj	$\langle \pi_0 h, \pi_1 h\rangle = h,$

and for all $h\colon C \wedge B \Rightarrow A$, $k\colon C \Rightarrow B \to A$:

evcur	$\mathrm{ev}^{A,B} \circ \langle \mathrm{cur}(h) \circ \pi_0^{C,A}, \pi_1^{C,A}\rangle = h,$
curev	$\mathrm{cur}(\mathrm{ev}^{A,B} \circ \langle k \circ \pi_0^{C,A}, \pi_1^{C,A}\rangle) = k.$ ⊠

REMARK. If we write $f \wedge g$ for $\langle f \circ \pi_0, g \circ \pi_1\rangle$, the last set of equations reads

$$\mathrm{ev} \circ (\mathrm{cur}(h) \wedge \mathrm{id}) = h,\ \ \mathrm{cur}(\mathrm{ev} \circ (k \wedge \mathrm{id})) = k.$$

From this last equation follows $\mathrm{cur}(\mathrm{ev}) = \mathrm{id}^{A\to B}$. The equation $\mathrm{cur}(\mathrm{ev}^{A,B}) = \mathrm{id}^{A\to B}$ together with $\mathrm{cur}(h)\circ k = \mathrm{cur}(h\circ(k\wedge\mathrm{id})$ again yields $\mathrm{cur}(\mathrm{ev}\circ(k\wedge\mathrm{id})) = k$.

8.1.4. EXAMPLES. (a) There are numerous examples of categories; they abound in mathematics. A very important example is the category **Set**, with as objects all sets, and as arrows from A to B all set-theoretic mappings from A to B. Composition is the usual function composition, and id^A is the identity mapping on A. **Set** is made into a CCC by choosing a singleton set as terminal object, say $\{\emptyset\}$. It is then obvious what to take for tr^A, since there is no choice; for the product $A \wedge B$ we choose a fixed representation of the cartesian product, say $\{\langle a, 0\rangle : a \in A\} \cup \{\langle b, 1\rangle : b \in B\}$ where $0 = \emptyset$, $1 = \{\emptyset\}$, and $\langle x, y\rangle$ is the set-theoretic ordered pair of x and y. $A \to B$ is the set of all functions from A to B. We leave it as an exercise to complete the definition. Note that in this example elements of a set A are in bijective correspondence with the arrows $f : \top \Rightarrow A$, and that there is a bijective correspondence between the *arrows* from A to B and the *elements* of the object $A \to B$.

(b) Given a directed graph $\mathcal{A} \equiv (\mathcal{A}_0, \mathcal{A}_1)$, we obtain a *free category over* $\mathcal{A}$, $\mathrm{Cat}(\mathcal{A})$, by taking as objects the objects of $\mathcal{A}$, and as arrows the equivalence classes of arrows in the free deduction graph over $\mathcal{A}$. That is to say the equality between arrows is the least equivalence $\approx$ satisfying

$$t \circ (s \circ r) \approx (t \circ s) \circ r, \;\; t \circ \mathrm{id} \approx t, \mathrm{id} \circ t \approx t,$$

and congruence:

$$\text{if } t \approx t',\;\; s \approx s' \text{ then } t \circ s \approx t' \circ s'.$$

In other words, equality between arrows can be proved from the axioms $t \circ (s \circ r) = (t \circ s) \circ r$, $t \circ \mathrm{id} = t$, $\mathrm{id} \circ t = t$, $t = t$, by means of the rules

$$\frac{t = t' \qquad t' = t''}{t = t''} \qquad \frac{t = t'}{t' = t} \qquad \frac{t = t' \qquad s = s'}{t \circ s = t' \circ s'}$$

Similarly we construct the free $\mathrm{CCC}(\mathcal{A})$ over $\mathcal{A}$, by first constructing the free tci-deduction graph and then imposing as equivalence on arrows the least equivalence relation $\approx$ satisfying

$$\begin{array}{l} t \circ (s \circ r) \approx (t \circ s) \circ r, \;\; t \circ \mathrm{id} \approx t, \;\; \mathrm{id} \circ t \approx t, \\ f \approx \mathrm{tr}^A \text{ for all } f : A \Rightarrow \top, \\ \pi_0 \circ \langle t, s\rangle \approx t, \;\; \pi_1 \circ \langle t, s\rangle \approx s, \;\; \langle \pi_0 \circ t, \pi_1 \circ t\rangle \approx t, \\ \mathrm{ev} \circ \langle \mathrm{cur}(s) \circ \pi_0, \pi_1\rangle \approx s, \;\; \mathrm{cur}(\mathrm{ev} \circ \langle t \circ \pi_0, \pi_1\rangle) \approx t, \end{array}$$

and the congruences

$$\begin{array}{l} \text{if } t \approx t',\;\; s \approx s' \text{ then } t \circ s \approx t' \circ s', \\ \text{if } t \approx t',\;\; s \approx s' \text{ then } \langle t, s\rangle \approx \langle t', s'\rangle, \\ \text{if } t \approx t' \text{ then } \mathrm{cur}(t) \approx \mathrm{cur}(t'). \end{array}$$

Again this can be reformulated as: equality $=$ between arrows can be proved from axioms $t = t$, $\langle \pi_0 \circ t, \pi_1 \circ t \rangle = t$, ... by means of the rules

$$\frac{t = t' \qquad t' = t''}{t = t''} \qquad \frac{t = t'}{t' = t} \qquad \frac{t = t' \qquad s = s'}{\langle t, s \rangle = \langle t', s' \rangle}, \text{ etc.}$$

An important special case of this construction is CCC($\mathcal{PV}$), where $\mathcal{PV}$, the collection of proposition variables, is interpreted as a discrete category with no arrows and with $\mathcal{PV}$ as its objects.

8.1.4A. ♠ Complete the definition of Set as a CCC.

Remark. Free categories, graphs and CCC's may also be defined by certain universal properties, or via the left adjoint of a forgetful functor; but since we concentrate here on the proof-theoretic aspects, while minimizing the category-theoretic apparatus, we have chosen to define the free objects by an explicit construction; cf. Lambek and Scott [1986, section I.4].

8.1.5. Lemma. *The following identities hold in any CCC and will be frequently used later on:*

$$\begin{aligned}
&\langle f, g \rangle \circ h = \langle f \circ h, g \circ h \rangle,\\
&\mathrm{ev}\langle \mathrm{cur}(f), g \rangle = f\langle \mathrm{id}, g \rangle,\\
&\mathrm{cur}(f) \circ g' = \mathrm{cur}(f \circ \langle g' \circ \pi_0, \pi_1 \rangle),
\end{aligned}$$

where $f\colon A \wedge B \Rightarrow C$, $g\colon A \Rightarrow B$, $g'\colon A' \Rightarrow A$. ⊠

8.1.5A. ♠ Prove this lemma.

8.1.6. Definition. Two objects A, B in a category $\mathcal{A} \equiv (\mathcal{A}_0, \mathcal{A}_1)$ are *isomorphic*, if there are arrows $f : A \Rightarrow B$, $g : B \Rightarrow A$ such that $g \circ f = \mathrm{id}^A$, $f \circ g = \mathrm{id}^B$. We write $A \simeq B$ for "A and B are isomorphic". ⊠

One readily sees that if $A \simeq A'$, $B \simeq B'$, then there is a bijective correspondence between the arrows from A to B and the arrows from A' to B'.

For example, the isomorphism $A \to \top \simeq \top$ holds in every CCC. This is verified as follows: $\mathrm{tr}^{A \to \top} : A \to \top \Rightarrow \top$ and $\mathrm{cur}(\mathrm{tr}^{\top \wedge A}) : \top \Rightarrow A \to \top$ are inverse to each other; $\mathrm{tr}^{A \to \top} \circ f = \mathrm{tr}^{\top} = \mathrm{id}^{\top}$ for every $f : \top \Rightarrow A \to \top$; and $\mathrm{cur}(\mathrm{tr}^{\top \wedge A}) \circ \mathrm{tr}^{A \to \top} = \mathrm{cur}(\mathrm{tr}^{\top \wedge A} \circ (\mathrm{tr}^{A \to \top} \wedge \mathrm{id}^A)) = \mathrm{cur}(\mathrm{tr}^{(A \to \top) \wedge A}) = \mathrm{cur}(\mathrm{ev}^{A \to \top, A}) = \mathrm{id}^{A \to \top}$. We shall have occasion to use this isomorphism, together with others mentioned in 8.1.6A below.

8.1.6A. ♠ Show that in any CCC the following isomorphisms hold: $A \wedge B \simeq B \wedge A$, $\top \wedge A \simeq A$, $(A \wedge B) \wedge C \simeq A \wedge (B \wedge C)$, $\top \to A \simeq A$, $A \to \top \simeq \top$, $(C \to A \wedge B) \simeq (C \to A) \wedge (C \to B)$, $(B \wedge C \to A) \simeq B \to (C \to A)$.

We note the following:

8.1.7. PROPOSITION. *(One-to-one correspondence between $\mathcal{A}_1(A, B)$ and $\mathcal{A}_1(\top, A \to B)$) In any tci-deduction graph there are operators $\ulcorner\ \urcorner$ and $^\sim$ such that for $f : A \Rightarrow B$, $g : \top \Rightarrow A \to B$:*

$$\ulcorner f \urcorner : \top \Rightarrow A \to B,\ \ g^\sim : A \Rightarrow B.$$

Moreover, in a CCC they satisfy:

$$\ulcorner f \urcorner^\sim = f,\ \ \ulcorner g^\sim \urcorner = g.$$

PROOF. Take

$$\ulcorner f \urcorner := \mathrm{cur}(f \circ \pi_1^{\top,A}),\ \ g^\sim := \mathrm{ev}^{A,B} \circ \langle g \circ \mathrm{tr}^A, \mathrm{id}^A \rangle.$$

8.1.7A. ♠ Verify the required properties for $\ulcorner\ \urcorner$ and $^\sim$ as defined.

8.2 Lambda terms and combinators

In this section we construct a one-to-one correspondence between the arrows of the free cartesian closed category over a countable discrete graph $\mathrm{CCC}(\mathcal{PV})$ and the closed terms of the extensional typed lambda calculus with function types and product types, $\boldsymbol{\lambda\eta}_{\to\wedge}$. This is the term calculus for $\to\wedge$-**Ni** with two extra conversions, namely η-conversion and surjectivity of pairing, in prooftree form

$$\dfrac{\dfrac{\begin{array}{c}\mathcal{D}\\ A \to B\end{array} \quad A^x}{B}}{A \to B}\,x \quad \text{cont} \quad \begin{array}{c}\mathcal{D}\\ A \to B\end{array} \qquad\qquad \dfrac{\dfrac{\begin{array}{c}\mathcal{D}\\ A_1 \wedge A_2\end{array}}{A_1} \quad \dfrac{\begin{array}{c}\mathcal{D}\\ A_1 \wedge A_2\end{array}}{A_2}}{A_1 \wedge A_2} \quad \text{cont} \quad \begin{array}{c}\mathcal{D}\\ A_1 \wedge A_2\end{array}$$

We have already encountered the inverses of these transformations (so-called expansions) in section 6.7.

8.2.1. DEFINITION. (*The calculus $\boldsymbol{\lambda\eta}_{\to\wedge\top}$*) We introduce an extension of $\boldsymbol{\lambda}_\to$ with product types (conjunction types) and a type of truth $\top$. The type structure $\mathcal{T}_{\to\wedge\top}$ corresponds with the objects of the free cartesian closed category $\mathrm{CCC}(\mathcal{PV})$ over the countable discrete graph $\mathcal{PV}$:

(i) $\top \in \mathcal{T}$, $P \in \mathcal{T}_{\to\wedge\top}$ for all $P \in \mathcal{PV}$.

(ii) If $A, B \in \mathcal{T}_{\to\wedge\top}$, then $(A \to B)$, $(A \wedge B) \in \mathcal{T}_{\to\wedge\top}$.

Terms are given by the clauses:

(iii) There is a countably infinite supply of variables of each type;

(iv) $*: \top$;

(v) if $t: A_0 \wedge A_1$, then $\mathbf{p}_i t: A_i$ $(i \in \{0,1\})$;

(vi) if $t: A \to B$, $s: A$, then $ts: B$;

(vii) if $t: A$, $s: B$, then $\mathbf{p}(t,s): A \wedge B$; for $\mathbf{p}(t,s)$ we also write $\langle t, s\rangle$.

Axioms and rules for $\beta\eta$-reduction are

true	$t^{\top} \succeq *$,
βcon	$(\lambda x^A.t^B)s^A \succeq t^B[x/s]$,
ηcon	$\lambda x^A.tx \succeq t$ $(x \notin \mathrm{FV}(t))$,
proj	$\mathbf{p}_i\langle t_0, t_1\rangle \succeq t_i$ $(i \in \{0,1\})$,
surj	$\langle \mathbf{p}_0 t, \mathbf{p}_1 t\rangle \succeq t$.

In addition we have

$$(\text{refl})\ t \succeq t \qquad\qquad (\text{repl})\ \frac{t \succeq s}{t[x/r] \succeq s[x/r]}$$

where x in "repl" denotes any occurrence, free or bound, of the variable x. ⊠

The rule repl may be split into special cases:

$$\frac{t \succeq t' \quad s \succeq s'}{\mathbf{p}(t,s) \succeq \mathbf{p}(t',s')} \qquad \frac{t \succeq t' \quad s \succeq s'}{ts \succeq t's'}$$

$$\frac{t \succeq t'}{\lambda x.t \succeq \lambda x.t'} \qquad \frac{t \succeq t'}{\mathbf{p}_i t \succeq \mathbf{p}_i t'}\ (i \in \{0,1\})$$

Below we shall prove strong normalization for $\boldsymbol{\lambda\eta}_{\to\wedge}$, extending the proof of section 6.8 for $\boldsymbol{\lambda}_{\to}$; by Newman's lemma (1.2.8) uniqueness of normal form readily follows.

8.2.2. NOTATION. Below we shall very often drop $\circ$ in the notation for composition, and simply write tt' for $t \circ t'$, whenever this can be done without impairing readability. ⊠

DEFINITION. An expression denoting an arrow in $\mathrm{CCC}(\mathcal{PV})$ is called a *combinator*. An $\wedge$-*combinator* is a combinator in $\mathrm{CCC}(\mathcal{PV})$ constructed without ev, cur. ⊠

8.2.3. THEOREM. *Let A, B be conjunctions of propositional variables and $\top$, such that no propositional variable occurs twice in A and such that each variable in B occurs in A. Then there is a unique $\wedge$-combinator*

$$\xi_{AB}: A \Rightarrow B.$$

Moreover, if A and B have the same variables, and also in B no variable occurs twice, then ξ_{AB} is an isomorphism with inverse ξ_{BA}.

PROOF. We use a main induction on the depth $|B|$ of B. If $B \equiv B_0 \wedge B_1$, we put $\xi_{AB} := \langle \xi_{AB_0}, \xi_{AB_1} \rangle$. For the basis case, B is either $\top$ or a propositional variable. If B is $\top$, take $\xi_{A\top} := \mathrm{tr}^A$. If B is a $P \in \mathcal{PV}$, then B occurs as a component of A and may be extracted by repeated use of $\mathbf{p}_0, \mathbf{p}_1$. This may be shown by a subinduction on $|A|$: suppose $A \equiv A_0 \wedge A_1$, then P occurs in A_i for $i = 0$ or $i = 1$; take $\xi_{AP} := \xi_{A_iP}\pi_i$.

The uniqueness of the $\wedge$-combinator from A to B may be shown as follows. Again, the main induction is on $|B|$. Suppose $\beta: A \Rightarrow B$ is another $\wedge$-combinator. If $B \equiv B_0 \wedge B_1$, then $\pi_i \circ \xi_{AB} : A \Rightarrow B_i$. By the IH it follows that $\pi_i \circ \xi_{AB} = \pi_i \circ \beta$, but then $\xi_{AB} = \langle \pi_0 \circ \xi_{AB}, \pi_1 \circ \xi_{AB} \rangle = \langle \pi_0 \circ \beta, \pi_1 \circ \beta \rangle = \beta$.

So we are left with the case of B prime. $B \equiv \top$ is trivial; so consider $B \equiv P$ for some variable P. We may assume that β is written in such a way that no associative regrouping β' of β (i.e. β' obtained from β by repeated use of the associativity for arrow composition) contains a subterm of one of the forms $\pi_i \circ \langle t, s \rangle$, $\mathrm{id} \circ t$, $t \circ \mathrm{id}$. Let β be written as $(\ldots(t_0 t_1)\ldots t_n)$ such that no t_i is of the form $t' \circ t''$. t_n cannot be of the form $\langle t', t'' \rangle$, for then B could not be a variable, since this would require somewhere a projection π_i to appear in front of some $\langle s', s'' \rangle$ (use induction on n).

So either $n = 1$ and then $\beta = \mathrm{id} = \xi_{AB}$, or $n > 1$, and then $t_n \equiv \pi_0$ or π_1, depending on whether, if $A \equiv A_0 \wedge A_1$, P occurs in A_0 or in A_1. Suppose $t_n \equiv \pi_0$, then $(\ldots(t_0 t_1)\ldots t_{n-1})$ is an $\wedge$-combinator of $A_0 \Rightarrow P$; now apply the IH, and it follows that $(\ldots(t_0 t_1)\ldots t_{n-1}) = \xi_{A_0P}$, hence $\beta = \xi_{A_0P} \circ \pi_0 = \xi_{AP}$. ⊠

8.2.4. DEFINITION. Let

$$\Theta \equiv x_1: A_1, \ldots, x_n: A_n, \quad \Theta' \equiv y_1: B_1, \ldots, y_m: B_m$$

be lists of typed variables such that in Θ all variables are distinct, and such that all variables in Θ' occur in Θ. Put

$$\bar{\Theta} := (\ldots(P_1 \wedge P_2)\ldots P_n), \quad \bar{\Theta}' := (\ldots(Q_1 \wedge Q_2)\ldots Q_n)$$

such that $Q_j = P_i$ iff $y_j = x_i$. By theorem 8.2.3 there is a unique $\wedge$-combinator $\xi_{\bar{\Theta}\bar{\Theta}'}$. Now define $\beta_{\Theta\Theta'}$ as the combinator obtained by replacing P_i by A_i in the types of the components of $\xi_{\bar{\Theta}\bar{\Theta}'}$. Furthermore put, for any list Θ:

$$\Theta^\circ := (\ldots((\top \wedge A_1) \wedge A_2) \ldots \wedge A_n).$$ ⊠

The following lemma is then readily checked.

8.2.5. LEMMA. *Let Θ, Θ' be as above, Θ'' a permutation of Θ.*

(i) $\beta_{\Theta,\Theta} = \mathrm{id}$; $\beta_{\Theta,\Theta''}\beta_{\Theta'',\Theta} = \mathrm{id}$; $\beta_{\Theta'',\Theta'} = \beta_{\Theta,\Theta'}\beta_{\Theta'',\Theta}$;

(ii) $\beta_{\Theta x,\Theta' x} = \langle \beta_{\Theta,\Theta'}\pi_0, \pi_1\rangle$ *if* $x \notin \Theta\Theta'$;

(iii) $\beta_{\Theta\Psi,\Theta'} = \beta_{\Theta,\Theta'} \circ \pi_0^\Psi$, *where in* $\Theta\Psi$ *each variable occurs exactly once, and where* π_0^Ψ *is a string of operators* π_0 *of the appropriate types, of a length equal to the length of* Ψ;

(iv) $\beta_{\Theta,\Theta'\Psi} = \pi_0^\Psi \circ \beta_{\Theta,\Theta'}$, *where each variable in* $\Theta'\Psi$ *also occurs in* Θ, *and where* π_0^Ψ *is a string of operators* π_0 *of the appropriate types, of a length equal to the length of* Ψ. ⊠

8.2.5A. ♠ Prove the lemma.

8.2.6. DEFINITION. (*Mapping* τ *from arrows in* CCC($\mathcal{PV}$) *to typed terms*) The mapping τ assigns a typed term $t^B[x^A]$ to each arrow from A to B in the CCC($\mathcal{PV}$). $\tau(t)$ has at most one free variable. We write $\tau_y(t)$ to indicate that y is the letter chosen to represent the variable free in the expression $\tau(t)$.

$$\begin{array}{ll}
\tau_x(\mathrm{id}^A) & := x^A, \\
\tau_x(\mathrm{tr}^A) & := *, \\
\tau_x(\pi_i^{A,B}) & := \mathbf{p}_i x^{A\wedge B}, \\
\tau_x(\langle t, s\rangle) & := \langle \tau_x t, \tau_x s\rangle \equiv \mathbf{p}(\tau_x t, \tau_x s), \\
\tau_x(\mathrm{ev}^{A,B}) & := (\mathbf{p}_0 x^{(A\to B)\wedge A})(\mathbf{p}_1 x^{(A\to B)\wedge A}), \\
\tau_x(\mathrm{cur}(t^{A\wedge B\to C})) & := \lambda z^B.(\lambda y^{A\wedge B}.\tau_y t)\langle x^A, z^B\rangle, \\
\tau_x(t \circ s) & := (\lambda y.\tau_y t)\tau_x s.
\end{array}$$

One should think of the variable x in $\tau_x(t)$ as being bound, i.e. what really matters is $\lambda x.\tau_x(t)$. However, the use of $\lambda x.\tau_x(t)$ is less convenient in computations. ⊠

8.2.7. DEFINITION. (*Mapping* σ *from typed terms to arrows*) Let t^B be a typed lambda term with its free variables contained in a list Θ. By induction on $|t|$ we define a combinator $\sigma_\Theta(t^B) : \Theta^\circ \Rightarrow B$.

$$
\begin{array}{ll}
\sigma_{\Theta x^A \Theta'}(x^A) & := \beta_{y^\top \Theta x^A \Theta', x^A}, \\
\sigma_\Theta(t^{A\to B} s^A) & := \mathrm{ev}^{A,B}\langle \sigma_\Theta t, \sigma_\Theta s\rangle, \\
\sigma_\Theta(\langle t, s\rangle) & := \langle \sigma_\Theta t, \sigma_\Theta s\rangle, \\
\sigma_\Theta(\lambda x^A.t^B) & := \mathrm{cur}(\sigma_{\Theta x^A}(t^B)) \text{ (w.l.o.g. } x^A \notin \Theta), \\
\sigma_\Theta(\mathbf{p}_i t^{A\wedge B}) & := \pi_i \sigma_\Theta(t) \ (i \in \{0,1\}), \\
\sigma_\Theta(*) & := \mathrm{tr}^{\Theta^\circ}.
\end{array}
$$
⊠

8.2.8. Lemma. *(τ preserves equality between combinators) For terms t, s denoting arrows in $\mathrm{CCC}(\mathcal{PV})$ such that $t = s$ we have $\tau_x(t) = \tau_x(s)$.*

Proof. The proof is by induction on the depth of a prooftree for $t = s$. We have to verify the statement for instances of the axioms, and show that the property is transmitted by the rules. The latter is straightforward. For the basis case of the axioms, we shall verify two complicated cases.

Case 1. The axiom evcur. Let $t\colon A \wedge B \Rightarrow C$.

$$
\begin{array}{l}
\tau_x(\mathrm{ev}\langle \mathrm{cur}(t)\pi_0, \pi_1\rangle) = \ (\text{def. of } \tau) \\
(\lambda y.\tau_y(\mathrm{ev}))\langle \tau_x(\mathrm{cur}(t)\pi_0), \tau_x\pi_1\rangle = \ (\text{def. of } \tau) \\
(\lambda y.\mathbf{p}_0 y(\mathbf{p}_1 y))\langle \tau_x(\mathrm{cur}(t)\pi_0), \tau_x\pi_1\rangle = \ (\beta\text{con, proj}) \\
(\tau_x(\mathrm{cur}(t)\pi_0))(\tau_x\pi_1) = \ (\text{def. of } \tau) \\
(\lambda y.\tau_y(\mathrm{cur}(t))\tau_x\pi_0)(\mathbf{p}_1 x^{A\wedge B}) = \ (\text{def. of } \tau) \\
(\lambda y^A \lambda z^B.((\lambda u^{A\wedge B}.\tau_u(t))\langle y^A, z^B\rangle))(\mathbf{p}_0 x^{A\wedge B})(\mathbf{p}_1 x^{A\wedge B}) = \ (\beta\text{con}) \\
(\lambda u^{A\wedge B}.\tau_u t)\langle \mathbf{p}_0 x, \mathbf{p}_1 x\rangle = \ (\text{surj}) \\
(\lambda u^{A\wedge B}.\tau_u t)x^{A\wedge B} = \ (\beta\text{con}) \ \tau_u t.
\end{array}
$$

Case 2. The axiom curev.

$$
\begin{array}{l}
\tau_x(\mathrm{cur}(\mathrm{ev}\langle t\pi_0, \pi_1\rangle)) = \ (\text{def. of } \tau) \\
\lambda z^B.(\lambda y^{A\wedge B}.\tau_y(\mathrm{ev}\langle t\pi_0, \pi_1\rangle))\langle x^A, z^B\rangle, \ \text{abbreviated } (*).
\end{array}
$$

We observe that

$$
\begin{array}{l}
\tau_y(\mathrm{ev}\langle t\pi_0, \pi_1\rangle) = \ (\text{def. of } \tau) \\
\lambda u^{(B\to C)\wedge B}.\tau_u(\mathrm{ev})\tau_y(\langle t\pi_0, \pi_1\rangle) = \ (\text{def. of } \tau) \\
(\lambda u^{(B\to C)\wedge B}.(\mathbf{p}_0 u)(\mathbf{p}_1 u))\langle \tau_y(t\pi_0), \tau_y\pi_1\rangle = \ (\beta\text{con, proj}) \\
\tau_y(t\pi_0)\tau_y\pi_1 = \ (\text{def. of } \tau) \ (\lambda x^A.\tau_x t)(\mathbf{p}_0 y^{A\wedge B})(\mathbf{p}_1 y^{A\wedge B}).
\end{array}
$$

Therefore

$$
\begin{array}{l}
\tau_x(\mathrm{cur}(\mathrm{ev}\langle t\pi_0, \pi_1\rangle)) = (*) = \\
\lambda z^B.(\lambda y^{A\wedge B}.(\lambda x^A.\tau_x t)(\mathbf{p}_0 y)(\mathbf{p}_1 y))\langle x^A, z^B\rangle = \ (\beta\text{con, proj}) \\
\lambda z^B.(\lambda x^A.\tau_x t)x^A z^B = \ (\beta\text{con}) \ \lambda z^B.\tau_x(t)z^B = \ (\eta\text{con}) \ \tau_x t.
\end{array}
$$
⊠

8.2.8A. ♠ Complete the proof.

8.2.9. LEMMA. *Let Θ, Θ' be lists of typed variables without repetitions, such that all variables of Θ are in Θ', and* FV(t) *is contained in Θ. Then*

$$\sigma_{\Theta'}(t) = \sigma_\Theta(t)\beta_{y^\top\Theta',y^\top\Theta}.$$

PROOF. By induction on $|t|$. We check one case. Let $t \equiv \lambda x^A.s$. Then

$$\begin{aligned}
&\sigma_{\Theta'}(t) = \mathrm{cur}(\sigma_{\Theta' x^A}(s)) = \text{ (IH)}\\
&\mathrm{cur}(\sigma_{\Theta x^A}(s)\beta_{y^\top\Theta' x^A,y^\top\Theta x^A}) = \text{ (8.1.5)}\\
&\mathrm{cur}(\sigma_{\Theta x^A}(s))\langle(\beta_{y^\top\Theta',y^\top\Theta})\pi_0,\pi_1\rangle = \text{ (8.1.5)}\\
&\mathrm{cur}(\sigma_{\Theta x^A}(s))\beta_{y^\top\Theta',y^\top\Theta} = \sigma_\Theta(\lambda x^A.s)\beta_{y^\top\Theta',y^\top\Theta} \text{ (def. of } \sigma).
\end{aligned}$$

⊠

8.2.9A. ♠ Complete the proof of the lemma.

8.2.10. LEMMA. *(σ_Θ respects β-conversion)*

$$\sigma_\Theta((\lambda x^A.t^B)s^A) = \sigma_\Theta(t^B[x^A/s^A]).$$

PROOF. The proof proceeds by induction on $|t|$. We first note that

$$\sigma_\Theta((\lambda x^A.t^B)s^A) = \mathrm{ev}\langle\mathrm{cur}(\sigma_\Theta(t^B)),\sigma_\Theta s\rangle = \sigma_\Theta(t)\langle\mathrm{id},\sigma_\Theta s\rangle.$$

Hence we only need to show

$$\sigma_{\Theta x}(t)\gamma = \sigma_\Theta(t[x/s]) \text{ where } \gamma := \langle\mathrm{id},\sigma_\Theta(s)\rangle.$$

If x is not free in t, then $t[x/s] \equiv t$, and $\sigma_{\Theta x}\gamma = \sigma_\Theta\pi_0\langle\mathrm{id},\sigma_\Theta(s)\rangle = \sigma_\Theta(t) = \sigma_\Theta(t[x/s])$. So we may concentrate on cases where x actually occurs free in t.

We check the most complicated case of the induction step, and leave the other cases to the reader.

Case 1. Let $t \equiv \lambda y^D.t'$.

$$\begin{aligned}
&\sigma_{\Theta x}(t)\gamma = \text{ (def. of } \sigma)\\
&\mathrm{cur}(\sigma_{\Theta xy}(t'))\gamma = \text{ (8.1.5)}\\
&\mathrm{cur}(\sigma_{\Theta xy}(t')\langle\gamma\pi_0,\pi_1\rangle) =\\
&\qquad \text{(since } \beta_{z^\top\Theta xy,z^\top\Theta yx} = \langle\langle\pi_0\pi_0,\pi_1\rangle,\pi_1\pi_0\rangle, \text{ and 8.2.9)}\\
&\mathrm{cur}(\sigma_{\Theta yx}(t')\langle\langle\pi_0\pi_0,\pi_1\rangle,\pi_1\pi_0\rangle)\langle\gamma\pi_0,\pi_1\rangle.
\end{aligned}$$

Now

$$\begin{aligned}
&\sigma_{\Theta yx}(t')\langle\langle\pi_0\pi_0,\pi_1\rangle,\pi_1\pi_0\rangle\langle\gamma\pi_0,\pi_1\rangle = \ (8.1.5)\\
&\sigma_{\Theta yx}(t')\langle\langle\pi_0\pi_0,\pi_1\rangle\langle\gamma\pi_0,\pi_1\rangle,\pi_1\pi_0\langle\gamma\pi_0,\pi_1\rangle\rangle = \ (8.1.5,\text{ proj})\\
&\sigma_{\Theta yx}(t')\langle\langle\pi_0\pi_0\langle\gamma\pi_0,\pi_1\rangle,\pi_1\langle\gamma\pi_0,\pi_1\rangle\rangle,\pi_1\gamma\pi_0\rangle = \ (\text{proj})\\
&\sigma_{\Theta yx}(t')\langle\langle\pi_0\gamma\pi_0,\pi_1\rangle,\pi_1\gamma\pi_0\rangle = \ (\text{def. of } \gamma)\\
&\sigma_{\Theta yx}(t')\langle\langle\pi_0\langle\text{id},\sigma_\Theta s\rangle\pi_0,\pi_1\rangle,\pi_1\langle\text{id},\sigma_\Theta s\rangle\pi_0\rangle = \ (\text{proj})\\
&\sigma_{\Theta yx}(t')\langle\langle\pi_0,\pi_1\rangle,\sigma_\Theta(s)\pi_0\rangle = \ (\text{surj})\\
&\sigma_{\Theta yx}(t')\langle\text{id},\sigma_\Theta(s)\pi_0\rangle = \ (\text{since } \sigma_\Theta(s)\pi_0 = \sigma_{\Theta y}(s) \text{ and } 8.2.5)\\
&\sigma_{\Theta yx}(t')\langle\text{id},\sigma_{\Theta y}(s)\rangle = \ (\text{by IH})\ \sigma_{\Theta y}(t'[x/s]),\ \text{hence}\\
&\sigma_\Theta(t[x/s]) = \text{cur}(\sigma_{\Theta y}t'[x/s]) = \sigma_\Theta(t)[x/s].
\end{aligned}$$

⊠

8.2.10A. ♠ Do the remaining cases in the proof of the lemma.

8.2.11. LEMMA. *(σ preserves $\beta\eta$-equality) If for lambda terms t, t' we have $t = t'$, then for all appropriate Θ, $\sigma_\Theta t = \sigma_\Theta t'$.*

PROOF. We prove this by induction on the depth of a deduction of $t = t'$. Preservation under the rules is easy; we also have to check that the property holds for axioms. For example, if $x \notin \text{FV}(t)$,

$$\begin{aligned}
&\sigma_\Theta(\lambda x.tx) = \ (\text{def. of } \sigma)\ \text{cur}(\sigma_{\Theta x}(tx) = \ (\text{def. of } \sigma)\\
&\text{cur}(\text{ev}\langle\sigma_{\Theta x}t,\sigma_{\Theta x}x\rangle) = \ (8.2.9)\ \text{cur}(\text{ev}\langle(\sigma_\Theta t)\pi_0,\pi_1\rangle) = \ (8.1.5)\ \sigma_\Theta t.
\end{aligned}$$

Also $\sigma_\Theta((\lambda x.t)s) = \sigma_\Theta(t[x/s])$ by lemma 8.2.10. Other cases are left to the reader. ⊠

8.2.12. LEMMA. *(σ_x is inverse to τ_x modulo a projection)*

$$\sigma_{x^A}(\tau_x(t)) = t\pi_1^{\top,A}.$$

PROOF. By induction on $|t|$. We check two difficult cases of the induction step and leave the others to the reader.
Case 1. Let $t \equiv \text{cur}(s)$, $s: A \wedge B \Rightarrow C$.

$$\begin{aligned}
&\sigma_x\tau_x(\text{cur}(s)) = \ (\text{def. of } \tau)\\
&\sigma_x(\lambda z^B.(\lambda y^{A\to B}.\tau_y s)\langle x^A, z^B\rangle) = \ (\text{def. of } \sigma)\\
&\text{cur}(\sigma_{xz}(\lambda y^{A\to B}.\tau_y s)\langle x^A, z^B\rangle)) = \ (\text{def. of } \sigma)\\
&\text{cur}(\text{ev}\langle\sigma_{xz}(\lambda y^{A\to B}.\tau_y s),\sigma_{xz}\langle x^A, z^B\rangle\rangle) = \ (\text{def. of } \sigma)\\
&\text{cur}(\text{ev}\langle\text{cur}(\sigma_{xzy}\tau_y s),\langle\pi_1\pi_0,\pi_1\rangle\rangle).
\end{aligned}$$

Now observe that

$$\begin{aligned}
&\sigma_{xzy}(\tau_y s) = \ (8.2.9)\ (\sigma_y\tau_y s)\langle\pi_0\pi_0\pi_0,\pi_1\rangle = \ (\text{IH})\\
&s\pi_1\langle\pi_0\pi_0\pi_0,\pi_1\rangle = \ (\text{proj})\ s\pi_1.
\end{aligned}$$

Hence

$$\begin{aligned}
&\mathrm{cur}(\mathrm{ev}\langle \mathrm{cur}(\sigma_{xzy}\tau_y s), \langle \pi_1\pi_0, \pi_1\rangle\rangle) = \\
&\mathrm{cur}(\mathrm{ev}\langle \mathrm{cur}(s\pi_1), \langle \pi_1\pi_0, \pi_1\rangle\rangle) = \ (8.1.5)\\
&\mathrm{cur}(s\pi_1\langle \mathrm{id}, \langle \pi_1\pi_0, \pi_1\rangle\rangle) = \ (\mathrm{proj})\\
&\mathrm{cur}(s\langle \pi_1\pi_0, \pi_1\rangle) = \ (8.1.5)\ \mathrm{cur}(s)\pi_1.
\end{aligned}$$

Case 2. Let $t \equiv s \circ s'$, $s\colon C \Rightarrow B$, $s'\colon A \Rightarrow C$.

$$\begin{aligned}
&\sigma_{x^A}(\tau_x(ss')) = \ (\text{def. of } \tau)\\
&\sigma_{x^A}((\lambda y^C.\tau_y s)\tau_x s') = \ (\text{def. of } \sigma)\\
&\mathrm{ev}^{C,B}\langle \sigma_x(\lambda y.\tau_y s), \sigma_x\tau_x s'\rangle = \ (\text{def. of } \sigma)\\
&\mathrm{ev}^{C,B}\langle \mathrm{cur}(\sigma_{xy}\tau_y s), \sigma_x\tau_x s'\rangle, \ \text{abbreviated } (*).
\end{aligned}$$

Observe that

$$\begin{aligned}
&\sigma_{xy}\tau_y s = \ (8.2.9)\ \sigma_y\tau_y s\langle \pi_0\pi_0, \pi_1\rangle = \ (\mathrm{IH})\\
&s\pi_1\langle \pi_0\pi_0, \pi_1^{\top\wedge A,C}\rangle = \ (\mathrm{proj})\ s\pi_1^{\top\wedge A,C},
\end{aligned}$$

hence $(*)$ is

$$\begin{aligned}
&\mathrm{ev}^{C,B}\langle \mathrm{cur}(\sigma_{xy}\tau_y s), \sigma_x\tau_x s'\rangle = \ (8.1.5)\\
&\mathrm{ev}^{C,B}\langle \mathrm{cur}(s\pi_1^{\top\wedge A,C}, s'\pi_1^{\top\wedge A,C}\rangle = \ (8.1.5)\\
&\mathrm{ev}^{C,B}\langle \mathrm{cur}(s), s'\rangle\pi_1^{\top\wedge A,C} = \ (8.1.5)\ s \circ s' \circ \pi_1^{\top\wedge A,C}.
\end{aligned}$$

⊠

8.2.12A. ♠ Complete the proof.

8.2.13. THEOREM. *For combinators t, s, $t = s$ iff $\tau_x(t) = \tau_x(s)$.*

PROOF. If $t = s$ then $\tau_x(t) = \tau_x(s)$ by lemma 8.2.8. If for two combinators $t, s\colon A \Rightarrow B$ we have $\tau_x(t) = \tau_x(s)$, we have by lemma 8.2.11 $\sigma_x(\tau_x(t)) = \sigma_x(\tau_x(s))$, hence $t\pi_1 = s\pi_1\colon \top \wedge A \Rightarrow B$ (lemma 8.2.12). Therefore $t \equiv t\pi_1\langle \mathrm{tr}^A, \mathrm{id}^A\rangle = s\pi_1\langle \mathrm{tr}^A, \mathrm{id}^A\rangle = s$. ⊠

8.3 Decidability of equality

We consider here the question of decidability of equality between combinators. First of all, we note that we can restrict attention to the case where A and B are objects of CCC($\mathcal{PV}$) which do not contain $\top$. To see this, one observes that any object B constructed from $\mathcal{PV}$, $\top$ by means of $\wedge, \rightarrow$ is either isomorphic to $\top$ or to an object B' not containing $\top$. This follows from the isomorphisms (cf. 8.1.6A):

$$\top \wedge A \simeq A \wedge \top \simeq A; \ A \rightarrow \top \simeq \top; \ \top \rightarrow A \simeq A.$$

By the results of the preceding section, the problem is equivalent to comparing closed terms in $\boldsymbol{\lambda\eta}_{\to\wedge}$. By proving strong normalization and uniqueness of normal form for $\boldsymbol{\lambda\eta}_{\to\wedge}$ we obtain decidability between the lambda terms, hence between the combinators. (In the calculus $\boldsymbol{\lambda\eta}_{\to\wedge\top}$ uniqueness of normal form fails, see 8.3.6C.)

8.3.1. We extend the results of section 6.8 in a straightforward way; the method uses a predicate "Comp" as before. We use $\to$, $\forall$ and $\wedge$ also on the metalevel, as in the following definition.

DEFINITION. For each formula (type) A in the $\wedge\to$-fragment we define the computability predicate Comp_A, by induction on the complexity of A; for prime A and $A \equiv B \to C$ the definition is as before (6.8.2), for $A \equiv B \wedge C$ we put

$$\begin{aligned}&\mathrm{Comp}_{B\wedge C}(t) := \\ &\qquad \mathrm{SN}(t) \wedge \forall t't''(t \succeq \mathbf{p}(t', t'') \to \mathrm{Comp}_B(t') \wedge \mathrm{Comp}_C(t'')).\end{aligned}$$ ⊠

8.3.2. LEMMA. *The properties* C1–4 *of lemma 6.8.3 remain valid for* Comp_A.

PROOF. The proof proceeds by induction on the complexity of A, as before. We have to consider one extra case for the induction step, namely where $A \equiv B \wedge C$.

C1 is trivial in this case, since included in the definition.

C2. If $t \succeq t^*$, then obviously $\mathrm{Comp}_{B\wedge C}(t^*)$ since $\mathrm{SN}(t^*)$ follows from $\mathrm{SN}(t)$ and if $t^* \succeq \mathbf{p}(t', t'')$, then also $t \succeq \mathbf{p}(t', t'')$.

C3. Assume $\forall t_1 \prec_1 t(\mathrm{Comp}_{B\wedge C}(t_1))$, and t non-introduced, that is to say not of the form $\mathbf{p}(t', t'')$. It is obvious from the assumption that $\mathrm{SN}(t)$. Assume $t \succeq \mathbf{p}(t', t'')$ in n steps. $n = 0$ is excluded since t is non-introduced; so there is a t_1 such that $t \succ_1 t_1 \succeq \mathbf{p}(t', t'')$; it now follows that $\mathrm{Comp}_B(t')$, $\mathrm{Comp}_C(t'')$.

For the induction step where $A \equiv B \to C$, we have to take η-conversion into account, but this does not change the proof. ⊠

8.3.3. LEMMA. $\forall s \in \mathrm{Comp}_A(\mathrm{Comp}_B(t[x/s]) \to \mathrm{Comp}_{A\to B}(\lambda x.t))$.

PROOF. We extend the proof of lemma 6.8.4. There is only one extra case if $(\lambda x.t)s \succ_1 t''$, namely

- $t \equiv t'x$, $t'' \equiv t's$.

But also in this case $\mathrm{Comp}_B(t'')$ by the assumption. ⊠

8.3.4. Lemma. $\text{Comp}_B(t) \wedge \text{Comp}_C(t') \to \text{Comp}_{B\wedge C}(\mathbf{p}(t,t'))$.

Proof. $\mathbf{p}(t,t')$ is strongly normalizable. For consider any reduction sequence of $\mathbf{p}(t,t')$:

$$t'' \equiv \mathbf{p}(t,t') \succ_1 t''_0 \succ_1 t''_1 \succ_1 t''_2 \ldots$$

If there is no step in the sequence where

$$(1) \qquad t''_n \equiv \mathbf{p}(t_n, t'_n) \equiv \mathbf{p}(\mathbf{p}_0 t''_{n+1}, \mathbf{p}_0 t''_{n+1}) \succ_1 t''_{n+1}$$

occurs, the reduction sequence is essentially a combination of reduction sequences of t and t' and hence bounded in length by $h_t + h_{t'}$. If on the other hand there is a first step as in (1), then the length of the reduction sequence is bounded by $h_t + h_{t'} + 1$, since the reduction tree of t''_n is embedded in the reduction tree of t.

Now assume $\mathbf{p}(t,t') \succeq \mathbf{p}(s,s')$. Then either this reduction is obtained by reducing t to s and t' to s', and then $\text{Comp}_B(s)$, $\text{Comp}_C(s')$ follow; or the reduction sequence proceeds as

$$\mathbf{p}(t,t') \succeq \mathbf{p}(\mathbf{p}_0 t'', \mathbf{p}_1 t'') \succ t'' \succeq \mathbf{p}(s,s'),$$

where $t \succeq \mathbf{p}_0 t''$, $t' \succeq \mathbf{p}_1 t''$. But in this case, also $t \succeq \mathbf{p}_0 t'' \succeq \mathbf{p}_0\mathbf{p}(s,s') \succ_1 s$, $t' \succeq \mathbf{p}_0 t'' \succeq \mathbf{p}_1\mathbf{p}(s,s') \succ_1 s'$, and therefore $\text{Comp}_B(s)$, $\text{Comp}_C(s')$. ⊠

8.3.5. Theorem. *All terms of* $\boldsymbol{\lambda\eta}_{\to\wedge}$ *are strongly computable under substitution, and hence strongly normalizable.*

Proof. The proof extends the proof of section 6.8 by considering some extra cases.

Case 4. $t \equiv \mathbf{p}_0 t'$. We have to show that for any substitution * with computable terms, t^* is computable. $t^* = \mathbf{p}_0(t'^*)$, and by the induction hypothesis t'^* is computable. So it suffices to show that whenever $\text{Comp}_{B\wedge C}(s)$, then $\text{Comp}_B(\mathbf{p}_0 s)$ for arbitrary s of type $B \wedge C$. This is done by induction on h_s. Since $\mathbf{p}_0 s$ is non-introduced, it suffices to prove $\text{Comp}_B(t)$ for all $t \prec_1 \mathbf{p}_0 s$. If $t \prec_1 \mathbf{p}_0 s$, then *either*

- $t \equiv \mathbf{p}_0(s')$ with $s' \prec_1 s$; then $\text{Comp}_B(t)$ by induction hypothesis, *or*
- $s \equiv \mathbf{p}(s', s'')$, $t \equiv s'$; $\text{Comp}_B(t)$ is now immediate from $\text{Comp}_{B\wedge C}(s)$.

Case 5. $t \equiv \mathbf{p}_1 t'$ is treated symmetrically.

Case 6. $t \equiv \mathbf{p}(t_0, t_1)$. We have to show that for a substitution * with computable terms, $t^* \equiv \mathbf{p}(t_0^*, t_1^*)$ is computable; but this now follows immediately from the induction hypothesis plus lemma 8.3.4. ⊠

8.3.6. PROPOSITION. $\succ_1$ *is weakly confluent and hence normal forms in* $\boldsymbol{\lambda\eta}_{\to\wedge}$ *are unique.*

PROOF. The proof of this fact proceeds as for $\boldsymbol{\lambda}_{\to}$ (cf. 1.2.11). ⊠

8.3.6A. ♠ Prove the confluence of terms in $\boldsymbol{\lambda\eta}_{\to\wedge}$.

8.3.6B. ♠ Show that in $\boldsymbol{\lambda\eta}_{\to\wedge}$ long normal forms (cf. 6.7.2) are unique. *Hint.* Show by induction on the complexity of t that if t, s are expanded normal and $t \succeq s$, then $t \equiv s$.

8.3.6C. ♠ Give an example showing failure of confluence in $\boldsymbol{\lambda\eta}_{\to\wedge\top}$. *Hint.* Consider $\mathbf{p}(\mathbf{p}_0 t, \mathbf{p}_1 t)$ with $\mathbf{p}_0 t$ or $\mathbf{p}_1 t$ of type $\top$, or $\lambda x^A.tx$ with tx of type $\top$, $x \notin \mathrm{FV}(t)$.

8.4 A coherence theorem for CCC's

Coherence theorems are theorems of the following type: given a free category $\mathcal{C}$ of a certain type (here cartesian closed) and objects A, B of $\mathcal{C}$ satisfying suitable conditions, there is exactly one combinator from A to B (modulo equality of arrows in a category of the given type). In this section we prove a coherence theorem for $\mathrm{CCC}(\mathcal{PV})$; the proof is based on Mints [1992e].

8.4.1. NOTATION. In this section we write $\vdash \Gamma \Rightarrow A$ if A is deducible from open assumptions Γ in $\wedge\to$-**Nm** (cf. 2.1.8). In term notation we write $\vec{x}{:}\,\Gamma \Rightarrow t{:}\,A$, or sometimes more compactly $t[\vec{x}{:}\,\Gamma]{:}\,A$ or $t[\vec{x}^\Gamma]{:}\,A$. We write $[x{:}\,B], \vec{y}{:}\,\Gamma \Rightarrow A$, or $[x^B], \vec{y}{:}\,\Gamma \Rightarrow A$ to indicate that the assumption x^B may be either present or absent. ⊠

8.4.2. DEFINITION. A $\wedge\to$-formula A is *balanced* if no propositional variable occurring in A has more than one occurrence with the same sign (so there are at most two occurrences of any propositional variable, and if there are two, they have opposite signs). A sequent $\Gamma \Rightarrow A$ is *balanced* if the formula $\bigwedge\Gamma \to A$ is balanced. ⊠

The theorem which we shall prove is

8.4.3. THEOREM. *(Coherence theorem) Let* A, B *be objects of* $\mathrm{CCC}(\mathcal{PV})$ *not containing* $\top$. *If* $A \Rightarrow B$ *is balanced, there is at most one combinator in* $\mathrm{CCC}(\mathcal{PV})$ *from* A *to* B.

The theorem is obviously false if we drop the condition of balance: in $\mathcal{A}$ there are two distinct arrows $f, g : P_0 \wedge P_0 \Rightarrow P_0 \wedge P_0$, namely $\mathrm{id}^{P_0\wedge P_0} \equiv \langle\pi_0, \pi_1\rangle$, and $\langle\pi_1, \pi_0\rangle$. The theorem will first be proved for so-called 2-sequents.

8.4.4. DEFINITION. A *2-sequent* is a sequent of the form $\Gamma \Rightarrow R$, where each formula in Γ has one of the following forms:

$$P \to P',\ (P \to P') \to P'',\ P'' \to (P \to P'),\ P'' \to P \wedge P'$$

for distinct $P, P', P'' \in \mathcal{PV}$. ⊠

8.4.5. PROPOSITION. *Let $\vec{x}{:}\,\Gamma \Rightarrow t{:}\,R$ be a $\beta\eta$-normal deduction term for a 2-sequent. Then t has one of the following forms:*

(a) x^R,

(b) $x^{P \to (Q \to R)} t_1^P t_2^Q$,

(c) $x^{Q \to R} t_1^Q$,

(d) $x^{(P \to Q) \to R} y^{P \to Q}$,

(e) $x^{(P \to Q) \to R} (\lambda y^P . t_1^Q)$,

(f) $x^{(P \to Q) \to R} (y^{P' \to (P \to Q)} t_1^{P'})$,

(g) $\mathbf{p}_0 (x^{P \to R \wedge Q} t_1^P)$ or $\mathbf{p}_1 (x^{P \to Q \wedge R} t_1^P)$,

where t_1, t_2 are again normal.

PROOF. If t is not a variable (case (a)), t represents a deduction ending in an E-rule; a main branch must therefore begin in an open assumption, corresponding to a variable. Since $\Gamma \Rightarrow R$ is a 2-sequent, we have the following possibilities. In the first three cases we assume that the final rule is $\to$E.

(1) The open assumption is of the form $x^{P \to (Q \to R)}$; this yields case (b).

(2) The open assumption is of the form $x^{P \to R}$ (case (c)).

(3) The open assumption is of the form $x^{(P \to Q) \to R}$, and t has the form $x^{(P \to Q) \to R} t_0^{P \to Q}$. t_0 may either be a variable (case (d)), or end with $\to$I (case (e)), or end with an E-rule. In the third subcase, the main branch of t_0 must start in an open assumption, i.e. a variable $y^{P' \to (P \to Q)}$; this yields case (f) of the proposition.

(4) Now assume that the final E-rule is $\wedge$E. The open assumption is of the form $P \to R \wedge Q$ or $P \to Q \wedge R$; this yields case (g). ⊠

8.4.6. LEMMA. *If $\Gamma \Rightarrow R$, $R \in \mathcal{PV}$ is derivable, then R has a positive occurrence among the formulas of Γ.*

PROOF. $\Gamma \Rightarrow R$ is intuitionistically true, hence classically true; if R has no positive occurrences in Γ, we can take a valuation v with $\mathrm{v}(R) = \bot$, $\mathrm{v}(Q) = \top$ for all Q distinct from R; this would make Γ true and R false. ⊠

8.4.7. LEMMA. *Let $\vec{x}{:}\Gamma \Rightarrow t{:}R$ be a $\beta\eta$-normal deduction term for a 2-sequent. Then any propositional variable occurring negatively in Γ also occurs positively in Γ.*

PROOF. By induction on $|t|$, using proposition 8.4.5 to distinguish cases. We shall leave most of the cases to the reader.
Case (b). $t \equiv x^{P\to(Q\to R)}t_1^P t_2^Q$, t_1, t_2 also normal. For negative occurrences of propositional variables in the sequents corresponding to t_1, t_2 we can therefore apply the IH. Q in the type of x is a negative occurrence; but $\vec{y}{:}\Gamma' \Rightarrow t_2{:}Q$ plus the preceding lemma shows that $\Gamma' \subset \Gamma$ necessarily contains a positive occurrence of Q. ⊠

8.4.7A. ♠ Do the remaining cases of the proof.

8.4.8. DEFINITION. Let $\mathcal{S} : \Gamma \Rightarrow R$ be a balanced 2-sequent. An $\mathcal{S}$-sequent is a sequent $\Gamma'\Delta \Rightarrow Q$ where $\Gamma' \subset \Gamma$, Q occurs positively in $\mathcal{S}$, and Δ is a list of propositional variables such that if $P \in \Delta$ then a formula $(P \to P_1) \to P_2$ occurs in Γ. ⊠

8.4.9. LEMMA. *Let $\mathcal{S} : \Gamma \Rightarrow R$ be a balanced 2-sequent, and let $A \in \Gamma$ be a formula with a strictly positive occurrence of R. Then all positive occurrences of R in the antecedent of any $\mathcal{S}$-sequent are in a single occurrence of A.*

PROOF. Strictly positive occurrence of R in A, A occurring in the antecedent of a 2-sequent, means that A has one of the forms $P \to R$, $P' \to (P \to R)$, $P \to R \wedge P'$, $P \to P' \wedge R$. We argue by contradiction. Let $\Gamma'\Delta \Rightarrow Q$ be an $\mathcal{S}$-sequent containing a positive occurrence of R not in A.

The occurrence cannot be in Γ', because then $\mathcal{S}$ would be unbalanced, since $\Gamma' \subset \Gamma$. Also, the occurrence can not be in Δ, because then Γ would contain $(R \to P_1) \to P_2$ for certain P_1, P_2, again making $\mathcal{S}$ unbalanced. ⊠

8.4.10. DEFINITION. An occurrence $A \in \Gamma$, $\Gamma \Rightarrow R$ a 2-sequent, is called *redundant* in the sequent if A contains a negative occurrence of R (so A is of one of the forms $R \to P_2$, $P_1 \to (R \to P_2)$, $(P_1 \to R) \to P_2$, $R \to (P_1 \to P_2)$, $R \to P_1 \wedge P_2$). ⊠

8.4.11. LEMMA. *Let $\mathcal{S}$ be a balanced 2-sequent, $\Gamma \Rightarrow R$ an $\mathcal{S}$-sequent, $\vec{x}{:}\Gamma \Rightarrow t{:}R$ a normal deduction term. Then Γ contains no redundant elements.*

PROOF. We use induction on $|t|$, arguing by cases according to 8.4.5.
Case (a). $t \equiv x^R$. $\Gamma \equiv R$ has no negative occurrences, hence no redundant members.

Case (b). $t \equiv x^{P\to(Q\to R)}t_1^P t_2^Q$. Then $[x^{P\to(Q\to R)}], \vec{y}{:}\, \Gamma_1 \Rightarrow t_1{:}\, P$ and $[x^{P\to(Q\to R)}], \vec{z}{:}\, \Gamma_2 \Rightarrow t_2{:}\, Q$ represent deductions of $\mathcal{S}$-sequents, hence x must be absent, by the IH. $P \to (Q \to R)$ contains a strictly positive occurrence of R, so the preceding lemma applies and we see that Γ_1, Γ_2 contain no positive occurrences of R. And thus by lemma 8.4.7 there is also no negative occurrence of R.

The other cases are left to the reader. ⊠

8.4.11A. ♠ Do the remaining cases of the proof.

8.4.12. PROPOSITION. *(Coherence for 2-sequents) Let $\Gamma \Rightarrow R$ be a balanced 2-sequent, $\vec{x}{:}\, \Gamma' \Rightarrow t{:}\, R$, $\vec{y}{:}\, \Gamma'' \Rightarrow s{:}\, R$ normal deduction terms such that $\Gamma', \Gamma'' \subset \Gamma$, $(\vec{x}{:}\, \Gamma') \cup (\vec{y}{:}\, \Gamma'') = \vec{z}{:}\, \Gamma$ for suitable $\vec{z}$. Then $\Gamma' = \Gamma'' = \Gamma$ and $t \equiv s$.*

PROOF. By induction on $|t| + |s|$. Assume $|t| \geq |s|$. We use a case distinction according to proposition 8.4.5.
Case (a). t is a variable; then $t \equiv x^R \equiv s$.
Case (b). Suppose $t \equiv x^{P\to(Q\to R)}t_1^P t_2^Q$. The term s cannot have one of the other forms, since this would yield that there were two distinct positive occurrences of R in Γ. Therefore $s \equiv x^{P\to(Q\to R)}s_1^P s_2^Q$. Furthermore

$$[x^{P\to(Q\to R)}], \vec{y}_1{:}\, \Gamma_1' \Rightarrow t_1{:}\, P, \quad [x^{P\to(Q\to R)}], \vec{y}_2{:}\, \Gamma_2' \Rightarrow t_2{:}\, Q,$$
$$[x^{P\to(Q\to R)}], \vec{z}_1{:}\, \Gamma_1'' \Rightarrow s_1{:}\, P, \quad [x^{P\to(Q\to R)}], \vec{z}_2{:}\, \Gamma_2'' \Rightarrow s_2{:}\, Q.$$

$P \to (Q \to R), \Gamma_1' \Rightarrow P$ is a $(\Gamma' \Rightarrow R)$-sequent. In this sequent $P \to (Q \to R)$ is redundant, since it contains a negative occurrence of P. Then $\vec{y}_1{:}\, \Gamma_1' \Rightarrow t_1{:}\, P$, and similarly

$$\vec{y}_2{:}\, \Gamma_2' \Rightarrow t_2{:}\, Q,\ \vec{z}_1{:}\, \Gamma_1'' \Rightarrow s_1{:}\, P,\ \vec{z}_2{:}\, \Gamma_2'' \Rightarrow s_2{:}\, Q.$$

It follows that $\vec{y}_1{:}\, \Gamma_1' \cup \vec{z}_1{:}\, \Gamma_1''$ and $\vec{y}_2{:}\, \Gamma_2' \cup \vec{z}_2{:}\, \Gamma_2''$ are again balanced, and we can apply the induction hypothesis.
Cases (c) and (d). Similar, and left to the reader.
Case (e). $t \equiv x^{(P\to Q)\to R}(\lambda y^P.t_1^Q)$. The possibility that s falls under one of the cases (a)-(c), (g) is readily excluded. For (e) we can argue as in case (b) above, using the IH. The possibilities (d) and (f) can both be excluded; since the argument is similar in both cases, we restrict ourselves to (f). So we have

$$s \equiv x^{(P\to Q)\to R}(y^{P'\to(P\to Q)}(s')^{P'}).$$

The terms t and s correspond to deductions

$$\dfrac{x{:}\, (P \to Q) \to R \qquad \dfrac{t_1{:}\, Q}{\lambda y.t_1{:}\, P \to Q}}{t \equiv x(\lambda y.t_1){:}\, R} \qquad \dfrac{x{:}\, (P \to Q) \to R \qquad \dfrac{z : P' \to (P \to Q) \qquad s'{:}\, P'}{zs'{:}\, P \to Q}}{s \equiv x(zs'){:}\, R}$$

$s' : P'$ is a normal deduction of

$$[(P \to Q) \to R], [P' \to (P \to Q)], \Gamma_1'' \Rightarrow P'.$$

$z : P' \to (P \to Q)$ is redundant, hence absent in s'. But then $x : (P \to Q) \to R$ is also absent, since Q is a negative occurrence in Γ, while another positive occurrence outside $R' \to (P \to Q)$ ought to be present but cannot be present since this would make $\Gamma \Rightarrow R$ unbalanced.

Among the assumptions of t_1 there must be positive occurrences of Q; also

$$\Gamma \subset \{(P \to Q) \to R, P' \to (P \to Q), \Gamma_1''\} \cup \{(P \to Q) \to R, \Gamma_1'\}.$$

But then these must coincide with $P' \to (P \to Q)$ in Γ_1' (otherwise there would be two positive occurrences of Q in Γ); and since Q is strictly positive in $P' \to (P \to Q)$, this is the only occurrence in Γ_1'. This would give as the only possibility for t_1

$$\cfrac{\cfrac{z : P' \to (P \to Q) \qquad t_2 : P'}{zt_2 : P \to Q} \qquad t_3 : P}{t_1 \equiv (zt_2)t_3 : Q}$$

But again from the balancedness of $\Gamma' \Rightarrow R$ it follows that apart from the occurrence of P in $(P \to Q) \to R$ there can be no other positive occurrence of P in Γ_1'. Hence $t_3 : P$ must be the bound assumption variable y^P, which contradicts the η-normality of t (for if t_3 is not a variable, the main branch of the subdeduction of $t_3 : P$ must contain another positive occurrence of P).

The cases (f) and (g) are left to the reader. ⊠

8.4.12A. ♠ Do the remaining cases of the proof.

8.4.13. *Reduction to 2-sequents.* Let us assign to a formula a "deviation" which measures the deviation of the formula from a form which is acceptable as antecedent formula in a 2-sequent:

$$\text{dev}(B \to C) := |B| + |C|, \ \text{dev}(B \wedge C) := |B| + |C| + 1.$$

For a multiset Γ, let n be the maximum of the deviations of formulas in Γ, and let m be the number of formula occurrences in Γ with deviation n. If we replace, in a multiset with formulas of deviation > 0, a formula A of maximum deviation by one or two formulas Γ_A according to a suitably chosen line in the table below, either n is lowered, or n stays the same and m is lowered.

DEFINITION. *(Deviation reductions)* A formula A which has a logical form not permitted in the antecedent of a 2-sequent may be replaced by a set Γ_A

with at most two elements; we also define for each replacement a term ϕ_A, according to the following table. The propositional variable P is fresh.

	A	Γ_A	ϕ_A
(a)	$B \to C$	$B \to P, P \to C$	$\lambda u^B.z^{P\to C}(y^{B\to P}u^B)$
(b)	$(D \to B) \to C$	$B \to P$, $(D \to P) \to C$	$\lambda u^{D\to B}.z^{(D\to P)\to C}(\lambda v^D.y^{B\to P}(u^{D\to B}v^D))$
(c)	$(B \to C) \to D$	$P \to B$, $(P \to C) \to D$	$\lambda u^{B\to C}.z^{(P\to C)\to D}(\lambda v^P.u^{B\to C}(y^{P\to B}v^P))$
(d)	$B \wedge C \to D$	$B \to (C \to D)$	$\lambda u^{B\wedge C}.y^{B\to(C\to D)}(\mathbf{p}_0 u)(\mathbf{p}_1 u)$
(e)	$C \to (D \to B)$	$C \to (D \to P)$, $P \to B$	$\lambda u^C\lambda v^D.z^{P\to B}(y^{C\to(D\to P)}u^C v^D)$
(f)	$B \wedge C$	B, C	$\langle y^B, z^C\rangle$
$(\text{g})_1$	$D \to B \wedge C$	$D \to P \wedge C$, $P \to B$	$\lambda u^D\langle y^{P\to B}(\mathbf{p}_0(z^{D\to(P\wedge C)}u^D)), \mathbf{p}_1(z^{D\to(P\wedge C)}u^D)\rangle$
$(\text{g})_2$	$D \to C \wedge B$	$D \to C \wedge P$, $P \to B$	$\lambda u^D\langle \mathbf{p}_0(z^{D\to(C\wedge P)}u^D), y^{P\to B}(\mathbf{p}_1(z^{D\to(C\wedge P)}u^D))\rangle$

Γ_A is not uniquely determined by A, but by A together with a subformula occurrence B in A, which is represented in Γ_A by a new propositional variable P. In all cases except (d) the free variables of the ϕ_A are $\{y, z\}$; in case (d) y is the only free variable. Note that by a suitable substitution for the new propositional variable P in Γ_A we get A back.

Repeated application of appropriate deviation reductions to a multiset Γ results ultimately in a multiset Δ, from which Γ is recoverable by a suitable substitution in the propositional variables, and such that Δ consists of formulas of one of the following forms only: P, $P \to P'$, $P \to (P' \to P'')$, $(P \to P') \to P''$, $P \to P' \wedge P''$. Thus Δ is almost the antecedent for a 2-sequent, except that P, P', P'' are not necessarily distinct. But further transformation to a 2-sequent can be achieved, again using the replacements in the table above; for example, $(P \to P') \to P'$ may be replaced by $(P \to Q) \to P'$, $Q \to P'$, with Q a fresh propositional variable. Summing up, we have shown:

LEMMA. *Repeated replacement of formulas A by Γ_A for a suitable choice of reduction rule from the table above in a sequent $\Delta \Rightarrow R$ terminates in a 2-sequent.* ⊠

8.4.14. LEMMA. *Let A, Γ_A and ϕ_A correspond according to one of the lines of the table in the preceding definition. Let P be a propositional variable not occurring in $A\Delta \Rightarrow B$. Let $\vec{u}$ be the new variables appearing as free variables in ϕ_A. Then*

(i) If $A\Delta \Rightarrow B$ is balanced, so is $\Gamma_A\Delta \Rightarrow B$.

(ii) *If $x\colon A, \vec{y}\colon \Delta \Rightarrow t\colon B$, then $\vec{u}\colon \Gamma_A, \vec{y}\colon \Delta \Rightarrow t[x/\phi_A]\colon B$.*

(iii) *If $x\colon A, \vec{y}\colon \Delta \Rightarrow t\colon B$ and $x\colon A, \vec{y}\colon \Delta \Rightarrow s\colon B$, then $t =_{\beta\eta} s$ iff $t[x/\phi_A] =_{\beta\eta} s[x/\phi_A]$, and x is $\beta\eta$-equal to some substitution instance of $\phi_{A[P/E]}$ for a suitable formula E.*

Here $\phi_{A[P/E]}$ is obtained from ϕ_A by substituting in the the types of all variables (free and bound) of ϕ_A the formula E for the proposition variable P. In fact, it will turn out that for E we can always take B as appearing in the table of the preceding definition.

PROOF. (i) by direct verification. (ii) is obtained by combining $\vec{u}\colon \Gamma_A$ (here $\vec{u}$ are the free variables of ϕ_A) with $x\colon A, \vec{y}\colon \Delta \Rightarrow t\colon B$ (using closure under substitution for prooftrees in N-systems).

(iii) Assume $x\colon A, \vec{y}\colon \Delta \Rightarrow t\colon B$ and $x\colon A, \vec{y}\colon \Delta \Rightarrow s\colon B$. Suppose first $t =_{\beta\eta} s$. Then $t[x/\phi_A] =_{\beta\eta} s[x/\phi_A]$ holds because $\beta\eta$-conversion is compatible with substitution.

Conversely, suppose $t[x/\phi_A] =_{\beta\eta} s[x/\phi_A]$. Choose a substitution of a formula E for the propositional variable P in ϕ_A such that A becomes logically equivalent to the conjunction of $\Gamma_{A[P/E]}$ (in fact, for E we can always take B in the table of the preceding definition), and a substitution $[\vec{u}/\vec{r}]$ in $\phi_{A[P/E]}$, such that $\phi_{A[P/E]}[\vec{u}/\vec{r}] =_{\beta\eta} x$. Then this shows that $t =_{\beta\eta} s$. The appropriate substitution $[\vec{u}/\vec{r}]$ is defined according to the cases of the table in the foregoing definition.

Case (a). $\phi_{A[P/B]} \equiv \lambda u^B.z^{B\to C}(y^{B\to B}u^B)$. Substitution $[y, z/\lambda v^A.v, x^{A\to B}]$.
Case (b). $\phi_{A[P/B]} \equiv \lambda u^{D\to B}.z^{(D\to B)\to C}(\lambda v^D.y^{B\to B}(u^{D\to B}v^B))$.
Substitution $[y, z/\lambda w^B.w, x^{(D\to B)\to C}]$.
Case (c). Similar to the preceding case, and left to the reader.
Case (d). There is no propositional substitution.
Substitution $[y/\lambda v^B\lambda w^C.x^{B\wedge C\to D}\mathbf{p}(v^B, w^C)]$.
Case (e). Substitution $[y, z/x^{C\to(D\to B)}, \lambda w^B.w]$.
Case (f). Substitution $[y, z/\mathbf{p}_0 x^{B\wedge C}, \mathbf{p}_1 x^{B\wedge C}]$.
Case (g)$_1$. Substitution $[y, z/\lambda w^B.w, x^{D\to C\wedge B}]$ for y, z.
Case (g)$_2$. Similar to the preceding case. ⊠

8.4.14A. ♠ Check the missing details.

8.4.15. LEMMA. *For any sequent $\Delta \Rightarrow B$, with $\Delta = A_1, \ldots, A_n$, there are a 2-sequent $\Gamma \Rightarrow R$, a sequence of terms $\vec{s} = s_1, \ldots, s_n$ and a term r, in variables $\vec{u}\colon \Gamma$, such that*

(i) *If $\vec{x}\colon \Delta \Rightarrow t\colon B$, then $\vec{u}\colon \Gamma \Rightarrow r(t[\vec{x}/\vec{s}])\colon R$.*

(ii) *If $\vec{x}\colon \Delta \Rightarrow t\colon B$, $\vec{x}\colon \Delta \Rightarrow t'\colon B$, then $t =_{\beta\eta} s$ iff $r(t[\vec{x}/\vec{s}]) =_{\beta\eta} r(t'[\vec{x}/\vec{s}])$.*

(iii) If $\Delta \Rightarrow B$ is balanced, so is $\Gamma \Rightarrow R$.

PROOF. We consider the sequent $\Delta B \to R \Rightarrow R$ (R a fresh propositional variable). If $\vec{x}: \Delta \Rightarrow t: B$ then $\vec{x}: \Delta, v: B \to R \Rightarrow vt: R$. Conversely, if $\vec{x}: \Delta, v: B \to R \Rightarrow vt: R$, we can substitute B for P, $\lambda w^B.w^B$ for v, and we get $\vec{x}: \Delta \Rightarrow t: B$ back, using the fact that $\lambda w^B.w^B: B \to B$ and the fact that the deduction terms are closed under substitution. Now repeatedly apply the preceding lemma to $\vec{x}: \Delta, v: B \to R \Rightarrow t: R$ until we have found a 2-sequent as in the statement of the lemma of 8.4.13. ⊠

8.4.16. *Proof of the coherence theorem.* This is now almost immediate: an arbitrary combinator from A to B in CCC($\mathcal{PV}$) with $A \to B$ balanced is representable by a lambda term; this may be seen as deduction term of a sequent. Lambda terms for balanced $\Gamma \Rightarrow B$ are in bijective correspondence with arrows for a suitable 2-sequent by the preceding lemma, and for arrows between 2-sequents the coherence theorem was proved in proposition 8.4.12.

8.5 Notes

8.5.1. The presentation in the first section is inspired by Lambek and Scott [1986], but in the next section, in the treatment of the correspondence between terms in the $\boldsymbol{\lambda\eta}$-calculus and arrows in the free cartesian closed category we have followed Mints [1992b]. The treatment of strong normalization for typed $\boldsymbol{\lambda\eta}_{\to\wedge}$ extends the earlier proof for $\boldsymbol{\lambda}_{\to}$. The first proof of the extension was due to de Vrijer [1987]; a proof by a quite different method, reducing the $\to\wedge$-case to the implication case, is given in Troelstra [1986], see also Troelstra and van Dalen [1988, 9.2.16].

8.5.2. *Coherence for CCC's.* Mints [1979] indicated a short proof of the coherence theorem for cartesian closed categories for objects constructed with $\to$ alone; Babaev and Solovjov [1979] proved by a different method the coherence theorem for $\to\wedge$-objects. Mints [1992d] then observed that the properties of the depth-reducing transformations (our lemmas 8.4.14, 8.4.15) established in Solovjov [1979] could be used to give a simplified proof (similar to his proof for the $\to$-case) for this result. The depth-reducing transformation as such was already known to Wajsberg [1938], and to Rose [1953]. Our treatment follows Mints [1992d], with a correction in proposition 8.4.12. (The correction was formulated after exchanges between Mints, Solovjov and the authors.)

There are obvious connections between these results and certain results on so-called BCK-logic. In particular, it follows from the $\to\wedge$-coherence theorem that for balanced implications $\to\wedge$-**I** is conservative over the corresponding fragment of BCK-logic. See, for example, Jaśkowski [1963], Hirokawa [1992],

Hindley [1993]. In the direction of simple type theory, this "ramifies" into theorems counting the number of different deductions of a formula (type).

8.5.3. *Other coherence theorems.* There is a host of coherence theorems for various kinds of categories; some use, as for the result sketched here, proof-theoretical methods, by reduction to a logical language (usually a fragment of the typed lambda calculus), for example Mints [1977], Babaev and Solovjov [1990]; others use very different methods, for example, Joyal and Street [1991].

In Kelly and Mac Lane [1971] the presentation of a coherence result for closed categories is entirely in terms of categories. However, inspired by Lambek [1968,1969], the proof uses an essential ingredient from proof theory, namely cut elimination (cf. Kelly and Mac Lane [1971, p. 101]). The results of Kelly and Mac Lane [1971] are extended and strengthened in Voreadou [1977] and Solovjov [1997].

Lambek [1968,1969,1972] systematically explores the relationships between certain types of categories and certain deductive systems, in the spirit of our first section in this chapter. The first of these papers deals with categories corresponding to the so-called Lambek calculus, the second paper deals with closed categories, and the third paper with cartesian closed categories; it contains a sketch of the connection between extensional typed combinatory logic (which is equivalent to $\boldsymbol{\lambda\eta}$) and arrows in the free CCC($\mathcal{PV}$). A different treatment was given in Lambek [1974], Lambek and Scott [1986].

The link between (a suitable equivalence relation on) natural deduction proofs and arrows in free cartesian closed categories was made more precise in Mann [1975].

Curien [1985,1986] also investigated the connection between the typed lambda calculus and CCC's; in these publications an intermediate system has been interpolated between extensional $\boldsymbol{\lambda\eta}$ and the categorical combinators. Hardin [1989] investigates Church–Rosser properties for (fragments of) these calculi.

Chapter 9

Modal and linear logic

Another possible title for this chapter might have been "some non-standard logics", since its principal aim is to illustrate how the methods we introduced for the standard logics **M**, **I** and **C** are applicable in different settings as well.

For the illustrations we have chosen two logics which are of considerable interest in their own right: the wellknown modal logic **S4**, and linear logic. For a long time modal logic used to be a fairly remote corner of logic. In recent times the interest in modal and tense logics has increased considerably, because of their usefulness in artificial intelligence and computer science. For example, modal logics have been used (1) in modelling epistemic notions such as belief and knowledge, (2) in the modelling of the behaviour of programs, (3) in the theory of non-monotonic reasoning.

The language of modal logics is an extension of the language of first-order predicate logic by one or more propositional operators, *modal operators* or *modalities*. Nowadays modal logics are extremely diverse: several primitive modalities, binary and ternary operators, intuitionistic logic as basis, etc. We have chosen **S4** as our example, since it has a fairly well-investigated proof theory and provides us with the classic example of a modal embedding result: intuitionistic logic can be faithfully embedded into **S4** via a so-called modal translation, a result presented in section 9.2.

Linear logic is one of the most interesting examples of what are often called "substructural logics" – logics which in their Gentzen systems do not have all the structural rules of Weakening, Exchange and Contraction. Linear logic has connections with **S4** and is useful in analyzing the "fine-structure" of deductions in **C** and **I**. In section 9.3 we introduce linear logic as a sequent calculus **Gcl** and sketch a proof of cut elimination, and demonstrate its expressive power by showing that intuitionistic linear logic can be faithfully embedded in **Gcl**. The next section shows for the case of intuitionistic implication logic how linear logic may be used to obtain some fine-structure in our standard logics. Finally we leave the domain of Gentzen-system techniques and discuss the simplest case of proofnets for linear logic, which brings graph-theoretic notions into play (section 9.5). Proofnets have been devised so as to exploit as fully as possible the symmetries present in (classical) linear logic.

9.1 The modal logic S4

The modal theory **S4** discussed in this section includes quantifiers; for its propositional part we write **S4p**. The language of **S4** is obtained by adding to the language of first-order predicate logic a unary propositional operator $\Box$. $\Box A$ may be read as *"necessarily A"* or *"box A"*. The dual of $\Box$ is $\Diamond$; $\Diamond A$ is pronounced as *"possibly A"* or *"diamond A"*; $\Diamond A$ may be defined in **S4** as $\neg\Box\neg A$.

9.1.1. DEFINITION. (*Hilbert system* **Hs** *for the modal logic* **S4**) A Hilbert system for **S4** is obtained by adding to the axiom schemas and rules for classical logic **Hc** the schemas

$\Box A \to A$ (T-*axioms*)
$\Box(A \to B) \to (\Box A \to \Box B)$ (K-*axioms*, or *normality axioms*)
$\Box A \to \Box\Box A$ (4-*axioms*)

and the *necessitation rule*:

$\Box$I If $\vdash A$ then $\vdash \Box A$.

The notion of a *deduction from a set of assumptions* Γ may be defined as follows.

The sequence $A_1, \ldots, A_n$ is a deduction of $\Gamma \vdash A$ (A from assumptions Γ) if $A \equiv A_n$, and for all A_i either

1. $A_i \in \Gamma$, or
2. there are $j, k < i$ such that $A_k \equiv A_j \to A_i$, or
3. $A_i \equiv \forall x A_j$, $j < i$, x not free in Γ, or
4. $A_i \equiv \Box A_j$, $j < i$, and there is a subsequence of $A_1, \ldots, A_j$ which is a derivation of $\vdash A_j$. ⊠

An alternative formulation of a "deduction from assumptions" is as follows: a deduction is a tree constructed starting from (Γ, Δ *sets* of formulas)

$$\Gamma \vdash A \ (A \in \Gamma) \qquad\qquad \vdash A \ (A \text{ axiom})$$

by means of rules.

$$\frac{\Gamma \vdash A}{\Gamma, B \vdash A}\,\text{W} \qquad\qquad \frac{\vdash A}{\vdash \Box A}\,\Box\text{I}$$

$$\frac{\Gamma \vdash A \to B \qquad \Gamma \vdash A}{\Gamma \vdash B}\,\to\!\text{E} \qquad\qquad \frac{\Gamma \vdash A}{\Gamma \vdash \forall x A}\,\forall\text{I}\ (x \notin \text{FV}(\Gamma))$$

Remark. Instead of $\to$E, $\Box$I one usually talks about "modus ponens" and "necessitation" (or "rule N") when discussing Hilbert systems. The designation "axiom K" or "K-axiom" ("K" from "Kripke") is standard in the literature on modal logics. "Axiom K" is also used in the literature for the axiom (schema) $A \to (B \to A)$, because of the connection with the combinator called K (in standard notation for combinatory logic) via the formulas-as-types parallel. In order to avoid confusion we have used in this text **k** for the combinator, and **k**-axiom for an axiom of the form $A \to (B \to A)$.

9.1.2. Lemma. *The deduction theorem*

$$\text{If } \Gamma, A \vdash B \text{ then } \Gamma \vdash A \to B$$

and the generalization of $\Box$I

$$\text{If } \Box B_1, \ldots, \Box B_n \vdash A \text{ then } \Box B_1, \ldots, \Box B_n \vdash \Box A$$

are derived rules in **Hs**.

Proof. We leave the proof of the deduction theorem to the reader. The generalization of the necessitation rule is proved by induction on n. Suppose we have already derived for all $\Box\Gamma$ of length n that $\Box\Gamma \vdash A \Rightarrow \Box\Gamma \vdash \Box A$. Now let $\Box\Gamma, \Box A \vdash B$. Then

$$\begin{array}{l} \Box\Gamma \vdash \Box A \to B \text{ (deduction theorem)} \\ \Box\Gamma \vdash \Box(\Box A \to B) \text{ (induction hypothesis)} \\ \Box\Gamma \vdash \Box\Box A \to \Box B \text{ (normality axiom, modus ponens)} \\ \Box\Gamma \vdash \Box A \to \Box B \text{ (4-axiom, transitivity of } \to\text{).} \end{array}$$

Hence with modus ponens and $\Box\Gamma, \Box A \vdash \Box A$ we find $\Box\Gamma, \Box A \vdash \Box B$. ⊠

Remark. One often finds in the literature the notion of a deduction from assumptions formulated without restriction on the rule $\Box$I, i.e.

$$\frac{\Gamma \vdash A}{\Gamma \vdash \Box A}$$

Let us write $\vdash^*$ for the notion of deducibility with this more liberal rule of necessitation. For $\vdash^*$ we can prove only a "*modal deduction theorem*", namely

$$\text{If } \Gamma, A \vdash^* B, \text{ then } \Gamma \vdash^* \Box A \to B.$$

In fact,

$$\Gamma \vdash^* B \text{ iff } \Box\Gamma \vdash^* B \text{ iff } \Box\Gamma \vdash B.$$

Of course, the generalized $\Box$I of the preceding lemma is now trivial. We have chosen the definition in 9.1.1, because it is more convenient in proving equivalences with Gentzen systems and systems of natural deduction.

In the notion of "deduction from assumptions" in first-order predicate logic we formulated the rule $\forall$I as: "If $\Gamma \vdash A$, then $\Gamma \vdash \forall x A$, provided $x \notin \mathrm{FV}(\Gamma)$". The variable restriction on $\forall$I is comparable to the restriction on $\Box$I (namely that the premise of $\Box$I is derived without assumptions). An alternative version of $\forall$I, "If $\Gamma \vdash A$, then $\Gamma \vdash \forall x A$" *without* the condition "$x \notin \mathrm{FV}(\Gamma)$" is analogous to the strong version of $\Box$I considered above. The unrestricted $\forall$I also entails a restriction on the deduction theorem, which now has to be modified as "If $\Gamma, A \vdash B$ then $\Gamma \vdash \forall A \to B$" where $\forall A$ is the universal closure of A.

9.1.2A. ♠ Defining "deduction from assumptions" as in the preceding remark, prove the modal deduction theorem and the equivalences $\Gamma \vdash^* B$ iff $\Box\Gamma \vdash^* B$ iff $\Box\Gamma \vdash B$.

9.1.2B. ♠ Prove modal replacement for **S4p** in the following form:

$$\Box(A \leftrightarrow B) \to (F[A] \leftrightarrow F[B])$$

for arbitrary contexts $F[*]$. Refine this result similarly to 2.1.8H.

9.1.3. Definition. (*The Gentzen system* **G1s**) This calculus is based on the language of first-order predicate logic with two extra operators, $\Box$ and $\Diamond$. To the sequent rules of **G1c** we add

$$\mathrm{L}\Box\ \frac{\Gamma, A \Rightarrow \Delta}{\Gamma, \Box A \Rightarrow \Delta} \qquad \mathrm{R}\Box\ \frac{\Box\Gamma \Rightarrow B, \Diamond\Delta}{\Box\Gamma \Rightarrow \Box B, \Diamond\Delta}$$

$$\mathrm{L}\Diamond\ \frac{\Box\Gamma, A \Rightarrow \Diamond\Delta}{\Box\Gamma, \Diamond A \Rightarrow \Diamond\Delta} \qquad \mathrm{R}\Diamond\ \frac{\Gamma \Rightarrow A, \Delta}{\Gamma \Rightarrow \Diamond A, \Delta}$$

From these rules we easily prove $\Diamond A \leftrightarrow \neg\Box\neg A$, as follows:

$$\dfrac{\dfrac{\dfrac{\dfrac{\dfrac{A \Rightarrow A \quad \bot \Rightarrow}{A, \neg A \Rightarrow}}{A, \Box\neg A \Rightarrow}}{\Diamond A, \Box\neg A \Rightarrow}}{\Diamond A, \Box\neg A \Rightarrow \bot}}{\Diamond A \Rightarrow \neg\Box\neg A} \qquad \dfrac{\dfrac{\dfrac{\dfrac{A \Rightarrow A \quad \bot \Rightarrow}{\Rightarrow \neg A, A}}{\Rightarrow \neg A, \Diamond A}}{\Rightarrow \Box\neg A, \Diamond A} \quad \bot \Rightarrow}{\neg\Box\neg A \Rightarrow \Diamond A}$$

So it suffices to add L$\Box$, R$\Box$ (with $\Delta = \emptyset$ in R$\Box$), or L$\Diamond$, R$\Diamond$ (with $\Gamma = \emptyset$ in L$\Diamond$). ⊠

9.1.3A. ♠ Show the equivalence of the Hilbert system **Hs** with **G1s** in the following sense: treat $\Diamond$ as defined and show $\Gamma \vdash A$ in **Hs** iff **G1s** $\vdash \Gamma \Rightarrow A$, for sets Γ.

9.1.3B. ♠ Formulate a one-sided sequent calculus for **S4**.

9.1.4. DEFINITION. (*The Gentzen system* **G3s**) A version of the sequent calculus where Weakening and Contraction have been built into the other rules is obtained by extending **G3c** with the following rules:

$$\text{L}\Box\ \frac{\Gamma, A, \Box A \Rightarrow \Delta}{\Gamma, \Box A \Rightarrow \Delta} \qquad\qquad \text{R}\Box\ \frac{\Box\Gamma \Rightarrow A, \Diamond\Delta}{\Gamma', \Box\Gamma \Rightarrow \Box A, \Diamond\Delta, \Delta'}$$

$$\text{L}\Diamond\ \frac{\Box\Gamma, A \Rightarrow \Diamond\Delta}{\Gamma', \Box\Gamma, \Diamond A \Rightarrow \Diamond\Delta, \Delta'} \qquad\qquad \text{R}\Diamond\ \frac{\Gamma \Rightarrow A, \Diamond A, \Delta}{\Gamma \Rightarrow \Diamond A, \Delta} \qquad \boxtimes$$

In order to retain the symmetry both $\Box$ and $\Diamond$ have been adopted as primitives.

9.1.4A. ♠ Prove that Contraction is derivable for **G3s**.

9.1.5. THEOREM. *Cut elimination holds for* **G3s**, **G1s**.

PROOF. For **G3s**, we can follow the model of the proof for **G3c**. Let us consider the case where the cut formula $\Box A$ is principal in both premises of the Cut rule, so the deduction with a critical cut as last inference terminates in

$$\dfrac{\dfrac{\begin{array}{c}\mathcal{D}_1\\ \Box\Gamma' \rightarrow A, \Diamond\Delta'\end{array}}{\Gamma'', \Box\Gamma' \Rightarrow \Box A, \Diamond\Delta', \Delta''} \qquad \dfrac{\begin{array}{c}\mathcal{D}_2\\ \Gamma, A, \Box A \Rightarrow \Delta\end{array}}{\Gamma, \Box A \Rightarrow \Delta}}{\Gamma, \Gamma'', \Box\Gamma' \Rightarrow \Delta, \Delta'', \Diamond\Delta'}$$

This is transformed into

$$\dfrac{\begin{array}{c}\mathcal{D}_1\\ \Box\Gamma' \Rightarrow A, \Diamond\Delta'\end{array} \qquad \dfrac{\dfrac{\begin{array}{c}\mathcal{D}_1\\ \Box\Gamma' \Rightarrow A, \Diamond\Delta'\end{array}}{\Gamma'', \Box\Gamma' \Rightarrow \Box A, \Diamond\Delta', \Delta''} \qquad \begin{array}{c}\mathcal{D}_2\\ \Gamma, A, \Box A \Rightarrow \Delta\end{array}}{\Gamma'', \Gamma, \Box\Gamma', A \Rightarrow \Delta, \Diamond\Delta', \Delta''}\,\text{Cut}}{\Gamma, \Gamma'', \Box\Gamma', \Box\Gamma' \Rightarrow \Delta, \Diamond\Delta', \Diamond\Delta', \Delta''}\,\text{Cut}$$

The extra cut is of lower degree, and the rank of the subdeduction ending in the cut of maximal degree is lower than in the original deduction. By the induction hypothesis we know how to eliminate cuts from this deduction; then we use closure under Contraction to obtain a cutfree deduction of the original conclusion.

Cut elimination for **G1s** does not work directly, but via introduction of the "Multicut" rule, as in 4.1.9.

9.1.5A. ♠ Complete the proof of the theorem and describe cut elimination for **G1s** with the help of "Multicut".

9.1.5B. ♠ Formulate a G3-type system for the logic **K**, defined as **S4**, but with only K-axioms, no T- and 4-axioms. Prove a cut elimination theorem for this system.

9.2 Embedding intuitionistic logic into S4

9.2.1. DEFINITION. (*The modal embedding*). The embedding exists in several variants. We describe a variant $^\circ$, and a more familiar variant $^\Box$. The definition is by induction on the depth of formulas (P atomic, not $\bot$):

$$\begin{array}{lll@{\qquad}lll}
P^\circ & := & P & P^\Box & := & \Box P \\
\bot^\circ & := & \bot & \bot^\Box & := & \bot \\
(A \wedge B)^\circ & := & A^\circ \wedge B^\circ & (A \wedge B)^\Box & := & A^\Box \wedge B^\Box \\
(A \vee B)^\circ & := & \Box A^\circ \vee \Box B^\circ & (A \vee B)^\Box & := & A^\Box \vee B^\Box \\
(A \to B)^\circ & := & \Box A^\circ \to B^\circ & (A \to B)^\Box & := & \Box(A^\Box \to B^\Box) \\
(\exists x A)^\circ & := & \exists x \Box A^\circ & (\exists x A)^\Box & := & \exists x A^\Box \\
(\forall x A)^\circ & := & \forall x A^\circ & (\forall x A)^\Box & := & \Box \forall x A^\Box
\end{array}$$

⊠

9.2.2. PROPOSITION. *The two versions of the modal embedding are equivalent in the following sense:* $\mathbf{S4} \vdash \Box A^\circ \leftrightarrow A^\Box$ *and hence* $\mathbf{S4} \vdash \Box\Gamma^\circ \Rightarrow A^\circ$ *iff* $\mathbf{S4} \vdash \Gamma^\Box \Rightarrow A^\Box$. ⊠

9.2.2A. ♠ Prove $\Box A^\circ \leftrightarrow A^\Box$ by induction on the depth of A.

9.2.3. THEOREM. *The embeddings* $^\Box$ *and* $^\circ$ *are sound, i.e. preserve deducibility.*

PROOF. It is completely straightforward to show, by induction on the length of a deduction in **G3i**, that if $\mathbf{G3i} \vdash \Gamma \Rightarrow A$, then $\mathbf{G3s} \vdash \Gamma^\Box \Rightarrow A^\Box$. We consider two typical cases of the induction step, namely where the last rule in the **G3i**-deduction is L→, R→ respectively.

L→: the deduction terminates with

$$\frac{\Gamma, A \to B \Rightarrow A \qquad \Gamma, B \Rightarrow C}{\Gamma, A \to B \Rightarrow C}$$

which is transformed into

$$\cfrac{\cfrac{\Gamma^\Box, \Box(A^\Box \to B^\Box) \Rightarrow A^\Box \qquad \cfrac{\Gamma^\Box, B^\Box \Rightarrow C^\Box}{\Gamma^\Box, \Box(A^\Box \to B^\Box), B^\Box \Rightarrow C^\Box}}{\Gamma^\Box, \Box(A^\Box \to B^\Box), A^\Box \to B^\Box \Rightarrow C^\Box}}{\Gamma^\Box, \Box(A^\Box \to B^\Box) \Rightarrow C^\Box}$$

The transition marked by the dashed line is justified by a weakening transformation applied to the proof of $\Gamma^\Box, B^\Box \Rightarrow C^\Box$.

R$\rightarrow$: We use as a lemma, that $\Box A^\Box \leftrightarrow A^\Box$ for all A (proof by induction on $|A|$). The deduction in **G3i** terminates in

$$\frac{\Gamma, A \Rightarrow B}{\Gamma \Rightarrow A \rightarrow B}$$

which is transformed into

$$\dfrac{\dfrac{\dfrac{\dfrac{\Gamma^\Box, A^\Box \Rightarrow B^\Box}{\Gamma^\Box \Rightarrow A^\Box \rightarrow B^\Box}}{\Box\Gamma^\Box \Rightarrow A^\Box \rightarrow B^\Box}\text{Cuts}}{\Box\Gamma^\Box \Rightarrow \Box(A^\Box \rightarrow B^\Box)}}{\Gamma^\Box \Rightarrow \Box(A^\Box \rightarrow B^\Box)}\text{Cuts}$$

where the "Cuts" are cuts with standard deductions of $\Box C^\Box \Rightarrow C^\Box$, $C^\Box \Rightarrow \Box C^\Box$ $(C \in \Gamma)$.

So apart from these cuts, the transformation of deductions is straightforward. A direct proof of soundness for $^\circ$ requires more complicated cuts. ⊠

9.2.3A. ♠ Complete the proof.

9.2.4. More interesting is the proof of faithfulness of the embeddings. We prove the faithfulness of $^\circ$ via a number of lemmas.

DEFINITION. Let $\mathcal{F}$ be the fragment of **S4** based on $\wedge$, $\vee$, $\rightarrow$, $\bot$, $\Box$. We assign to each deduction $\mathcal{D}$ in the $\mathcal{F}$-fragment of **G3s** a *grade* $\rho(\mathcal{D})$ which counts the applications of rules in $\mathcal{D}$ other than R$\rightarrow$, R$\wedge$, R$\forall$. More precisely, the assignment may be read off from the following schemas, where ρ, ρ' are grades assigned to the premises, and the expression to the left of the conclusion gives the grade of the conclusion, expressed in terms of the grade(s) of the premises.

$\rho(\mathcal{D}) = 0$ for an axiom $\mathcal{D}$

$$\frac{\rho : \Gamma \Rightarrow \Delta}{\rho + 1 : \Gamma' \Rightarrow \Delta'} \text{ for L}\wedge, \text{R}\vee, \text{L}\Box, \text{R}\Box, \text{L}\forall$$

$$\frac{\rho : \Gamma, A \Rightarrow \Delta \qquad \rho' : \Gamma, B \Rightarrow \Delta}{\rho + \rho' + 1 : \Gamma, A \vee B \Rightarrow \Delta} \qquad \frac{\rho : \Gamma \Rightarrow \Delta, A \qquad \rho' : \Gamma, B \Rightarrow \Delta}{\rho + \rho' + 1 : \Gamma, A \rightarrow B \Rightarrow \Delta}$$

$$\frac{\rho : \Gamma, A \Rightarrow B, \Delta}{\rho : \Gamma \Rightarrow A \rightarrow B, \Delta} \qquad \frac{\rho : \Gamma \Rightarrow A[x/y], \forall x A, \Delta}{\rho : \Gamma \Rightarrow \forall x A, \Delta}$$

$$\frac{\rho : \Gamma \Rightarrow A, \Delta \qquad \rho' : \Gamma \Rightarrow B, \Delta}{\rho + \rho' : \Gamma \Rightarrow A \wedge B, \Delta}$$

9.2.5. LEMMA. *(Inversion lemma) Let $\vdash_n \Gamma \Rightarrow \Delta$ mean that $\Gamma \Rightarrow \Delta$ can be proved by a* **G3s**-*derivation $\mathcal{D}$ with $\rho(\mathcal{D}) \leq n$, where the axioms have only atomic principal formulas. Then*

(i) $\vdash_n \Gamma \Rightarrow A \to B, \Delta$ *iff* $\vdash_n \Gamma, A \Rightarrow B, \Delta$;

(ii) $\vdash_n \Gamma \Rightarrow A \wedge B, \Delta$ *iff* $\vdash_n \Gamma \Rightarrow A, \Delta$ *and* $\vdash_n \Gamma \Rightarrow B, \Delta$;

(iii) $\vdash_n \Gamma \Rightarrow \forall x A, \Delta$ *iff* $\vdash_n \Gamma \Rightarrow A[x/y], \Delta$ *(y not free in $\Gamma, \Delta, \forall x A$).*

From this it follows that a cutfree derivation $\mathcal{D}$ of $\Gamma \Rightarrow \Delta$ in the $\mathcal{F}$-fragment may be assumed to consist of derivations of a number of sequents $\Gamma_i \Rightarrow \Delta_i$, with all formulas in Δ_i atomic, disjunctions, existential or modal, followed by applications of R$\wedge$, R$\to$ and R$\forall$ only. ⊠

9.2.5A. ♠ Prove the inversion lemma.

9.2.6. DEFINITION. A formula C is *primitive* if C is either atomic, or a disjunction, or starts with an existential quantifier. ⊠

LEMMA. *Let $^\circ$ be the embedding of 9.2.1. Suppose we have derivations of either*

(a) $\Box\Gamma^\circ, \Delta^\circ \Rightarrow \Box\Lambda^\circ$, *or*

(b) $\Box\Gamma^\circ, \Delta^\circ \Rightarrow \Box\Lambda^\circ, B^\circ$ *(B° primitive).*

Then there are derivations of these sequents where any sequent with more than one formula in the succedent has one of the following two forms:

(i) $\Box\Sigma^\circ, \Delta^\circ \Rightarrow \Box\Theta^\circ, A^\circ$ *with* $|\Theta| \geq 1$, *A° primitive, or*

(ii) $\Box\Sigma^\circ, \Delta^\circ \Rightarrow \Box\Theta^\circ$, *with* $|\Theta| \geq 2$.

PROOF. We apply induction on $\rho(\mathcal{D})$. If $\rho(\mathcal{D}) = 0$, $\mathcal{D}$ is necessarily an axiom and the result is trivial.

Suppose $\rho(\mathcal{D}) > 0$. If a sequent of type (a) is the consequence of a right rule, the rule must be R$\Box$; in this case $\Delta = \emptyset$, $|\Lambda| = 1$, hence the deduction ends with

$$\frac{\Box\Gamma^\circ \Rightarrow A^\circ}{\Box\Gamma^\circ \Rightarrow \Box A^\circ}$$

The premise has been obtained by application of R$\wedge$, R$\to$ and R$\forall$ from deductions $\mathcal{D}_i$ of $\Box\Gamma_i^\circ \Rightarrow A_i^\circ$, A_i° primitive. Now apply the induction hypothesis to the $\mathcal{D}_i$.

If a sequent of type (b) is the conclusion of a right rule, the rule must be R$\vee$ or R$\exists$. Consider the case where the rule is R$\vee$. Then the final rule application is

$$\frac{\Box\Gamma^\circ \Rightarrow \Box\Lambda^\circ, \Box B_0^\circ, \Box B_1^\circ}{\Box\Gamma^\circ \Rightarrow \Box\Lambda^\circ, \Box B_0^\circ \vee \Box B_1^\circ}$$

We can then apply the IH to the premise. Similarly if the last rule applied was R$\exists$.

Now assume that a sequent of type (a) or (b) was obtained by a left rule, for example

$$\frac{\Delta^\circ, \Box\Gamma^\circ \Rightarrow \Box A^\circ, \Box\Lambda^\circ \qquad \Delta^\circ, \Box\Gamma^\circ, B^\circ \Rightarrow \Box\Lambda^\circ}{\Delta^\circ, \Box\Gamma^\circ, \Box A^\circ \to B^\circ \Rightarrow \Box\Lambda^\circ}$$

Then we can apply the IH. Other cases are left to the reader. ⊠

9.2.7. DEFINITION. Let (a), (b), (i), (ii) be as in the preceding lemma. *Standard* derivations are cutfree derivations with conclusions of type (a) or (b) and each sequent with more than one formula in the succedent of types (i) or (ii) only. ⊠

LEMMA. *If* **G3s** $\vdash \Box\Gamma^\circ \Rightarrow A^\circ$, *then there is a cutfree derivation such that all its applications of* R$\to$, R$\forall$ *have at most one formula in the succedent.*

PROOF. By the inversion lemma, $\Box\Gamma^\circ \Rightarrow A^\circ$ can be obtained from deductions $\mathcal{D}_i$ of $\Box\Gamma_i^\circ \Rightarrow A_i^\circ$ with A_i° primitive, using only R$\to$, R$\wedge$, R$\forall$, and by the preceding lemma the $\mathcal{D}_i$ may be assumed to be standard. So if we have somewhere an application of R$\to$ or R$\forall$ with more than one formula in the succedent of the conclusion, the conclusion must be of one of the forms (i), (ii) in the statement of the preceding lemma, which is obviously impossible. ⊠

9.2.8. THEOREM. *(Faithfulness of the modal embedding)*

$$\mathbf{G3s} \vdash \Box\Gamma^\circ \Rightarrow A^\circ \ \textit{iff}\ \mathbf{G3s} \vdash \Gamma^\Box \Rightarrow A^\Box \ \textit{iff}\ \mathbf{Ip} \vdash \bigwedge\Gamma \to A.$$

PROOF. The first equivalence has already been established, so it suffices to show that if $\Box\Gamma^\circ \Rightarrow A^\circ$ then **Ip** $\vdash \bigwedge\Gamma \to A$. Now delete all modalities in a standard derivation of $\Box\Gamma^\circ \Rightarrow A^\circ$; then we find a derivation in a classical sequent calculus in which all applications of R$\to$ and R$\forall$ have one formula in the succedent; it is easy to see that then all derivable sequents are intuitionistically valid, by checking that, whenever $\Gamma \Rightarrow \Delta$ occurs in such a deduction, then $\Gamma \Rightarrow \bigvee\Delta$ is provable in **Ip** (cf. 3.2.1A). ⊠

9.3 Linear logic

There are several possible reasons for studying logics which are, relative to our standard logics, "substructural", that is to say, when formulated as Gentzen systems, do not have all the structural rules.

For example, in systems of relevance logic, one studies notions of formal proof where in the proof of an implication the premise has to be used in an essential way (the premise has to be *relevant* to the conclusion). In such a logic we cannot have Weakening, since Weakening permits us to conclude $B \to A$ from A; B does not enter into the argument at all. On the other hand, Contraction is retained.

In the calculus of Lambek [1958], designed to model certain features of the syntax of natural language, not only Weakening and Contraction are absent, but Exchange as well. Therefore the Lambek calculus has two analogues of implication, A/B and $A\backslash B$. The rules for these operators are (Γ, Δ, Δ' sequences, antecedents of sequents always inhabited)

$$\text{L}\backslash\ \frac{\Gamma \Rightarrow A \qquad \Delta B \Delta' \Rightarrow C}{\Delta \Gamma (A\backslash B) \Delta' \Rightarrow C} \qquad \text{R}\backslash\ \frac{A\Gamma \Rightarrow B}{\Gamma \Rightarrow A\backslash B}$$

$$\text{L}/\ \frac{\Gamma \Rightarrow A \qquad \Delta B \Delta' \Rightarrow C}{\Delta (A/B) \Gamma \Delta' \Rightarrow C} \qquad \text{R}/\ \frac{\Gamma A \Rightarrow B}{\Gamma \Rightarrow A/B}$$

In so-called BCK-logic, Weakening and Exchange are permitted, but Contraction is excluded (see, for example, Ono and Komori [1985]).

If we think of formulas, not as representing propositions, but as types of information, and each occurrence of a formula A as representing a bit of information of type A, we naturally may want to keep track of the use of information; and then several occurrences of A are not equivalent to a single occurrence. These considerations lead to linear logic, introduced in Girard [1987a]. In the Gentzen system for linear logic we have neither Weakening nor Contraction, but Exchange is implicitly assumed since in the sequents $\Gamma \Rightarrow \Delta$ considered the Γ, Δ are multisets, not sequences.

From the viewpoint of structural proof theory, the most interesting aspect of linear logic is that it can be used to obtain more insight into the systems for the standard logics. This is illustrated in a simple case (intuitionistic implication logic) in the next section. See in addition the literature mentioned in the notes

9.3.1. *Conjunction*

In order to set the stage, we consider the rules for conjunction. For both the left and the right rule two obvious possibilities present themselves:

$$\text{L}\wedge\ \frac{\Gamma, A_i \Rightarrow \Delta}{\Gamma, A_0 \wedge A_1 \Rightarrow \Delta}\ (i = 0, 1) \qquad \text{L}\wedge'\ \frac{\Gamma, A_0, A_1 \Rightarrow \Delta}{\Gamma, A_0 \wedge A_1 \Rightarrow \Delta}$$

$$\text{R}\wedge\ \frac{\Gamma \Rightarrow A_0, \Delta \quad \Gamma \Rightarrow A_1, \Delta}{\Gamma \Rightarrow A_0 \wedge A_1, \Delta} \qquad \text{R}\wedge'\ \frac{\Gamma_0 \Rightarrow A_0, \Delta_0 \quad \Gamma_1 \Rightarrow A_1, \Delta_1}{\Gamma_0, \Gamma_1 \Rightarrow A_0 \wedge A_1, \Delta_0, \Delta_1}$$

In the absence of the rules W and C, we have to consider four possible combinations: (L∧,R∧), (L∧′,R∧), (L∧,R∧′), (L∧′,R∧′). However, the second and the third combination yield undesirable results: versions of Weakening and/or Contraction become derivable. For example, combining L∧′ with R∧ permits us to derive Contraction by one application of the Cut rule:

$$\frac{\dfrac{A \Rightarrow A \quad A \Rightarrow A}{A \Rightarrow A \wedge A} \qquad \dfrac{\Gamma, A, A \Rightarrow \Delta}{\Gamma, A \wedge A \Rightarrow \Delta}}{\Gamma, A \Rightarrow \Delta}$$

Therefore only the first and the fourth combination remain. It is easy to see that for each of these combinations the crucial step in cut elimination, where the cut formula has been active in both premises, is possible:

$$\frac{\dfrac{\Gamma \Rightarrow A_0, \Delta \quad \Gamma \Rightarrow A_1, \Delta}{\Gamma \Rightarrow A_0 \wedge A_1, \Delta} \qquad \dfrac{\Gamma', A_0 \Rightarrow \Delta'}{\Gamma', A_0 \wedge A_1 \Rightarrow \Delta'}}{\Gamma, \Gamma' \Rightarrow \Delta, \Delta'}\ \text{Cut}$$

becomes

$$\frac{\Gamma \Rightarrow A_0, \Delta \quad \Gamma', A_0 \Rightarrow \Delta'}{\Gamma, \Gamma' \Rightarrow \Delta, \Delta'}\ \text{Cut}$$

and

$$\frac{\dfrac{\Gamma \Rightarrow A_0, \Delta \quad \Gamma' \Rightarrow A_1, \Delta'}{\Gamma, \Gamma' \Rightarrow A_0 \wedge A_1, \Delta, \Delta'} \qquad \dfrac{\Gamma'', A_0, A_1 \Rightarrow \Delta''}{\Gamma'', A_0 \wedge A_1 \Rightarrow \Delta''}}{\Gamma, \Gamma', \Gamma'' \Rightarrow \Delta, \Delta', \Delta''}\ \text{Cut}$$

becomes

$$\frac{\Gamma' \Rightarrow A, \Delta' \qquad \dfrac{\Gamma \Rightarrow A_0, \Delta \quad \Gamma'', A_0, A_1 \Rightarrow \Delta'}{\Gamma, \Gamma'', A_1 \Rightarrow \Delta, \Delta''}\ \text{Cut}}{\Gamma, \Gamma', \Gamma'' \Rightarrow \Delta, \Delta', \Delta''}\ \text{Cut}$$

"Context-free" or "context-sharing" becomes visible only in the case of rules with several premises. But a secondary criterion is whether rules "mesh together" in proving cut elimination. That is why L∧′ is to be regarded as "context-free": its natural counterpart R∧′ is context-free.

9.3.1A. ♠ Investigate the consequences of the second and third combinations of conjunction rules in more detail; can you get full Weakening and Contraction?

9.3.2. *Gentzen systems for linear logic*

In linear logic Weakening and Contraction are absent. As we have seen, in the absence of these rules, the combinations L$\wedge$, R$\wedge$ and L$\wedge'$, R$\wedge'$ characterize two distinct analogues of conjunction: the context-free *tensor* (notation $\star$), with rules corresponding to (L$\wedge$,R$\wedge$), and the context-sharing *and* (notation $\sqcap$) with rules corresponding to (L$\wedge'$,R$\wedge'$). In a similar way $\vee$ splits into context-free *par* ($+$) and context-sharing *or* ($\sqcup$), $\to$ into context-free *linear implication* ($\multimap$) and context-sharing *additive implication* ($\rightsquigarrow$). There is a single, involutory, *negation* ($\sim$).

The logical constants $\top$ (true) and $\bot$ (false) also split each into two operators, namely $\top$ (true), $\mathbf{1}$ (unit) and $\bot$ (false), $\mathbf{0}$ (zero) respectively. This yields the following set of rules for the "pure" propositional part of classical linear logic:

Logical axiom and Cut rule:

$$\text{Ax}\ A \Rightarrow A \qquad\qquad \text{Cut}\ \frac{\Gamma \Rightarrow A, \Delta \qquad \Gamma', A \Rightarrow \Delta'}{\Gamma, \Gamma' \Rightarrow \Delta, \Delta'}$$

Rules for the propositional constants:

$$\text{L}{\sim}\ \frac{\Gamma \Rightarrow A, \Delta}{\Gamma, {\sim}A \Rightarrow \Delta} \qquad\qquad \text{R}{\sim}\ \frac{\Gamma, A \Rightarrow \Delta}{\Gamma \Rightarrow {\sim}A, \Delta}$$

$$\text{L}\sqcap\ \frac{\Gamma, A_i \Rightarrow \Delta}{\Gamma, A_0 \sqcap A_1 \Rightarrow \Delta}\,(i = 0, 1) \qquad \text{R}\sqcap\ \frac{\Gamma \Rightarrow A, \Delta \qquad \Gamma \Rightarrow B, \Delta}{\Gamma \Rightarrow A \sqcap B, \Delta}$$

$$\text{L}\star\ \frac{\Gamma, A, B \Rightarrow \Delta}{\Gamma, A \star B \Rightarrow \Delta} \qquad\qquad \text{R}\star\ \frac{\Gamma \Rightarrow A, \Delta \qquad \Gamma' \Rightarrow B, \Delta'}{\Gamma, \Gamma' \Rightarrow A \star B, \Delta, \Delta'}$$

$$\text{L}\sqcup\ \frac{\Gamma, A \Rightarrow \Delta \qquad \Gamma, B \Rightarrow \Delta}{\Gamma, A \sqcup B \Rightarrow \Delta} \qquad \text{R}\sqcup\ \frac{\Gamma \Rightarrow A_i, \Delta}{\Gamma \Rightarrow A_0 \sqcup A_1, \Delta}\,(i = 0, 1)$$

$$\text{L}+\ \frac{\Gamma, A \Rightarrow \Delta \qquad \Gamma', B \Rightarrow \Delta'}{\Gamma, \Gamma', A + B \Rightarrow \Delta, \Delta'} \qquad \text{R}+\ \frac{\Gamma \Rightarrow A, B, \Delta}{\Gamma \Rightarrow A + B, \Delta}$$

$$\text{L}\multimap\ \frac{\Gamma \Rightarrow A, \Delta \qquad \Gamma', B \Rightarrow \Delta'}{\Gamma, \Gamma', A \multimap B \Rightarrow \Delta, \Delta'} \qquad \text{R}\multimap\ \frac{\Gamma, A \Rightarrow B, \Delta}{\Gamma \Rightarrow A \multimap B, \Delta}$$

$$\text{L}\rightsquigarrow\ \frac{\Gamma \Rightarrow A, \Delta \quad \Gamma, B \Rightarrow \Delta}{\Gamma, A \rightsquigarrow B \Rightarrow \Delta} \qquad \text{R}\rightsquigarrow\ \frac{\Gamma \Rightarrow B, \Delta}{\Gamma \Rightarrow A \rightsquigarrow B, \Delta}\ \frac{\Gamma, A \Rightarrow \Delta}{\Gamma \Rightarrow A \rightsquigarrow B, \Delta}$$

$$\text{L}\mathbf{1}\ \frac{\Gamma \Rightarrow \Delta}{\Gamma, \mathbf{1} \Rightarrow \Delta} \qquad\qquad \text{R}\mathbf{1}\ \ \Rightarrow \mathbf{1}$$

(no L$\top$) $\qquad$ R$\top$ $\Gamma \Rightarrow \top, \Delta$

L0 $\mathbf{0} \Rightarrow$ $\qquad$ $\text{R}\mathbf{0}\ \dfrac{\Gamma \Rightarrow \Delta}{\Gamma \Rightarrow \Delta, \mathbf{0}}$

L$\bot$ $\Gamma, \bot \Rightarrow \Delta$ $\qquad$ (no R$\bot$)

It is routine to add rules for the quantifiers; in behaviour they are rather like the context-sharing operators (no good context-free versions are known), even if this is not "visible" since there are no multi-premise quantifier rules.

Rules for the quantifiers (y not free in Γ, Δ*):*

$$\text{L}\forall\ \frac{\Gamma, A[x/t] \Rightarrow \Delta}{\Gamma, \forall x\, A \Rightarrow \Delta} \qquad \text{R}\forall\ \frac{\Gamma \Rightarrow A[x/y], \Delta}{\Gamma \Rightarrow \forall x\, A, \Delta}$$

$$\text{L}\exists\ \frac{\Gamma, A[x/y] \Rightarrow \Delta}{\Gamma, \exists x\, A \Rightarrow \Delta} \qquad \text{R}\exists\ \frac{\Gamma \Rightarrow A[x/t], \Delta}{\Gamma \Rightarrow \exists x\, A, \Delta}$$

More interesting is the addition of two operators ! ("of course" or "storage") and ? ("why not" or "reuse") which behave like the two **S4**-modalities $\Box, \Diamond$. These operators re-introduce Weakening and Contraction, but in a controlled way, for specific formulas only. Intuitively "$!A$" means something like "A can be used zero, one or more times", and "$?A$" means "A can be obtained zero, one or more times". The rules express that for formulas $!A$ we have Weakening and Contraction on the left, and for $?A$ we have Weakening and Contraction on the right.

Rules for the exponentials:

$$\text{W!}\ \frac{\Gamma \Rightarrow \Delta}{\Gamma, !A \Rightarrow \Delta} \quad \text{L!}\ \frac{\Gamma, A \Rightarrow \Delta}{\Gamma, !A \Rightarrow \Delta} \quad \text{R!}\ \frac{!\Gamma \Rightarrow A, ?\Delta}{!\Gamma \Rightarrow !A, ?\Delta} \quad \text{C!}\ \frac{\Gamma, !A, !A \Rightarrow \Delta}{\Gamma, !A \Rightarrow \Delta}$$

$$\text{W?}\ \frac{\Gamma \Rightarrow \Delta}{\Gamma \Rightarrow ?A, \Delta} \quad \text{L?}\ \frac{!\Gamma, A \Rightarrow ?\Delta}{!\Gamma, ?A \Rightarrow ?\Delta} \quad \text{R?}\ \frac{\Gamma \Rightarrow A, \Delta}{\Gamma \Rightarrow ?A, \Delta} \quad \text{C?}\ \frac{\Gamma \Rightarrow ?A, ?A, \Delta}{\Gamma \Rightarrow ?A, \Delta}$$

DEFINITION. We denote the Gentzen system of *classical linear logic* by **Gcl**. *Intuitionistic linear logic* **Gil** is the subsystem of **Gcl** where all succedents contain at most one formula, and which does not contain ?, +, **0**, $\sim$. We use **CL**, **IL** for the sets of sequents provable in **Gcl** and **Gil** respectively. ⊠

Examples. We give two examples of deductions in **Gil**:

$$\dfrac{\dfrac{\dfrac{\dfrac{\dfrac{A \Rightarrow A}{A \sqcap B \Rightarrow A}}{!(A \sqcap B) \Rightarrow A}}{!(A \sqcap B) \Rightarrow !A} \quad \dfrac{\dfrac{\dfrac{B \Rightarrow B}{A \sqcap B \Rightarrow B}}{!(A \sqcap B) \Rightarrow B}}{!(A \sqcap B) \Rightarrow !B}}{!(A \sqcap B), !(A \sqcap B) \Rightarrow !A \star !B}}{!(A \sqcap B) \Rightarrow !A \star !B}\,\mathrm{C}!$$

$$\dfrac{\dfrac{\dfrac{\dfrac{\dfrac{A \Rightarrow A \quad C \Rightarrow C}{A \multimap C, A \Rightarrow C}}{(A \multimap C) \sqcap (B \multimap C), A \Rightarrow C} \quad \dfrac{\dfrac{B \Rightarrow B \quad C \Rightarrow C}{B \multimap C, B \Rightarrow C}}{(A \multimap C) \sqcap (B \multimap C), B \Rightarrow C}}{(A \multimap C) \sqcap (B \multimap C), A \sqcup B \Rightarrow C}}{(A \multimap C) \sqcap (B \multimap C) \Rightarrow (A \sqcup B) \multimap C}}{\Rightarrow (A \multimap C) \sqcap (B \multimap C) \multimap ((A \sqcup B) \multimap C)}$$

9.3.3. *De Morgan dualities*

There is a great deal of redundancy in the very symmetric calculus **Gcl**: there is a set of "De Morgan dualities" which permits the elimination of many constants. Using $\vdash A \Leftrightarrow B$ as abbreviation for "$\vdash A \Rightarrow B$ and $\vdash B \Rightarrow A$", we have

$$\begin{array}{llll}
\vdash A \star B \Leftrightarrow {\sim}({\sim}A + {\sim}B), & & \vdash A + B \Leftrightarrow {\sim}({\sim}A \star {\sim}B), \\
\vdash A \sqcap B \Leftrightarrow {\sim}({\sim}A \sqcup {\sim}B), & & \vdash A \sqcup B \Leftrightarrow {\sim}({\sim}A \sqcap {\sim}B), \\
\vdash A \multimap B \Leftrightarrow {\sim}A + B, & & \vdash A \leadsto B \Leftrightarrow {\sim}A \sqcup B, \\
\vdash \forall x A \Leftrightarrow {\sim}\exists x {\sim}A, & & \vdash \exists x A \Leftrightarrow {\sim}\forall x {\sim}A, \\
\vdash !A \Leftrightarrow {\sim}?{\sim}A, & & \vdash ?A \Leftrightarrow {\sim}!{\sim}A.
\end{array}$$

The operators $\star, +, \sqcap$ and $\sqcup$ are associative and commutative modulo $\Leftrightarrow$, and further $\sqcap$ and $\sqcup$ are idempotent modulo $\Leftrightarrow$. The constants $\mathbf{0}, \mathbf{1}, \bot$ and $\top$ behave as neutral elements w.r.t. $+, \star, \sqcup$ and $\sqcap$ respectively (i.e. $\vdash A \star \mathbf{1} \Leftrightarrow A$, $\vdash A \sqcup \bot \Leftrightarrow A$, etc.).

Exploiting the symmetries of the classical calculus, we can introduce a system **GScl** which has one-sided sequents only, just like the **GS**-systems for ordinary classical logic (9.3.3G).

9.3.3A. ♠ Prove in **Gcl** the De Morgan dualities listed above.

9.3.3B. ♠ Prove in **Gcl** for the relevant operators the associativity, idempotency, commutativity, and neutral-element properties. Which of these properties also hold in **Gil**?

9.3.3C. ♠ Prove in **Gil** $!A \star !B \Rightarrow !(A \sqcap B)$ and $(A \sqcup B) \star C \Rightarrow (A \star C) \sqcup (B \star C)$. Show that $(A \sqcup B) \sqcap C \Rightarrow (A \sqcap C) \sqcup (B \sqcap C)$ is not derivable in **Gcl**.

9.3.3D. ♠ A natural deduction system for $\multimap$**IL** is given by the axioms and rules (1) $A \Rightarrow A$, (2) If $\Gamma, A \Rightarrow B$ then $\Gamma \Rightarrow A \multimap B$, (3) If $\Gamma \Rightarrow A \multimap B$ and $\Delta \Rightarrow A$ then $\Gamma\Delta \Rightarrow B$ (Γ, Δ finite multisets). Prove equivalence with **Gil**.

9.3.3E. ♠ Show that a Hilbert system for $\multimap$**IL** is obtained by taking the axiom schemas $(A \multimap B) \multimap ((B \multimap C) \multimap (A \multimap C))$ and $[A \multimap (B \multimap C)] \multimap [B \multimap (A \multimap C)]$, with modus ponens as deduction rule (Troelstra [1992a]). *Hint.* Use the preceding exercise, and derive a deduction theorem for the Hilbert system.

9.3.3F. ♠ The positive ($\mathcal{P}$) and negative ($\mathcal{N}$) contexts in the language of linear logic are defined as for the language of ordinary first-order logic, that is to say that

$$\begin{array}{rcl} \mathcal{P} & = & * \mid A \sqcap \mathcal{P} \mid \mathcal{P} \sqcap A \mid A \star \mathcal{P} \mid \mathcal{P} \star A \mid A + \mathcal{P} \mid \mathcal{P} + A \mid A \sqcup \mathcal{P} \\ & & \mid \mathcal{P} \sqcup A \mid A \multimap \mathcal{P} \mid \mathcal{N} \multimap A \mid A \rightsquigarrow \mathcal{P} \mid \mathcal{N} \rightsquigarrow A \mid \sim\mathcal{N} \\ \mathcal{N} & = & A \sqcap \mathcal{N} \mid \mathcal{N} \sqcap A \mid A \star \mathcal{N} \mid \mathcal{N} \star A \mid A + \mathcal{N} \mid \mathcal{N} + A \mid A \sqcup \mathcal{N} \\ & & \mid \mathcal{N} \sqcup A \mid A \multimap \mathcal{N} \mid \mathcal{P} \multimap A \mid A \rightsquigarrow \mathcal{N} \mid \mathcal{P} \rightsquigarrow A \mid \sim\mathcal{P} \end{array}$$

Let $F[*], G[*]$ be a positive and a negative context respectively, and let $\vec{z}$ be a list of variables free in B or C but bound in $F[B]$ or $F[C]$; similarly, let $\vec{u}$ be a list of variables free in B or C but bound in $G[B]$ or $G[C]$. Prove that in **Gcl**, **Gil** without exponentials $!, ?$ (Troelstra [1992a, 3.10])

$$\begin{array}{l} \vdash \mathbf{1} \sqcap \forall \vec{z}(B \multimap C) \Rightarrow F[B] \multimap F[C], \\ \vdash \mathbf{1} \sqcap \forall \vec{u}(B \multimap C) \Rightarrow G[C] \multimap G[B]. \end{array}$$

In the full theories we only have

$$\text{If } \vdash (B \multimap C) \text{ then } \vdash F[B] \multimap F[C] \text{ and } G[C] \multimap G[B].$$

9.3.3G. ♠ Write down a system **GScl** with one-sided sequents for **CL**, similar to the GS-systems for **C**.

9.3.3H. ♠ (Approximation theorem, Girard [1987a]) Prove the following result. Let

$$\begin{array}{l} !_n A := (\mathbf{1} \sqcap A) \star \ldots (n \text{ times }) \ldots \star (\mathbf{1} \sqcap A) \\ ?_n A := (\mathbf{0} \sqcup A) + \ldots (n \text{ times }) \ldots + (\mathbf{0} \sqcup A). \end{array}$$

Suppose that we have shown **GScl** $\vdash \Gamma$ (**GScl** as in the preceding exercise), and assume each occurrence α of $!$ in Γ to have been assigned a label $n(\alpha) \in \mathbb{N} \setminus \{0\}$. Then we can assign to each occurrence β of $?$ a label $n(\beta) \in \mathbb{N} \setminus \{0\}$ such that if Γ' is obtained from Γ by replacing every occurrence α of $!$ and every occurrence β of $?$ by $!_{n(\alpha)}, ?_{n(\beta)}$ respectively, then in **GScl** without $!$, $?$ we have $\vdash \Gamma'$. *Hint.* Use induction on the depth of deductions, starting from *atomic* instances of the axioms, and use the monotonicity laws of exercise 9.3.3F.

9.3.4. THEOREM. *Cut elimination holds for* **Gcl** *and* **Gil**.

PROOF. The proof is more or less standard, and similar to the proof for **G3s**. However, we cannot "absorb contraction", so we need an analogue of the Multicut rule. For example, one would like to replace

$$\dfrac{\dfrac{\begin{array}{c}\mathcal{D}\\ !\Gamma \Rightarrow A, ?\Delta\end{array}}{!\Gamma \Rightarrow !A, ?\Delta} \quad \dfrac{\begin{array}{c}\mathcal{D}'\\ \Gamma', !A, !A \Rightarrow \Delta'\end{array}}{\Gamma', !A \Rightarrow \Delta'}}{!\Gamma, \Gamma' \Rightarrow ?\Delta, \Delta'}\,\text{Cut}$$

by

$$\dfrac{\dfrac{\dfrac{\begin{array}{c}\mathcal{D}\\ !\Gamma \Rightarrow A, ?\Delta\end{array}}{!\Gamma \Rightarrow !A, ?\Delta} \quad \dfrac{\dfrac{\begin{array}{c}\mathcal{D}\\ !\Gamma \Rightarrow A, ?\Delta\end{array}}{!\Gamma \Rightarrow !A, ?\Delta} \quad \begin{array}{c}\mathcal{D}'\\ \Gamma', !A, !A \Rightarrow \Delta'\end{array}}{!A, !\Gamma, \Gamma' \Rightarrow ?\Delta, \Delta'}\,\text{Cut}}{\Gamma, !\Gamma, \Gamma' \Rightarrow ?\Delta, ?\Delta, \Delta'}\,\text{Cut}}{!\Gamma, \Gamma' \Rightarrow ?\Delta, \Delta'}\,\text{C!,C?}$$

where the double line stands for a number of C!- and C?-applications. This does not work because the lower cut will have the same height as the original one. The solution is to permit certain derivable generalizations of the Cut rule, similar to "Multicut":

$$\dfrac{\Gamma \Rightarrow \Delta, !A \qquad \Gamma, (!A)^n, \Rightarrow !A, \Delta'}{\Gamma, \Gamma' \Rightarrow \Delta, \Delta'}\,\text{Cut! } (n > 1)$$

$$\dfrac{\Gamma \Rightarrow (?A)^n, \Delta \qquad \Gamma', ?A \Rightarrow \Delta'}{\Gamma, \Gamma' \Rightarrow \Delta, \Delta'}\,\text{Cut? } (n > 1)$$

Let us write "Cut*" for either Cut?, Cut!, or Cut. Now we can transform the deduction above into

$$\dfrac{\dfrac{\begin{array}{c}\mathcal{D}\\ !\Gamma \Rightarrow A, ?\Delta\end{array}}{!\Gamma \Rightarrow !A, ?\Delta} \quad \begin{array}{c}\mathcal{D}'\\ \Gamma', !A, !A \Rightarrow \Delta'\end{array}}{!\Gamma, \Gamma' \Rightarrow ?\Delta, \Delta'}\,\text{Cut!}$$

If the cutformula is not principal in at least one of the premises of the terminal cut, we have to permute Cut* upwards over a premise where the cut formula is not principal. This works as usual, except where the Cut* involved is Cut! or Cut? and the multiset $(!A)^n$ or $(?A)^n$ removed by Cut! or Cut? is derived from *two* premises of a context-free rule (R$\star$, L+, L$\multimap$ or Cut*); a representative example is

$$\dfrac{\Gamma'' \Rightarrow !A, \Delta'' \qquad \dfrac{\Gamma, (!A)^p \Rightarrow \Delta, B \quad \Gamma', (!A)^q \Rightarrow \Delta', C}{\Gamma, \Gamma', (!A)^{p+q} \Rightarrow \Delta, \Delta', B \star C}\,\text{R}\star}{\Gamma, \Gamma', \Gamma'' \Rightarrow \Delta, \Delta', \Delta'', B \star C}\,\text{Cut}^*$$

If either $p = 0$ or $q = 0$, there is no difficulty in permuting Cut! upwards on the right. But if $p, q > 0$, then in this case cutting $\Gamma'' \Rightarrow !A, \Delta''$ with both the upper sequents on the right, followed by R$\star$, leaves us with duplicated Γ'', Δ''. To get out of this difficulty, we look at the premise on the left. There are two possibilities. If $!A$ is not principal in the left hand premise of the Cut!, we can permute Cut! upwards on the left. The obstacle which prevented permuting with the right premise does not occur here, since only a single occurrence of $!A$ is involved. On the other hand, if $!A$ is principal in the left hand premise, we must have $\Gamma'' \equiv !\Gamma'''$, $\Delta'' \equiv ?\Delta'''$ for suitable Γ''', Δ''', and we may cut with the upper sequents on the right, followed by contractions of $!\Gamma''', !\Gamma'''$ into $!\Gamma'''$ and of $?\Delta''', ?\Delta'''$ into $?\Delta'''$, and an application of the multiplicative rule (R$\star$ in our example). ⊠

9.3.4A. ♠ Complete the proof (cf. Lincoln et al. [1992]).

9.3.4B. ♠ Devise a variant of **Gcl** where the W- and C-rules for the exponentials have been absorbed into the other rules, and which permits cut elimination. *Hint.* Cf. **G3s**.

9.3.5. PROPOSITION. *A fragment of* **CL** *determined by a subset* $\mathcal{L}$ *of* $\{\star, \multimap, \sqcap, \sqcup, \mathbf{1}, \bot, \top, \forall, \exists, !\}$ *is conservative over* **IL** *(i.e. if* **CL** *restricted to* $\mathcal{L}$ *proves* A*, then so does* **IL** *restricted to* $\mathcal{L}$*) iff* $\mathcal{L}$ *does not include both* $\multimap$ *and* $\bot$.

PROOF. $\Leftarrow$ Suppose that $\bot$ is not in $\mathcal{L}$, let $\mathcal{D}$ be a cutfree deduction of $\Gamma \Rightarrow A$, and assume that $\mathcal{D}$ contains a sequent with a succedent consisting of more than one formula. This can happen only if there is an application of L$\multimap$ of the form

$$\frac{\Gamma \Rightarrow A, C \qquad \Gamma', B \Rightarrow}{\Gamma, \Gamma', A \multimap B \Rightarrow C}$$

We can then find a branch in the deduction tree with empty succedents only. This branch must end in an axiom with empty succedent, which can only be L0 or L$\bot$. The first possibility is excluded since the whole deduction is carried out in a sublanguage of **Gil**; the second possibility is excluded by assumption.

If $\multimap$ is not in $\mathcal{L}$, we can prove, by a straightforward induction on the length of a cutfree deduction of a sequent $\Delta \Rightarrow B$, that all sequents in the deduction have a single consequent.

$\Rightarrow$ As to the converse, in the fragment $\{\multimap, \bot\}$ of **Gcl** we can prove (P, Q, R, S atomic)

$$P \multimap ((\bot \multimap Q) \multimap R),\ (P \multimap S) \multimap \bot \Rightarrow R$$

This sequent does not have a cutfree deduction in **Gil**. ⊠

9.3.5A. ♠ Construct a **Gcl**-derivation for the sequent mentioned in the proof.

9.3.6. DEFINITION. The *Girard embedding* of **I** into **CL** is defined by

$$\begin{array}{lll} P^* & := & P \ (P \text{ atomic}), \\ \bot^* & := & \bot, \\ (A \wedge B)^* & := & A^* \sqcap B^*, \\ (A \vee B)^* & := & !A^* \sqcup !B^*, \\ (A \to B)^* & := & !A^* \multimap B^*, \\ (\exists x A)^* & := & \exists x !A^*, \\ (\forall x A)^* & := & \forall x A^*. \end{array}$$

⊠

CL is at least as expressive as the standard logics, as shown by the following result:

9.3.7. THEOREM. *If* $\mathbf{Gcl} \vdash \Gamma^* \Rightarrow A^*$, *then* $\mathbf{G1i} \vdash \Gamma \Rightarrow A$.

PROOF. Almost immediate from the observations that (1) if we replace $\sqcap, \sqcup, !$ by $\wedge, \vee, \Box$ in the *-clauses, and we read $^\circ$ for *, we obtain the faithful embedding of **I** into **S4** of 9.2.1, and (2) ! behaves like $\Box$ in **S4**, and all the laws for $\forall, \exists, \multimap, \sqcap, +, \bot$ in **Gcl** are also laws for their analogues $\forall, \exists, \to, \wedge, \vee, \bot$ in **S4**. This second fact has as a consequence that if $\mathbf{Gcl} \vdash \Gamma^* \Rightarrow A^*$, then $\mathbf{S4} \vdash \Gamma^\circ \Rightarrow A^\circ$ for the modal embedding $^\circ$, hence $\mathbf{G1i} \vdash \Gamma \Rightarrow A$. ⊠

9.4 A system with privileged formulas

In this section we illustrate how linear logic may be used to encode "fine structure" of **G1c** and **G1i**. We consider a version of the intuitionistic Gentzen system for implication, rather similar to →h-**GKi** of 6.3.5, in which at most one of the antecedent formulas is "privileged". That is to say, sequents are of the form

$$\Pi; \Gamma \Rightarrow A, \text{ with } |\Pi| \leq 1.$$

So Π contains at most one formula. To improve readability we write $-;\Gamma$ ($\Pi; -$) for a $\Pi;\Gamma$ with empty Π (empty Γ). The rules are suggested by thinking of $\Pi;\Gamma \Rightarrow A$ as $\Pi^*, !\Gamma^* \Rightarrow A^*$ in linear logic, where * is the Girard embedding.

9.4.1. DEFINITION. The system **IU** ("U" from "universal") is given by
Axioms

$$A; - \Rightarrow A$$

Logical rules

$$\text{L}\!\to \frac{-;\Gamma \Rightarrow A \qquad B;\Gamma' \Rightarrow C}{A \to B; \Gamma, \Gamma' \Rightarrow C} \qquad \text{R}\!\to \frac{\Pi; \Gamma, A \Rightarrow B}{\Pi; \Gamma \Rightarrow A \to B}$$

Structural rules

$$\text{LW}\ \frac{\Pi;\Gamma\Rightarrow A}{\Pi;\Gamma,B\Rightarrow A}\qquad \text{LC}\ \frac{\Pi;\Gamma,B,B\Rightarrow A}{\Pi;\Gamma,B\Rightarrow A}\qquad \text{D}\ \frac{B;\Gamma\Rightarrow A}{-;B,\Gamma\Rightarrow A}\qquad\boxtimes$$

REMARK. An equivalent system is obtained by taking the context-sharing version of L→. Observe also that $A\to B$ can never be introduced by L→, if B has been introduced by Weakening or Contraction. Thus deleting semicolons in all sequents of an **IU**-proof, we have, modulo some repetitions caused by D, a restricted kind of intuitionistic deduction.

9.4.2. PROPOSITION. *The system* **IU** *is closed under the rules "Headcut" (Cut_h) and "Midcut" (Cut_m):*

$$\text{Cut}_\text{h}\ \frac{\Pi;\Gamma\Rightarrow A\qquad A;\Gamma'\Rightarrow B}{\Pi;\Gamma,\Gamma'\Rightarrow B}\qquad \text{Cut}_\text{m}\ \frac{-;\Gamma\Rightarrow A\qquad \Pi;A,\Gamma'\Rightarrow B}{\Pi;\Gamma,\Gamma'\Rightarrow B}\qquad\boxtimes$$

9.4.2A. ♠ Show that **IU** is closed under Cut_h and Cut_m.

9.4.3. THEOREM. *If* $\mathbf{G1i}\vdash\Gamma\Rightarrow A$ *then* $\mathbf{IU}\vdash -;\Gamma\Rightarrow A$.

PROOF. By induction on the length of deductions $\mathcal{D}$ of $\Gamma\Rightarrow A$ in **G1i**. Problems can arise only when the last rule in $\mathcal{D}$ is L→ or L∀. So suppose $\mathcal{D}$ to end with

$$\frac{\Gamma\Rightarrow A\qquad \Gamma,B\Rightarrow C}{\Gamma,A\to B\Rightarrow C}$$

By IH we have in **IU**

$$-;\Gamma\Rightarrow A\quad\text{and}\quad -;\Gamma,B\Rightarrow C$$

So we cannot apply L→ directly, since this would require a premise $B;\Gamma\Rightarrow C$. However, we can derive $-;A\to B,A\Rightarrow B$:

$$\cfrac{\cfrac{\cfrac{A;-\Rightarrow A}{-;A\Rightarrow A}\qquad B;-\Rightarrow B}{A\to B;A\Rightarrow B}}{-;A\to B,A\Rightarrow B}$$

and then we may construct a derivation with the help of cuts

$$\cfrac{\cfrac{\cfrac{-;\Gamma\Rightarrow A\qquad -;A,A\to B\Rightarrow B}{-;\Gamma,A\to B\Rightarrow B}\text{Cut}_\text{m}\qquad -;\Gamma,B\Rightarrow C}{-;\Gamma,\Gamma,A\to B\Rightarrow C}\text{Cut}_\text{m}}{-;\Gamma,A\to B\Rightarrow C}\text{LC}\qquad\boxtimes$$

As noted above, the rule L$\to$ of **IU** is motivated by the Girard embedding: if we read $B; \Gamma \Rightarrow C$ as $B^*, !\Gamma^* \Rightarrow C^*$, it precisely corresponds to

$$\dfrac{\dfrac{!\Gamma^* \Rightarrow A^*}{!\Gamma^* \Rightarrow !A^*} \quad !\Delta^*, B^* \Rightarrow C^*}{!\Gamma^*, !\Delta^*, !A^* \multimap B^* \Rightarrow C^*}$$

It is easy to see that the soundness of * for **IU** can be proved by a straightforward step-by-step deduction transformation, without the need for auxiliary cuts.

This contrasts with the proof of soundness of * for **G1i**. In that case we need cuts to correct matters (cf. the proof of 9.2.3). For example,

$$\dfrac{C \Rightarrow C \quad \dfrac{A \Rightarrow A}{B, A \Rightarrow A}}{C, C \to B, A \Rightarrow A}$$

leads to

$$\dfrac{\dfrac{\dfrac{\dfrac{\dfrac{!C^* \Rightarrow !C^* \quad B^* \Rightarrow B^*}{!C^*, !C^* \multimap B^* \Rightarrow B^*}}{!(!C^* \multimap B^*), !C^* \Rightarrow B^*}}{!(!C^* \multimap B^*), !C^* \Rightarrow !B^*}}{!(!C^* \multimap B^*) \Rightarrow !C^* \to !B^*} \quad \dfrac{!C^* \Rightarrow !C^* \quad \dfrac{\dfrac{A^* \Rightarrow A^*}{!A^* \Rightarrow A^*}}{!B^*, !A^* \Rightarrow A^*}}{!C^*, !C^* \multimap !B^*, !A^* \Rightarrow A^*}\,\text{L}\multimap}{!C^*, !(!C^* \multimap B^*), !A^* \Rightarrow A^*}\,\text{Cut}$$

After translating and eliminating the auxiliary cut we are left with

$$\dfrac{\dfrac{A^* \Rightarrow A^*}{!A^* \Rightarrow A^*}}{!C^*, !(!C^* \multimap B^*), !A^* \Rightarrow A^*}\,\text{W}$$

deriving from a deduction with weakenings alone, not requiring auxiliary cuts under translation. This suggests that translated deductions of **G1i** after elimination of the auxiliary cuts will correspond to deductions in **IU**; this impression is confirmed by the following result.

9.4.4. PROPOSITION. *If $\mathcal{D}$ is a deduction in the $!\multimap$-fragment of* **Gcl** *of a sequent $\Pi^*, !\Gamma^* \Rightarrow !\Sigma^*, \Delta^*$, where all cutformulas are of the form A^* or $!A^*$, and all identity axioms of the form $A^* \Rightarrow A^*$, then the skeleton of $\mathcal{D}$, $\mathrm{sk}(\mathcal{D})$, obtained by replacing sequents $\Pi^*, !\Gamma^* \Rightarrow !\Sigma^*, \Delta^*$ by $\Pi; \Gamma \Rightarrow \Sigma, \Delta$, is an* **IU***-deduction modulo possible repetitions of sequents.*

PROOF. We prove the statement of the theorem, together with $|\Pi \cup \Sigma| \leq 1$ and $|\Sigma \cup \Delta| \leq 1$ simultaneously by induction on the length of $\mathcal{D}$.
Basis. For an axiom $A^* \Rightarrow A^*$, with skeleton $-; A \Rightarrow A$, all conditions are met.

Induction step. Case 1. Suppose the last step is L⊸:

$$\frac{\Pi_0^*, !\Gamma_0^* \Rightarrow !A^*, !\Sigma_0^*, \Delta_0^* \qquad B^*, \Pi_1^* \Rightarrow !\Sigma_1^*, \Delta_1^*}{\Pi_0^*, \Pi_1^*, !A^* \multimap B^*, !\Gamma_0^*, !\Gamma_1^* \Rightarrow !\Sigma_o^*, !\Sigma_1^*, \Delta_0^*, \Delta_1^*}$$

With the IH we see that

$$\Pi_i = \Sigma_0 = \Delta_0 = \emptyset, \ \ |\Sigma_1 \cup \Delta_1| = 1.$$

Case 2. Suppose the last rule is L!, so $\mathcal{D}$ ends with

$$\frac{\Pi^*, A^*, !\Gamma^* \Rightarrow !\Sigma^*, \Delta^*}{\Pi^*, !A^*, !\Gamma^* \Rightarrow !\Sigma^*, \Delta^*}$$

So by IH $|\Sigma \cup \Delta| = 1$, $|\Sigma| = 0$, hence $|\Delta| = 1$, and hence $A; \Gamma \Rightarrow \Delta$; then $-; A, \Gamma \Rightarrow \Delta$ by dereliction D.
Case 3. The last step is R!

$$\frac{!\Gamma^* \Rightarrow A^*}{!\Gamma^* \Rightarrow !A^*}$$

By the IH we can derive $-; \Gamma \Rightarrow A$ which is the skeleton of the conclusion. ⊠

9.4.4A. ♠ Extend **IU** so as to cover the other propositional operators as well.

9.5 Proofnets

9.5.1. In this section we present Girard's notion of proofnet for the context-free ("multiplicative") fragment containing only $+, \star$ as operators, and $\sim$ as a defined operation; formulas are constructed from *positive literals* $P, Q, P', \dots$ and *negative literals* $\sim P, \sim Q, \sim P', \dots$ by means of $+$ and $\star$, using the De Morgan symmetries of 9.3.3 for defining $\sim A$ for compound A (as with the GS-calculi for **C**). As our starting point we take the one-sided Gentzen system for this fragment, with axioms and rules

$$A, \sim A \qquad \frac{A, \Gamma \qquad B, \Delta}{A \star B, \Gamma, \Delta} \qquad \frac{A, B, \Gamma}{A + B, \Gamma} \qquad \frac{\Gamma, A \qquad \Delta, \sim A}{\Gamma, \Delta}\,\text{Cut}$$

Just as in the Gentzen systems for our standard logics, cutfree sequent proofs may differ in the order of the application of the rules, e.g. the two deductions

$$\cfrac{\cfrac{\cfrac{A, \sim A \quad B, \sim B}{A \star B, \sim A, \sim B} \quad C{\sim}C}{(A \star B) \star C, \sim A, \sim B, \sim C}}{(A \star B) \star C, \sim A + \sim B, \sim C} \qquad \cfrac{\cfrac{\cfrac{A, \sim A \quad B, \sim B}{A \star B, \sim A, \sim B}}{A \star B, \sim A + \sim B} \quad C{\sim}C}{(A \star B) \star C, \sim A + \sim B, \sim C}$$

represent "essentially" the same proof: only the order of the application of the rules differs. The proofs also exhibit a lot of redundancy inasmuch as the inactive formulas are copied many times. There is also a good deal of non-determinism in the process of cut elimination, due to the possibilities for permuting cut upwards either to the left or to the right.

Proofnets were introduced by Girard [1987a] in order to remove such redundancies and the non-determinism in cut elimination, and to find a unique representative for equivalent deductions in **Gcl**. In the same manner as **Ni** improves upon **G1i**, proofnets improve on **Gcl**. For the full calculus **Gcl** proofnets are complicated, but for the $+\star$-fragment there is a simple and satisfying theory. Proofnets are labelled graphs, and therefore we define in the next subsection some graph-theoretic notions for later use.

9.5.2. *Notions from graph theory*

DEFINITION. A *graph* is a pair $G \equiv (X, R)$, where X is a set, and R is an irreflexive symmetric binary relation on X (R irreflexive means $\forall x \neg Rxx$, and R symmetric means $\forall xy(Rxy \to Ryx)$). The elements of X are the *nodes* or *vertices* of the graph, and the pairs in R are the *edges* of the graph. (X', R') is a *subgraph* of (X, R) if $X' \subset X$ and $R' \subset R$.

We shall use lower case letters $x, y, z, \ldots$ for nodes; an edge (x, y) is simply written as xy or yx.

A node is of *order* k if it belongs to exactly k different edges. ⊠

In the usual way finite graphs may be represented as diagrams with the nodes as black dots; dots x, y are connected by a line segment if Rxy. For example the picture below is a graph with five nodes and seven edges (so the crossing of the two diagonal lines does not count as a node).

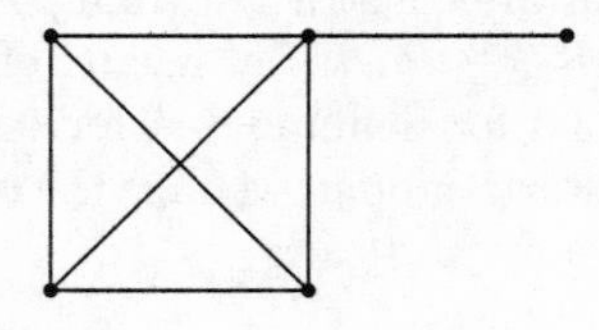

DEFINITION. A *path* $x_1 x_2 \ldots x_n$ in a graph G is a sequence of nodes x_1, x_2, $\ldots$, x_n with $n > 1$, and $x_i x_{i+1}$ an edge for $1 \leq i < n$. A path $x_1 x_2 \ldots x_n$ is said to be a path *from* x_1 *to* x_n. A *subpath* of a path $x_1 x_2 \ldots x_n$ is a path $x_i x_{i+1} x_{i+2} \ldots x_j$ with $1 \leq i < j \leq n$.

A *cycle* is a path of the form $x_1 x_2 \ldots x_n x_1$, such that for no $i \neq j$ $x_i = x_j$ (so a cycle cannot contain a proper subpath which is a cycle).

For the *concatenation* $x_1 \ldots x_n x_{n+1} \ldots x_m$ of paths $\alpha = x_1 \ldots x_n$ and $\beta = x_{n+1} \ldots x_m$ we write $\alpha \frown \beta$. ⊠

REMARK. A cycle is sometimes defined as a path of the form $x_1x_2 \dots x_nx_1$; cycles satisfying our extra condition above are then called *simple cycles.*

DEFINITION. A graph is *connected* if for each pair of nodes x, y there is a path from x to y. The *component* of a node x in G is the largest connected subgraph containing x. A *tree* (-graph) is a graph which is connected and contains no cycles. ⊠

9.5.3. *Proof structures*

A proof structure with hypotheses is a graph with nodes labelled by formulas or by the symbol "cut", built from the following *components*, each consisting of 1, 2 or 3 labelled nodes with 0, 1 or 2 edges between them:

- single nodes labelled with a formula H (a *hypothesis*); the conclusion and premise of this component are H.
- *axiom links* A——$\sim A$, with conclusions $A, \sim A$ and no premises;
- *cut links* A——cut——$\sim A$ with premise $A, \sim A$ and no conclusions;
- logical links with premises A, B, namely *tensor links* or $\star$-links A——$A \star B$——B, with conclusion $A \star B$, and *par links* or $+$-links A——$A + B$——B with conclusion $A + B$.

So edges with cut, $A \star B$ or $A + B$ always appear in pairs. More precisely we define proof structures as follows:

DEFINITION. The notion of a *proof structure* and the notion of the *set of conclusions of a proof structure* are defined simultaneously. We write $\nu, \mu, \dots$ for proof structures. CON(ν) is the set of conclusions of ν, a subset of the labelled nodes of ν. Let $\nu, A, B, \dots D$ indicate a proof structure (PS) ν with some of its conclusions labelled $A, B, \dots D$. We shall indulge in a slight abuse of notation in frequently using the labels of the nodes to designate the nodes themselves. Proof structures are generated by:

(i) (hypothesis clause) a single node labelled with a formula H is a PS with conclusion H;

(ii) (axiom clause) A——$\sim A$ is a PS (axiom), with conclusions $A, \sim A$;

(iii) (join clause) if ν, μ are PS's, then so is $\nu \cup \mu$, with $\text{CON}(\nu \cup \mu) = \text{CON}(\nu) \cup \text{CON}(\mu)$;

(iv) (cut clause) if $\nu, A, \sim A$ is a PS, so is the graph obtained by adding edges and the symbol "cut" A——cut——$\sim A$; the new conclusions are $\text{CON}(\nu) \cup \{\text{cut}\} \setminus \{A, \sim A\}$ (i.e. conclusions of ν except $A, \sim A$, and "cut" added);

(v) ($\star$-clause, $+$-clause) if ν, A, B is a PS, then so are the graphs obtained by adding two edges and a node A —— $A \star B$ —— B ($\star$-link) or A —— $A + B$ —— B ($+$-link). The conclusions are CON(ν) with A, B omitted and $A \star B$, respectively $A + B$ added.

The hypotheses of a PS are simply all the nodes which went into the construction by the hypothesis clause. Sometimes we shall use the expression "the conclusions of a PS" also for the multiset of labels corresponding to the set of conclusions. A notation for PS which is closer to our usual deduction notation is obtained by the following version of the definition (with the obvious clauses for terminal nodes):

(i)′ $\overline{H}$ is a hypothesis for any multiplicative formula H;

(ii)′ $\overline{A \qquad {\sim}A}$ is a PS (axiom link);

(iii)′ the union of two PS's is a PS;

(iv)′ connecting conclusions A, B in a PS by

$$\frac{A \quad B}{A \star B} \text{ or } \frac{A \quad B}{A + B}$$

gives a new PS (adding a $\star$-link and a $+$-link respectively);

(v)′ connecting terminal nodes A, ${\sim}A$ in a PS by

$$\frac{A \quad {\sim}A}{\text{cut}}$$

gives a new PS. ⊠

REMARK. Alternatively, a PS may be globally characterized as a finite graph, with nodes labelled with formulas or "cut"; every node is either a hypothesis or conclusion of a unique link, and is the premise of at most one link. A formula which is not the premise of another link is a *conclusion*.

A hypothesis which is a conclusion is of degree 0, otherwise of degree 1; axioms are of degree 1 when conclusions of the PS, otherwise of degree 2; and conclusions of $+, \star$-links are of degree 2 when conclusions of the PS, otherwise of degree 3.

9.5.3A. ♠ Show that the global characterization of a PS in the remark is equivalent to the inductive characterization in the definition of PS.

EXAMPLE. The two deductions at the beginning of this section are both represented by the following PS without hypotheses, and with conclusions $\sim A + \sim B$, $(A \star B) \star C$, $\sim C$.

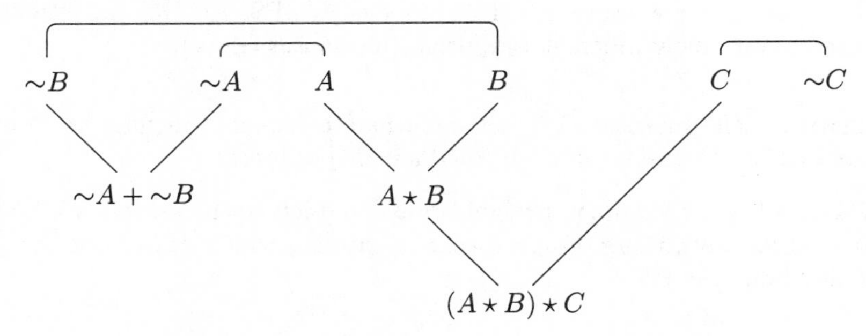

If we delete any single edge entering one of the nodes $\sim A + \sim B$, $A \star B$, or $(A \star B) \star C$, the result is no longer a PS. If we delete both edges in $\sim A + \sim B$, or any of the axiom links, the result is again a PS, but now with hypotheses.

Another graphic representation of the PS exhibited above:

$$\dfrac{\sim A \qquad \sim B}{\sim A + \sim B} \qquad \dfrac{\dfrac{A \qquad B}{A \star B} \qquad \overline{C \qquad \sim C}}{(A \star B) \star C}$$

A certain subset of the PS's, the set of *inductive PS's* corresponds in an obvious way to sequent deductions.

9.5.4. DEFINITION. *Inductive PS's (IPS's)* are obtained by the clauses:

(i) A single node with label H (a *hypothesis*) is an IPS;

(ii) A —— $\sim A$ is an IPS (axiom), with $A, \sim A$ as conclusions;

(iii) if ν, A and $\nu', \sim A$ are IPS's, then so is ν, A —— cut —— $\sim A, \nu'$ (cut link: two new edges and a node labelled "cut"), the conclusions are $(\mathrm{CON}(\nu) \cup \mathrm{CON}(\nu') \cup \{\mathrm{cut}\}) \setminus \{A, \sim A\}$;

(iv) if ν, A and ν', B are IPS's, then so is ν, A —— $A \star B$ —— B, ν' (two new edges and a node $A \star B$), the conclusions are $(\mathrm{CON}(\nu) \cup \mathrm{CON}(\nu') \cup \{A \star B\}) \setminus \{A, B\}$;

(v) if ν, A, B is an IPS, then so is

$$\begin{array}{c} \nu, A, B \\ \backslash\,/ \\ A + B \end{array}$$

(two new edges and a node $A + B$), conclusions as in the corresponding clause of the preceding definition. ⊠

N.B. The example above of a PS is in fact an IPS. An IPS can usually be generated in many different ways from the clauses (i)–(v).

LEMMA. *The multiset* Γ *is derivable in the sequent calculus iff* Γ *is the multiset of conclusions of an IPS without hypotheses.* ⊠

The proof of the lemma is straightforward and left to the reader. We are now looking for an intrinsic, graph-theoretic criterion which singles out the IPS's from among the PS's.

9.5.5. DEFINITION. A *switching* of a PS is a graph obtained by omitting one of the two edges of every +-link of the PS. We call a PS a *proofnet* if every switching is a tree. ⊠

We shall now establish that a PS is a proofnet iff it is an IPS. One side is easy:

9.5.6. LEMMA. *Each IPS is a proofnet.* ⊠

9.5.6A. ♠ Prove the lemma by induction on the generation of an IPS.

In order to to prove the converse we move to a more abstract setting and prove a graph-theoretic theorem. In this connection it is useful to observe that we need not consider proofnets with cut links: if we replace in a PS a cut link A —— cut —— $\sim A$ by a $\star$-link, A —— $A \star \sim A$ —— $\sim A$, the result is again a PS which is a proofnet iff the original PS was a proofnet.

9.5.7. *Abstract proof structures*

DEFINITION. An *abstract proof structure* (APS) $S \equiv (N, E, P)$ is a triple with (N, E) a finite graph, P set of pairs (xy, xz) of edges with $y \neq z$; x is the *basis* of the pair. A node can be the basis of at most one pair in P. An element of P is said to be a *pair of* S. The graph terminology introduced above is also applied, with the obvious meaning, to an APS.

A *sub-APS* (N', E', P') of an APS (N, E, P) is an APS with (N', E') a subgraph of (N, E), and such that P' is P restricted to $E \times E$.

A *switching* of an APS (N, E, P) is a subgraph (N, E') of (N, E) such E' is obtained by omitting one edge from each pair of the APS. (So if the APS has n pairs, there are 2^n switchings).

An APS is a *net*, if all its switchings are trees. ⊠

EXAMPLE. The following picture shows an APS on the left (with the pairs marked by $*$ at the bases of the pairs), with its four switchings on the right. Clearly this APS is a net.

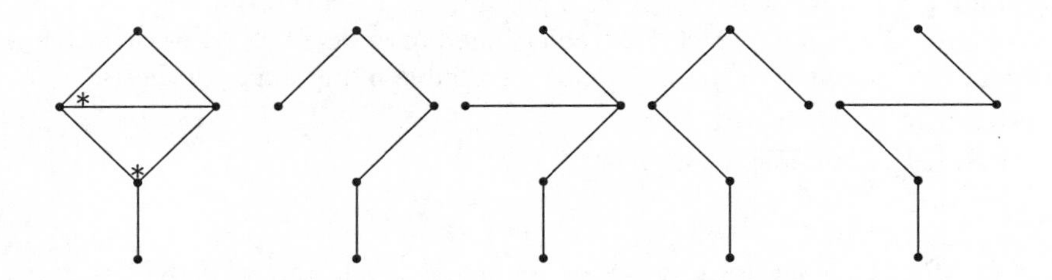

9.5.7A. ♠ Is there a PS with this APS as underlying graph, the pairs corresponding to +-links?

9.5.8. DEFINITION. Let $S \equiv (N, E, P)$ be an APS, $c \in P$ a pair with edges xy, xz and basis x. Let $c(S) := S \setminus \{xy, xz\}$.

S_c^- is the component of x in $c(S)$, S_c^+ is $c(S) \setminus S_c^-$. The pair c is a *section* if S_c^- does not contain y or z; that is to say, each path in S from x to y or from x to z passes through one of the edges xy, xz. ⊠

9.5.9. LEMMA. *Let S be an APS, and c a section of S. Then S is a net if S_c^- and S_c^+ are nets.*

PROOF. Each switching of S decomposes into switchings of S_c^-, S_c^+ connected by a single edge. ⊠

9.5.10. DEFINITION. A pair of edges $\{xy, xz\}$ is *free* in the APS S if xy, xz are the only edges with x as endpoint (in other words, x is of degree 2).

A path *passes through* a pair if it contains both its edges.

A *free path (free cycle)* is a path (cycle) which passes only through free pairs, and passes through at least one pair.

A path γ is *nice* if for each node $x \notin \gamma$, there are two free paths from x to γ, distinct in x (i.e. starting with distinct edges in x), each intersecting γ in a single node. ⊠

N.B. Since a cycle does not contain smaller cycles, it follows that a cycle passes through a pair $\{xy, xz\}$ iff the edges appear consecutively in the cycle (as $\ldots yxz \ldots$).

9.5.11. PROPOSITION. *Let S be a net with an inhabited set of pairs P, then one of the pairs is a section.*

PROOF. We shall derive a contradiction from the following three assumptions:

(H1) S is a net,

(H2) P is inhabited,

(H3) no pair of P is a section.

Assuming H1–3 we shall construct a properly increasing sequence S_0, S_1, S_2, ... of sub-APS's of S, which contradicts the finiteness of S. In our construction of the sequence we use the following double induction hypothesis:

Hi1$[n]$ S_n is a sub-APS of S;

Hi2$[n]$ S_n has a free, nice cycle D_n.

Basis. S must contain a cycle S_0; for if not, each pair of S is a section, contradicting H2–3. Then $D_0 := S_0$ satisfies Hi1[0], Hi2[0]. (Observe that *relative to* S_0 each pair through which S_0 passes is necessarily free.) Note also that we can assume S_0 to have at least one pair; for if all cycles did not contain pairs, there would be switchings containing cycles.

Induction step. Now assume Hi1$[n]$, Hi2$[n]$ for S_n, D_n. D_n is free in S_n, and must contain at least one pair – otherwise there would be a switching of S containing a cycle.

Let $\{yx, xz\}$ be a free pair in D_n; it cannot be a section of S, hence there is a path in S from x to y or from x to z, not passing through yx, xz. Starting from x, let $u \in S_n$ be the first edge after x that the path has in common with S_n. Let α be this path from x to u; α intersects S_n in x and u only. We put $S'_n := S_n \cup \alpha$. S_{n+1} will be S'_n with a new set of pairs added, namely all edges paired in S, and not yet paired in S_n. We note

(a) Either such a pair entirely belongs to α, and then it is free, or the pair has u as basis, and then one edge belongs to α;

(b) The only possible non-free new pair has basis u;

(c) The only old pairs possibly becoming non-free have basis x or u.

We now have to check Hi1$[n+1]$, Hi2$[n+1]$ for the new S_{n+1}. Hi1$[n+1]$ is automatically guaranteed.

Construction of D_{n+1}. By Hi2$[n]$ for S_n, if $u \notin D_n$, there exist two free paths δ_1, δ_2 in S_n from u to D_n, distinct in u, each intersecting D_n in a single point only. We call these intersection points u_1, u_2 respectively (see fig. a). $x = u_1$ and $x = u_2$ are excluded since x has degree 2 in S_n, and u_1, u_2 have degree 3.

One of the paths $\alpha \frown \delta_1$, $\alpha \frown \delta_2$ is free. For δ_1, δ_2 are free in S_n (by hypothesis), hence also free in S_{n+1}; α is also free, and $\alpha \frown \delta_1$, and $\alpha \frown \delta_2$ intersect in u only. δ_1 and δ_2 differ in u, so $\alpha \frown \delta_1$, $\alpha \frown \delta_2$ cannot *both* pass through a possible new pair with basis u; say $\alpha \frown \delta_1$ does not do so. Then $\alpha \frown \delta_1$ is free.

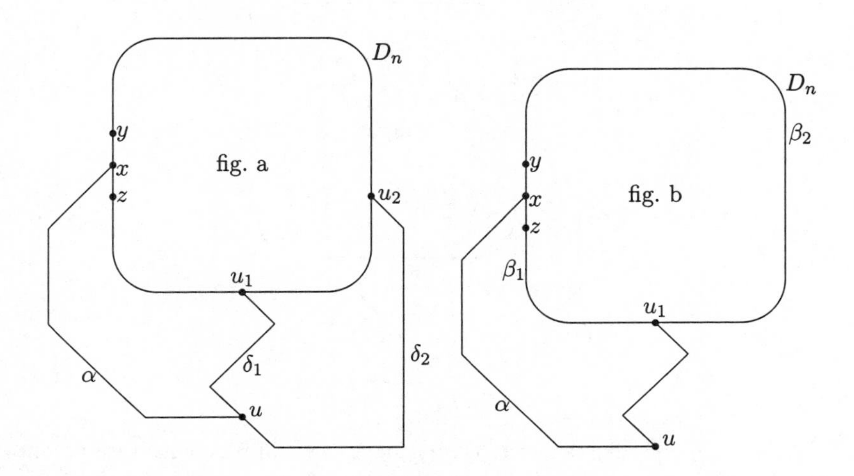

There are two proper subpaths β_1, β_2 of D_n connecting u_1 with x. Then either $\alpha \frown \delta_1 \frown \beta_1$ or $\alpha \frown \delta_1 \frown \beta_2$ is free. To see this, note (fig. b) $\alpha \frown \delta_1$ and β_1, $\alpha \frown \delta_1$ and β_2 respectively intersect in $\{u_1, x\}$ only; β_1, β_2 are free in S_{n+1} (since they are free in S_n) and do not contain pairs which can become non-free going from S_n to S_{n+1}. β_1 and β_2 are distinct in u_1, and hence one of the two joins in u_1 is correct, i.e. does not pass through a new pair added in u. And since the unique pair with basis x belongs to D_n, the junctions $\beta_1 \frown \alpha$, $\beta_2 \frown \alpha$ do not pass through this pair.
Suppose, say, $\alpha \frown \delta_1 \frown \beta_1$ is free; then this will be our D_{n+1}.

If $u \in D_n$, then argue as above, with $u_1 = u$.

Niceness of D_{n+1}. Finally we have to show that D_{n+1} is nice. Take some $v^* \in S_{n+1} \setminus D_{n+1}$. Then by the construction $v^* \in S_n$.

(i) If $v^* \in D_n$, take the two proper subpaths of D_n connecting v with x and with u_1.

(ii) If $v^* \notin D_n$, there are two free paths ϵ_1, ϵ_2 from v ending in D_n, distinct from v, by Hi2[n]. Two possible situations for $(v^*, \epsilon_1, \epsilon_2)$ are pictured in fig. c: $(v, \epsilon_1, \epsilon_2)$ and $(v', \epsilon_1', \epsilon_2')$.

Consider ϵ_1. If it does not intersect D_{n+1}, one can continue it to x or to u_1 such that it is free.

Or if it does intersect D_{n+1}, then the piece of ϵ_1 from v to the first intersection with D_{n+1} is correct. The two free paths ϵ_1, ϵ_2 are distinct in v^* since they always agree on the first edge with ϵ_1, ϵ_2 respectively, where they must be different. ⊠

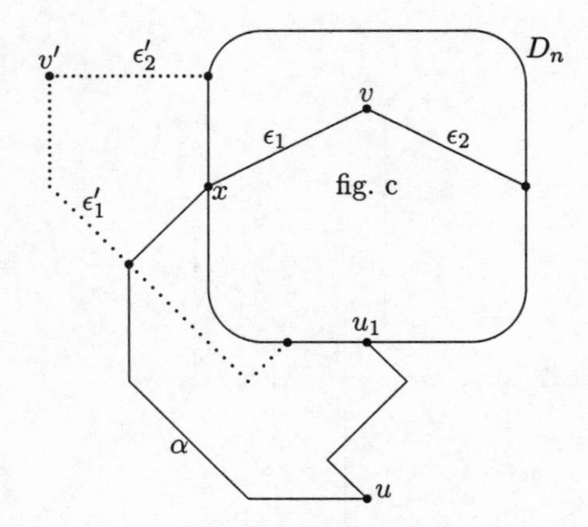

9.5.11A. ♠ The preceding abstract theory for APS's still functions if we permit that a pair has the form (xy, xy), but then in the definition of free pair we must restrict attention to pairs with x, y, z distinct. We consider two contractions on such APS's when finite: (a) deleting an edge xy not belonging to a pair and identifying x and y, and deleting the edges of a pair (xy, xy) and identifying x and y. This yields a notion of reduction which is terminating and confluent. Show that an APS S is a net iff the normal form with respect to this reduction is a single point (Danos [1990]).

9.5.12. *Equating IPS's and proofnets*

We are now ready to prove that inductive proof structures and nets coincide. We have already seen that inductive proof structures are proofnets (9.5.6). It remains to show the converse.

THEOREM. *Every proofnet is an IPS.*

PROOF. Let ν be a given proofnet, and $S = (N, E, P)$ the corresponding APS, where (N, E) is the graph underlying ν, and P consists of the pairs of edges corresponding to $+$-links in ν. S is a net, since ν is a proofnet. We apply induction on the number of $+$-links in ν.

Basis. There are no $+$-links in ν; we apply a subinduction on the number of $\star$-links in ν.

Subbasis. If there is no $\star$-link, ν has been obtained from axiom links, hypotheses and join (cf. 9.5.3). But join is excluded, since ν is connected; hence ν is an axiom or a hypothesis, hence an IPS by definition.

Subinduction step. Suppose there is at least one $\star$-link in ν. The last clause applied in an inductive construction of ν can never have been a join, since

this would conflict with ν being connected. So let the last clause applied in an inductive construction of ν be a tensor clause, with conclusion $A \star B$ and premises A, B, and let ν_A, ν_B be the components of A and B in ν after deletion of the tensor link. ν_A, ν_B are connected (if not, ν would not be connected), are disjoint (otherwise ν would contain a cycle) and together with the tensor link make up all of ν (otherwise ν would not be connected). Hence, by IH, ν_A, ν_B are IPS's, and so is ν.
Induction step. ν has +-links, so S has pairs. Hence we can find a pair c in S which is a section. Then S_c^- and S_c^+ are nets. The proof structure ν_c^+ corresponding to S_c^+ has conclusions A, B and is an IPS by the IH; add a +-link to obtain an IPS ν^* with conclusion $A + B$. S_c^- corresponds to ν_c^- which has $A + B$ as hypothesis and which is also an IPS by the IH; substituting ν^* for the hypothesis $A + B$ in ν_c^- yields ν as an IPS. ⊠

COROLLARY. *A PS ν is an IPS iff ν is a proofnet.* ⊠

9.6 Notes

For a general introduction to modal logics, the reader may consult, for example, Bull and Segerberg [1984], Hughes and Cresswell [1968], Fitting [1983, 1993], Mints [1992d].

9.6.1. *The modal logic* **S4**. The first axiomatization of **S4** is due to C. I. Lewis (in Lewis and Langford [1932]), but the Hilbert system given here is due to Gödel [1933a]. For some background information on this paper see the "Introductory Note to *1933f*" in Gödel [1986].

An early reference for a Gentzen system for **S4** with a proof of cut elimination is Curry [1952a]. A form of **G3s** with □ as the only primitive modality appears in Kanger [1957].

Some examples of modal logics which in many respects can be treated in the same way as **S4** are the following sublogics of **S4** (cf. the definition of **S4** in 9.1.1), based on the same classical basis and the same deduction rules as **S4**, but with only some of the modal axioms: **K** (K-axioms only), **K4** (K- and 4-axioms only), **T** (K- and T-axioms). These systems are examples of *normal* modal logics, i.e. they have a modal-logic Kripke semantics of the standard type.

In Mints [1990,1994b] resolution calculi for **S4** and some closely related logics are developed, in the spirit of sections 7.4, 7.5.

9.6.2. *Interpolation for modal logics.* Fitting [1983] and Rautenberg [1983] contain proofs of interpolation for many modal formalisms. In Fitting [1983, 3.9, 7.13] it is shown that in **S4** (and some closely related modal logics such as **K**, **K4**, **T**) one can impose an additional condition on the interpolant of $A \to B$.

In order to define this extra condition, let us extend the notion of positive and negative formula occurrence by adding to the clauses for positive and negative contexts in 1.1.4: $\Box B^+$, $\Diamond B^+ \in \mathcal{P}$ and $\Box B^-$, $\Diamond B^- \in \mathcal{N}$. A formula is said to be of *□-type* (*◇-type*) if all occurrences of □ occur positively (negatively) and all occurrences of ◇ occur negatively (positively). Then we can require for the interpolant C of $A \to B$

(i) if B is of ◇-type, then C is of ◇-type;

(ii) if A is of □-type, then C is of □-type;

(iii) if A is of □-type and B is of ◇-type, then C is non-modal, that is to say contains neither ◇ nor □.

9.6.3. *Modal embedding of* **I** *into* **S4**. There are many, slightly different, embeddings of **I** into **S4**. Gödel [1933a] proved correctness and conjectured faithfulness for a particular embedding and a slight variant of this, in the case of **Ip** and **S4p**.

The original motivation in Gödel [1933a] for the embedding was provided by reading $\Box A$ as "A is (intuitionistically) provable", where "provable" is not to be read as "formally provable in a specific recursively axiomatizable system" (as Gödel was careful to point out), but rather as "provable by any intuitionistically acceptable argument", or "provable in the sense of the BHK-interpretation" (2.5.1).

For some variant embeddings, see McKinsey and Tarski [1948], Maehara [1954]. The $^\circ$ mapping is copied from the Girard embedding (9.3.6, Girard [1987a]) of intuitionistic logic into classical linear logic. For $^\Box$ (the mapping used by Rasiowa and Sikorski [1953,1963], but for the fact that we have $\bot$ instead of $\neg$ as a primitive) the proof of faithfulness is somewhat easier to give. The proof of faithfulness given here is an adaptation of the proof by Schellinx [1991] of faithfulness for the original Girard embedding of **I** into classical linear logic. Our proof is rather similar to the argument in Maehara [1954]; as in Maehara's proof, a sequent calculus for **I** with finite multisets in the succedent plays a role (cf. exercise 3.2.1A).

A very elegant alternative method for proving faithfulness is found in Flagg and Friedman [1986]; this method works not only for predicate logic, but also for other formalisms based on intuitionistic logic on the one hand, and **S4** on the other hand (for example, intuitionistic arithmetic and so-called epistemic arithmetic; proof-theoretic arguments for the faithfulness of the embedding in this case had been given before in Goodman [1984] and Mints [1978].

It should be pointed out, however, that **S4** is by no means the strongest system for which such an embedding works (cf. again the Introductory Note to *1933f* in Gödel [1986]).

9.6.4. *Semantic tableaux for modal logics.* Semantic tableaux (cf. 4.9.7) have been widely used in the study of modal logics (for example, see Fitting [1983,1988], Goré [1992] to obtain completeness proofs relative to a suitable Kripke semantics. These proofs as a rule then establish completeness for a cutfree sequent calculus, with closure under Cut as a by-product.

Direct proofs of cut elimination by a Gentzen-type algorithm are less numerous (e.g. Curry [1952a], Valentini [1986]).

Just as for intuitionistic logic, semantic tableaus for modal logics cannot have all rules strictly cumulative. Thus we have for the modalities in **S4** the rules

$$\frac{\Theta, \mathbf{t}A}{\Theta^*, \mathbf{t}\Diamond A} \qquad \frac{\Theta, \mathbf{f}A}{\Theta^*, \mathbf{f}\Box A}$$

where $\Theta^* := \{\mathbf{t}\Box A : \mathbf{t}\Box A \in \Theta\} \cup \{\mathbf{f}\Diamond A : \mathbf{f}\Diamond A \in \Theta\}$. More flexibility is achieved by considering tableaus with indexed sequents, also called "prefixed tableau systems" (Fitting [1983, chapter 8]); the idea for such systems (but without the indices explicitly appearing) goes back to Kripke [1963]. The indices correspond to "worlds" in the Kripke semantics for the logic under consideration.

Mints [1994a] gives a cut elimination algorithm for a whole group of such systems. He also shows that there is a close relationship between these sequent calculi with indexed sequents, and systems (Wansing [1994]) in the display-logic style of Belnap [1982]. N- and G- systems with a linearly ordered set of "levels" are considered in Martini and Masini [1993], Masini [1992,1993].

9.6.5. *Linear logic.* For a first introduction beyond the present text, the reader may consult Troelstra [1992a]. Full bibliographical information on linear logic may be found in electronic form under `http://www.cs.cmu.edu/~carsten/linearbib/linearbib.html`.

The system of linear logic in its present form was introduced by Girard [1987a]. As already noted in the preamble of section 9.3, this was not the first study of Gentzen systems in which (some of) the structural rules had been dropped; see Došen [1993] and the references given there. The novel idea of Girard was to reintroduce Weakening and Contraction in a controlled form by means of the exponential operators (namely !, ?).

Some of our symbols for linear logic deviate from the ones used in Girard [1987a] and many other papers on linear logic: we use $\sqcap, \sqcup, \star, +, \bot, \mathbf{0}, \sim$ for Girard's $\&, \oplus, \otimes, \wp, 0, \bot, {}^{\bot}$ (note the interchange between $0, \bot$); $\rightsquigarrow$ does not appear in Girard's paper; the other symbols coincide. $\wp$ is also printed as an upside-down &. For the reasons behind our choice of notation, see Troelstra [1992a, 2.7].

The embeddability of **I** into **Gcl** is stated in Girard [1987a] and completely proved in Schellinx [1991]. A one-sided version of **G3s** is embedded into classical linear logic **Gcl** in Martini and Masini [1994]. Combining this with the embedding of **I** into **S4** yields another proof of the embeddability of **I** into **Gcl**. The results in 9.4 are taken from Schellinx [1994]. For much farther-reaching results in this direction, see Danos et al. [1997,1995].

Interpolation for fragments of **CL** is treated in Roorda [1994]. For resolution calculi for linear logic, see Mints [1993].

Natural deduction formulations of **IL** are discussed in Benton et al. [1992], Mints [1995], Ronchi della Rocca and Roversi [1994], Troelstra [1995].

Proofnets were introduced in Girard [1987a]. The original criterion for a proof structure to be a proofnet was formulated as the so-called "longtrip condition". It is not hard to see that this criterion is equivalent to the switching criterion as given here (the terminology of "switching" was suggested by the notion of a longtrip). Later Girard [1991] gave a much simpler proof, also covering the case of quantifiers, reproduced (without the quantifiers) in Troelstra [1992a, chapter 17]. Another proof, of independent interest, was given in the thesis of Danos [1990]; this is the proof presented here. For yet another proof see Metayer [1994]. An important early paper on proofnets is Danos and Regnier [1989]. In the meantime, the concept of proofnets has been extended (without the "boxes" of Girard [1987a]) to cover quantifiers and context-sharing operators $\sqcap, \sqcup$ as well (Girard [1996]).

Basic logic of Sambin et al. [1997], Sambin and Faggian [1998] goes one step beyond linear logic; here also the role of the contexts is isolated. Among the extensions of basic logic we find linear logic and quantum logic, and the cut elimination procedure for basic logic extends to these systems.

9.6.6. *Computational content of classical logic.* For a long time the quest after a form of "computational content" in classical logic, comparable to the computational content in **I** provided by the formulas-as-types parallel, seemed hopeless. But recently the picture has changed. See for example Danos et al. [1997], Joinet et al. [1998], Danos et al. [1999], where methods for "constructivizing" classical logic, in the sense just referred to, are being studied via a classification of possible methods for embedding Gentzen's system LK into linear logic.

Chapter 10

Proof theory of arithmetic

This chapter presents an example of the type of proof theory inspired by Hilbert's programme and the Gödel incompleteness theorems. The principal goal will be to offer an example of a true mathematically meaningful principle not derivable in first-order arithmetic. Some experience with formal proofs in arithmetic and the first elements of recursion theory will facilitate understanding for the reader, even if most sections (the last two excepted) are essentially self-contained.

The main tool for proving theorems in arithmetic is clearly the induction schema

$$\mathrm{Ind}(A,x) \qquad A[x/0] \to \forall x(A \to A[x/\mathrm{S}x]) \to \forall x A.$$

Here A is an arbitrary formula. An equivalent form of this schema is "cumulative" induction

$$\mathrm{Ind}(A,x)^* \quad \forall y{<}x\,(A[x/y] \to A) \to \forall x A.$$

$\mathrm{Ind}(A,x)$ and $\mathrm{Ind}(A,x)^*$ refer to the standard ordering of the natural numbers. Now it is tempting to try to strengthen arithmetic by allowing more general induction schemas, e.g. with respect to the lexicographical ordering of $\mathbb{N} \times \mathbb{N}$. More generally, we might pick an arbitrary well-ordering $\lhd$ over $\mathbb{N}$, i.e. a linear ordering without infinite descending sequences. Then the following schema of *transfinite induction* holds.

$$\mathrm{TI}_{\lhd}(A,x) \quad \forall x(\forall y{\lhd}x\, A[x/y] \to A) \to \forall x A.$$

This can be read as follows. Suppose the property $A(x)$ is "progressive", i.e. from the validity of $A(y)$ for all $y \lhd x$ we can always conclude that $A(x)$ holds. Then $A(x)$ holds for all x.

To see the validity of this schema consider the set of all x such that $A(x)$ does not hold. If this set is not empty, then by the well-foundedness of $\lhd$ it must contain a smallest element x_0. But by the choice of x_0 we have $\forall y{\lhd}x_0\, A(y)$ and hence a contradiction against the assumed progressiveness of $A(x)$.

One might wonder whether this schema of transfinite induction actually strengthens arithmetic. We will prove here a classic result of Gentzen [1943] which in a sense answers this question completely. However, in order to state the result we have to be more explicit about the well-orderings used. This is done in the following section 10.1.

10.1 Ordinals below ε_0

From elementary set theory we know that there are particular well-ordered sets called *ordinals* such that any well-ordered set is isomorphic to an ordinal, and the ordinals themselves form a well-ordered class. Here we restrict ourselves to a countable set of relatively small ordinals, traditionally called ordinals below ε_0. Moreover, we equip these ordinals with an extra structure (a kind of algebra). It is then customary to speak of *ordinal notations*. These ordinal notations can be introduced without any set theory in a purely formal, combinatorial way. Our treatment is based on the Cantor normal form for ordinals; for detailed information we refer to Bachmann [1955]. We also introduce some elementary relations and operations for such ordinal notations, which will be used later. For brevity we from now on use the word "ordinal" instead of "ordinal notation".

10.1.1. DEFINITION. We define the two notions α is an *ordinal* and $\alpha < \beta$ for ordinals α, β simultaneously by induction:

- If $\alpha_m, \ldots, \alpha_0$ are ordinals, $m \geq -1$ and $\alpha_m \geq \ldots \geq \alpha_0$ (where $\alpha \geq \beta$ means $\alpha > \beta$ or $\alpha = \beta$), then

$$\omega^{\alpha_m} + \cdots + \omega^{\alpha_0}$$

  is an ordinal. Note that the empty sum denoted by 0 is allowed here.

- If $\omega^{\alpha_m} + \cdots + \omega^{\alpha_0}$ and $\omega^{\beta_n} + \cdots + \omega^{\beta_0}$ are ordinals, then

$$\omega^{\alpha_m} + \cdots + \omega^{\alpha_0} < \omega^{\beta_n} + \cdots + \omega^{\beta_0}$$

  iff there is an $i \geq 0$ such that $\alpha_{m-i} < \beta_{n-i}$, $\alpha_{m-i+1} = \beta_{n-i+1}, \ldots, \alpha_m = \beta_n$, or else $m < n$ and $\alpha_m = \beta_n, \ldots, \alpha_0 = \beta_{n-m}$.

For proofs by induction on ordinals it is convenient to introduce the notion of *level* of an ordinal by the stipulations (a) if α is the empty sum 0, level(α) = 0, and (b) if $\alpha = \omega^{\alpha_m} + \ldots + \omega^{\alpha_0}$ with $\alpha_m \geq \ldots \geq \alpha_0$, then level($\alpha$) = level($\alpha_m$) + 1. ⊠

For ordinals of level k, $\omega_k \leq \alpha < \omega_{k+1}$, where $\omega_0 = 0$, $\omega_1 = \omega$, $\omega_{k+1} = \omega^{\omega_k}$.

NOTATION. We shall use the notation 1 for ω^0, a for $\omega^0 + \cdots + \omega^0$ with a copies of ω^0 and $\omega^\alpha a$ for $\omega^\alpha + \cdots + \omega^\alpha$ again with a copies of ω^α. ⊠

Note that limit ordinals (ordinals $\neq 0$ not having an immediate predecessor) are written as $\alpha + \omega^\alpha \cdot a$ for $\alpha > 0$, $a > 0$.

10.1.1A. ♠ Prove (by induction on the levels in the inductive definition) that $<$ is a linear order with 0 as the smallest element. Show that the ordering is decidable.

10.1.2. DEFINITION. (*Addition*) We now define addition for ordinals:

$$(\omega^{\alpha_m} + \cdots + \omega^{\alpha_0}) + (\omega^{\beta_n} + \cdots + \omega^{\beta_0}) := \omega^{\alpha_m} + \cdots + \omega^{\alpha_i} + \omega^{\beta_n} + \cdots + \omega^{\beta_0}$$

where i is minimal such that $\alpha_i \geq \beta_n$; if there is no such i, take $i = m + 1$ (i.e. $\omega^{\beta_n} + \cdots + \omega^{\beta_0}$). ⊠

10.1.2A. ♠ Prove that $+$ is an associative operation which is strictly monotonic in the second argument and weakly monotonic in the first argument. Note that $+$ is not commutative: $1 + \omega = \omega \neq \omega + 1$.

10.1.2B. ♠ There is also a commutative version of addition: the *natural* (or *Hessenberg*) sum of two ordinals is defined by

$$(\omega^{\alpha_m} + \cdots + \omega^{\alpha_0}) \# (\omega^{\beta_n} + \cdots + \omega^{\beta_0}) := \omega^{\gamma_{m+n}} + \cdots + \omega^{\gamma_0},$$

where $\gamma_{m+n}, \ldots, \gamma_0$ is a decreasing permutation of $\alpha_m, \ldots, \alpha_0, \beta_n, \ldots, \beta_0$. Prove that $\#$ is associative, commutative and strictly monotonic in both arguments.

10.1.3. We will also need to know how ordinals of the form $\beta + \omega^\alpha$ can be approximated from below. First note that

$$\delta < \alpha \to \beta + \omega^\delta a < \beta + \omega^\alpha.$$

Furthermore, for any $\gamma < \beta + \omega^\alpha$ with $\alpha > 0$ we can find a $\delta < \alpha$ and an a such that

$$\gamma < \beta + \omega^\delta a.$$

10.1.3A. ♠ Prove this, and describe an algorithm that, when given $\alpha > 0$, β, γ such that $\gamma < \beta + \omega^\alpha$, produces δ, a with $\gamma < \beta + \omega^\delta a$.

10.1.4. DEFINITION. We now define 2^α for ordinals α. Let $\alpha_m \geq \cdots \geq \alpha_0 \geq \omega > k_n \geq \cdots \geq k_1 > 0$. Then (writing $\exp_2(\alpha), \exp_\omega(\alpha)$ for $2^\alpha, \omega^\alpha$)

$$\begin{aligned}&\exp_2(\omega^{\alpha_m} + \cdots + \omega^{\alpha_0} + \omega^{k_n} + \cdots + \omega^{k_1} + \omega^0 a)\\ &\quad := (\exp_\omega(\omega^{\alpha_m} + \cdots + \omega^{\alpha_0} + \omega^{k_n - 1} + \cdots + \omega^{k_1 - 1}))2^a.\end{aligned}$$ ⊠

10.1.4A. ♠ Prove that $2^{\alpha+1} = 2^\alpha + 2^\alpha$ and that 2^α is strictly monotonic in α.

10.1.5. In order to work with ordinals in a purely arithmetical system we set up some effective bijection between our ordinals $< \varepsilon_0$ and non-negative integers (i.e. a Gödel numbering). For its definition it is useful to refer to ordinals in the form

$$\omega^{\alpha_m} k_m + \cdots + \omega^{\alpha_0} k_0 \quad \text{with } \alpha_m > \cdots > \alpha_0 \text{ and } k_i \neq 0 \ (m \geq -1).$$

(By convention, $m = -1$ corresponds to the empty sum.)

DEFINITION. For any ordinal α we define its Gödel number $\ulcorner \alpha \urcorner$ inductively by

$$\ulcorner \omega^{\alpha_m} k_m + \cdots + \omega^{\alpha_0} k_0 \urcorner := \left(\prod_{i \leq m} p_{\ulcorner \alpha_i \urcorner}^{k_i} \right) - 1,$$

where p_n is the n-th prime number starting with $p_0 := 2$. For any non-negative integer x we define its corresponding ordinal notation $\mathrm{o}(x)$ inductively by

$$\mathrm{o}\left(\left(\prod_{i \leq \ell} p_i^{q_i} \right) - 1 \right) := \sum_{i \leq \ell} \omega^{\mathrm{o}(i)} q_i,$$

where the sum is to be understood as the natural sum. ⊠

10.1.6. LEMMA. *(i)* $\mathrm{o}(\ulcorner \alpha \urcorner) = \alpha$, *(ii)* $\ulcorner \mathrm{o}(x) \urcorner = x$.

PROOF. This can be proved easily by induction. ⊠

Hence we have a simple bijection between ordinals and non-negative integers. Using this bijection we can transfer our relations and operations on ordinals to computable relations and operations on non-negative integers.

10.1.7. NOTATION. We use the following abbreviations.

$$\begin{array}{lcl} x \prec y & := & \mathrm{o}(x) < \mathrm{o}(y), \\ \omega^x & := & \ulcorner \omega^{\mathrm{o}(x)} \urcorner, \\ x \oplus y & := & \ulcorner \mathrm{o}(x) + \mathrm{o}(y) \urcorner, \\ xk & := & \ulcorner \mathrm{o}(x) k \urcorner, \\ \omega_k & := & \ulcorner \omega_k \urcorner, \end{array}$$

where $\omega_0 := 1$, $\omega_{k+1} := \omega^{\omega_k}$. ⊠

We leave it to the reader to verify that $\prec$, $\lambda x.\omega^x$, $\lambda xy.x \oplus y$, $\lambda xk.xk$ and $\lambda k.\ulcorner \omega_k \urcorner$ are all primitive recursive.

10.2 Provability of initial cases of TI

We now derive initial cases of the principle of transfinite induction in arithmetic, i.e. of

$$\mathrm{TI}_{\prec a}(P) \qquad \forall x(\forall y{\prec}x\, Py \to Px) \to \forall x{\prec}a\, Px$$

for some number a and a predicate symbol P, where $\prec$ is the standard order of order type ε_0 defined in the preceding section. In section 10.4 we will see that our results here are optimal in the sense that for our full system of ordinals $< \varepsilon_0$ the principle

$$\mathrm{TI}_{\prec}(P) \qquad \forall x(\forall y{\prec}x\, Py \to Px) \to \forall x Px$$

of transfinite induction is underivable. All these results are due to Gentzen [1943].

10.2.1. DEFINITION. By an *arithmetical system* **Z** we mean a theory based on minimal logic $\forall\!\to\!\bot$-**M** (including equality axioms) with the following properties. The language of **Z** consists of a fixed (possibly countably infinite) supply of function and relation constants which are assumed to denote fixed functions and relations on the non-negative integers for which a computation procedure is known. Among the function constants there must be a constant S for the successor function and 0 for (the 0-place function) zero. Among the relation constants there must be a constant $=$ for equality and $\prec$ for the ordering of type ε_0 of the natural numbers, as introduced in section 10.1. In order to formulate the general principle of transfinite induction we also assume that a unary relation symbol P is present, which acts like a free set variable.

Terms are built up from object variables x, y, z by means of $f(t_1, \dots, t_m)$, where f is a function constant. We identify closed terms which have the same value; this is a convenient way to express in our formal systems the assumption that for each function constant a computation procedure is known. Terms of the form $\mathrm{S}(\mathrm{S}(\dots \mathrm{S}(0)\dots))$ are called *numerals*. We use the notation $\mathrm{S}^n 0$ or $\overline{n}$ or (only in this chapter) even n for them. *Formulas* are built up from $\bot$ and atomic formulas $R(t_1, \dots, t_m)$, with R a relation constant or a relation symbol, by means of $A \to B$ and $\forall x A$. Recall that we abbreviate $A \to \bot$ by $\neg A$.

The *axioms* of **Z** will always include the *Peano axioms*, i.e. the universal closures of

PA1	$\mathrm{S}x = \mathrm{S}y \to x = y$,
PA2	$\mathrm{S}x = 0 \to A$,
$\mathrm{Ind}(A, x)$	$A[x/0] \to \forall x(A \to A[x/\mathrm{S}x]) \to \forall x A$,

with A an arbitrary formula. We express our assumption that for any relation constant R a decision procedure is known by adding the axiom $R\vec{n}$ whenever $R\vec{n}$ is true, and $\neg R\vec{n}$ whenever $R\vec{n}$ is false. Concerning $\prec$ we require irreflexivity and transitivity for $\prec$ as axioms, and also – following Schütte – the universal closures of

ord1 $x \prec 0 \to A$,
ord2 $z \prec y \oplus \omega^0 \to (z \prec y \to A) \to (z = y \to A) \to A$,
ord3 $x \oplus 0 = x$,
ord4 $x \oplus (y \oplus z) = (x \oplus y) \oplus z$,
ord5 $0 \oplus x = x$,
ord6 $\omega^x 0 = 0$,
ord7 $\omega^x(\mathrm{S}y) = \omega^x y \oplus \omega^x$,
ord8 $z \prec y \oplus \omega^{\mathrm{S}x} \to z \prec y \oplus \omega^{\mathrm{e}(x,y,z)}\mathrm{m}(x,y,z)$,
ord9 $z \prec y \oplus \omega^{\mathrm{S}x} \to \mathrm{e}(x,y,z) \prec \mathrm{S}x$,

where $\oplus$, $\lambda xy.\omega^x y$, e and m denote the appropriate function constants and A is any formula. (The reader should check that e, m can be taken to be primitive recursive.) These axioms are formal counterparts to the properties of the ordinal notations observed in the preceding section; for example, ord8 correponds to the remark in 10.1.3. We also allow an arbitrary supply of true formulas $\forall\vec{x}A$ with A quantifier-free and without P as axioms. Such formulas are called Π_1-*formulas* (in the literature also Π^0_1-formulas).

Moreover, we may also add an *ex-falso-quodlibet schema* or even a *stability schema* for A:

Efq $\bot \to A$,
Stab $\neg\neg A \to A$.

Addition of Efq leads to an intuitionistic arithmetical system (the $\forall\to\bot$-fragment of a version of Heyting arithmetic **HA**, cf. 6.6.2) and addition of Stab to a classical arithmetical system (a version of Peano arithmetic **PA**; see 10.5). Note that in our $\forall\to\bot$-fragment of minimal logic these schemas are derivable from their instances

Efq_R $\forall\vec{x}(\bot \to R\vec{x})$,
Stab_R $\forall\vec{x}(\neg\neg R\vec{x} \to R\vec{x})$,

with R a relation constant or the special relation symbol P. The proof uses theorem 2.3.6 and the first half of exercise 2.3.6A. Note also that when the stability schema is present, we can replace PA2, ord1 and ord2 by their more familiar classical versions

$\mathrm{PA2^c}$ $\mathrm{S}x \neq 0$,
$\mathrm{ord1^c}$ $x \not\prec 0$,
$\mathrm{ord2^c}$ $z \prec y \oplus \omega^0 \to z \neq y \to z \prec y$.

We will also consider *restricted* arithmetical systems $\mathbf{Z}_k$. They are defined like $\mathbf{Z}$, but with the induction schema $\mathrm{Ind}(A, x)$ restricted to formulas A of level $\mathrm{lev}(A) \leq k$. The *level* of a formula A is defined by

$$\begin{aligned} \mathrm{lev}(R\vec{t}) &:= \mathrm{lev}(\bot) := 0, \\ \mathrm{lev}(A \to B) &:= \max(\mathrm{lev}(A) + 1, \mathrm{lev}(B)), \\ \mathrm{lev}(\forall x A) &:= \max(1, \mathrm{lev}(A)). \end{aligned}$$

However, the trivial special case of induction $A[x/0] \to \forall x A[x/\mathrm{S}x] \to \forall x A$, which amounts to case distinction, is allowed for arbitrary A. (This is needed in the proof of theorem 10.2.3 below; in the full language with $\vee$ this is equivalent to adding $\forall x(x = 0 \vee \exists y(x = Sy))$.) ⊠

10.2.2. THEOREM. *(Provable initial cases of TI in* $\mathbf{Z}$*) Transfinite induction up to* ω_n*, i.e. for arbitrary* $A(x)$

$$\forall x(\forall y{\prec}x\, A(y) \to A(x)) \to \forall x{\prec}\omega_n\, A(x)$$

is derivable in $\mathbf{Z}$.

PROOF. To any formula $A(x)$ we assign a formula $A^+(x)$ (with respect to a fixed variable x) by

$$A^+(x) := \forall y(\forall z{\prec}y\, A(z) \to \forall z{\prec}y \oplus \omega^x\, A(z)).$$

We first show

If $A(x)$ is progressive, then $A^+(x)$ is progressive,

where "$B(x)$ is *progressive*" means $\forall x(\forall y{\prec}x\, B(y) \to B(x))$. So assume that $A(x)$ is progressive and

(1) $\forall y{\prec}x\, A^+(y)$.

We have to show $A^+(x)$. So assume further

(2) $\forall z{\prec}y\, A(z)$

and $z \prec y \oplus \omega^x$. We have to show $A(z)$.

Case $x = 0$. Then $z \prec y \oplus \omega^0$. By ord2 it suffices to derive $A(z)$ from $z \prec y$ as well as from $z = y$. If $z \prec y$, then $A(z)$ follows from (2), and if $z = y$, then $A(z)$ follows from (2) and the progressiveness of $A(x)$.

Case $\mathrm{S}x$. From $z \prec y \oplus \omega^{\mathrm{S}x}$ we obtain $z \prec y \oplus \omega^{\mathrm{e}(x,y,z)}\mathrm{m}(x, y, z)$ by (ord8) and $\mathrm{e}(x, y, z) \prec \mathrm{S}x$ by ord9. From (1) we obtain $A^+(\mathrm{e}(x, y, z))$. By the definition of $A^+(x)$ we get

$$\forall u{\prec}y \oplus \omega^{\mathrm{e}(x,y,z)}v\, A(u) \to \forall u{\prec}(y \oplus \omega^{\mathrm{e}(x,y,z)}v) \oplus \omega^{\mathrm{e}(x,y,z)}\, A(u)$$

and hence, using ord4 and ord7

$$\forall u{\prec}y \oplus \omega^{\mathrm{e}(x,y,z)}v\, A(u) \to \forall u{\prec}y \oplus \omega^{\mathrm{e}(x,y,z)}\mathrm{S}(v)\, A(u).$$

Also from (2) and ord6, ord3 we obtain

$$\forall u{\prec}y \oplus \omega^{\mathrm{e}(x,y,z)}0\, A(u).$$

Using an appropriate instance of the induction schema we can conclude that

$$\forall u{\prec}y \oplus \omega^{\mathrm{e}(x,y,z)}\mathrm{m}(x,y,z)\, A(u)$$

and hence $A(z)$.

We now show, by induction on n, how for an arbitrary formula $A(x)$ we can obtain a derivation of

$$\forall x(\forall y{\prec}x\, A(y) \to A(x)) \to \forall x{\prec}\omega_n\, A(x).$$

So assume the left hand side, i.e. assume that $A(x)$ is progressive.
Case 0. Then $x \prec \omega^0$ and hence $x \prec 0 \oplus \omega^0$ by ord5. By ord2 it suffices to derive $A(x)$ from $x \prec 0$ as well as from $x = 0$. Now $x \prec 0 \to A(x)$ holds by ord1, and $A(0)$ then follows from the progressiveness of $A(x)$.
Case $n+1$. Since $A(x)$ is progressive, by what we have shown above $A^+(x)$ is also progressive. Applying the IH to $A^+(x)$ yields $\forall x{\prec}\omega_n\, A^+(x)$, and hence $A^+(\omega_n)$ by the progressiveness of $A^+(x)$. Now the definition of $A^+(x)$ (together with ord1 and ord5) yields $\forall z{\prec}\omega^{\omega_n}\, A(z)$. ⊠

Note that in the induction step of this proof we have derived transfinite induction up to ω_{n+1} for $A(x)$ from transfinite induction up to ω_n for a formula of level higher than the level of $A(x)$.

10.2.3. We now want to refine the preceding theorem to a corresponding result for the subsystems $\mathbf{Z}_k$ of $\mathbf{Z}$.

THEOREM. *(Provable initial cases of* TI *in* $\mathbf{Z}_k$*) Let* $1 \leq \ell \leq k$*. Then in* $\mathbf{Z}_k$ *we can derive transfinite induction for any formula* $A(x)$ *of level* $\leq \ell$ *up to* $\omega_{k-\ell+2}[m]$ *for arbitrary* m*, i.e.*

$$\forall x(\forall y{\prec}x\, A(y) \to A(x)) \to \forall x{\prec}\omega_{k-\ell+2}[m]\, A(x),$$

where $\omega_1[m] := m$*,* $\omega_{i+1}[m] := \omega^{\omega_i[m]}$*.*

PROOF. Note first that if $A(x)$ is a formula of level $\ell \geq 1$, then the formula $A^+(x)$ constructed in the proof of the preceding theorem has level $\ell + 1$, and for the proof of

If $A(x)$ is progressive, then $A^+(x)$ is progressive,

we have used induction with an induction formula of level ℓ.

Now let $A(x)$ be a fixed formula of level $\leq \ell$, and assume that $A(x)$ is progressive. Define $A^0 := A$, $A^{i+1} := (A^i)^+$. Then $\text{lev}(A^i) \leq \ell + i$, and hence in $\mathbf{Z}_k$ we can derive that $A^1, A^2, \dots A^{k-\ell+1}$ are all progressive. Now from the progressiveness of $A^{k-\ell+1}(x)$ we obtain $A^{k-\ell+1}(0)$, $A^{k-\ell+1}(1)$, $A^{k-\ell+1}(2)$ and generally $A^{k-\ell+1}(m)$ for any m, i.e. $A^{k-\ell+1}(\omega_1[m])$. But since

$$A^{k-\ell+1}(x) = (A^{k-\ell})^+(x) = \forall y(\forall z{\prec}y\, A^{k-\ell}(z) \to \forall z{\prec}y \oplus \omega^x\, A^{k-\ell}(z))$$

we first get (with $y = 0$) $\forall z{\prec}\omega_2[m]\, A^{k-\ell}(z)$ and then $A^{k-\ell}(\omega_2[m])$ by the progressiveness of $A^{k-\ell}$. Repeating this argument we finally obtain

$\forall z{\prec}\omega_{k-\ell+2}[m]\, A^0(z)$. ⊠

Our next aim is to prove that these bounds are sharp. More precisely, we will show that in $\mathbf{Z}$ (no matter how many true Π_1-formulas we have added as axioms) one cannot derive "purely schematic" transfinite induction up to ε_0, i.e. one cannot derive the formula

$$\forall x(\forall y{\prec}x\, Py \to Px) \to \forall x Px$$

with a relation symbol P, and that in $\mathbf{Z}_k$ one cannot derive transfinite induction up to ω_{k+1}, i.e. the formula

$$\forall x(\forall y{\prec}x\, Py \to Px) \to \forall x{\prec}\omega_{k+1}\, Px.$$

This will follow from the method of normalization applied to arithmetical systems, which we have to develop first.

10.3 Normalization with the omega rule

We will show in theorem 10.4.12 that a normalization theorem does not hold for arithmetical systems $\mathbf{Z}$, in the sense that for any formula A derivable in $\mathbf{Z}$ there is a derivation of the same formula A in $\mathbf{Z}$ which only uses formulas of a level bounded by the level of A. The reason for this failure is the presence of induction axioms, which can be of arbitrary level.

Here we remove that obstacle against normalization in a somewhat drastic way: we leave the realm of proofs as finite combinatory objects and replace the induction axioms by a rule with infinitely many premises, the so-called ω-rule (suggested by Hilbert and studied by Lorenzen, Novikov and Schütte), which allows us to conclude that $\forall x A(x)$ from $A(0), A(1), A(2), \dots$, i.e.

$$\frac{\begin{matrix}\mathcal{D}_0 & \mathcal{D}_1 & & \mathcal{D}_n & \\ A(0) & A(1) & \cdots & A(n) & \cdots\end{matrix}}{\forall x A(x)}\,\omega$$

So derivations can be viewed as labelled infinite (countably branching) trees. As in the finitary case a label consists of the derived formula and the name of the rule applied. Since we define derivations inductively, any such derivation tree must be well-founded, i.e. must not contain an infinite descending path.

Clearly this ω-rule can also be used to replace the rule $\forall$I. As a consequence we do not need to consider free individual variables.

It is plain that any derivation in an arithmetical system **Z** can be translated into an infinitary derivation with the ω-rule; this will be carried out in lemma 10.3.5 below. The resulting infinitary derivation has a noteworthy property: in any application of the ω-rule the cutranks of the infinitely many immediate subderivations $\mathcal{D}_n$ are bounded, and also their sets of free assumption variables are bounded by a finite set. Here the cutrank of a derivation is as usual the least number $\geq$ the level of any subderivation obtained by $\to$I as the main premise of $\to$E or by the ω-rule as the main premise of $\forall$E, where the *level of a derivation* is the level of its type as a term, i.e. of the formula it derives. Clearly a derivation is called normal iff its cutrank is zero, and we will prove below that any (possibly infinite) derivation of finite cutrank can be transformed into a derivation of cutrank zero. The resulting normal derivation will continue to be infinite, so the result may seem useless at first sight. However, we will be able to bound the depth of the resulting derivation in an informative way, and this will enable us in 10.4 to obtain the desired results on unprovable initial cases of transfinite induction. Let us now carry out this programme.

N.B. The standard definition of cutrank in predicate logic measures the depth of formulas; here one uses the level, as in section 6.10.

10.3.1. DEFINITION. We introduce the systems $\mathbf{Z}^\infty$ of ω-arithmetic as follows. $\mathbf{Z}^\infty$ has the same language and – apart from the induction axioms – the same axioms as **Z**. Derivations in $\mathbf{Z}^\infty$ are infinite objects. It is useful to employ a term notation for these, and we temporarily use d, e, f to denote such (infinitary) derivation terms. For the term corresponding to the deduction obtained by applying the ω-rule to d_i, $i \in \mathbb{N}$ we write $\langle d_i \rangle_{i<\omega}$. However, for our purposes here it suffices to only consider derivations whose depth is bounded below ε_0. ⊠

In the present chapter we will also regard the term t in $\forall$E as a "minor premise", as mentioned in 2.1.6, remark (v). The notion of a track (see 6.2.2) is adapted accordingly.

DEFINITION. We define the notion "d is a *derivation of depth* $\leq \alpha$" (written $|d| \leq \alpha$) inductively as follows (i ranges over numerals).

(A) Any assumption variable u^A with A a closed formula and any axiom ax^A is a derivation of depth $\leq \alpha$, for any α.

($\to$I) If d^B is a derivation of depth $\leq \alpha_0 < \alpha$, then $(\lambda u^A.d^B)^{A\to B}$ is a derivation of depth $\leq \alpha$.

($\to$E) If $d^{A\to B}$ and e^A are derivations of depths $\leq \alpha_i < \alpha$ (i=1,2), then $(d^{A\to B}e^A)^B$ is a derivation of depth $\leq \alpha$.

(ω) For all $A(x)$, if $d_i^{A(i)}$ are derivations of depths $\leq \alpha_i < \alpha$ $(i < \omega)$, then $(\langle d_i^{A(i)}\rangle_{i<\omega})^{\forall x A}$ is a derivation of depth $\leq \alpha$.

($\forall$E) For all $A(x)$, if $d^{\forall x A}$ is a derivation of depth $\leq \alpha_0 < \alpha$, then, for all i, $(d^{\forall x A} i)^{A(i)}$ is a derivation of depth $\leq \alpha$. ⊠

NOTATION. We will use $|d|$ to denote the least α such that $|d| \leq \alpha$. ⊠

Note that in ($\forall$E) it suffices to use numerals as minor premises. The reason is that we only need to consider closed terms, and any such term is in our setup identified with a numeral.

10.3.2. DEFINITION. The *cutrank* $\mathrm{cr}(d)$ of a derivation d is defined by

$$\begin{aligned}
\mathrm{cr}(u^A) &:= \mathrm{cr}(\mathrm{ax}^A) := 0,\\
\mathrm{cr}(\lambda u.d) &:= \mathrm{cr}(d),\\
\mathrm{cr}(d^{A\to B}e^A) &:= \begin{cases} \max(\mathrm{lev}(A\to B), \mathrm{cr}(d), \mathrm{cr}(e)) & \text{if } d = \lambda u.d',\\ \max(\mathrm{cr}(d), \mathrm{cr}(e)) & \text{otherwise,}\end{cases}\\
\mathrm{cr}(\langle d_i\rangle_{i<\omega}) &:= \sup\nolimits_{i<\omega} \mathrm{cr}(d_i),\\
\mathrm{cr}(d^{\forall x A} j) &:= \begin{cases} \max(\mathrm{lev}(\forall x A), \mathrm{cr}(d)) & \text{if } d = \langle d_i\rangle_{i<\omega},\\ \mathrm{cr}(d) & \text{otherwise.}\end{cases}
\end{aligned}$$
⊠

Clearly $\mathrm{cr}(d) \in \mathbb{N} \cup \{\omega\}$ for all d. For our purposes it will suffice to consider only derivations with finite cutranks (i.e. with $\mathrm{cr}(d) \in \mathbb{N}$) and with finitely many free assumption variables.

10.3.3. LEMMA. *If d is a derivation of depth $\leq \alpha$, with free assumption variables among $u, \vec{u}$ and of cutrank $\mathrm{cr}(d) = k$, and e is a derivation of depth $\leq \beta$, with free assumption variables among $\vec{u}$ and of cutrank $\mathrm{cr}(e) = \ell$, then $d[u/e]$ is a derivation with free assumption variables among $\vec{u}$, of depth $|d[u/e]| \leq \beta + \alpha$ and of cutrank $\mathrm{cr}(d[u/e]) \leq \max(\mathrm{lev}(e), k, \ell)$.*

PROOF. Straightforward induction on the depth of d. ⊠

10.3.3A. ♠ Give the proof in some detail.

10.3.4. Using this lemma we can now embed our systems $\mathbf{Z}_k$ (i.e. arithmetic with induction restricted to formulas of level $\leq k$) and hence $\mathbf{Z}$ into $\mathbf{Z}^\infty$. In this embedding we refer to the number $n_{\mathrm{I}}(d)$ of nested applications of the induction schema within a $\mathbf{Z}_k$-derivation d.

DEFINITION. The *nesting* of applications of induction in d, $\mathrm{n_I}(d)$, is defined by induction on d, as follows.

$$\begin{aligned}
\mathrm{n_I}(u) &:= \mathrm{n_I}(\mathrm{ax}) := 0 \text{ (axioms and assumption variables)},\\
\mathrm{n_I}(\mathrm{Ind}) &:= 1,\\
\mathrm{n_I}(\mathrm{Ind}\,de) &:= \max(\mathrm{n_I}(d), \mathrm{n_I}(e)+1),\\
\mathrm{n_I}(de) &:= \max(\mathrm{n_I}(d), \mathrm{n_I}(e)), \text{ if } d \text{ is not of the form } \mathrm{Ind}\,d_0,\\
\mathrm{n_I}(\lambda u.d) &:= \mathrm{n_I}(d) \text{ (case of } \to\mathrm{I}).\\
\mathrm{n_I}(\lambda x.d) &:= \mathrm{n_I}(d) \text{ (case of } \forall\mathrm{I}).\\
\mathrm{n_I}(dt) &:= \mathrm{n_I}(d) \text{ (case of } \forall\mathrm{E}).
\end{aligned}$$

⊠

10.3.5. LEMMA. *Let a $\mathbf{Z}_k$-derivation in long normal form (see 6.7.2) be given with $\leq m$ nested applications of the induction schema, i.e. of*

$$\mathrm{Ind}(A,x) \qquad A[x/0] \to \forall x(A \to A[x/\mathrm{S}x]) \to \forall x A.$$

all with $\mathrm{lev}(A) \leq k$. We consider subderivations d^B not of the form $\mathrm{Ind}\,\vec{t}$ or $\mathrm{Ind}\,\vec{t}d_0$. For every such subderivation and closed substitution instance $B\sigma$ of B we construct $(d^\infty_\sigma)^{B\sigma}$ in $\mathbf{Z}^\infty$ with free assumption variables $u^{C\sigma}$ for u^C free assumption of d, such that $|d^\infty_\sigma| < \omega^{m+1}$ and $\mathrm{cr}(d^\infty_\sigma) \leq k$, and moreover such that d is obtained by $\to$I iff d^∞_σ is, and d is obtained by $\forall$I or of the form $\mathrm{Ind}\,\vec{t}d_0e$ iff d^∞_σ is obtained by the ω-rule.

PROOF. By recursion on such subderivations d.
Case u^C or ax. Take $u^{C\sigma}$ or ax.
Case $\mathrm{Ind}\,\vec{t}de'$. Since the deduction is in long normal form, $e' = \lambda xv.e$. By IH we have d^∞_σ and e^∞_σ. (Note that neither d nor e can have one of the forbidden forms $\mathrm{Ind}\,\vec{t}$ and $\mathrm{Ind}\,\vec{t}d_0$, since both are in long normal form). Write $e^\infty_\sigma(t,f)$ for $e^\infty_\sigma[x,v/t,f]$, and let

$$(\mathrm{Ind}\,\vec{t}d(\lambda xv.e))^\infty_\sigma := \langle d^\infty_\sigma, e^\infty_\sigma(0,d^\infty_\sigma), e^\infty_\sigma(1, e^\infty_\sigma(0,d^\infty_\sigma)), \ldots\rangle.$$

By IH $|e^\infty_\sigma| \leq \omega^{m-1}{\cdot}p$ and $|d^\infty_\sigma| \leq \omega^m{\cdot}q$ for some $p, q < \omega$. By lemma 10.3.3 we obtain

$$\begin{aligned}
&|e^\infty_\sigma(0,d^\infty_\sigma)| \leq \omega^m{\cdot}q + \omega^{m-1}{\cdot}p,\\
&|e^\infty_\sigma(1,e^\infty_\sigma(0,d^\infty_\sigma))| \leq \omega^m{\cdot}q + \omega^{m-1}{\cdot}2p
\end{aligned}$$

and so on, and hence

$$|(\mathrm{Ind}\,d(\lambda xv.e))^\infty_\sigma| \leq \omega^m{\cdot}(q+1).$$

Concerning the cutrank we have by IH $\mathrm{cr}(d^\infty_\sigma), \mathrm{cr}(e^\infty_\sigma) \leq k$. Therefore

$$\begin{aligned}
&\mathrm{cr}(e^\infty_\sigma(0,d^\infty_\sigma)) \leq \max(\mathrm{lev}(A(0)), \mathrm{cr}(d^\infty_\sigma), \mathrm{cr}(e^\infty_\sigma)) \leq k,\\
&\mathrm{cr}(e^\infty_\sigma(1,e^\infty_\sigma(0,d^\infty_\sigma))) \leq \max(\mathrm{lev}(A(1)), k, \mathrm{cr}(e^\infty_\sigma)) = k,
\end{aligned}$$

and so on, and hence

$$\mathrm{cr}((\mathrm{Ind}\, d(\lambda xv.e))^\infty_\sigma) \le k.$$

Case $\lambda u^C.d^B$. By IH, we have $(d^\infty_\sigma)^{B\sigma}$ with possibly free assumptions $u^{C\sigma}$. Take $(\lambda u.d)^\infty_\sigma := \lambda u^{C\sigma}.d^\infty_\sigma$.

Case de, with d not of the form $\mathrm{Ind}\,\vec{t}$ or $\mathrm{Ind}\,\vec{t}d_0$. By IH we have d^∞_σ and e^∞_σ. Since de is subderivation of a normal derivation we know that d and hence also d^∞_σ is not obtained by $\to$I. Therefore $(de)^\infty_\sigma := d^\infty_\sigma e^\infty_\sigma$ is normal and $\mathrm{cr}(d^\infty_\sigma e^\infty_\sigma) = \max(\mathrm{cr}(d^\infty_\sigma), \mathrm{cr}(e^\infty_\sigma)) \le k$. Also we clearly have $|d^\infty_\sigma e^\infty_\sigma| < \omega^{m+1}$.

Case $(\lambda x.d)^{\forall x B(x)}$. By IH for every i and substitution instance $B(i)\sigma$ we have $d^\infty_{\sigma,i}$. Take $(\lambda x.d)^\infty_\sigma := \langle d^\infty_{\sigma,i}\rangle_{i<\omega}$.

Case $(dt)^{B[x/t]}$. By IH, we have $(d^\infty_\sigma)^{(\forall x B)\sigma}$. Since dt is a subderivation of a normal derivation, d is not obtained by $\forall$I, hence d^ω_σ is not obtained by the ω-rule. Therefore we may take $(dt)^\infty_\sigma := d^\infty_\sigma i$, where i is the numeral with the same value as $t\sigma$. ⊠

10.3.6. DEFINITION. A derivation is called *convertible* or a *redex* if it is of the form $(\lambda u.d)e$ or else $\langle d_i\rangle_{i<\omega} j$, which can be converted into $d[u/e]$ or d_j, respectively. A derivation is called *normal* if it does not contain a convertible subderivation. Note that a derivation is normal iff it is of cutrank 0.

Call a derivation a *simple application* if it is of the form $d_0 d_1 \dots d_m$ with d_0 an assumption variable or an axiom. ⊠

10.3.7. We want to define an operation which by repeated conversions transforms a given derivation into a normal one with the same end formula and no additional free assumption variables. The usual methods to achieve such a task have to be adapted properly in order to deal with the new situation of infinitary derivations. Here we give a particularly simple argument due to Tait [1965].

LEMMA. *For any derivation d^A of depth $\le \alpha$ and cutrank $k+1$ we can find a derivation $(d^k)^A$ with free assumption variables contained in those of d, which has depth $\le 2^\alpha$ and cutrank $\le k$.*

PROOF. By induction on α. The only case which requires some argument is when the derivation is of the form de with $|d| \le \alpha_1 < \alpha$ and $|e| \le \alpha_2 < \alpha$, but is not a simple application. We first consider the subcase where $d^k = \lambda u.d_1(u)$ and $\mathrm{lev}(d) = k+1$. Then $\mathrm{lev}(e) \le k$ by the definition of level (recall that the level of a derivation was defined to be the level of the formula it derives), and hence $d_1[u/e^k]$ has cutrank $\le k$ by lemma 10.3.3. Furthermore, also by lemma 10.3.3, $d_1[u/e^k]$ has depth $\le 2^{\alpha_2} + 2^{\alpha_1} \le 2^{\max(\alpha_2,\alpha_1)+1} \le 2^\alpha$. Hence we can take $(de)^k$ to be $d_1[u/e^k]$.

In the subcase where $d^k = \langle d_i \rangle_{i<\omega}$, $\text{lev}(d) = k+1$ and $e^k = j$ we can take $(de)^k$ to be d_j, since clearly d_j has cutrank $\leq k$ and depth $\leq 2^\alpha$. If we are not in the above subcases, we can simply take $(de)^k$ to be $d^k e^k$. This derivation clearly has depth $\leq 2^\alpha$. Also it has cutrank $\leq k$, which can be seen as follows. If $\text{lev}(d) \leq k+1$ we are done. But $\text{lev}(d) \geq k+2$ is impossible, since we have assumed that de is not a simple application. In order to see this, note that if de is not a simple application, it must be of the form $d_0 d_1 \dots d_n e$ with d_0 not an assumption variable or axiom and d_0 not itself of the form $d'd''$; then d_0 must end with an introduction $\to$I or ω, hence there is a cut of a degree exceeding $k+1$, which is excluded by assumption. ⊠

10.3.7A. ♠ Complete the proof.

As an immediate consequence we obtain

10.3.8. THEOREM. *(Normalization for $\mathbf{Z}^\infty$) For any derivation d^A of depth $\leq \alpha$ and cutrank $\leq k$ we can find a normal derivation $(d^*)^A$ with free assumption variables contained in those of d, which has depth $\leq 2^\alpha_k$, where $2^\alpha_0 := \alpha$, $2^\alpha_{m+1} := 2^{2^\alpha_m}$.* ⊠

As in section 6.2 we can now analyze the structure of normal derivations in $\mathbf{Z}^\infty$. In particular we obtain

10.3.9. THEOREM. *(Subformula property for $\mathbf{Z}^\infty$) Let $\mathcal{D}$ be a normal deduction in $\mathbf{Z}^\infty$ for $\Gamma \vdash A$. Then each formula in $\mathcal{D}$ is a subformula of a formula in $\Gamma \cup \{A\}$.*

PROOF. We prove this for tracks (see 6.2.2) of order n, by induction on n. ⊠

10.3.9A. ♠ Complete the proof.

10.4 Unprovable initial cases of TI

We now apply the technique of normalization for arithmetic with the ω-rule to obtain a proof that transfinite induction up to ε_0 is underivable in $\mathbf{Z}$, i.e. a proof of

$$\mathbf{Z} \nvdash \forall x(\forall y \prec x\, Py \to Px) \to \forall x Px$$

with a relation symbol P, and that transfinite induction up to ω_{k+1} is underivable in $\mathbf{Z}_k$, i.e. a proof of

$$\mathbf{Z}_k \nvdash \forall x(\forall y \prec x\, Py \to Px) \to \forall x \prec \omega_{k+1}\, Px.$$

It clearly suffices to prove this for arithmetical systems based on classical logic. Hence we may assume that we have used only the classical versions PA2^c, ord1^c and ord2^c of the axioms from section 10.2.

Our proof is based on an idea of Schütte, which consists in adding a so-called *progression rule* to the infinitary systems. This rule allows us to conclude Pj (where j is any numeral) from all Pi for $i \prec j$.

10.4.1. DEFINITION. More precisely, we define the notion of a derivation in $\mathbf{Z}^\infty + \text{Prog}(P)$ of depth $\leq \alpha$ by the inductive clauses of definition 10.3.1 and the additional clause Prog(P):

(Prog) If for all $i \prec j$ we have derivations d_i^{Pi} of depths $\leq \alpha_i < \alpha$, then $\langle d_i^{Pi}\rangle_{i\prec j}^{Pj}$ is a derivation of depth $\leq \alpha$.

We also define $\text{cr}(\langle d_i\rangle_{i\prec j}) := \sup_{i\prec j} \text{cr}(d_i)$. ⊠

Since this progression rule only deals with derivations of atomic formulas, it does not affect the cutranks of derivations. Hence the proof of normalization for $\mathbf{Z}^\infty$ carries over unchanged to $\mathbf{Z}^\infty + \text{Prog}(P)$. In particular we have

10.4.2. LEMMA. *For any derivation d^A in $\mathbf{Z}^\infty + \text{Prog}(P)$ of depth $\leq \alpha$ and cutrank $\leq k+1$ we can find a derivation $(d^k)^A$ in $\mathbf{Z}^\infty + \text{Prog}(P)$ with free assumption variables contained in those of d, which has depth $\leq 2^\alpha$ and cutrank $\leq k$.* ⊠

10.4.3. We now show that from the progression rule for P we can easily derive the progressiveness of P.

LEMMA. *We have a normal derivation of $\forall x(\forall y{\prec}x\, Py \to Px)$ in $\mathbf{Z}^\infty + \text{Prog}(P)$ with depth* 5.

PROOF.

$$\dfrac{\cdots \quad \dfrac{\dfrac{\cdots \quad \dfrac{\dfrac{\forall y{\prec}j\, Py}{i \prec j \to Pi}\,\forall\text{E} \qquad i \prec j}{Pi}\,{\to}\text{E} \quad \cdots \quad (\text{all } i \prec j)}{Pj}\,\text{Prog}}{\forall y{\prec}j\, Py \to Pj}\,{\to}\text{I} \quad \cdots \quad (\text{all } j)}{\forall x(\forall y{\prec}x\, Py \to Px)}\,\omega$$

⊠

10.4.4. The crucial observation now is that a normal derivation of $P\ulcorner\beta\urcorner$ must essentially have a depth of at least β. However, to obtain the right estimates for the subsystems $\mathbf{Z}_k$ we cannot apply lemma 10.4.2 down to cutrank 0 (i.e. to normal form) but must stop at cutrank 1. Such derivations, i.e. those of cutrank ≤ 1, will be called *quasi-normal*; they can also be analyzed easily.

10.4.5. We begin by showing that a quasi-normal derivation of a quantifier-free formula can always be transformed without increasing its cutrank or its depth into a quasi-normal derivation of the same formula which

1. does not use the ω-rule, *and*

2. contains $\forall$E only in the initial part of a track starting with an axiom.

Recall that our axioms are of the form $\forall \vec{x} A$ with A quantifier-free.

DEFINITION. (*Quasi-subformula*) The quasi-subformulas of a formula A are given by the following clauses.

(i) A, B are quasi-subformulas of $A \to B$;

(ii) $A(i)$ is a quasi-subformula of $\forall x A(x)$, for all numerals i;

(iii) if A is a quasi-subformula of B, and C is an atomic formula, then $C \to A$ and $\forall x A$ are quasi-subformulas of B;

(iv) "... is a quasi-subformula of ..." is a reflexive and transitive relation. ⊠

EXAMPLE. $Q \to \forall x(P \to A)$, P, Q atomic, is a quasi-subformula of $A \to B$.

We now transfer the subformula property for normal derivations (theorem 10.3.9) to a quasi-subformula property for quasi-normal derivations.

10.4.6. THEOREM. *(Quasi-subformula property) Let $\mathcal{D}$ be a quasi-normal deduction in $\mathbf{Z}^\infty + \mathrm{Prog}(P)$ for $\Gamma \vdash A$. Then each formula in $\mathcal{D}$ is a quasi-subformula of a formula in $\Gamma \cup \{A\}$.*

PROOF. We prove this for tracks of order n, by induction on n. ⊠

10.4.6A. ♠ Prove this in detail.

10.4.7. COROLLARY. *Let $\mathcal{D}$ be a quasi-normal deduction in $\mathbf{Z}^\infty + \mathrm{Prog}(P)$ of a formula $\forall \vec{x} A$ with A quantifier-free from quantifier-free assumptions. Then any track in $\mathcal{D}$ of positive order ends with a quantifier-free formula.*

PROOF. If not, then the major premise of the $\to$E whose minor premise is the offending end formula of the track, would contain a quantifier to the left of $\to$. This contradicts theorem 10.4.6. ⊠

Our next aim is to eliminate the ω-rule. For this we need the notion of an *instance* of a formula.

10.4.8. DEFINITION. (*Instance*) The instances of a formula are given by the following clauses.

(i) If B' is an instance of B and A is quantifier-free, then $A \to B'$ is an instance of $A \to B$;

(ii) $A(i)$ is an instance of $\forall x A(x)$, for all numerals i;

(iii) The relation "... is an instance of ..." is reflexive and transitive. ⊠

10.4.9. LEMMA. *Let $\mathcal{D}$ be a quasi-normal deduction in $\mathbf{Z}^\infty + \text{Prog}(P)$ of a formula A without $\forall$ to the left of $\to$ from quantifier-free assumptions. Then for any quantifier-free instance A' of A we can find a quasi-normal derivation $\mathcal{D}'$ of A' from the same assumptions such that*

(i) $\mathcal{D}'$ does not use the ω-rule,

(ii) $\mathcal{D}'$ contains $\forall$E only in the initial elimination part of a track starting with an axiom, and

(iii) $|\mathcal{D}'| \le |\mathcal{D}|$.

PROOF. By induction on the depth of $\mathcal{D}$. We distinguish cases according to the last rule in $\mathcal{D}$.
Case $\to$E.

$$\frac{A \to B \quad A}{B} \to\text{E}$$

By the quasi-subformula property 10.4.6 A must be quantifier-free. Let B' be a quantifier-free instance of B. Then by definition $A \to B'$ is a quantifier-free instance of $A \to B$. The claim now follows from the IH.
Case $\to$I.

$$\frac{B}{A \to B} \to\text{I}$$

Any instance of $A \to B$ has the form $A \to B'$ with B' an instance of B. Hence the claim follows from the IH.
Case $\forall$E.

$$\frac{\forall x A(x) \quad i}{A(i)} \forall\text{E}$$

Then any quantifier-free instance of $A(i)$ is also a quantifier-free instance of $\forall x A(x)$, and hence the claim follows from the IH.
Case ω.

$$\frac{\dots \quad A(i) \quad \dots \quad (\text{all } i < \omega)}{\forall x A(x)} \omega$$

Any quantifier-free instance of $\forall x A(x)$ has the form $A(i)'$ with $A(i)'$ a quantifier-free instance of $A(i)$. Hence the claim again follows from the IH. ⊠

10.4.9A. ♠ Do the remaining cases.

DEFINITION. A derivation d in $\mathbf{Z}^\infty + \mathrm{Prog}(P)$ is called a *$P\vec{\alpha}, \neg P\vec{\beta}$-refutation* if $\vec{\alpha}$ and $\vec{\beta}$ are disjoint and d derives a formula $\vec{A} \to B := A_1 \to \cdots \to A_k \to B$ with $\vec{A}$ and the free assumptions in d among $P\ulcorner\alpha_1\urcorner, \dots, P\ulcorner\alpha_m\urcorner, \neg P\ulcorner\beta_1\urcorner, \dots, \neg P\ulcorner\beta_n\urcorner$ or true quantifier-free formulas without P, and B a false quantifier-free formula without P or else among $P\ulcorner\beta_1\urcorner, \dots, P\ulcorner\beta_n\urcorner$. ⊠

(So, classically, a $P\vec{\alpha}, \neg P\vec{\beta}$-refutation shows $\bigwedge_i P\ulcorner\alpha_i\urcorner \to \bigvee_j P\ulcorner\beta_j\urcorner$.)

10.4.10. LEMMA. *Let d be a quasi-normal $P\vec{\alpha}, \neg P\vec{\beta}$-refutation. Then*

$$\min(\vec{\beta}) \le |d| + \mathrm{lh}(\vec{\alpha}'),$$

where $\vec{\alpha}'$ is the sublist of $\vec{\alpha}$ consisting of all $\alpha_i < \min(\vec{\beta})$, and $\mathrm{lh}(\vec{\alpha}')$ denotes the length of the list $\vec{\alpha}'$.

PROOF. By induction on $|d|$. By lemma 10.4.9 we may assume that d does not contain the ω-rule, and contains $\forall$E only in a context where leading universal quantifiers of an axiom are removed. We distinguish cases according to the last rule in d.

Case $\to$I. By our definition of refutations the claim follows immediately from the IH.

Case $\to$E. Then $d = f^{C \to (\vec{A} \to B)} e^C$. If C is a true quantifier-free formula without P or of the form $P\ulcorner\gamma\urcorner$ with $\gamma < \min(\vec{\beta})$, the claim follows from the IH for f:

$$\min(\vec{\beta}) \le |f| + \mathrm{lh}(\vec{\alpha}') + 1 \le |d| + \mathrm{lh}(\vec{\alpha}').$$

If C is a false quantifier-free formula without P or of the form $P\ulcorner\gamma\urcorner$ with $\min(\vec{\beta}) \le \gamma$, the claim follows from the IH for e:

$$\min(\vec{\beta}) \le |e| + \mathrm{lh}(\vec{\alpha}') + 1 \le |d| + \mathrm{lh}(\vec{\alpha}').$$

It remains to consider the case when C is a quantifier-free implication involving P. Then $\mathrm{lev}(C) \ge 1$, hence $\mathrm{lev}(C \to (\vec{A} \to B)) \ge 2$ and therefore (since $\mathrm{cr}(d) \le 1$) f must be a simple application (10.3.6) starting with an axiom. Now our only axioms involving P are Eq_P: $\forall x, y(x = y \to Px \to Py)$ and Stab_P: $\forall x(\neg\neg Px \to Px)$, and of these only Stab_P has the right form. Hence $f = \mathrm{Stab}_P \ulcorner\gamma\urcorner$ and therefore e: $\neg\neg P\ulcorner\gamma\urcorner$. Now from $\mathrm{lev}(\neg\neg P\ulcorner\gamma\urcorner) = 2$, the assumption $\mathrm{cr}(e) \le 1$ and again the form of our axioms involving P, it follows that e must end with $\to$I, i.e. $e = \lambda u^{\neg P\ulcorner\gamma\urcorner}.e_0^{\perp}$. So we have

$$\dfrac{\dfrac{f}{\neg\neg P\ulcorner\gamma\urcorner \to P\ulcorner\gamma\urcorner} \qquad \dfrac{\begin{matrix}[u\colon \neg P\ulcorner\gamma\urcorner] \\ e_0 \\ \perp\end{matrix}}{\neg\neg P\ulcorner\gamma\urcorner}}{P\ulcorner\gamma\urcorner}$$

The claim now follows from the IH for e_0.

Case $\forall$E. By assumption we then are in the initial part of a track starting with an axiom. Since d is a $P\vec{\alpha}, \neg P\vec{\beta}$-refutation, that axiom must contain P. It cannot be the equality axiom $\mathrm{Eq}_P\colon \forall x, y(x = y \to Px \to Py)$, since $\ulcorner\gamma\urcorner = \ulcorner\delta\urcorner \to P\ulcorner\gamma\urcorner \to P\ulcorner\delta\urcorner$ can never be (whether $\gamma = \delta$ or $\gamma \neq \delta$) the end formula of a $P\vec{\alpha}, \neg P\vec{\beta}$-refutation. For the same reason it can not be the stability axiom $\mathrm{Stab}_P\colon \forall x(\neg\neg Px \to Px)$. Hence the case $\forall$E cannot occur.

Case Prog(P). Then $d = \langle d_\delta^{P\ulcorner\delta\urcorner}\rangle_{\delta<\gamma}^{P\ulcorner\gamma\urcorner}$. By assumption on d, γ is in $\vec{\beta}$. We may assume $\gamma = \beta_i := \min(\vec{\beta})$, for otherwise the premise deduction $d_{\beta_i}\colon P\ulcorner\beta_i\urcorner$ would be a quasi-normal $P\vec{\alpha}, \neg P\vec{\beta}$-refutation, to which we could apply the IH.

If there are no $\alpha_j < \gamma$, the argument is simple: every d_δ is a $P\vec{\alpha}, \neg P\vec{\beta}, \neg P\delta$-refutation, so by IH, since also no $\alpha_j < \delta$,

$$\min(\vec{\beta}, \delta) = \delta \leq |d_\delta|,$$

hence $\gamma = \min(\vec{\beta}) \leq |d|$.

To deal with the situation that some α_j are less than γ, we observe that there can be at most finitely many α_j immediately preceding γ; so let ε be the least ordinal such that

$$\forall\delta(\varepsilon \leq \delta < \gamma \to \delta \in \vec{\alpha}).$$

Then $\varepsilon, \varepsilon + 1, \dots, \varepsilon + k - 1 \in \vec{\alpha}$, $\varepsilon + k = \gamma$. ε is either a successor or a limit. If $\varepsilon = \varepsilon' + 1$, it follows by the IH that since $d_{\varepsilon'}$ is a $P\vec{\alpha}, \neg P\vec{\beta}, \neg P(\varepsilon - 1)$-refutation,

$$\varepsilon - 1 \leq |d_{\varepsilon-1}| + \mathrm{lh}(\vec{\alpha}') - k,$$

where $\vec{\alpha}'$ is the sequence of $\alpha_j < \gamma$. Hence $\varepsilon \leq |d| + \mathrm{lh}(\vec{\alpha}') - k$, and so

$$\gamma \leq |d| = \mathrm{lh}(\vec{\alpha}').$$

If ε is a limit, there is a sequence $\langle\delta_{f(n)}\rangle_n$ with limit ε, and with all $\alpha_j < \varepsilon$ below $\delta_{f(0)}$, and so by IH

$$\delta_{f(n)} \leq |d_{f(n)}| + \mathrm{lh}(\vec{\alpha}') - k,$$

and hence $\varepsilon \leq |d_{f(n)}| + \mathrm{lh}(\vec{\alpha}') - k$, so $\gamma \leq |d| + \mathrm{lh}(\vec{\alpha}')$. ⊠

10.4.11. THEOREM. *Transfinite induction up to ε_0 is underivable in* $\mathbf{Z}$, *i.e.*

$$\mathbf{Z} \nvdash \forall x(\forall y{\prec}x\, Py \to Px) \to \forall x Px$$

with a relation symbol P, and for $k \geq 3$ transfinite induction up to ω_{k+1} is underivable in $\mathbf{Z}_k$, i.e.

$$\mathbf{Z}_k \nvdash \forall x(\forall y{\prec}x\, Py \to Px) \to \forall x{\prec}\omega_{k+1}\, Px.$$

Proof. We restrict ourselves to the second part. So assume that transfinite induction up to ω_{k+1} is derivable in $\mathbf{Z}_k$. Then by the embedding of $\mathbf{Z}_k$ into $\mathbf{Z}^\infty$ (lemma 10.3.5) and the normal derivability of the progressiveness of P in $\mathbf{Z}^\infty + \mathrm{Prog}(P)$ with finite depth (lemma 10.4.3) we can conclude that $\forall x{\prec}\omega_{k+1}\, Px$ is derivable in $\mathbf{Z}^\infty + \mathrm{Prog}(P)$ with depth $< \omega^{m+1}$ and cutrank $\leq k$. (Note that here we need $k \geq 3$, since the formula expressing transfinite induction up to ω_{k+1} has level 3). Now $k-1$ applications of lemma 10.4.2 yield a derivation of the same formula $\forall x{\prec}\omega_{k+1}\, Px$ in $\mathbf{Z}^\infty + \mathrm{Prog}(P)$ with depth $\gamma < 2_{k-1}^{\omega^{m+1}} < \omega_{k+1}$ and cutrank ≤ 1.

Hence there is also a quasi-normal derivation of $P\ulcorner\gamma+3\urcorner$ in $\mathbf{Z}^\infty + \mathrm{Prog}(P)$ with depth $\gamma+2$ and cutrank ≤ 1, of the form

$$\dfrac{\dfrac{\begin{matrix}\mathcal{D}\\ \forall x{\prec}\omega_{k+1} Px\end{matrix}}{\ulcorner\gamma+3\urcorner \prec \omega_{k+1} \to P\ulcorner\gamma+3\urcorner} \qquad \begin{matrix}\mathcal{D}'\\ \ulcorner\gamma+3\urcorner \prec \omega_{k+1}\end{matrix}}{P\ulcorner\gamma+3\urcorner}$$

where $\mathcal{D}'$ is a deduction of finite depth (it may even be an axiom, depending on the precise choice of axioms for $\mathbf{Z}$); this contradicts lemma 10.4.10. ⊠

10.4.12. *Normalization for arithmetic is impossible*

The normalization theorem for first-order logic applied to one of our arithmetical systems $\mathbf{Z}$ is not particularly useful since we may have used in our derivation induction axioms of arbitrary complexity. Hence it is tempting to first eliminate the induction schema in favour of an induction rule allowing us to conclude $\forall x A(x)$ from a derivation of $A(0)$ and a derivation of $A(\mathrm{S}x)$ with an additional assumption $A(x)$ to be cancelled at this point (note that this rule is equivalent to the induction schema), and then to try to normalize the resulting derivation in the new system $\mathbf{Z}$ with the induction rule. We will apply theorems 10.4.11 and 10.2.2 to show that even a very weak form of the normalization theorem cannot hold in $\mathbf{Z}$ with the induction rule.

Theorem. *The following weak form of a normalization theorem for* $\mathbf{Z}$ *with the induction rule is false: "For any derivation* d^B *with free assumption variables among* $\vec{u}^{\vec{A}}$ *for formulas* $\vec{A}, B$ *of level* $\leq \ell$ *there is a derivation* $(d^*)^B$, *with free assumption variables contained in those of* d*, which contains only formulas of level* $\leq k$*, where* k *depends on* ℓ *only."*

Proof. Assume that such a normalization theorem holds. Consider the formula

$$\forall x(\forall y{\prec}x\, Py \to Px) \to \forall x{\prec}\omega_{n+1}\, Px$$

expressing transfinite induction up to ω_{n+1}, which is of level 3. By theorem 10.2.2 it is derivable in $\mathbf{Z}$. Now from our assumption it follows that there exists a derivation of this formula containing only formulas of level $\leq k$, for some k independent of n. Hence $\mathbf{Z}_k$ derives transfinite induction up to ω_{n+1} for any n. But this clearly contradicts theorem 10.4.11. ⊠

10.5 TI for non-standard orderings

The results proved up to now in this chapter all refer to the standard definition of a well-ordering $\prec$ of order type ε_0 in section 10.1. We now consider the question whether these results can be transferred to orderings defined in a less standard way. It will turn out that all our attempts in this direction fail.

The results in this section require classical logic, but apart from that are to a large extent independent of the particular formulation of an arithmetical system. However, in 10.5.4 it will be convenient to be more specific about the function constants allowed. Therefore we assume that we have constants for all primitive recursive functions; clearly it then suffices to have a single relation constant $=$ for equality. As non-logical axioms we take the defining equations for all primitive recursive functions (and of course the equality axioms) plus the Peano axioms PA1, PA2^c and the induction schema from 10.2.1. The resulting formalism is called *Peano arithmetic* **PA**.

10.5.1. We first consider the schema $\mathrm{TI}_{\lhd}(A, x)$, where $\lhd$ is a primitive recursive (non-standard) definition of a well-ordering of order type $< \varepsilon_0$. By means of a counterexample we will see that in general $\mathrm{TI}_{\lhd}(A, x)$ is unprovable in **PA**, even if the ordering defined by $\lhd$ is the standard ordering $<$ of the natural numbers.

Let $\forall x A(x)$ be an arbitrary universal formula of arithmetic. We may assume that $x < y \wedge A(y) \to A(x)$ is provable; otherwise take $\forall z{<}x\, A(z)$ instead of $A(x)$. Depending on A we define an ordering $<_A$ by

$$n <_A m := \begin{cases} n < m \text{ and } A(n), \text{ or} \\ m < n \text{ and } \neg A(n). \end{cases}$$

Then $n <_A m$ or $m <_A n$ whenever $n \neq m$. To see this, suppose $n < m$. Then in case $A(n)$ we have $n <_A m$, and in case $\neg A(n)$ we also have $\neg A(m)$ by our assumption on A, hence $m <_A n$.

If $\forall x A(x)$ is true, $<_A$ defines the standard ordering $<$ of $\mathbb{N}$. Otherwise there is a minimal k such that $\neg A(k)$ holds; the ordering may then be visualized as

$$(1) \qquad 0, 1, 2, 3, \ldots, k-1, \ldots k+3, k+2, k+1, k,$$

i.e. the initial segment $[0, k-1]$ ordered by $<$, followed by the segment $(\omega, k]$ ordered by $>$. So the linear ordering $<_A$ is a well-ordering iff $\forall x A(x)$ holds.

We now show that we can formally derive $\forall x A(x)$ from an instance of transfinite induction on $<_A$ within **PA**, or more precisely

$$(2) \qquad \mathbf{PA} + \mathrm{TI}_{<_A}(A, x) \vdash \forall x A.$$

To prove (2), recall that $\mathrm{TI}_{<_A}(A, x)$ is

$$\forall y(\forall x{<_A}y\, A(x) \to A(y)) \to \forall x A(x).$$

It clearly suffices to prove the premise $\forall y(\forall x{<_A}y\, A(x) \to A(y))$. So let y be given and assume

(3) $\forall x{<_A}y\, A(x)$.

We have to show $A(y)$. So assume $\neg A(y)$. Then $k \leq y$ for the minimal k such that $\neg A(k)$. Because of the form (1) of the ordering $<_A$ we can conclude that y is in the non-well-founded part of $<_A$, hence $y+1 <_A y$. Therefore $A(y+1)$ by (3), contradicting $\neg A(k)$ and $k \leq y < y+1$.

Note that if A is quantifier-free, then the ordering $<_A$ is primitive recursive (and the above argument may be recast in minimal logic). Now since there is a true formula $\forall x(fx = 0)$ with f primitive recursive that is unprovable in **PA** (e.g. the formula expressing the consistency of **PA**), we have

PROPOSITION. *There is a primitive recursive definition* $\lhd$ *of the standard ordering of* $\mathbb{N}$ *such that* **PA** *does not derive the schema* $\mathrm{TI}_{\lhd}(A, x)$ *of transfinite induction with respect to* $\lhd$. ⊠

10.5.2. We now ask ourselves whether $\mathbf{PA} \vdash \mathrm{TI}_{\lhd}(A, x)$ for an ordering $\lhd$ and all arithmetical A implies that $\lhd$ defines a well-ordering of order type less than ε_0. The answer is no, in a strong sense: it is not even true that such a $\lhd$ must be well-founded, even if we require it to be primitive recursive. We now prove this by means of a counterexample. The idea for this example is based on the properties of an implicit truth function for arithmetic. As a first step we need Tarski's classic result on the undefinability of a truth predicate for arithmetic.

For the rest of this chapter we presuppose some familiarity with Gödel numberings of arithmetic. We shall assume that a standard numbering for terms, formulas and formal proofs of classical first-order arithmetic **PA** has been given.

We use $\mathbf{p}$ for some standard primitive recursive bijective coding of pairs of natural numbers onto the natural numbers, with inverses $\mathbf{p}_0, \mathbf{p}_1$.

Furthermore there is some coding of finite sequences from $\mathbb{N}$ into $\mathbb{N}$, such that the primitive recursive extraction function $\lambda xy.(x)_y$ given by

$$(n)_y = \begin{cases} x_y \text{ if } y < u, \\ 0 \text{ otherwise,} \end{cases}$$

for an n coding the sequence $\langle x_0, x_1, \ldots, x_{u-1}\rangle$, and a primitive recursive length function lth such that

$$\mathrm{lth}(n) = u.$$

For the Gödel number of an expression $\mathcal{E}$ we write $\ulcorner\mathcal{E}\urcorner$.

10.5.3. THEOREM. *(Undefinability of a truth predicate for* **PA***) A truth predicate for arithmetic is a predicate T such that $T(\ulcorner A \urcorner) \leftrightarrow A$ is true for all arithmetical sentences A. There is no arithmetically definable truth predicate for arithmetic.*

PROOF. Let Sub_z be the arithmetical operation such that

$$\mathrm{Sub}_z(\ulcorner t \urcorner, \ulcorner A \urcorner) = \ulcorner A[z/t] \urcorner,$$

where z is a fixed variable, t a term, and $[z/t]$ indicates substitution of t for the variable z. Assume now T to be arithmetically definable.

$$S(x, y) := T(\mathrm{Sub}_z(\ulcorner \overline{x} \urcorner, y)),$$

where $\overline{0} := 0, \overline{n+1} := S\overline{n}$. Then we have

$$S(\overline{n}, \ulcorner A \urcorner) \leftrightarrow A[z/\overline{n}].$$

To see this, note

$$S(\overline{n}, \ulcorner A \urcorner) \leftrightarrow S(n, \ulcorner A \urcorner) \leftrightarrow T(\mathrm{Sub}_z(\ulcorner \overline{n} \urcorner, \ulcorner A \urcorner)) \leftrightarrow T(\ulcorner A[z/\overline{n}] \urcorner) \leftrightarrow A[z/\overline{n}].$$

Now a contradiction follows. Let

$$A := \neg S(z, z), \quad \text{and } \overline{n} = \ulcorner A \urcorner.$$

Then

$$S(\overline{n}, \overline{n}) \leftrightarrow S(\overline{n}, \ulcorner A \urcorner) \leftrightarrow A[z/\overline{n}] = \neg S(\overline{n}, \overline{n}). \qquad \boxtimes$$

10.5.4. *Implicit truth function*

Let f be the characteristic function of the truth predicate for arithmetical sentences, i.e. for sentences A:

$$f(\ulcorner A \urcorner) = 0 \leftrightarrow A.$$

(For definiteness, we may assume fn to be any value, say 0, on n which are not the Gödel number of a formula). f is not arithmetically definable, as we have seen, but in any case it must satisfy a number of conditions with respect to the logical operators:

(i) $f(\ulcorner s_1 = s_2 \urcorner) = 0 \leftrightarrow \mathrm{Val}(\ulcorner s_1 \urcorner) = \mathrm{Val}(\ulcorner s_2 \urcorner)$,

(ii) $f(c(\ulcorner A \urcorner, \ulcorner B \urcorner)) = 0 \leftrightarrow f(\ulcorner A \urcorner) = 0 \wedge f(\ulcorner B \urcorner) = 0$, where $c(\ulcorner A \urcorner, \ulcorner B \urcorner) = \ulcorner A \wedge B \urcorner$,

(iii) $f(n(\ulcorner A \urcorner)) = 0 \leftrightarrow f(\ulcorner A \urcorner) = 1$, where $n(\ulcorner A \urcorner) = \ulcorner \neg A \urcorner$,

(iv) $f(\ulcorner \forall v_i A(v_i) \urcorner) = 0 \leftrightarrow \forall n (f(\mathrm{Sub}_{v_i}(\ulcorner \overline{n} \urcorner, \ulcorner A \urcorner)) = 0)$,

where Val is a function computing the value of the closed terms. The conditions (i)–(iv) completely determine f on the set Sent of Gödel numbers of sentences.

LEMMA. *f satisfies a Π^0_2-condition of the form*

$$\forall x \exists y R(x, y, f),$$

where R is a primitive recursive predicate.

PROOF. We recast the conditions above. For the components $(n)_i$ of n, when n is viewed as the code of a finite sequence, we write n_i in this proof. For definiteness, we assume that if $n \in \text{Sent}$, then n_0 describes the main operator of a formula given as a Gödel number: $n_0 = 0$ for $\ulcorner A \urcorner = n$ prime, $n_0 = 1, 2, 3$ for conjunctions, negations, and universal quantifiers respectively. From (i)–(iv) we obtain

$$\begin{aligned}
\forall n {\in} \text{Sent}((fn = 0 \to \{&(n_0 = 0 \to \text{Val}(n_1) = \text{Val}(n_2)) \wedge \\
&(n_0 = 1 \to fn_1 = 0 \wedge fn_2 = 0) \wedge \\
&(n_0 = 2 \to fn_1 \neq 0) \wedge \\
&(n_0 = 3 \to \forall m(f(\text{Sub}(n_1, n_2, m)) = 0))\}) \wedge \\
(fn \neq 0 \to \{&(n_0 = 0 \to \text{Val}(n_1) \neq \text{Val}(n_2)) \wedge \\
&(n_0 = 1 \to fn_1 \neq 0 \vee fn_2 \neq 0) \wedge \\
&(n_0 = 2 \to fn_1 = 0) \wedge \\
&(n_0 = 3 \to \exists m'(f(\text{Sub}(n_1, n_2, m')) \neq 0))\})).
\end{aligned}$$

Here $\text{Sub}(n_1, n_2, m) = \text{Sub}_{v_{n_2}}(\ulcorner \overline{m} \urcorner, n_1)$ and if $n_0 = 3$, n codes $\ulcorner \forall v_{n_2} B \urcorner$. Note that $\text{Val}(n_1) = \text{Val}(n_2)$ can be written in the form $\exists k Q(n_1, n_2, k)$ with primitive recursive Q. Intuitively, $Q(n_1, n_2, k)$ says that k codes two terminating computations yielding the values of the closed terms coded by n_1, n_2, and that the values are equal. Similarly, $\text{Val}(n_1) \neq \text{Val}(n_2)$ can be written as $\exists k' Q'(n_1, n_2, k')$.

By moving the quantifiers $\forall m$, $\exists m'$, $\exists k$, $\exists k'$ outwards we obtain a quantifier-free formula R preceded by $\forall n \forall m \exists m' \exists k \exists k'$, and the truth of the lemma is now obvious. ⊠

10.5.4A. ♠ Describe the construction of Q mentioned in the proof.

10.5.5. REMARK. At this point one can conclude easily that Beth's definability theorem 4.4.2B and hence also the interpolation theorem 4.4.2 cannot hold for Peano arithmetic **PA**. To see this, note that as in 10.5.4 one can give an implicit definition of the truth predicate T for arithmetic (here for definiteness we assume that Tn is false for n which are not the Gödel number of a formula), in the form $A(P) := \forall x \exists y R(x, y, P)$ with a unary predicate symbol P and a primitive recursive predicate R. Clearly P is uniquely determined by $A(P)$, i.e. $A(P) \wedge A(Q) \to \forall x(Px \leftrightarrow Qx)$ is derivable in **PA**. Now if Beth's theorem would hold, then there would be a formula C not involving P such that $A(P) \to \forall x(C \leftrightarrow Px)$ in derivable in **PA** and hence true in the standard model. This contradicts Tarski's undefinability theorem 10.5.3.

10.5.6. Definition. Let $f'x := \min_y R(x, y, f)$, where f is the characteristic function of the truth predicate, and R is as above. ⊠

Note that $\forall x R(x, f'x, f)$ holds. Let $\langle f', f\rangle$ be the encoding of f' and f into a single sequence; this sequence is obviously not arithmetical, but satisfies the Π^0_1-condition just mentioned.

10.5.7. We now want to argue that any pair f', f satisfying $\forall x R(x, f'x, f)$ is such that f is the characteristic function of the truth predicate for arithmetical sentences. To see this, observe that from $\forall x R(x, f'x, f)$ we obtain $\forall x \exists y R(x, y, f)$ and hence the condition given in the proof of the lemma in 10.5.4. But as already noted this determines f on the set Sent of Gödel numbers of sentences.

10.5.8. Let us now make use of this insight to produce the counterexample we are aiming at. As a first step we will rewrite $\forall x R(x, f'x, f)$ in the form $\forall x Q(\overline{\alpha}x)$, or more precisely construct a primitive recursive predicate Q such that $\{\langle f', f\rangle \mid \forall x R(x, f'x, f)\} = \{\alpha \mid Q(\overline{\alpha}x)\}$.

The first step consists in replacing each subterm ft in $R(x, f'x, f)$ by $\mathbf{p}_1(\alpha t)$ and $f'x$ by $\mathbf{p}_0(\alpha x)$, yielding the form $\forall x R_1(x, \alpha)$. We now argue that this can be rewritten as

$$\forall x \forall x_1 \ldots \forall x_n (x_1 = \alpha t_1 \wedge \cdots \wedge x_n = \alpha t_n \rightarrow R_2(x, x_1, \ldots, x_n))$$

where $t_1, \ldots, t_n$ are terms without α. This can be proved easily by induction on the number of nestings of α; e.g. $\forall x(\alpha(\alpha x) = 0)$ is rewritten as $\forall x \forall x_1 \forall x_2 (x_1 = \alpha x \wedge x_2 = \alpha x_1 \rightarrow x_2 = 0)$. Now this can be further rewritten as

$$\begin{aligned}\forall y \forall x, x_1, \ldots, x_n, y_1, \ldots, y_n {\leq} y\, (y_1 = t_1 \wedge \cdots \wedge y_n = t_n \wedge \\ x_1 = \alpha y_1 \wedge \cdots \wedge x_n = \alpha y_n \rightarrow R_2(x, x_1, \ldots, x_n)).\end{aligned}$$

Finally αy_i can be replaced by $(\overline{\alpha}y)_{y_i}$ and y by $\mathrm{lth}(\overline{\alpha}y)$. This yields the form $\forall x Q(\overline{\alpha}x)$ with a primitive recursive Q.

10.5.9. Definition. Let α be a variable ranging over functions in $\mathbb{N} \rightarrow \mathbb{N}$. With any Π^1_1-sentence $\forall \alpha \exists x \neg Q(\overline{\alpha}x)$ we associate a tree

$$\mathrm{Tr}_Q := \{\overline{\alpha}x \mid \forall y {\leq} x\, Q(\overline{\alpha}y)\}.$$

The converse of the partial ordering $\lhd$ on Tr_Q (initial segment ordering of finite sequences) can be extended to a linear ordering $\lhd^*$ (the *Brouwer–Kleene ordering* of Tr_Q) by defining the converse

$$\begin{aligned}\overline{\alpha}x \rhd^* \overline{\beta}y \text{ if either } & (\overline{\alpha}x \lhd \overline{\beta}y) \text{ or} \\ & (\overline{\alpha}z = \overline{\beta}z \text{ and } \alpha z > \beta z \text{ for some } z < x, y).\end{aligned}$$ ⊠

10.5.10. LEMMA. *If* Tr_Q *is well-founded under the converse of* $\lhd$, *then* $\lhd^*$ *is a well-ordering.*

PROOF. Consider any infinite sequence within Tr_Q

$$\overline{\alpha}_0 x_0 \rhd^* \overline{\alpha}_1 x_1 \rhd^* \overline{\alpha}_2 x_2 \rhd^* \cdots ;$$

from this we find an increasing sequence in $\lhd$, within Tr_Q, that is to say, an infinite sequence α in Tr_Q. For $\alpha 0$ we can take (assuming $x_0 > 0$)

$$\alpha 0 = \lim_{i \to \infty} \alpha_i 0.$$

This limit is determined, since $\overline{\alpha}_1 x_1$ is either a prolongation of $\overline{\alpha}_0 x_0$, or $\overline{\alpha}_0 z = \overline{\alpha}_1 z$ and $\alpha_1 z < \alpha_0 z$; so either $\alpha_0 0 = \alpha_1 0$ or $\alpha_1 0 < \alpha_0 0$; the value of α_i at 0 can go down at most finitely often, then $\alpha_i 0$ remains fixed. And so on for $\alpha 1, \alpha 2$ etc. ⊠

10.5.11. THEOREM. *The Brouwer–Kleene ordering* $\lhd^*$ *of the* Π^1_1*-sentence* $\forall \alpha \exists x \neg Q(\bar{\alpha} x)$, Q *as in 10.5.8, is primitive recursive, not well-founded, but well-founded w.r.t. all arithmetical sequences. It follows that* $\mathrm{TI}_{\lhd^*}(A, x)$ *holds for any arithmetical formula* $A(x)$, *i.e.*

$$\forall y (\forall x \lhd^* y A(x) \to A(y)) \to \forall x A(x).$$

PROOF. If the Brouwer–Kleene ordering $\lhd^*$ were well-founded, it would mean that there was no truth definition; in other words, the encoding of arithmetical truth $\langle f', f \rangle$ provides an infinite sequence on the Brouwer–Kleene ordering.

On the other hand, we have observed in 10.5.7 that any $\alpha = \langle f', f \rangle$ satisfying $\forall x Q(\bar{\alpha} x)$ and hence $\forall x R(x, f'x, f)$ is such that f is the characteristic function of the truth predicate for arithmetical sentences. Hence there is no arithmetically definable such α, i.e. $\lhd^*$ is arithmetically well-founded.

Finally let $A(x)$ be any arithmetical formula and assume that $\mathrm{TI}_{\lhd^*}(A, x)$ does not hold. Then the premise $\forall y (\forall x \lhd^* y A(x) \to A(y))$ is true but the conclusion $\forall x A(x)$ is false, hence $\neg A(x_0)$ for some x_0. From the premise for x_0 we obtain an $x_1 \lhd^* x_0$ such that $\neg A(x_1)$, then an $x_2 \lhd^* x_1$ such that $\neg A(x_2)$, and so on. If we pick the smallest x_0 such that $\neg A(x_0)$, then the smallest $x_1 \lhd^* x_0$ such that $\neg A(x_1)$ and so on, we obtain an arithmetically definable sequence $x_0, x_1, x_2, \ldots$ such that $x_{i+1} \lhd^* x_i$ for all i, contradicting our previous observation. ⊠

10.6 Notes

Some important general references on the branch of proof theory generated by (modifications of) Hilbert's programme (cf. remarks in the preface) are

Schütte [1960], Kreisel [1977], Schütte [1977], Takeuti [1987], Girard [1987b], Pohlers [1989], Buchholz et al. [1981]. Girard [1987b] covers more than just proof theory in the Hilbert programme tradition. In the appendices to Takeuti [1987] some leading proof-theorists have given their views on proof theory.

10.6.1. *Gentzen's consistency proofs.* The proof in Gentzen [1936] starts from a formalism for arithmetic based on natural deduction. Gentzen defines a notion of "reduction step" for deductions, which preserves correctness. If no reduction step is possible, the conclusion must be a sequent the truth of which is immediately decidable. He then assigns ordinal notations less than ε_0 to derivations and shows that suitable reduction steps lower the ordinal (notation) assigned to a derivation, ultimately producing a derivation to which no reduction step is applicable. A fully reduced derivation of $1 = 2$ is impossible, and from this it may be concluded that arithmetic is consistent. In fact, Gentzen's argument uses $\mathrm{TI}_{\prec}(A, x)$ for a quantifier-free A in his consistency proof.

Originally Gentzen had a different version of the proof, not based on transfinite induction, but on a notion of "reduction rule" instead. A reduction rule is something like a strategy for reducing derivations to correct derivations. Objections raised to this proof because of a supposed use of Brouwer's fan theorem induced Gentzen to shortcircuit this discussion by using transfinite induction instead. On this early version of Gentzen's proof, see Bernays [1970].

Gentzen's first proof is not easy to follow, and in Gentzen [1938] he presents a second version, based on a "Gentzen system" (as it is called in this book), which is more perspicuous, even if the formalism of arithmetic itself is somewhat less natural.

Finally, in Gentzen [1943], he proved the initial cases of transfinite induction along $\prec$ (10.2.2) and gave a direct proof of the underivability of $\mathrm{TI}_{\prec}(P)$ for a predicate letter P (10.4.11).

Schütte [1950a] showed that Gentzen's consistency proofs could be made more perspicuous using an infinitary proof system with ω-rule (such as $\mathbf{Z}$ in our exposition), and embedding standard arithmetic in the infinitary system.

Our exposition in section 10.4 takes Gentzen [1943] as point of departure, but incorporates Schütte's idea of using an infinitary system. Moreover, the logical basis is an N-system; for a similar exposition based on a Gentzen system, see Schwichtenberg [1977].

Gentzen's results on transfinite induction provided the first example of a true, mathematically meaningful statement not provable in first-order arithmetic, in contrast to the original incompleteness result of Gödel, where the unprovable statement was entirely motivated by metamathematical considerations. Still, transfinite induction up to ε_0 might be regarded as esoteric. The first example of a purely combinatorial statement, of "straightforwardly mathematical character" was found by Paris [1978]; see also Harrington and

Paris [1977]. After the first result of this type, many more followed; examples with references may be found in Takeuti [1987, section 12], Buchholz and Wainer [1987], Gallier [1991], Friedman and Sheard [1995].

10.6.2. *Subsystems of* $\mathbf{Z}$. Refinements of Gentzen's theorem 10.2.2 on provable and 10.4.11 on unprovable initial cases on TI in $\mathbf{Z}$ to corresponding results for the subsystems $\mathbf{Z}_k$ were first obtained by Mints [1971] and Parsons [1973].

10.6.3. *Continuous cut elimination.* An important version of cut elimination for infinitary systems is *continuous* cut elimination, due to Mints [1975]. Provided one permits a "repetition rule" (which simply repeats the premise as the conclusion), cut elimination may be defined as a *continuous* operation (in the usual tree topology) on prooftrees.

This technique has been successfully applied in, for example, Gordeev [1988] and Buchholz [1991]. Buchholz obtains a neat proof of the uniform reflection principle by this method. The uniform reflection principle may be stated as

$$\mathrm{Proof}_{\mathbf{PA}}(\ulcorner A(\dot{x})\urcorner \to A(x)) \qquad (\mathrm{FV}(A) \subset \{x\}),$$

where $\mathrm{Proof}_{\mathbf{S}}$ is the standard arithmetized proof predicate for system $\mathbf{S}$, and $\ulcorner A(\dot{x})\urcorner$ is the Gödel number of $A(S^x 0)$ as a function of x, i.e. $\ulcorner A(\dot{x})\urcorner$ is represented by a term containing at most x free.

For continuous normalization of infinite terms, see Schwichtenberg [1998].

10.6.4. *TI for non-standard orderings.* Both counterexamples in the text are due to Kreisel. The first one, in 10.5.1, was given by Kreisel in lectures on proof theory at UCLA (Kreisel [1968]). It is also mentioned in Kreisel [1977]. The second one, leading to theorem 10.5.11, is from Kreisel [1953]. In fact, we can find a primitive recursive ordering which is not well-founded, but well-founded w.r.t. all *hyper*arithmetical sequences. This follows from the characterization of Π^1_1-predicates as $\Sigma^1_{1,\mathrm{Hyp}}$-predicates given by Kleene [1955, Theorem XXVI].

This should be contrasted with the result of Friedman and Scedrov [1986] showing that matters for intuitionistic first-order arithmetic $\mathbf{HA}$ are different.

The remark on the failure of interpolation for arithmetic (10.5.5) we owe to G.E. Mints, who considers it to be folklore.

Chapter 11

Second-order logic

11.1 Intuitionistic second-order logic

There exists a close connection between intuitionistic second-order logic and the so-called *polymorphic lambda calculus* $\boldsymbol{\lambda 2}$. $\boldsymbol{\lambda 2}$ is an extension of $\boldsymbol{\lambda}_{\rightarrow}$ permitting abstraction over type variables, and under the formulas-as-types paradigm it may be regarded as isomorphic to the natural deduction system for intuitionistic second-order propositional logic $\rightarrow\forall^2\mathbf{Nip}^2$, that is to say, $\rightarrow\mathbf{Nmp}$ extended with quantification over propositions. $\boldsymbol{\lambda 2}$ is an important component of various type systems studied in computer science.

In this chapter we show how strong normalization for intuitionistic second-order predicate logic may be reduced to normalization for $\mathbf{Nmp}^2$, and we present a proof of strong normalization for $\rightarrow\forall^2\mathbf{Nip}^2$ ($=\boldsymbol{\lambda 2}$). Next we show that intuitionistic second-order arithmetic $\mathbf{HA}^2$ (formal intuitionistic analysis) can be represented in intuitionistic second-order logic, and show that every recursive function which is provably total in $\mathbf{HA}^2$ can be represented by a term of $\boldsymbol{\lambda 2}$. From the formal deduction establishing totality of the function being considered, one can read off an algorithm, encoded by a term of $\boldsymbol{\lambda 2}$, for computing the function.

11.1.1. *Description of* $\mathbf{Ni}^2$

To the first-order language, second-order quantifiers $\forall X^n, \exists X^n$ are added. If we wish to distinguish the symbols for second-order quantification from those for first-order quantification, we write $\forall^2, \exists^2$. The additional quantifier rules are given by

$$\frac{A}{\forall Y^n A[X^n/Y^n]}\,\forall^2\mathrm{I} \qquad\qquad \frac{\forall X^n A}{A[X^n/\lambda x_1 \ldots x_n.B]}\,\forall^2\mathrm{E}$$

$$\frac{A[X^n/\lambda x_1 \ldots x_n.B]}{\exists X^n A}\,\exists^2\mathrm{I} \qquad\qquad \frac{\exists X^n A \qquad \begin{matrix}[A[X^n/Y^n]]^u \\ \mathcal{D} \\ C\end{matrix}}{C}\,u, \exists^2\mathrm{E}$$

where

- in $\forall^2$I A does not depend on open hypotheses containing X^n free, and Y^n is free for X^n in A;

- $A[X^n/\lambda x_1 \dots x_n.B]$ is obtained from A by replacing each occurrence of a subformula $X^n t_1 \dots t_n$ in A by $B[x_1, \dots, x_n/t_1, \dots, t_n]$;

- in $\exists^2$E Y^n does not occur free in assumptions on which C depends exept $A[X^n/Y^n]$, nor does Y^n occur free in C.

These rules are the same as in 6.6.3, except that the restriction on B has been dropped.

11.1.2. *Restriction to the language with* $\to, \forall, \forall^2$

There is a good deal of redundancy in the operators of second-order logic, since we can define $\bot, \wedge, \vee, \exists, \exists^2$ from $\to, \forall, \forall^2$ as follows. In the definitions below X^0 is not free in A, B.

$$\begin{array}{lcl}
\bot & := & (\forall X^0)X^0, \\
A \wedge B & := & \forall X^0((A \to (B \to X)) \to X), \\
A \vee B & := & \forall X^0((A \to X) \to ((B \to X) \to X)), \\
\exists y A & := & \forall X^0(\forall y(A \to X) \to X), \\
\exists Y^n A & := & \forall X^0(\forall Y^n(A \to X) \to X)).
\end{array}$$

Under these definitions, the usual introduction and elimination rules for the defined operators become derived rules in the fragment based on $\to, \forall, \forall^2$. For example, the following deductions show that $\wedge$I, $\wedge\text{E}_\text{L}$ are derivable.

$$\dfrac{\dfrac{\dfrac{\dfrac{A\to(B\to X^0)^u \quad A}{B \to X^0} \quad B}{X^0}}{(A\to(B\to X^0))\to X^0}\,u}{\forall X^0((A\to(B\to X))\to X)} \qquad \dfrac{\dfrac{\forall X^0((A\to(B\to X))\to X)}{(A\to(B\to A))\to A} \quad \dfrac{\dfrac{A^u}{B \to A}}{A\to(B\to A)}\,u}{A}$$

and the proof of $\wedge\text{E}_\text{R}$ is similar to the proof of $\wedge\text{E}_\text{L}$. Note that the distinction between minimal and intuitionistic logic disappears in second-order logic.

Henceforth we shall assume $\mathbf{I}^2$ to be formulated with primitives $\to, \forall, \forall^2$ only, unless expressly indicated otherwise.

11.1.2A. ♠ Derive the rules for the other defined operators.

11.1.3. *Normal deductions and normalization for* $\mathbf{Ni}^2$

We can formulate notions of conversion, reduction and normal form in the same way as for **Ni**. In particular, to the detour conversions of the $\to\forall$-fragment of **Ni** we add $\forall^2$-detour conversions:

$$\dfrac{\dfrac{\begin{matrix}\mathcal{D}\\ A\end{matrix}}{\forall Y^n A[x^n/y^n]}}{A[X^n/\lambda\vec{x}.B]} \quad \text{cont} \quad \begin{matrix}\mathcal{D}[X^n/\lambda\vec{x}.B]\\ A[X^n/\lambda\vec{x}.B]\end{matrix}$$

Since we restrict attention to the $\to\forall\forall^2$-language, there is no need for considering permutative conversions and immediate simplifications. But it is of interest to note that the detour conversions for the operators $\wedge, \vee, \bot, \exists, \exists^2$ correspond under the definitions to transformations of prooftrees which result from $\to\forall\forall^2$-detour conversions. On the other hand, permutative conversions correspond after translation under the definitions to transformations not generated by the $\to\forall\forall^2$-detour conversions.

11.1.4. Proposition. *Strong normalization w.r.t. detour conversions of* $\mathbf{Ni}^2$ *in the full language is reducible to strong normalization w.r.t. detour conversions for the* $\to\forall\forall^2$*-fragment of* $\mathbf{Ni}^2$.

11.1.4A. ♠ Prove the preceding proposition in detail.

11.1.4B. ♠ Give an example of a permutation conversion in the full second-order language which does not translate into a series of conversions relative to $\to, \forall$ and $\forall^2$.

There is no meaningful subformula property for second-order logic. Any $A[X^n/\lambda\vec{x}.B]$ ought to count as a subformula of $\forall X^n A$, but the logical complexity of the subformula may be very much larger than the complexity of $\forall X^n A$. This is also the reason why it does not seem to be worthwhile to strive for normalization relative to the full language, including permutative conversions.

Nevertheless, some useful conclusions can be drawn from the fact that a derivation can be brought into normal form.

For one thing, a normal derivation of a first-order formula (i.e. a formula without $\forall^2$) does not contain second-order quantifiers. Another example is given by the following (cf. 6.2.7D):

11.1.5. Proposition. *A normal derivation without open assumptions ends with an introduction.*

Proof. Let $\mathcal{D}$ be a normal derivation without open assumptions. Assume that $\mathcal{D}$ ends with an elimination rule. Follow a main branch (defined as in

6.2.5) starting from the conclusion. We pass through eliminations only; at the top we find a formula which cannot be an open assumption (there are none, by hypothesis), but also cannot be discharged by $\to$I since there are no introductions below this formula; contradiction. ⊠

11.1.6. PROPOSITION. *Let $\mathcal{D}$ be a normal derivation in* $\mathbf{Ni}^2$ *of* $A \vee B$ *without open hypotheses, i.e.* $\mathcal{D}$ *derives* $\forall X^0((A \to X) \to (B \to X) \to X)$. *Then* $\mathbf{Ni}^2 \vdash A$ *or* $\mathbf{Ni}^2 \vdash B$.

PROOF. **First proof.** There is an immediate proof from the preceding proposition, if we consider normal deduction w.r.t. the full language, so that $\vee$ appears as a primitive.

However, it is instructive to see how we have to argue if $\vee$ is defined and we are reasoning about the $\to\forall\forall^2$-fragment.

Second proof. Let us follow the main branch (defined as in 6.2.5) of $\mathcal{D}$, going upwards from the conclusion.

If the final rule applied in $\mathcal{D}$ is an E-rule, then the main branch passes through eliminations only and its topmost formula must be an open assumption, which is impossible. So the final rule applied is $\forall^2$I. The immediate subdeduction $\mathcal{D}_1$ of $\mathcal{D}$ therefore terminates with $(A \to X) \to (B \to X) \to X$ and has no open assumptions. Again, the final rule of $\mathcal{D}_1$ must have been an introduction, so the premise $(B \to X) \to X$ is the conclusion of a subdeduction $\mathcal{D}_2$ from open assumptions of the form $A \to X$.

Case 1. $A \to X$ does actually occur as the top formula of the main branch. Then the first rule applied to $A \to X$ has to be an elimination rule, otherwise $A \to X$ or $\forall X(A \to X)$ would have to occur as a subformula of the conclusion in the strict sense. We see that this is impossible, if we keep in mind that X does not occur in A or B. So $\to$E is applied at the top –

$$\dfrac{A \to X \qquad \begin{matrix}\mathcal{D}' \\ A\end{matrix}}{X}$$

– and then introductions must follow in the main branch. $\mathcal{D}'$ may use $A \to X, B \to X$ as open assumptions. But if in $\mathcal{D}'$ we replace the X everywhere by $A \vee B$, these assumptions become derivable and we have found a deduction for A.

Case 2. If there is no formula $A \to X$ at the top of the main branch, it must be the case that the final rule of $\mathcal{D}_2$ is an introduction discharging $B \to X$ at the top of the main branch and possibly other places. The argument is similar to the preceding case, and we find a deduction of B. ⊠

11.2 Ip^2 and $\lambda 2$

11.2.1. Ni^2 *and polymorphic lambda calculus*

In a more or less routine fashion we can reformulate **Ni**^{2} as a calculus of typed terms. We shall not give here the description for the full calculus, but only for intuitionistic second-order *propositional* logic **Ip**^{2}. In **Ip**^{2} only propositional variables $X^0, Y^0, Z^0, \ldots$ occur; we drop the superscript ^{0}. As in the full system, $\bot, \wedge, \vee, \exists, \exists^2$ are definable.

The new clauses describing the formation of terms for deductions in propositional **Ni**^{2} are given by

$$\frac{t\colon A}{\Lambda X.t\colon (\forall X)A} \qquad \frac{t\colon (\forall X)A}{t^{(\forall X)A}B\colon A[X/B]}$$

There is an obvious condition to be met in the case of $\forall^2$I: t may not contain free individual variables with a type in which X occurs free. We have added new operators of *type abstraction*, ΛX, and *type application* (application of a term to a type (= formula). The resulting calculus of typed terms is also known as the *polymorphic lambda calculus* or *system* F (Girard [1971,1972]) or $\boldsymbol{\lambda}$**2**.

Henceforth *term* will be used for first-order terms; for second-order terms we use *type* or *formula.*

The conversions for $\to$ and $\forall^2$ correspond in term notation to

$$\begin{array}{l}(\lambda x^A.t^B)(s^A) \ \text{ cont } \ t[x^A/s^A]\colon B\\ (\Lambda X.t^A\colon (\forall X)B)A \ \text{ cont } \ t[X/A]\colon B[X/A]\end{array}$$

Observe that we do *not* have application on the level of types: instead of having $((\forall X)A)B$ convert to $A[X/B]$, we simply identify $((\forall X)A)B$ with $A[X/B]$, so that we can in fact dispense with the notation $((\forall X)A)B$.

For this system we shall prove strong normalization and uniqueness of normal form in the next section. The remainder of this section is devoted to computational aspects of $\boldsymbol{\lambda}$**2**, in preparation for section 11.5.

11.2.2. *Computational content*

DEFINITION. Let X be a fixed propositional variable in $\boldsymbol{\lambda}_{\to}$. We put

$$\mathrm{N}'_X := X \to ((X \to X) \to X).$$

This is the *type* of *(variant) natural numbers over* X. The so-called *(variant) Church numerals* are terms of type N'_X in normal form:

$$\bar{n}'_X := \lambda x^X f^{X \to X}.f^n(x) : \mathrm{N}'_X,$$

where as before

$$f^0(x) := x; \quad f^{n+1}(x) := f(f^n(x)).$$

We shall drop the subscript X in the sequel, since it will be kept fixed. ⊠

Of course we can also define $\bar{n}'_A$ and N'_A for an arbitrary type A in the same manner.

NOTATION. In the remainder of this chapter we shall simply write $\mathrm{N}_A, \bar{n}_A$ for $\mathrm{N}'_A, \bar{n}'_A$ respectively. ⊠

11.2.2A. ♠ Show that the variant Church numerals of type X are the only terms in normal form of type $\mathrm{N} \equiv \mathrm{N}_X$.

11.2.2B. ♠ All extended polynomials are representable in the variant Church numerals. *Hint.* Take as representing terms

$$\begin{array}{lcl} F_+ & := & \lambda x^N y^N z^X f^{X\to X}.x(yzf)f \\ F_\times & := & \lambda x^N y^N z^X f^{X\to X}.xz(\lambda u^X.yuf) \\ F_{\mathbf{p}_i^k} & := & \lambda x_1^N \ldots x_k^N.x_i \\ F_{\mathbf{c}_n} & := & \lambda x^N.\bar{n} \\ F_{\mathrm{sg}} & := & \lambda x^N y^X f^{X\to X}.xy(\lambda z^X.fy) \\ F_{\overline{\mathrm{sg}}} & := & \lambda x^N y^X f^{X\to X}.x(fy)(\lambda z^X.y). \end{array}$$

As we have seen before in 1.2.21, the class of representable functions becomes larger, if one considers N_A for arbitrary A, and permits representing terms where the types of the input numerals and output numerals may differ.

In $\boldsymbol{\lambda}_\to$ there is arbitrariness in the choice of the type X in N_X. This arbitrariness is removed in $\boldsymbol{\lambda}\mathbf{2}$. There we put:

11.2.3. DEFINITION.

$$\mathrm{N} := \forall X(X \to ((X \to X) \to X)),$$
$$\bar{n} := \Lambda X \lambda x^X f^{X\to X}.f^n(x).$$
⊠

The results on representability of extended polynomials carry over to $\boldsymbol{\lambda}\mathbf{2}$, modulo small adaptations. But as will become clear from the exercises, many more functions are representable in $\boldsymbol{\lambda}\mathbf{2}$ than are representable in $\boldsymbol{\lambda}_\to$. In particular, the functions representable in $\boldsymbol{\lambda}\mathbf{2}$ are closed under recursion.

11.2.3A. ♠ Show that the following operators $\mathbf{p}, \mathbf{p}_0, \mathbf{p}_1$ may be taken as pairing with inverses for types U, V with a defined product type $U \wedge V$:

$$\mathbf{p} := \lambda u^U v^V \Lambda X \lambda x^{U\to(V\to X)}.xuv,$$
$$\mathbf{p}_0 := \lambda x^{U\wedge V}.xU(\lambda y^U z^V.y), \quad \mathbf{p}_1 := \lambda x^{U\wedge V}.xV(\lambda y^U z^V.z).$$

N.B. These terms encode the deductions exhibited in 11.1.2.

11.2.3B. ♠ Let $\mathsf{S} \equiv \lambda z^N \Lambda X \lambda x^X y^{X\to X}.y(zXxy)$, $\mathsf{It} \equiv \Lambda X.\lambda u^X f^{X\to X} z^N.zXuf$. Show that

$$\begin{aligned} &\mathsf{It}\,Xu^X f^{X\to X}\bar{0} &&= u, \\ &\mathsf{It}\,Xu^X f^{X\to X}(\mathsf{S}t) &&= f(\mathsf{It}\,Xu^X f^{X\to X}t). \end{aligned}$$

11.2.3C. ♠ Define with the help of pairing and the iterator of the preceding exercise a *recursor* Rec such that

$$\begin{aligned} &\mathsf{Rec}\,Xu^X f^{X\to(N\to X)}\bar{0} = u^X, \\ &\mathsf{Rec}\,Xu^X f^{X\to(N\to X)}(\overline{\mathsf{S}n}) = f(\mathsf{Rec}\,uf\bar{n})\bar{n}. \end{aligned}$$

11.3 Strong normalization for $\mathbf{Ni}^2$

We first show that strong normalization for full second-order logic can be reduced to strong normalization for propositional $\mathbf{Ni}^2$ or $\boldsymbol{\lambda 2}$.

11.3.1. PROPOSITION. *Strong normalization for* $\mathbf{Ni}^2$ *is a consequence of strong normalization for propositional* $\mathbf{Ni}^2$.

PROOF. We define a mapping ϕ from formulas and deductions of $\mathbf{Ni}^2$ to formulas and deductions of propositional $\mathbf{Ni}^2$, as follows.

$$\phi(X^n t_1 \dots t_n) := X^*$$

where X^* is some propositional variable bijectively associated with the relation variable X^n, and for compound formulas we put

$$\begin{aligned} &\phi(A \to B) := \phi A \to \phi B, \\ &\phi(\forall X^n A) \;:= \forall X^* \phi A, \\ &\phi(\forall x A) \quad\;:= \forall x^* \phi A, \;\; x^* \text{ not free in } \phi A. \end{aligned}$$

Here x^* is a *propositional* variable associated to the individual variable x. Deductions are translated as follows. We write $\overset{\phi}{\mapsto}$ for "ϕ maps to".

- The single-node prooftree "A" is translated into the single-node prooftree "ϕA".
- For prooftrees of depth greater than one, we define inductively

$$\dfrac{\begin{matrix}\mathcal{D}\\ B\end{matrix}}{A \to B}\,{\to}\mathrm{I} \qquad \overset{\phi}{\mapsto} \qquad \dfrac{\begin{matrix}\phi\mathcal{D}\\ \phi B\end{matrix}}{\phi A \to \phi B}\,{\to}\mathrm{I}$$

$$\dfrac{\begin{matrix}\mathcal{D}\\ A \to B\end{matrix} \quad \begin{matrix}\mathcal{D}'\\ A\end{matrix}}{B}\,{\to}\mathrm{E} \qquad \overset{\phi}{\mapsto} \qquad \dfrac{\begin{matrix}\phi\mathcal{D}\\ \phi A \to \phi B\end{matrix} \quad \begin{matrix}\phi\mathcal{D}'\\ \phi A\end{matrix}}{\phi B}\,{\to}\mathrm{E}$$

$$\dfrac{\begin{matrix}\mathcal{D}\\ A[X^n]\end{matrix}}{\forall Y^n A[x^n/Y^n]}\forall^2\mathrm{I} \quad \overset{\phi}{\mapsto} \quad \dfrac{\begin{matrix}\phi\mathcal{D}\\ \phi(A[X^n])\end{matrix}}{\phi(\forall Y^* A[x^n/Y^n])}\forall^2\mathrm{I}$$

$$\dfrac{\begin{matrix}\mathcal{D}\\ \forall X^n A\end{matrix}}{A[X^n/\lambda\vec{x}.B]}\forall^2\mathrm{E} \quad \overset{\phi}{\mapsto} \quad \dfrac{\begin{matrix}\phi\mathcal{D}\\ \forall X^*\phi(A)\end{matrix}}{\phi(A)[X^*/\phi B]}\forall^2\mathrm{E}$$

$$\dfrac{\begin{matrix}\mathcal{D}\\ A[x]\end{matrix}}{\forall y A[x/y]}\forall\mathrm{I} \quad \overset{\phi}{\mapsto} \quad \dfrac{\begin{matrix}\phi\mathcal{D}\\ \phi(A[x])\end{matrix}}{\forall y^*\phi(A[x])}\forall^2\mathrm{I}$$

$$\dfrac{\begin{matrix}\mathcal{D}\\ \forall x A\end{matrix}}{A[x/t]}\forall\mathrm{E} \quad \overset{\phi}{\mapsto} \quad \dfrac{\begin{matrix}\phi\mathcal{D}\\ \forall x^*\phi A\end{matrix}}{\phi A}\forall^2\mathrm{E}$$

Note that for all *individual* variables $\vec{x}$ and terms $\vec{t}$

$$\phi A \equiv \phi(A[\vec{x}\,/\vec{t}\,]).$$

Furthermore,

$$\mathcal{D} \succ_1 \mathcal{D}' \Rightarrow \phi\mathcal{D} \succ \phi(\mathcal{D}'). \qquad \boxtimes$$

11.3.2. In contrast to the situation for $\boldsymbol{\lambda}_{\rightarrow}$, the propositional variables in $\mathbf{Ni}^2$ cannot simply be regarded as formulas of minimal complexity, since in the course of the normalization process quite complex formulas may be substituted for these variables. The notion of computability for $\boldsymbol{\lambda 2}$ has to reflect this; the idea is to assign "variable computability predicates" to the propositional variables. A "computability candidate" has to satisfy certain requirements. A straightforward generalization of 6.8.3 (on which the proof of strong normalization of $\boldsymbol{\lambda 2}$ in Girard et al. [1988] is based) is obtained by defining the notion of a computability candidate as follows.

DEFINITION. A term t is called *non-introduced* if t is not of the form $\lambda x.s$ or $\Lambda X.s$. A set of terms $\mathbf{X}$, all of the same type A, is a *computability candidate* (a *c.c.*) (*of type* A) iff

CC1 If $t \in \mathbf{X}$ then $\mathrm{SN}(t)$;

CC2 If $t \in \mathbf{X}$ and $t \succeq t'$ then $t' \in \mathbf{X}$;

CC3 If t is non-introduced, and $\forall t' \prec_1 t (t' \in \mathbf{X})$ then $t \in \mathbf{X}$. $\boxtimes$

As a corollary of CC3 we find

CC4 If $t : A$ is non-introduced and normal, then $t \in \mathbf{X}$.

Instead of this definition, obtained by transferring the properties C1–3 in 6.8.3 from computability predicates for function types to c.c.'s, we use the slightly different notion of a saturated set which is convenient for generalizations.

11.3.3. *Saturated sets.* To motivate our definition of saturated sets we first collect some properties of the set SN of strongly normalizable terms. In this section, $\vec{\varepsilon}$ will be used for a sequence of first- and second-order terms (types). We say that $\vec{\varepsilon}$ is in SN, if the first-order terms of $\vec{\varepsilon}$ are in SN.

LEMMA.

(i) If $\vec{\varepsilon} \in \mathrm{SN}$, *then for any variable* x *of the appropriate type* $x\vec{\varepsilon} \in \mathrm{SN}$.

(ii) If $t \in \mathrm{SN}$, *then* $\lambda x.t \in \mathrm{SN}$.

(iii) If $t[x/r]\vec{\varepsilon} \in \mathrm{SN}$ *and* $r \in \mathrm{SN}$, *then* $(\lambda x.t)r\vec{\varepsilon} \in \mathrm{SN}$.

(iv) If $t \in \mathrm{SN}$, *then* $\Lambda X.t \in \mathrm{SN}$.

(v) If $t[X/B]\vec{\varepsilon} \in \mathrm{SN}$, *then* $(\Lambda X.t)B\vec{\varepsilon} \in \mathrm{SN}$.

PROOF. (i) Immediate, since every reduction step must take place in a member of $\vec{\varepsilon}$.

(ii), (iv) are treated similarly.

(iii) (cf. lemma 6.8.4) Assume $t[x/r]\vec{\varepsilon} \in \mathrm{SN}$ and $r \in \mathrm{SN}$; we have to show $(\lambda x.t)r\vec{\varepsilon} \in \mathrm{SN}$. We use induction on $h_r + h_{\vec{\varepsilon}} + h_t$, the sum of the sizes of the reduction trees of r, $\vec{\varepsilon}$ and t. If $(\lambda x.t)r\vec{\varepsilon} \succ_1 t''$, then *either*

- $t'' \equiv (\lambda x.t)r'\vec{\varepsilon}$ with $r \succ_1 r'$, and by induction hypothesis $t'' \in \mathrm{SN}$; *or*
- $t'' \equiv (\lambda x.t)r\vec{\varepsilon}'$ with $\varepsilon_i \succ_1 \varepsilon_i'$, and by induction hypothesis $t'' \in \mathrm{SN}$; *or*
- $t'' \equiv (\lambda x.t')r\vec{\varepsilon}$ with $t \succ_1 t'$, and by induction hypothesis $t'' \in \mathrm{SN}$; *or*
- $t'' \equiv t[x/r]\vec{\varepsilon}$, and $t'' \in \mathrm{SN}$ holds by assumption.

(v) is treated similarly. ⊠

Parts (i), (iii) and (v) yield non-introduced terms. We call a set **A** of strongly normalizable terms of type A saturated if it is closed under (i), (iii) and (v):

DEFINITION. A set **A** of terms is said to be *saturated of type* A (notation **A**: A) if **A** consists of terms of type A such that

Sat-1 If $t \in \mathbf{A}$, then $t \in \mathrm{SN}$.

Sat-2 If $\vec{\varepsilon} \in \mathrm{SN}$, then for any variable x of the appropriate type $x\vec{\varepsilon} \in \mathbf{A}$.

Sat-3 If $t[x/r]\vec{\varepsilon} \in \mathbf{A}$ and $r \in \mathrm{SN}$, then $(\lambda x.t)r\vec{\varepsilon} \in \mathbf{A}$.

Sat-4 If $t[X/B]\vec{\varepsilon} \in \mathbf{A}$, then $(\Lambda X.t)B\vec{\varepsilon} \in \mathbf{A}$. ⊠

So in particular the set SN_A of strongly normalizable terms of type A is saturated.

11.3.4. DEFINITION. The following definition of a predicate of *strong computability* "Comp" extends the definition given before for predicate logic.

Let A be a formula with $\mathrm{FV}(A) \subset \vec{X}$, and let $\vec{B} \equiv B_1, \ldots, B_n$ be a sequence of formulas of the same length, and let $\vec{\mathbf{B}} \equiv \mathbf{B}_1, \ldots, \mathbf{B}_n$ be a sequence of saturated sets with $\mathbf{B}_i \colon B_i$. We define $\mathrm{Comp}_A[\vec{X}/\vec{B}]$ (*computability under assignment* of $\vec{\mathbf{B}}$ to $\vec{X}$) as follows:

(i) $\mathrm{Comp}_{X_i}[\vec{X}/\vec{\mathbf{B}}] := \mathbf{B}_i$,

(ii) $\mathrm{Comp}_{B\to C}[\vec{X}/\vec{\mathbf{B}}] := \{\, t{\in}\mathrm{SN} : \forall s{\in}\mathrm{Comp}_B[\vec{X}/\vec{\mathbf{B}}](ts \in \mathrm{Comp}_C[\vec{X}/\vec{\mathbf{B}}])\,\}$,

(iii) $\mathrm{Comp}_{(\forall Y)C}[\vec{X}/\vec{\mathbf{B}}] := \{\, t{\in}\mathrm{SN} : \forall D \forall \mathbf{D}\colon D(tD \in \mathrm{Comp}_C[\vec{X}, Y/\vec{\mathbf{B}}, \mathbf{D}])\,\}$
(D ranging over types). ⊠

NOTATION. In order to save on notation, we shall use in the remainder of this section a standard abbreviation: we write Comp^*_D for $\mathrm{Comp}_D[\vec{X}/\vec{\mathbf{B}}]$, where $\mathbf{B}_i \colon B_i$ are saturated sets $(1 \le i \le m)$, fixed in each proof. So $\mathrm{Comp}^*_D[Y/\mathbf{C}]$ stands for $\mathrm{Comp}_D[\vec{X}, Y/\vec{\mathbf{B}}, \mathbf{C}]$ etc. ⊠

11.3.5. LEMMA. *Comp^*_A is a saturated set of type $A[\vec{X}/\vec{B}]$.*

PROOF. We have to show Sat-1–4 with Comp for $\mathbf{A}$, i.e.
(i) If $r \in \mathrm{Comp}^*_A$ then $r \in \mathrm{SN}$.
(ii) If $\vec{\varepsilon} \in \mathrm{SN}$ then $x\vec{\varepsilon} \in \mathrm{Comp}^*_A$.
(iii) If $r[x/s]\vec{\varepsilon} \in \mathrm{Comp}^*_A$ and $s \in \mathrm{SN}$, then $(\lambda x.r)s\vec{\varepsilon} \in \mathrm{Comp}^*_A$.
(iv) If $r[Y/C]\vec{\varepsilon} \in \mathrm{Comp}^*_A$ then $(\Lambda Y.r)C\vec{\varepsilon} \in \mathrm{Comp}^*_A$.

(i) follows from the definition of Comp^*_A and the fact that every saturated set is a subset of SN.

(ii) is proved by induction on the depth of A. So assume $\vec{\varepsilon} \in \mathrm{SN}$. Note that this implies $x\vec{\varepsilon} \in \mathrm{SN}$, by the properties of SN.
Case (ii)1. $A \equiv X_i$. By Sat-2 for $\mathbf{B}_i$ we find $x\vec{\varepsilon} \in \mathbf{B}_i$, and hence $x\vec{\varepsilon} \in \mathrm{Comp}^*_{X_i}$.
Case (ii)2. $A \equiv A_1 \to A_2$. We have to show $x\vec{\varepsilon} \in \mathrm{Comp}^*_{A_1\to A_2}$. Assume $s \in \mathrm{Comp}^*_{A_1}$; we then have to show $x\vec{\varepsilon}s \in \mathrm{Comp}^*_{A_2}$. Since $s \in \mathrm{SN}$ by (i), we find that $x\vec{\varepsilon}s \in \mathrm{Comp}^*_{A_2}$ (using (ii) for A_2).
Case (ii)3. $A \equiv (\forall Y)A_1$. We have to show that $x\vec{\varepsilon} \in \mathrm{Comp}^*_{(\forall Y)A_1}$. So let $\mathbf{C}\colon C$ be a saturated set, then we must show that $x\vec{\varepsilon}C \in \mathrm{Comp}^*_{A_1}[Y/\mathbf{C}]$. But this is a consequence of (ii) for A_1.

(iii) is proved by induction on the depth of A. We assume $s \in \mathrm{SN}$, $r[x/s]\vec{\varepsilon} \in \mathrm{Comp}^*_A$. Note that this implies $(\lambda x.r)s\vec{\varepsilon} \in \mathrm{SN}$, by the properties of SN.
Case (iii)1. $A \equiv X_i$. We have to show $(\lambda x.r)s\vec{\varepsilon} \in \mathbf{B}_i$. But $s \in \mathrm{SN}$, so this follows from Sat-3.
Case (iii)2. $A \equiv A_1 \to A_2$. Let $t \in \mathrm{Comp}^*_{A_1}$. Then we must show $(\lambda x.r)s\vec{\varepsilon}t \in \mathrm{Comp}^*_{A_2}$. By IH for A_2 it suffices to show that $r[x/s]\vec{\varepsilon}t \in \mathrm{Comp}^*_{A_2}$, which holds by definition of Comp.

Case (iii)3. $A \equiv (\forall Y)A_1$. Let $\mathbf{C}{:}\,C$ be a saturated set. We must show $(\lambda x.r)s\vec{\varepsilon}C \in \mathrm{Comp}^*_{A_1}[Y/\mathbf{C}]$. By the IH for A_1 it suffices to show that $r[x/s]\vec{\varepsilon}C \in \mathrm{Comp}^*_{A_1}[Y/\mathbf{C}]$, which holds by definition.

(iv) is again proved by induction on the depth of A. Let $r[Y/C]\vec{\varepsilon} \in \mathrm{Comp}^*_A$. Note that this implies $(\Lambda Y.r)C\vec{\varepsilon} \in \mathrm{SN}$, by the properties of SN.

Cases (iv)1 and (iv)2. Left to the reader.

Case (iv)3. $A \equiv (\forall Z)A_1$. Let $\mathbf{D}{:}\,D$ be a saturated set. We must show $(\Lambda Y.r)C\vec{\varepsilon}D \in \mathrm{Comp}^*_{A_1}[Z/\mathbf{D}]$. By the IH it suffices to show $r[Y/C]\vec{\varepsilon}D \in \mathrm{Comp}^*_{A_1}[Z/\mathbf{D}]$, which is trivial. ⊠

11.3.5A. ♠ Supply the proofs of the missing cases for (iv) in the proof of the preceding lemma.

11.3.6. LEMMA. *Let $\vec{\mathbf{B}}{:}\,\vec{B}$ be a sequence of saturated sets as before. Then*

$$\mathrm{Comp}_{A[Y/C]}[\vec{X}/\vec{\mathbf{B}}] = \mathrm{Comp}_A[\vec{X}, Y/\vec{\mathbf{B}}, \mathrm{Comp}_C[\vec{X}/\vec{\mathbf{B}}]].$$

(The right side is well-defined by the preceding lemma.)

PROOF. By induction on the complexity of A. Let $A \equiv \forall Z.B$. Then

$$\begin{aligned}
&\mathrm{Comp}^*_{(\forall ZB)[Y/C]} := \\
&\{\, t{\in}\mathrm{SN} : \forall D \forall \mathbf{D}{:}D(tD \in \mathrm{Comp}^*_{B[Y/C]}[Z/\mathbf{D}])\,\} = \\
&\{\, t{\in}\mathrm{SN} : \forall D \forall \mathbf{D}{:}D(tD \in \mathrm{Comp}^*_B[Y, Z/\mathrm{Comp}^*_C, \mathbf{D}])\,\} = \\
&\mathrm{Comp}^*_A[Y/\mathrm{Comp}^*_C].
\end{aligned}$$

We leave the other cases to the reader. ⊠

11.3.6A. ♠ Complete the proof of the preceding lemma.

11.3.7. THEOREM. *Let $t[x_1{:}\,A_1, \ldots, x_n{:}\,A_n]{:}\,C$, and assume that the free second-order variables of $\{t, A_1, \ldots, A_n\}$ are contained in $\{X_1, \ldots, X_m\}$.*

For $1 \le i \le m$, $1 \le j \le n$ let $\mathbf{B}_i{:}\,B_i$ be saturated sets, $s_j{:}\,A_j[\vec{X}/\vec{B}]$, $s_j \in \mathrm{Comp}_{A_j}[\vec{X}/\vec{\mathbf{B}}]$. Then

$$t[\vec{X}/\vec{B}][\vec{x}/\vec{s}] \in \mathrm{Comp}_C[\vec{X}/\vec{\mathbf{B}}].$$

PROOF. Let us write t^* for $t[\vec{X}/\vec{B}][\vec{x}/\vec{s}]$. We prove the statement of the theorem by induction on t.

Case 1. $t \equiv x_i^{A_i}$. Immediate, since $s_i \in \mathrm{Comp}^*_{A_j}$ by assumption.

Case 2. $t \equiv rs$. By IH, r^*, s^* are strongly computable, hence so is r^*s^*.

Case 3. $t \equiv r^{(\forall Z)A}C{:}\, A[Z/C]$. By IH $r^* \in \mathrm{Comp}^*_{(\forall Z)A}$. By the definition of Comp^*, lemma 11.3.5 and lemma 11.3.6

$$r^*C[\vec{X}/\vec{B}] \in \mathrm{Comp}^*_A[Z/\mathrm{Comp}^*_C] = \mathrm{Comp}^*_{A[Z/C]}.$$

Case 4. $t \equiv \lambda z^{C_1}.r^{C_2}$. We have to show that $\lambda z^{C_1[\vec{X}/\vec{B}]}.r^* \in \mathrm{Comp}^*_{C_1 \to C_2}$. So assume $s \in \mathrm{Comp}^*_{C_1}$. Then we have to show $(\lambda z.r^*)s \in \mathrm{Comp}^*_{C_2}$. By Sat-3 it suffices to show that $r^*[x/s] \in \mathrm{Comp}^*_{C_2}$ (for $\mathrm{Comp}^*_{C_2}$ is saturated by lemma 11.3.5). But this follows by IH from

$$r^*[x/s] = r[\vec{X}/\vec{B}][\vec{x}, x/\vec{s}, s].$$

Case 5. $t \equiv \lambda Z.r$. t^* is of the form $\lambda Z.r^*$. We have to show $t^* \in \mathrm{Comp}^*_{(\forall Z)A_1}$. So let $\mathbf{C}{:}\, C$. Then we have to show $(\lambda Z.r^*)C \in \mathrm{Comp}^*_{A_1}[Z/\mathbf{C}]$. By Sat-4 we need only to show $r^*[Z/C] \in \mathrm{Comp}^*_{A_1}[Z/\mathbf{C}]$, which follows by IH. ⊠

COROLLARY. *All terms of* $\boldsymbol{\lambda 2}$ *belong to* SN, *hence they are strongly normalizable.* ⊠

11.3.8. *An auxiliary system*

In section 11.5 we should like to map deductions and formulas of second-order arithmetic to terms and types of $\boldsymbol{\lambda 2}$, in such a way that the notion of reducibility is preserved, and second-order propositions go to the corresponding type. For this it is necessary that $(\forall X)X$ is mapped to an inhabited type of $\boldsymbol{\lambda 2}$; but since the type $(\forall X)X$ is uninhabited in $\boldsymbol{\lambda 2}$, we introduce the auxiliary system $\boldsymbol{\lambda 2 \Omega}$. In the next section we shall show that in $\boldsymbol{\lambda 2 \Omega}$ precisely the same recursive functions are representable as in $\boldsymbol{\lambda 2}$; this is proved by a suitable encoding of $\boldsymbol{\lambda 2 \Omega}$ into $\boldsymbol{\lambda 2}$.

DEFINITION. Let $\boldsymbol{\lambda 2 \Omega}$ be obtained from $\boldsymbol{\lambda 2}$ by adding a constant $\Omega{:}\, \forall X.X$. There are no conversion rules involving Ω. ⊠

THEOREM.

(i) *The terms in* $\boldsymbol{\lambda 2 \Omega}$ *are strongly normalizable.*

(ii) *Normal forms in* $\boldsymbol{\lambda 2}$ *and* $\boldsymbol{\lambda 2 \Omega}$ *are unique.*

(iii) *If we include conjunction as a type-forming operator, strong normalization and uniqueness of normal form remain valid for* $\boldsymbol{\lambda 2}$ *and* $\boldsymbol{\lambda 2 \Omega}$.

PROOF. (i) As for $\boldsymbol{\lambda 2}$, adding some cases where necessary.

(ii) By extending the proof of weak confluence in 1.2.11. The rest of the proof is left to the reader. ⊠

11.3.8A. ♠ Supply the missing details in the proof of the preceding theorem.

11.4 Encoding of $\lambda 2\Omega$ into $\lambda 2$

11.4.1. PROPOSITION. *Let* $t\colon \mathrm{N} \to \mathrm{N}$ *be a [closed] term in* $\boldsymbol{\lambda 2\Omega}$*. Then there is a [closed] term* t' *in* $\boldsymbol{\lambda 2}$ *such that* $t\bar{n} \succeq \bar{m} \Rightarrow t'\bar{n} \succeq \bar{m}$.

PROOF. For the proof we define (1) an encoding $^\circ$ of types and terms of $\boldsymbol{\lambda 2\Omega}$ into $\boldsymbol{\lambda 2}$, for which it can be proved that if $t \succ_1 t'$, then $t^\circ \succeq (t')^\circ$, and (2) maps W, C such that $\mathsf{W}\bar{n} \succeq \bar{n}^\circ$, $\mathsf{C}\bar{n}^\circ \succeq \bar{n}$. From this the t' in the statement of the theorem is obtained as $\lambda z^{\mathrm{N}}.\mathsf{C}(t^\circ(\mathsf{W}z))$. Once the right definitions for $^\circ$, W, C have been given, the proofs become straightforward inductive arguments, which are left almost entirely to the reader.

We define the encoding map $^\circ$ on types by

$$\begin{array}{ll} X^\circ & := X, \\ (A \to B)^\circ & := A^\circ \to B^\circ, \\ (\forall X.A)^\circ & := \forall X(X \to A^\circ). \end{array}$$

It follows that

$$(A[X/B])^\circ \equiv A^\circ[X/B^\circ].$$

In order to define the encoding of terms, we assume that there is associated to each type variable X a "fresh" individual variable x of type X, not occurring free in the terms to be encoded. The correspondence between the X and the associated x^X is assumed to be one-to-one. Below, in the definition of the encoding, we shall use the tacit convention that the type variables X_i, X, Y have x_i, x, y associated to them.

For A with $\mathrm{FV}(A) = \{X_1, \ldots, X_p\}$, we define a term $\tau_A\colon A^\circ$ with free variables $x_1\colon X_1, \ldots, x_p\colon X_p, X_1, \ldots, X_p$ as follows:

$$\begin{array}{lll} \tau_{X_i} & := & x_i\colon X_i \\ \tau_{B\to C} & := & \lambda y\colon B^\circ.\tau_C \ (y \notin \mathrm{FV}(\tau_C)) \\ \tau_{\forall X.B} & := & \Lambda X \lambda x^X.\tau_B. \end{array}$$

For closed A, $\tau_A : A^\circ$ is closed. For example,

$$\tau_{(\forall X)X} \equiv \Lambda X \lambda x^X.x\colon \forall X(X \to X).$$

We now extend $^\circ$ to terms, and associate to each $t \in A$, with free variables $X_1, \ldots, X_p$, $y_1\colon A_1, \ldots, y_q\colon A_q$ a term $t^\circ\colon A^\circ$ with free variables $X_1, \ldots, X_p$, $x_1\colon X_1, \ldots, x_p\colon X_p$, $y_1\colon A_1^\circ, \ldots, y_q\colon A_q^\circ$, as follows:

$$\begin{array}{llll} \text{if } A \equiv A_i & \text{then } (y_i\colon A_i)^\circ & := & y_i\colon A^\circ; \\ \text{if } A \equiv B \to C & \text{then } (\lambda y\colon B.t\colon C)^\circ & := & \lambda y\colon B^\circ.t^\circ\colon C^\circ; \\ & ((t\colon B {\to} A)(s\colon B))^\circ & := & (t^\circ\colon B^\circ {\to} A^\circ)(s^\circ\colon B^\circ); \\ \text{if } A \equiv (\forall X)B & \text{then } (\Lambda X.t\colon \forall X.B)^\circ & := & \Lambda X \lambda x\colon X.t^\circ\colon B^\circ; \\ \text{if } A \equiv B[X/C] & \text{then } ((t\colon (\forall X)B)C)^\circ & := & (t^\circ\colon ((\forall X)B)^\circ)C^\circ\tau_C; \\ \text{if } A \equiv (\forall X)X & \text{then } \Omega^\circ & := & \tau_{\forall X.X} \equiv \Lambda X \lambda x\colon X.x. \end{array}$$

Then one verifies commutativity with substitution:

(i) $(t[y{:}\,A/s{:}\,A])^\circ \equiv t^\circ[y{:}\,A^\circ/s^\circ{:}\,A^\circ]$,
(ii) $\tau_{A[X/B]} \equiv \tau_A[X/B^\circ][x{:}\,B^\circ/\tau_B{:}\,B^\circ]$, and
(iii) $(t[X/B])^\circ \equiv t^\circ[X/B^\circ][x{:}\,B^\circ/\tau_B{:}\,B^\circ]$

by induction on t, A and t respectively, and uses this in verifying

$$\text{if } t \succ_1 t' \text{ then } t^\circ \succeq (t')^\circ.$$

For example, we have as one step in the induction for (ii):

$$\begin{aligned}&\tau_{(\forall Y)C[X/B]} \equiv \Lambda Y \lambda y^Y.\tau_{C[X/B]} \equiv\\ &\Lambda Y \lambda y^Y.(\tau_C[X/B^\circ][x{:}\,B^\circ/\tau_B{:}\,B^\circ]) \equiv\\ &(\Lambda Y \lambda y^Y.(\tau_C))[X/B^\circ][x{:}\,B^\circ/\tau_B{:}\,B^\circ] \equiv\\ &\tau_{(\forall Y)C}[X/B^\circ][x{:}\,B^\circ/\tau_B{:}\,B^\circ].\end{aligned}$$

Further, we note that

$$\begin{aligned}\mathrm{N}^\circ &\equiv \forall X(X \to (X \to ((X \to X) \to X)))\\ (\bar{n})^\circ &\equiv \Lambda X \lambda x^X y^X z^{X\to X}.z^n y.\end{aligned}$$

We can define operators W ("weaken") and C ("contract") such that

$$\mathsf{W}\bar{n} \succeq \bar{n}^\circ, \quad \mathsf{C}\bar{n}^\circ \succeq \bar{n};$$

simply take

$$\begin{aligned}\mathsf{W} &:= \lambda u^{\mathrm{N}} \Lambda X \lambda x^X y^X z^{X\to X}.uXxz,\\ \mathsf{C} &:= \lambda u^{\mathrm{N}^\circ} \Lambda X \lambda y^X z^{X\to X}.uXyyz.\end{aligned}$$

For the term t' of the proposition we may now take

$$t' := \lambda z^{\mathrm{N}}.\mathsf{C}(t^\circ(\mathsf{W}z)).$$ ⊠

11.4.1A. ♠ Fill in the missing details of this proof.

11.5 Provably recursive functions of $\mathbf{HA}^2$

11.5.1. DEFINITION. *Intuitionistic second-order arithmetic* $\mathbf{HA}^2$ (alternative notation **HAS**) is obtained taking the language of pure second-order logic, with a single individual constant 0 (*zero*), a single function constant S (*successor*), and a binary predicate symbol = for equality between individuals, with the axioms and rules of intuitionistic second-order logic, and the following axioms for equality and successor:

$$\begin{aligned}&\forall x(x = x),\\ &\forall X^1 \forall xy(x = y \wedge Xx \to Xy),\\ &\forall xy(\mathrm{S}x = \mathrm{S}y \to x = y),\\ &\forall x(\neg \mathrm{S}x = 0), \text{ i.e. } \forall x(\mathrm{S}x = 0 \to (\forall X^0)X),\end{aligned}$$

and the induction axiom:

$$\forall X^1(X0 \wedge \forall x(Xx \to X(\mathrm{S}x)) \to \forall y Xy).$$

Note that the second equality axiom implies

$$\forall xy(x = y \to y = x);$$
$$\forall xyz(x = y \wedge x = z \to y = z);$$
$$\forall xy(x = y \to \mathrm{S}x = \mathrm{S}y).$$

(For the first, take $Xz \equiv (z = x)$, then $x = y \wedge Xx \to Xy$ yields $x = y \wedge x = x \to y = x$, and since $x = x$ holds, $x = y \to y = x$; for the second take $Xx \equiv (x = z)$; for the third, take $Xz \equiv (\mathrm{S}x = \mathrm{S}z)$). ⊠

11.5.2. DEFINITION. A subsystem $\mathbf{HA}^{2*}$ based on $\mathbf{I}^2$ and equivalent to $\mathbf{HA}^2$ is the following. The language consists of $0, \mathrm{S}, =$ as before, but we define a predicate $\mathbb{N}$ by

$$x \in \mathbb{N} \equiv \mathbb{N}x := \forall X^1(X0 \to (\forall x(Xx \to X(\mathrm{S}x)) \to Xx)).$$

As axioms we include

$$\forall x(x = x),\ \forall xyz(x = y \to (x = z \to y = z)),$$
$$\forall xy(\mathrm{S}x = \mathrm{S}y \to x = y),\ \forall xy(x = y \to \mathrm{S}x = \mathrm{S}y),$$
$$\forall x(\mathrm{S}x = 0 \to (\forall X^0)X).$$

The *standard model* of $\mathbf{HA}^{2*}$ (and of $\mathbf{HA}^2$) has $\mathbb{N}$ and the powersets of $\mathbb{N}^k$ as domains of individuals and k-place relations respectively; 0, $\mathrm{S}, =$ get their usual interpretation. ⊠

11.5.3. DEFINITION. $X^n \in \mathsf{Ext} := \forall \vec{x}{\in}\mathbb{N}\forall \vec{y}{\in}\mathbb{N}(\vec{x} = \vec{y} \wedge X\vec{x} \to X\vec{y})$, where $\vec{x} \in \mathbb{N}$ stands for $x_1 \in \mathbb{N}, \ldots, x_n \in \mathbb{N}$ etc. ⊠

So "$X \in \mathsf{Ext}$" means that X behaves as an "extensional predicate" w.r.t. $\mathbb{N}$, or, perhaps more appropriately, satisfies *replacement* w.r.t. elements of $\mathbb{N}$. Note that $\mathbb{N} \in \mathsf{Ext}$.

11.5.4. DEFINITION. (*Interpretation of* $\mathbf{HA}^2$ *in* $\mathbf{HA}^{2*}$) Formulas of $\mathbf{HA}^2$ can now be interpreted in $\mathbf{HA}^{2*}$, by relativization of all individual variables to $\mathbb{N}$, and all relation variables to Ext; so $\forall x$, $\exists x$ go to $\forall x{\in}\mathbb{N}$, $\exists x{\in}\mathbb{N}$, and $\forall X^n, \exists X^n$ to $\forall X^n{\in}\mathsf{Ext}, \exists X^n{\in}\mathsf{Ext}$; with respect to the other logical operators the embedding is a homomorphism. ⊠

From now on we take as a matter of convenience the language based on $\to, \wedge, \forall, \forall^2, \exists$; from the results in 11.1.4 we know that we have strong normalization w.r.t. detour conversions for this language.

11.5.4A. ♠ Verify that the embedding of the definition (call it ψ) indeed satisfies for all closed A of $\mathbf{HA}^2$: $\mathbf{HA}^2 \vdash A \Rightarrow \mathbf{HA}^{2*} \vdash \psi(A)$.

11.5.5. Definition. (*Normal deduction* $\mathcal{D}[n]$ of $\mathrm{S}^n0 \in \mathbb{N}$ in $\mathbf{HA}^{2*}$) For $\mathcal{D}[0]$ we take

$$\dfrac{\dfrac{\dfrac{0 \in X^{1}}{\forall y(y \in X \to \mathrm{S}y \in X) \to 0 \in X}\,{\to}\mathrm{I}}{0 \in X \to (\forall y(y \in X \to \mathrm{S}y \in X) \to 0 \in X)}\,{\to}\mathrm{I},1}{0 \in \mathbb{N}}\,\forall^2\mathrm{I}$$

and for $\mathcal{D}[n]$ if $n > 0$:

$$\dfrac{\dfrac{\dfrac{\dfrac{\dfrac{\forall y(y \in X \to \mathrm{S}y \in X)^{1}}{\mathrm{S}^{n-1}0 \in X \to \mathrm{S}^n0 \in X}\,\forall\mathrm{E} \qquad \begin{matrix}[0 \in X]^2\\ \mathcal{D}_{n-1}\\ \mathrm{S}^{n-1}0 \in X\end{matrix}}{\mathrm{S}^n0 \in X}\,{\to}\mathrm{E}}{\forall y(y \in X \to \mathrm{S}y \in X) \to \mathrm{S}^n0 \in X}\,{\to}\mathrm{I},1}{0 \in X \to \forall y(y \in X \to \mathrm{S}y \in X) \to \mathrm{S}^n0 \in X}\,{\to}\mathrm{I},2}{\mathrm{S}^n0 \in \mathbb{N}}\,\forall^2\mathrm{I}$$

where $\mathcal{D}_{n-1}$ is either $0 \in X$ (for $n = 1$) or is of the form

$$\dfrac{\dfrac{\forall y(y \in X \to \mathrm{S}y \in X)}{\mathrm{S}^{n-2}0 \in X \to \mathrm{S}^{n-1}0 \in X}\,\forall\mathrm{E} \qquad \begin{matrix}[0 \in X]\\ \mathcal{D}_{n-2}\\ \mathrm{S}^{n-2}0 \in X\end{matrix}}{\mathrm{S}^{n-1}0 \in X}$$

⊠

It is easy to see that $\mathcal{D}[n]$ translates under the collapsing map into $\bar{n}$ of $\boldsymbol{\lambda}\mathbf{2}$.

11.5.6. Lemma. *$\mathcal{D}[n]$ is the unique normal deduction of* $\mathrm{S}^n0 \in \mathbb{N}$.

Proof. Let $\mathcal{D}$ be a normal deduction of $\mathrm{S}^n0 \in \mathbb{N}$. $\mathcal{D}$ cannot end with an E-rule. For assume $\mathcal{D}$ to end with an E-rule; if we follow a main branch from the conclusion upwards, the branch must start in a purely first-order axiom, such as $\forall x(x = x)$, or in $\forall x(\mathrm{S}x = 0 \to (\forall X^0)X)$. From a purely first-order axiom we can never arrive at the conclusion alone, passing through eliminations only. Elimination starting from $\forall x \neg \mathrm{S}x = 0$ must begin with

$$\dfrac{\dfrac{\forall x(\mathrm{S}x = 0 \to (\forall X)X)}{\mathrm{S}t = 0 \to (\forall X)X} \qquad \begin{matrix}\mathcal{D}'\\ \mathrm{S}t = 0\end{matrix}}{(\forall X)X}$$

$$\vdots$$

However, no open assumptions of $\mathcal{D}'$ can be discharged lower down; so $\mathcal{D}'$ must deduce $\mathrm{S}t = 0$ from the axioms. But in the standard model $\mathrm{S}t = 0$ is false (here we rely on the *consistency* of the system).

Therefore the deduction ends with an introduction, with subdeduction $\mathcal{D}_1$ of the premise $0 \in X \to (\forall y(y \in X \to \mathrm{S}y \in X) \to \mathrm{S}^n 0 \in X)$. Again, the final step of $\mathcal{D}_1$ must be an introduction, and the premise of the conclusion of $\mathcal{D}_1$ is $\forall y(y \in X \to \mathrm{S}y \in X) \to \mathrm{S}^n 0 \in X$, which is the conclusion of subdeduction $\mathcal{D}_2$ from assumptions $0 \in X$. $\mathcal{D}_2$ cannot end with an E-rule; for if it did, a main branch of $\mathcal{D}_2$ would have to terminate either in $0 \in X$ or in an axiom. The axioms are excluded as possibilities for the same reasons as before (assign $\mathbb{N}$ to the variable X in the "standard model" part of the argument). To $0 \in X$ no elimination rule is applicable. So the final step of $\mathcal{D}_2$ is an I-rule, and the immediate subdeduction $\mathcal{D}_3$ of $\mathcal{D}_2$ derives $\mathrm{S}^n 0 \in X$ from assumptions $0 \in X$ and $\forall y(y \in X \to \mathrm{S}y \in X)$. The last rule of $\mathcal{D}_3$ must be an elimination rule, and the main branch must terminate in $0 \in X$ or in $\forall y(y \in X \to \mathrm{S}y \in X)$. If it terminates in $0 \in X$ and $n = 0$, we are done. Terminating in $0 \in X$ while $n \neq 0$ is impossible. So assume the main branch to start in $\forall y(y \in X \to \mathrm{S}y \in X)$:

$$\dfrac{\dfrac{\forall y(y \in X \to \mathrm{S}y \in X)}{t \in X \to \mathrm{S}t \in X} \qquad \begin{array}{c}\mathcal{D}''\\ t \in X\end{array}}{\mathrm{S}t \in X}$$
$$\vdots$$

but then $(\mathrm{S}t \in X) \equiv (\mathrm{S}^n 0 \in X)$ etc. ⊠

11.5.7. DEFINITION. We define a *collapsing map* $[\![\ ]\!]$ from formulas and deductions of $\mathbf{HA}^{2*}$ to formulas and deductions of $\boldsymbol{\lambda 2\Omega}$. Let M be a fixed inhabited type of $\boldsymbol{\lambda 2\Omega}$, e.g. $M \equiv \forall X(X \to X)$, containing $T \equiv \Lambda X \lambda x.x^X$. To the relation variables X^n we let bijectively correspond propositional (type-) variables X^*. For formulas we take

$$\begin{array}{lcl}
[\![t = s]\!] & := & M, \\
[\![Xt_1 \dots t_n]\!] & := & X^*, \\
[\![A \to B]\!] & := & [\![A]\!] \to [\![B]\!], \\
[\![\forall x A]\!] & := & [\![A]\!], \\
[\![\exists x A]\!] & := & [\![A]\!], \\
[\![\forall X^n A]\!] & := & \forall X^* [\![A]\!].
\end{array}$$

Observe that

$$[\![A[x/t]]\!] \equiv [\![A]\!], \quad [\![t \in \mathbb{N}]\!] \equiv \mathrm{N}.$$

The definition is extended to deductions as follows. We have not before introduced a complete term calculus for $\mathbf{HA}^{2*}$, but the notations below will be self-explanatory. The definition proceeds by induction on construction of deduction terms, or what is the same, by induction on the length of deductions. First the basis case, assumptions and axioms:

$$\begin{array}{lll}
x_i : A_i & \mapsto & x_i\colon [\![A_i]\!], \\
\forall x(x = x) & \mapsto & T\colon M, \\
\forall xyz(x = y \to (x = z \to y = z)) & \mapsto & \lambda x^M y^M.T^M\colon M \to M \to M, \\
\forall xy(\mathrm{S}x = \mathrm{S}y \to x = y) & \mapsto & \lambda y^M.T^M\colon M \to M, \\
\forall xy(x = y \to \mathrm{S}x = \mathrm{S}y) & \mapsto & \lambda y^M.T^M\colon M \to M, \\
\forall x(\mathrm{S}x = 0 \to \forall X.X) & \mapsto & \lambda z^M.\Omega\colon M \to (\forall X)X.
\end{array}$$

For the rules we put

$$\begin{array}{lll}
\lambda x^A.t^B\colon A \to B & \mapsto & \lambda x^{[\![A]\!]}.[\![t]\!]^{[\![B]\!]}\colon [\![A \to B]\!], \\
t^{A\to B}s^A\colon B & \mapsto & [\![t]\!]^{[\![A\to B]\!]}[\![s]\!]^{[\![A]\!]}\colon [\![B]\!], \\
\lambda x^I.t\colon \forall xA & \mapsto & [\![t]\!]\colon [\![\forall x^I A]\!] (\text{note } [\![\forall x^I A]\!] \equiv [\![A[x/y]]\!]), \\
t^{\forall y^I A[x/y]}s^I\colon A[x/s] & \mapsto & [\![t]\!]\colon [\![A]\!] \ \ (\text{note } [\![A]\!] \equiv [\![\forall y A[x/y]]\!]), \\
\mathbf{p}(t_0, t_1^{A[x/t_0]})\colon \exists xA & \mapsto & [\![t_1]\!]\colon [\![A]\!] \ \ (\text{note } [\![A[x/t_0]]\!] = [\![A]\!]), \\
\mathrm{E}^{\exists}_{y,z}(t^{\exists xA}, s(y, z^{A[x/y]}))\colon C & \mapsto & [\![s(y, z^{A[x/y]})]\!][z/[\![t]\!]^{[\![A]\!]}]\colon [\![C]\!].
\end{array}$$

For the next two cases observe that $[\![A[X/\lambda\vec{x}.C]]\!] \equiv [\![A]\!][X^*/[\![C]\!]]$.

$$\begin{array}{lll}
\Lambda X^n.t^A\colon \forall X^n A & \mapsto & \Lambda X^*.[\![t]\!]^{[\![A]\!]}\colon \forall X^*[\![A]\!], \\
t^{\forall X^n A}(\lambda\vec{x}.B)\colon A[X^n/\lambda\vec{x}.B] & \mapsto & [\![t]\!]^{\forall X^*[\![A]\!]}[\![B]\!]\colon [\![A]\!][X^*/[\![B]\!]].
\end{array}$$

For the $\wedge$-rules we need on the right hand side defined operators:

$$\begin{array}{l}
\mathbf{p}^{A,B} := \lambda x^A y^B \Lambda X \lambda z^{A\to(B\to X)}.zxy, \\
\mathbf{p}_0^{A,B} := \lambda u^{A\wedge B}.uA(\lambda x^A y^B.x), \ \ \mathbf{p}_1^{A,B} := \lambda u^{A\wedge B}.uB(\lambda x^A y^B.y).
\end{array}$$

Then

$$\mathbf{p}(t^A, s^B)\colon A \wedge B \ \mapsto \ \mathbf{p}([\![t]\!]^{[\![A]\!]}, [\![s]\!]^{[\![B]\!]})\colon [\![A \wedge B]\!],$$

etc. ⊠

Proposition. *The collapsing map preserves reductions: if $\mathcal{D}$ reduces to $\mathcal{D}'$, then the collapse of $\mathcal{D}$ reduces to the collapse of $\mathcal{D}'$.* ⊠

11.5.7A. ♠ Verify this.

11.5.8. THEOREM. *The provably total recursive functions of* $\mathbf{HA}^2$ *are representable in* $\boldsymbol{\lambda}\mathbf{2}$.

PROOF. Let $\mathcal{D}$ be a deduction in $\mathbf{HA}^{2*}$, with conclusion,

$$(1) \qquad \vdash \forall x{\in}\mathbb{N}\, \exists y{\in}\mathbb{N} A(x,y),$$

or expanded

$$\vdash \forall x(x \in \mathbb{N} \to \exists y(y \in \mathbb{N} \wedge A(x,y))).$$

The collapsing map applied to $\mathcal{D}$ produces a term t such that

$$t\colon [\![\forall x(x \in \mathbb{N} \to \exists y(y \in \mathbb{N} \wedge A(x,y)))]\!] \equiv \mathrm{N} \to (\mathrm{N} \wedge [\![A]\!]).$$

Then $t^* := \lambda x^{\mathrm{N}}.\mathbf{p}_0(tx)$ encodes the recursive function f implicit in the proof of (1). This is seen as follows. If we specialize the deduction $\mathcal{D}$ to $x = \mathrm{S}^n 0$ we obtain a deduction $\mathcal{D}_n$ which must end with applications of $\forall$E, $\to$E along the main branch:

$$\dfrac{\dfrac{\begin{matrix}\mathcal{D}'_n \\ \forall x(x \in \mathbb{N} \to \exists y(y \in \mathbb{N} \wedge A(x,y)))\end{matrix}}{\mathrm{S}^n0 \in \mathbb{N} \to \exists y(y \in \mathbb{N} \wedge A(\mathrm{S}^n0,y))} \qquad \dfrac{\mathcal{D}[n]}{\mathrm{S}^n0 \in \mathbb{N}}}{\exists y(y \in \mathbb{N} \wedge A(\mathrm{S}^n0,y))}$$

which after normalization must end with an introduction:

$$\dfrac{\begin{matrix}\mathcal{D}''_n \\ \mathrm{S}^m0 \in \mathbb{N} \wedge A(\mathrm{S}^m0, \mathrm{S}^n0)\end{matrix}}{\exists y(y \in \mathbb{N} \wedge A(\mathrm{S}^n0,y))}$$

Applying $\wedge$E to $\mathcal{D}''_n$ we find a deduction $\mathcal{D}^+$ of $\mathrm{S}^m0 \in \mathbb{N}$; the collapsing map produces a term which must be equal to $t^*\bar{n}$. By lemma 11.5.6, the deduction $\mathcal{D}^+$ normalizes to $\mathcal{D}[m]$, so $t^*\bar{n}$ normalizes to $\bar{m}$. Combining this with the embedding of $\boldsymbol{\lambda}\mathbf{2\Omega}$ into $\boldsymbol{\lambda}\mathbf{2}$, we have the desired result. Since $A(\mathrm{S}^n0, \mathrm{S}^m0)$ is provable in $\mathbf{HA}^2$, it is true in the standard model, so $m = f(n)$. ⊠

REMARK. As observed in 11.6.4, the $\boldsymbol{\lambda}\mathbf{2}$-representable functions are in fact *exactly* the functions provably total recursive in $\mathbf{HA}^2$.

11.5.9. *The provably total recursive functions of* $\mathbf{PA}^2$

The provably total recursive functions of classical second-order arithmetic $\mathbf{PA}^2$ are in fact also provably total in $\mathbf{HA}^2$; $\mathbf{PA}^2$ is just $\mathbf{HA}^2$ with classical logic. This fact may be proved directly, but can also be obtained from the characterization of the provably recursive functions of $\mathbf{HA}^2$. In outline, the proof is as follows.

(a) In $\mathbf{HA}^2$ and $\mathbf{PA}^2$ we can conservatively extend the language and the system by adding symbols for all the primitive recursive functions. Then a provably total recursive function is given by a code number n such that $\vdash \forall x \exists y(\chi_T(\bar{n}, x, y) = 0)$, where χ_T is the characteristic function of Kleene's T-predicate.

(b) The Gödel-Gentzen negative translation ^{g} (cf. 2.3) embeds $\mathbf{PA}^2$ into $\mathbf{HA}^2$. Hence, if $\mathbf{PA}^2 \vdash \exists y(\chi_T(\bar{n}, x, y) = 0)$, then $\mathbf{HA}^2 \vdash \neg\neg\exists y(\chi_T(\bar{n}, x, y) = 0)$ (using $\neg\forall x \neg A \leftrightarrow \neg\neg\exists x A$).

(c) By a method due to Friedman [1978] and Dragalin [1979] we can show that $\mathbf{HA}^2$ is closed under "Markov's rule" in the form: "if $\vdash \neg\neg\exists x(t(x, \vec{y}) = 0)$ then $\vdash \exists x(t(x, \vec{y}) = 0)$" (an exposition of this result is found in Troelstra and van Dalen [1988, section 5.1]).

As to (a), we use the result on the conservativity of the addition for symbols for definable functions, applied to functions defined by primitive recursion (4.4.12). Therefore we need to show that in $\mathbf{HA}^2$ we can define graph predicates $H(\vec{x}, z)$ for each primitive recursive function h. The graph PRD for the predecessor function prd is given by

$$\mathrm{PRD}(x, z) := (x = 0 \wedge z = 0) \vee \exists y(x = \mathrm{S}y \wedge z = y)$$

The crucial step in the construction of these H goes as follows. If h is obtained from f and the number m (for notational simplicity we do not consider additional numerical parameters):

$$h(0) = m, \quad h(\mathrm{S}z) = f(z, h(z))$$

and we have already constructed F as the graph of f, we obtain the graph H as

$$H(z, u) := \forall X^2(A(z, X) \to X(z, u)),$$

where

$$A(z, X^2) := \forall z' \leq z([z' = 0 \to (Xz'v \leftrightarrow v = m)] \wedge$$
$$[z' \neq 0 \to (Xz'v) \leftrightarrow \exists v'z''(\mathrm{PRD}(z', z'') \wedge X(z'', v') \wedge F(z'', v', v))]).$$

Then one proves by induction on z

$$\exists X^2 A(z, X), \text{ and } \forall z' \leq z \exists! u(A(z, X) \to X(z', u)).$$

11.6 Notes

For a general introduction to higher-order logic, see Leivant [1994].

11.6.1. *Takeuti's conjecture.* Closure under Cut for second-order classical logic, known as "Takeuti's Conjecture", was first proved by Tait [1966] by a semantical argument, using classical metamathematical reasoning; this was extended to higher-order logic by Takahashi [1967].

Prawitz [1967] also gave a proof, extended to higher-order logic in Prawitz [1968], along the same lines as Takahashi's proof. Takahashi [1970] deals with type theory with extensionality.

Prawitz [1970] uses Beth models to obtain closure under Cut for a cutfree system of intuitionistic second-order logic (cf. 4.9.1).

11.6.2. *Normalization and strong normalization for* $\boldsymbol{\lambda 2}$. Girard [1971] was the first to prove a normalization theorem for a system of terms corresponding to intuitionistic second-order logic in $\wedge, \to, \forall, \forall^2, \exists^2$, based on his idea of "reducibility candidates" as a kind of variable computability predicate, as a method for extending the method of computability predicates of Tait [1967] to higher-order systems. Martin-Löf [1971b] proves normalization (not strong normalization) for $\mathbf{HA}^2$ with $\to, \forall, \forall^2$ in the form of deduction trees (not terms), using Girard's idea. Prawitz [1971], also inspired by Girard [1971], contains a proof of strong normalization, for intuitionistic first- and second-order logic, covering also permutation conversions for $\vee, \exists$. (Prawitz [1981] is a supplement, in particular for classical second-order logic.) Girard [1972] also proves strong normalization. Girard [1971] is also the first place where $\boldsymbol{\lambda 2}$ is defined (called system "F" by Girard). Not much later, $\boldsymbol{\lambda 2}$ was rediscovered by Reynolds [1974]. (Strong) normalization for $\boldsymbol{\lambda 2}$ and the associated logical systems has been re-proved many times, always in essence by the same method (for example, Osswald [1973], Tait [1975]). A smooth version is given in the recent Girard et al. [1988]; but the proof presented here is based on Matthes [1998], where the result has been generalized considerably.

The method of computability predicates by Tait, with its extension by Girard, has a semantical flavour; see, for example, Hyland and Ong [1993], Altenkirch [1993] and Gallier [1995].

As to the significance of cut elimination and normalization for second- and higher-order logic, see Kreisel and Takeuti [1974], Girard [1976], Päppinghaus [1983].

11.6.3. $\boldsymbol{\lambda 2}$ as a type theory has been further strengthened, for example in the "calculus of constructions" of Coquand and Huet [1988], but then the formulas-as-types parallel cannot any longer be viewed as an isomorphism. See for this aspect Geuvers [1993,1994]. For some information on extensions of $\boldsymbol{\lambda 2}$ see Barendregt [1992]. An elegant normalization proof for the calculus of constructions is found in Geuvers and Nederhof [1991].

The characterization of the provable recursive functions of $\mathbf{HA}^2$ as the functions representable in $\boldsymbol{\lambda 2}$ is due to Girard [1971,1972]; the proof given here follows Girard et al. [1988].

Leivant [1990] gave a proof of this characterization for classical second-order arithmetic, by another, more semantically inspired method.

11.6.4. *A converse theorem.* All $\boldsymbol{\lambda 2}$-representable functions are in fact provably recursive in $\mathbf{HA}^2$. The idea for the proof is based on (1) arithmetizing the syntax of $\boldsymbol{\lambda 2}$, and (2) observing that the proof of strong normalization for any given closed term t of $\boldsymbol{\lambda 2}$ can in fact be carried out in $\mathbf{HA}^2$ itself, and in particular that we can prove

$$\mathbf{HA}^2 \vdash \forall n \exists ! m \mathrm{Red}(\ulcorner t\bar{n} \urcorner, \ulcorner \bar{m} \urcorner),$$

where $\ulcorner s \urcorner$ is the code of the $\boldsymbol{\lambda 2}$-term s, and $\mathrm{Red}(t, t')$ expresses that the term with code t reduces to the term with code t'. The proof is carried out in Girard [1972]. For a proof with similar details, see, for example, the formalization of normalization of $\mathbf{Ni}^2$ in Troelstra [1973, section 4.4].

Solutions to selected exercises

2.1.8B.

$$
\dfrac{\dfrac{\dfrac{\dfrac{\neg\neg A \to \neg\neg B^{\,v} \quad \dfrac{\dfrac{\neg(A \to B)^{\,u} \quad \dfrac{\dfrac{\dfrac{\neg A^{\,w'} \quad A^{\,w}}{\bot}}{B}}{A \to B}\,w}{\bot}}{\neg\neg A}\,w'}{\neg\neg B} \quad \dfrac{\dfrac{\neg(A \to B)^{\,u} \quad \dfrac{B^{\,w''}}{A \to B}}{\bot}}{\neg B}\,w''}{\bot}}{\neg\neg(A \to B)}\,u}{(\neg\neg A \to \neg\neg B) \to \neg\neg(A \to B)}\,v
$$

2.1.8E.

$$
\dfrac{\dfrac{\dfrac{\neg(\neg A \land \neg B)^{\,u} \quad \dfrac{\dfrac{\dfrac{\neg(A \lor B)^{\,v} \quad \dfrac{A^{\,w}}{A \lor B}}{\bot}}{\neg A}\,w \quad \dfrac{\dfrac{\neg(A \lor B)^{\,v} \quad \dfrac{B^{\,w'}}{A \lor B}}{\bot}}{\neg B}\,w'}{\neg A \land \neg B}}{\bot}}{A \lor B}\,\bot_{\mathrm{c}}, v}{\neg(\neg A \land \neg B) \to A \lor B}\,u
$$

The other half of the equivalence is easier and holds even intuitionistically.

$$
\dfrac{\dfrac{\dfrac{\neg A^{\,v} \quad \dfrac{(A \to B) \to A^{\,w} \quad \dfrac{\dfrac{\dfrac{\neg A^{\,v} \quad A^{\,u}}{\bot}}{B}\,\bot_{\mathrm{i}}}{A \to B}\,u}{A}}{\bot}}{A}\,\bot_{\mathrm{c}}, v}{((A \to B) \to A) \to A}\,w
$$

2.1.8F. Write $\mathrm{P}_{X,Y}$ for $((X \to Y) \to X) \to X$.

$$\dfrac{\dfrac{\dfrac{\mathrm{P}_{C,A} \quad \dfrac{\dfrac{A\to C^v \quad \dfrac{\mathrm{P}_{A,B} \quad \dfrac{\dfrac{C\to A^w \quad \dfrac{(A\to B)\to C^u \quad A\to B^{w''}}{C}}{A}}{(A\to B)\to A}\,w''}{A}}{C}}{(C\to A)\to C}\,w}{C}}{(A\to C)\to C}\,v}{((A\to B)\to C)\to(A\to C)\to C}\,u$$

2.1.8G.

$$\dfrac{\dfrac{\mathrm{P}_{A,C} \quad \dfrac{\dfrac{\mathrm{P}_{A,B} \quad \dfrac{\dfrac{(A\to(B\wedge C))\to A^{w'} \quad \dfrac{\dfrac{\dfrac{A\to B^u \quad A^w}{B} \quad \dfrac{A\to C^v \quad A^w}{C}}{B\wedge C}}{A\to B\wedge C}\,w}{A}}{(A\to B)\to A}\,u}{A}}{(A\to C)\to A}\,v}{A}}{\mathrm{P}_{A,B\wedge C}}\,w'$$

2.1.8H. Proof by induction on the complexity of contexts. We do two cases. Let $G[*] \equiv F[*] \to C$.

$$\dfrac{\dfrac{\dfrac{\dfrac{F[B]\to C^u \quad \dfrac{\dfrac{\mathrm{IH} \quad \forall\vec{x}(A\to B)^v}{F[A]\to F[B]} \quad F[A]^w}{F[B]}}{C}}{F[A]\to C}\,w}{(F[B]\to C)\to F[A]\to C}\,u}{\forall\vec{x}(A\to B)\to(F[B]\to C)\to F[A]\to C}\,v$$

Let $F[*] \equiv \forall y F'[*]$, then

$$\dfrac{\dfrac{\dfrac{\dfrac{\dfrac{\mathrm{IH} \quad \dfrac{\forall y\vec{x}(A\to B)^u}{\forall\vec{x}(A\to B)}}{F'[A]\to F'[B]} \quad \dfrac{\forall yF'[A]^v}{F'[A]}}{F'[B]}}{\forall yF'[B]}}{\forall yF'[A]\to\forall F'[B]}\,v}{\forall y\vec{x}(A\to B)\to\forall yF'[A]\to\forall F'[B]}\,u$$

2.2.2A. The prooftree is exhibited with at each node

$$\mathit{term}\colon\mathit{type} \mid \mathrm{FV_a}(\mathit{term}) \mid \mathrm{FV_i}(\mathit{term})$$

Types of subterms have been omitted.

$$\dfrac{u\colon \exists x Rx \mid u \mid \emptyset \qquad \dfrac{\dfrac{v\colon \forall x(Rx \to R'y) \mid v \mid y}{vx\colon Rx \to R'y \mid v \mid x, y} \qquad w\colon Rx \mid w \mid x}{vxw\colon R'y \mid v, w \mid x, y}}{\dfrac{\mathrm{E}^{\exists}_{w,x}(u, vxw)\colon R'y \mid u, v \mid y}{\dfrac{\lambda u.\mathrm{E}^{\exists}_{w,x}(u, vxw)\colon \exists x Rx \to R'y \mid v \mid y}{\lambda vu.\mathrm{E}^{\exists}_{w,x}(u, vxw)\colon \forall x(Rx \to R'y) \to (\exists x Rx \to R'y) \mid \emptyset \mid y}\,v}\,u}\,w$$

2.3.8A. For the equivalence $\mathbf{I} \vdash A^{\mathrm{g}} \leftrightarrow A^{\mathrm{q}}$ let A^* be obtained from A by inserting $\neg\neg$ after each $\forall$, so $A^{\mathrm{q}} \equiv \neg\neg A^*$. Then prove by induction on the depth of A that $\neg\neg(A^* \leftrightarrow A^{\mathrm{g}})$.

2.4.2D. Let us call the alternative system $\mathbf{H}'$, and let us introduce abbreviations for the axioms mentioned: $\mathbf{k}^{A,B}$ (cf. 1.3.6), $\mathbf{w}^{A,B}$ (contraction), $\mathbf{e}^{A,B,C}$ (permutation) and $\mathbf{t}^{A,B,C}$ (near-transitivity).

We may prove the equivalence to $\to$**Hi** either directly, by giving a derivation of the axiom schema **s** in $\mathbf{H}'$, or by paralleling the proof that $\to$**Hi** is equivalent to $\to$**Ni**. For the latter strategy, we have to derive $A \to A$, and to show closure under $\to$I. The first is easy: take $(A \to A \to A) \to A \to A$ as instance of the contraction axiom, and detach the premise as an instance of **k**.

To show closure under $\to$I, suppose we have derived in $\mathbf{H}'$

$$\dfrac{\begin{matrix}\mathcal{D}_0 \\ A \to B\end{matrix} \qquad \begin{matrix}\mathcal{D}_1 \\ A\end{matrix}}{B}$$

where $\mathcal{D}_0, \mathcal{D}_1$ use assumptions from Γ, C. By the induction hypothesis there are $\mathcal{D}_0^*$, $\mathcal{D}_1^*$ deriving $C \to (A \to B)$ and $C \to A$ respectively from Γ. We combine these in a new proof

$$\dfrac{\mathbf{w}^{C,B} \qquad \dfrac{\dfrac{\mathbf{t}^{A,C\to B,C} \qquad \dfrac{\mathbf{e}^{C,A,B} \qquad \begin{matrix}\mathcal{D}_0^* \\ C \to A \to B\end{matrix}}{A \to C \to B}}{(C \to A) \to (C \to C \to B)} \qquad \begin{matrix}\mathcal{D}_1^* \\ C \to A\end{matrix}}{C \to C \to B}}{C \to B}$$

3.1.3C.

$$\dfrac{\dfrac{\dfrac{\dfrac{\dfrac{\dfrac{A \Rightarrow A \qquad \dfrac{\bot \Rightarrow}{A, \bot \Rightarrow}}{\neg A, A \Rightarrow}}{\neg A, A \Rightarrow B}}{\neg A \Rightarrow A \to B} \qquad \dfrac{\bot \Rightarrow \bot}{\neg A, \bot \Rightarrow \bot}}{\neg(A \to B), \neg A \Rightarrow \bot}}{\neg(A \to B) \Rightarrow \neg\neg A} \qquad \dfrac{\dfrac{\dfrac{\dfrac{\dfrac{B \to B}{A, B \Rightarrow B}}{B \Rightarrow A \to B} \qquad \dfrac{\bot \Rightarrow \bot}{B, \bot \Rightarrow \bot}}{\neg(A \to B), B \Rightarrow \bot}}{\neg(A \to B) \Rightarrow \neg B} \qquad \dfrac{\bot \Rightarrow \bot}{\neg(A \to B), \bot \Rightarrow \bot}}{\neg(A \to B), \neg\neg B \Rightarrow \bot}}{\dfrac{\neg\neg A \to \neg\neg B, \neg(A \to B) \Rightarrow \bot}{\dfrac{\neg\neg A \to \neg\neg B \Rightarrow \neg\neg(A \to B)}{\Rightarrow \neg\neg A \to \neg\neg B \to \neg\neg(A \to B)}}}$$

3.1.3D.

$$\dfrac{\dfrac{\dfrac{\dfrac{A \Rightarrow B}{A, A \Rightarrow B}}{\Rightarrow A, A \to B}}{\Rightarrow A, \exists x(A \to B)} \qquad \dfrac{\dfrac{\dfrac{\dfrac{B \Rightarrow B}{A, B \Rightarrow B}}{B \Rightarrow A \to B}}{B \Rightarrow \exists x(A \to B)}}{\exists x B \Rightarrow \exists x(A \to B)}}{A \to \exists x B \Rightarrow \exists x(A \to B)}$$

$$\dfrac{\dfrac{\dfrac{\dfrac{\dfrac{A, B \Rightarrow A, B}{A \Rightarrow B, B{\to}A}}{\Rightarrow A{\to}B, B{\to}A}}{A{\to}B, (A \to B) \vee (B \to A)}}{(A{\to}B) \vee (B{\to}A), (A{\to}B) \vee (B{\to}A)}}{(A{\to}B) \vee (B{\to}A)} \qquad \dfrac{\dfrac{\dfrac{\dfrac{\dfrac{\dfrac{Ax, Az \Rightarrow Az, \forall y Ay}{Ax \Rightarrow Az, Az{\to}\forall y Ay}}{Ax \Rightarrow Az, \exists x(Ax{\to}\forall y Ay)}}{Ax \Rightarrow \forall y Ay, \exists x(Ax{\to}\forall y Ay)}}{\Rightarrow Ax{\to}\forall y Ay, \exists x(Ax{\to}\forall y Ay)}}{\Rightarrow \exists x(Ax{\to}\forall y Ay), \exists x(Ax{\to}\forall y Ay)}}{\Rightarrow \exists x(Ax{\to}\forall y Ay)}$$

3.2.1A. For the proof of equivalence we have to appeal to closure under Cut for **G1i**. The easy direction is

$$\text{If } \mathbf{G1i} \vdash \Gamma \Rightarrow \bigvee \Delta \text{ then m-}\mathbf{G1i} \vdash \Gamma \Rightarrow \Delta.$$

The proof is perfectly straightforward, by an induction on the depth of deductions in **G1i**, provided we prove simultaneously

$$\text{If } \mathbf{G1i} \vdash \Gamma \Rightarrow \text{ then m-}\mathbf{G1i} \vdash \Gamma \Rightarrow .$$

In the induction step, it helps to distinguish cases according to whether Δ contains zero, one or more formulas. The difficult direction is to show

$$(*) \text{ If m-}\mathbf{G1i} \vdash \Gamma \Rightarrow \Delta \text{ then } \mathbf{G1i} \vdash \Gamma \Rightarrow \bigvee \Delta.$$

Using closure of **G1i** under Cut we first establish that

$$\mathbf{G1i} \vdash (A \vee B) \vee C \Rightarrow A \vee (B \vee C),$$
$$\mathbf{G1i} \vdash A \vee (B \vee C) \Rightarrow (A \vee B) \vee C,$$
$$\mathbf{G1i} \vdash A \vee B \Rightarrow B \vee A.$$

This permits us to disregard bracketing and order of the formulas in forming $\Gamma \Rightarrow \bigvee \Delta$. Now the proof proceeds by induction on the depth of deductions in m-**G1i**. We treat some crucial cases in the proof of $(*)$.

The last step in the proof in m-**G1i** was R$\wedge$:

$$\dfrac{\Gamma \Rightarrow \Delta, A \qquad \Gamma \Rightarrow \Delta, B}{\Gamma \Rightarrow \Delta, A \wedge B}$$

Let D be $\bigvee \Delta$. By induction hypothesis we have deductions of

$$\mathbf{G1i} \vdash \Gamma \Rightarrow D \vee A \text{ and } \mathbf{G1i} \vdash \Gamma \Rightarrow D \vee B$$

Hence we have

$$\mathbf{G1i} \vdash \Gamma \Rightarrow (D \vee A) \wedge (D \vee B)$$

We now prove $\mathbf{G1i} \vdash (D \vee A) \wedge (D \vee B) \Rightarrow D \vee (A \wedge B)$, and apply Cut.

Suppose now the last step in the proof in m-**G1i** was L$\to$:

$$\frac{\Gamma \Rightarrow A, \Delta \quad \Gamma, B \Rightarrow \Delta}{\Gamma, A \to B \Rightarrow \Delta}$$

By IH we have in **G1i** a proof $\mathcal{D}'$ of $\Gamma \Rightarrow A \vee D$ and a proof $\mathcal{D}''$ of $\Gamma, B \Rightarrow D$.

We get the required conclusion from the following deduction:

$$\dfrac{\begin{matrix}\mathcal{D}' \\ \Gamma \Rightarrow D \vee A\end{matrix} \quad \dfrac{\dfrac{D \Rightarrow D}{\Gamma, D, A \to B \Rightarrow D}\text{LW} \quad \dfrac{\dfrac{A \Rightarrow A \quad B \Rightarrow B}{A, A \to B \Rightarrow B}\text{L}\to \quad \begin{matrix}\mathcal{D}'' \\ \Gamma, B \Rightarrow D\end{matrix}}{\Gamma, A, A \to B \Rightarrow D}\text{Cut}}{\Gamma, D \vee A, A \to B \Rightarrow D}\text{L}\vee}{\dfrac{\Gamma\Gamma, A \to B \Rightarrow D}{\Gamma, A \to B \Rightarrow D}\text{LC}}\text{Cut}$$

3.3.3B. The equivalence between classical G-systems and classical N-systems for subsets of the operators not containing $\perp$.

(a) Applications of $\perp_c$ in **Nc** as on the left may be replaced by applications of Peirce's rule as on the right:

$$\begin{matrix}[A \to \perp]^u \\ \mathcal{D} \\ \dfrac{\perp}{A}u,\perp_c\end{matrix} \qquad \begin{matrix}[A \to \perp]^u \\ \mathcal{D} \\ \dfrac{\dfrac{\perp}{A}\perp_i}{A}\text{P},u\end{matrix}$$

Conversely, applications of P in **Nc**$'$ as on the left may be replaced by applications of $\perp_c$ as on the right:

$$\begin{matrix}[A \to B]^u \\ \mathcal{D} \\ \dfrac{A}{A}\text{P},u\end{matrix} \qquad \dfrac{\begin{matrix}\dfrac{\dfrac{\dfrac{\neg A^v \quad A^u}{\perp}}{B}\perp_i}{A \to B}u \\ \mathcal{D} \\ A\end{matrix} \quad \neg A^v}{\dfrac{\perp}{A}\perp_c, v}$$

(b) The equivalence between any two 1-equivalents is easily seen to hold by the following deduction steps:

$$\dfrac{\Gamma, A \to B \Rightarrow B, \Delta \quad \dfrac{A \Rightarrow B, A}{\Rightarrow A \to B, A}}{\Gamma \Rightarrow A, B, \Delta}\text{Cut} \qquad \dfrac{\Gamma \Rightarrow A, B, \Delta \quad \Gamma, A \Rightarrow A, \Delta}{\Gamma, A \to B \Rightarrow B, \Delta}\text{L}\to$$

By repeated application of these deduction steps, we see that from $\Gamma, \Delta \to B \Rightarrow B$ follows $\Gamma \Rightarrow B, \Delta$ and vice versa ($\Delta \to A$ is short for $D_1 \to A, \ldots, D_n \to A$ where $\Delta \equiv D_1, \ldots, D_n$).

(c) We give a proof by induction on the depth of deductions in **G2c**. A is said to be the *goal-formula* of the 1-equivalent $\Gamma, \Delta \to A \Rightarrow A$ of the sequent $\Gamma \Rightarrow A, \Delta$.

Let us consider a crucial case. Suppose the **G2c**-deduction ends with

$$\dfrac{\begin{matrix}\mathcal{D} \\ \Gamma, A \Rightarrow B, \Delta\end{matrix}}{\Gamma \Rightarrow A \to B, \Delta}$$

By induction hypothesis we have a deduction $\mathcal{D}^*$ in $\mathbf{Nc}'$ deriving B from $\Delta{\to}A$, A, Γ, $A \to B$. Let $\Delta \equiv D_1, \ldots, D_n$. Then

$$\dfrac{\dfrac{\dfrac{D_i \to (A \to B) \quad B_i^u}{A \to B} \qquad A^v}{\dfrac{B}{[D_i \to B]}\,u}\;(1 \le i \le n)}{}\qquad \begin{array}{c} \mathcal{D} \\ \dfrac{B}{A \to B}\,v \end{array}$$

which shows that in $\mathbf{Nc} \vdash \Gamma, \Delta \to (A \to B) \Rightarrow A \to B$. For our goal-formula we have chosen the principal formula.

Another case is when the deduction in **G2c** ends with

$$\dfrac{\Gamma \Rightarrow A, \Delta \qquad \Gamma \Rightarrow B, \Delta}{\Gamma, A \wedge B, \Delta}$$

By IH we have natural deductions of A from $\Delta \to A$ and of B from $\Delta \to B$. We construct a new deduction

$$\dfrac{\begin{array}{c} (1 \le i \le n)\ \dfrac{\dfrac{\dfrac{D_i \to A \wedge B \quad D_i^u}{A \wedge B}}{A}}{[D_i \to A]}\,u \\ \mathcal{D}_1 \\ A \end{array} \qquad \begin{array}{c} v\,\dfrac{\dfrac{\dfrac{D_i \to A \wedge B \quad D_i^v}{A \wedge B}}{B}}{[D_i \to B]}\ (1 \le i \le n) \\ \mathcal{D}_2 \\ B \end{array}}{A \wedge B}$$

Here again we have used the principal formula as goal-formula. But in this case the alternative, choosing a formula of the context as goal-formula, works at least as well, if not better. Let C be a formula from the context.

$$\begin{array}{c} \dfrac{\dfrac{A \wedge B \to C \qquad \dfrac{A^v \quad B^w}{A \wedge B}}{C}}{[A \to C]}\,v \\ \mathcal{D}_1 \\ \dfrac{C}{[B \to C]}\,w \\ \mathcal{D}_2 \\ C \end{array}$$

If on the other hand the context is empty, the case becomes completely straightforward.

If the final rule is an R$\vee$–

$$\dfrac{\Gamma \Rightarrow A, \Delta}{\Gamma \Rightarrow A \vee B, \Delta}$$

– it is definitely more advantageous to work with a 1-equivalent where $A \vee B$ is not the goal-formula. If Δ is empty, the case is trivial. If $\Delta \equiv C, \Delta'$, we construct a deduction as follows:

$$\dfrac{\dfrac{A \vee B \to C \quad \dfrac{A^v}{A \vee B}}{C}}{[A \to C]}\, v$$
$$\mathcal{D}$$
$$C$$

Here $\mathcal{D}$ is a proof given by the induction hypothesis. The treatment of the case where the final rule is R$\forall$ is similar to the case of R$\wedge$; and the case for R$\exists$ resembles the case for R$\vee$. The treatment of the cases where the deduction ends with a left rule is easy, since then the goal-formula is not affected.

3.3.4B. We show that in **G2i*** + Cut, that is **G2i** + Cut with sequents with inhabited succedent, we can define N such that $|\mathsf{N}(\mathcal{D})| < k|\mathcal{D}|$, where k can be taken to be 2 for the full system, and 1 for the system without $\bot$.

Axioms in **G2i*** + Cut are either of the form $\Gamma, A \Rightarrow A$ or of the form $\Gamma, \bot \Rightarrow A$. In the first case, the axiom is translated as a prooftree with a single node A, and $|\mathsf{N}(\mathcal{D})| = 0 < 2^{|\mathcal{D}|} = 2^0 = 1$. In the second case, the axiom is translated as $\frac{\bot}{A}$. Hence $|\mathsf{N}(\mathcal{D})| = 1 < 2{\cdot}2^{|\mathcal{D}|} = 2^1 = 2$. Now we consider the induction step. We check three typical cases. Let $d_i = |\mathcal{D}_i|$, $d'_i = |\mathcal{D}'_i|$, $\mathcal{D}'_i = \mathsf{N}(\mathcal{D}_i)$, $d = |\mathcal{D}|$, $d' = |\mathcal{D}'|$ for $i = 1, 2$.

Case 1. $\mathcal{D}$ ends with R$\wedge$, so $\mathcal{D}$ is as on the left, and is translated as on the right:

$$\dfrac{\begin{matrix}\mathcal{D}_1 \\ \Gamma \Rightarrow A\end{matrix} \quad \begin{matrix}\mathcal{D}_2 \\ \Gamma \Rightarrow B\end{matrix}}{\Gamma \Rightarrow A \wedge B} \qquad\qquad \dfrac{\begin{matrix}\mathcal{D}'_1 \\ A\end{matrix} \quad \begin{matrix}\mathcal{D}'_2 \\ B\end{matrix}}{A \wedge B}$$

Then $d' = \max(d'_1, d'_2) + 1 < \max(k2^{d_1}, k2^{d_2}) + 1 = k2^{\max(d_1,d_2)} + 1 \leq k2^{\max(d_1,d_2)+1} = k2^d$.

Case 2. $\mathcal{D}$ ends with L$\to$; $\mathcal{D}$ and $\mathcal{D}'$ have respectively the forms

$$\dfrac{\begin{matrix}\mathcal{D}_1 \\ \Gamma \Rightarrow A\end{matrix} \quad \begin{matrix}\mathcal{D}_2 \\ \Gamma, B \Rightarrow C\end{matrix}}{\Gamma, A \to B \Rightarrow C} \qquad\qquad \begin{matrix}\dfrac{A \to B \quad \begin{matrix}\mathcal{D}_1 \\ A\end{matrix}}{[B]} \\ \mathcal{D}_2 \\ C\end{matrix}$$

Now

$$d' \leq d_1 + d_2 + 1 < k2^{d_1} + (k2^{d_2} - 1) + 1$$

(using $d_1 < k2^{d_1}$, $d_2 \leq k2^{d_2} - 1$), hence

$$d' < k(2^{d_1} + 2^{d_2}) \leq k2{\cdot}2^{\max(d_1,d_2)} = k2^d.$$

Case 3. $\mathcal{D}$ ends with Cut; then $\mathcal{D}$ and $\mathcal{D}'$ have the forms

$$\dfrac{\begin{matrix}\mathcal{D}_1 \\ \Gamma \Rightarrow A\end{matrix} \quad \begin{matrix}\mathcal{D}_2 \\ \Gamma, A \Rightarrow B\end{matrix}}{\Gamma', \Gamma \Rightarrow B} \qquad\qquad \begin{matrix}\mathcal{D}_1 \\ [A] \\ \mathcal{D}_2 \\ B\end{matrix}$$

and

$$d' \leq d_1 + d_2 < k(2^{d_1} + 2^{d_2}) = k2{\cdot}2^{\max}(d_1, d_2) = k2^d.$$

3.5.7A. Let $B \equiv ((PQ)Q)P$, $A \equiv BQ$. We first show that $A, QP \Rightarrow Q$ is actually provable:

$$\dfrac{\dfrac{\dfrac{\dfrac{\dfrac{\dfrac{A, P, (PQ)Q, (PQ)Q \Rightarrow P}{A, P, (PQ)Q \Rightarrow B} \quad P, (PQ)Q, Q \Rightarrow Q}{A, P, (PQ)Q \Rightarrow Q} \quad (PQ)Q, P \Rightarrow P}{A, QP, (PQ)Q \Rightarrow P}}{A, QP \Rightarrow B} \quad QP, Q \Rightarrow Q}{A, QP \Rightarrow Q}$$

Now suppose we have a proof of $A, QP \Rightarrow Q$ with the restricted version of the rule L$\rightarrow$. Then we can show that all possibilities for constructing a proof bottom-up fail.

(a) If A was introduced by L$\rightarrow$, $A, QP \Rightarrow Q$ reduces to $QP \Rightarrow B$ and $QP, Q \Rightarrow Q$. $QP \Rightarrow B$ reduces to $QP, (PQ)Q \Rightarrow P$, which, if $(PQ)P$ was introduced by L$\rightarrow$, reduces to

(aa) $QP \Rightarrow PQ$ and $QP, Q \Rightarrow P$ – breakdown or, if QP was introduced by L$\rightarrow$, to

(ab) $P, (PQ)Q \Rightarrow P$ and $(PQ)Q \Rightarrow Q$, the latter in turn reduces to $Q \Rightarrow Q$ and $\Rightarrow PQ$ – breakdown.

(b) If QP was introduced by L$\rightarrow$, $A, QP \Rightarrow Q$ reduces to $A, P \Rightarrow Q$ and $A \Rightarrow Q$, hence to $A \Rightarrow Q$; this reduces to $Q \Rightarrow Q$ and $\Rightarrow B$, the latter sequent reduces to $(PQ)Q \Rightarrow P$ which is underivable.

3.5.11A. Let us assume that $\mathcal{D}$ proves $\vdash_n \Gamma, \forall x A \Rightarrow \Delta$; since Γ, Δ consist of quantifier-free formulas, Γ, Δ does not contain $\forall x A$. $\mathcal{D}$ contains instances α_i of L$\rightarrow$:

$$\frac{\forall x A, A[x/t_i], \Gamma_i \Rightarrow \Delta_i}{\forall x A, \Gamma_i \Rightarrow \Delta_i}$$

Let $t_0, \ldots, t_{n-1}$ be a complete list of the terms involved in these applications. (This sequence is possibly empty!) Replace in the deduction $\mathcal{D}$ the occurrences of $\forall x A$ by $A[x/t_0], \ldots, A[x/t_{n-1}]$. The result is a correct proof in **G3**[**mic**], except that the α_i are transformed into instances of contraction. (Why is the proof correct?) By closure under contraction of **G3**[**mic**], we can successively remove these instances of contraction. An alternative proof simply uses induction on the depth of $\mathcal{D}$.

The extra result for **G3c** may be proved similarly, or by reducing it to the preceding result using the definition of $\exists x A$ as $\neg\forall x \neg A$, and the possibility of shifting formulas from left to right and vice versa in **G3c**.

4.2.7A Assume **G3i** $\vdash \forall x(P \vee Rx) \rightarrow P \vee \forall x Rx$. By inversion, this is equivalent to having a proof $\mathcal{D}'$ of **G3i** $\vdash \forall x(P \vee Rx) \Rightarrow P \vee \forall x Rx$. In order to show that this is unprovable, we establish by induction something more general, namely that there is no deduction $\mathcal{D}''$ for $\forall x(P \vee Rx), Rt_0, Rt_1, \ldots, Rt_{n-1} \Rightarrow P \vee \forall x Rx$. This is proved by induction on the depth of $\mathcal{D}''$. Let us abbreviate the antecedent of the conclusion of $\mathcal{D}''$ as Δ_n. If the last rule applied in $\mathcal{D}''$ is R$\vee$, it would mean that $\Delta_n \Rightarrow P$ or $\Delta_n \Rightarrow \forall x Rx$ ought to be provable; but neither of these is even classically valid. If the last rule applied in $\mathcal{D}''$ is L$\forall$, it means that we have a proof $\mathcal{D}'''$ of depth less than $|\mathcal{D}''|$ showing $\Delta_n, P \vee Rt_n \Rightarrow P \vee \forall x Rx$, which by the inversion lemma means that there is a proof of no greater depth of $\Delta_n, P, Rt_n \Rightarrow P \vee \forall x Rx$. This is impossible by the induction hypothesis.

4.2.7B The only somewhat awkward case is the test for the sequent $BQ, QP \Rightarrow Q$ with $B \equiv ((PQ)Q)P$. (We have dropped $\to$ in the notation, since the sequent is purely implicational.) We put $A \equiv BQ$. The application of the algorithm leads in principle to many branches in the search tree, which all yield failure because of repetition of sequents. However, after some experimenting with branches, we may note the following.

Suppose we concentrate in the search tree on the left premises of applications of L$\to$, and forget about the right premises. Moreover, we always apply in reverse R$\to$, whenever possible. Then

$$\begin{array}{ll} A, \Gamma \Rightarrow C & \text{yields } A, \Gamma, (PQ)Q \to P, \\ QP, \Gamma \Rightarrow C & \text{yields } QP, \Gamma \Rightarrow Q, \\ (PQ)Q, \Gamma \Rightarrow C & \text{yields } (PQ)Q, \Gamma, P \Rightarrow Q. \end{array}$$

Hence, after repeatedly treating the initial sequent by L$\to$, always followed when applicable by R$\to$, we always find a sequent $\Gamma \Rightarrow A$ with A equal to P or Q, and Γ as a set a subset of $A, (PQ)Q, QP, P$.

Modulo contraction, applying L$\to$ (followed by R$\to$ when applicable) always again yields one of these sequents, so all branches produce repetitions. (Actually, the reader may note that the sequent considered is not even classically provable, by finding a suitable falsifying valuation.)

4.2.7D. Let **G1i[mic]**$^\circ$ be the system obtained from **G1i[mic]** by replacing L$\to$ by its context-free version (as in Gentzen's original system). Suppose a deduction $\mathcal{D}$ in **G1i[mic]**$^\circ$ establishes $\vdash_n B, A \to B, \Gamma \Rightarrow \Delta$; then we readily see by induction on n that there is also a deduction $\mathcal{D}'$ establishing $\vdash_n B, \Gamma \Rightarrow \Delta$, and $\mathcal{D}'$ is restricted if $\mathcal{D}$ is restricted. This suffices to prove the property by induction on the depth of proofs.

4.2.7F. Let $\mathcal{D}$ be a deduction in **G3i** with $|\mathcal{D}| \leq n$ of $\Gamma \Rightarrow C$. $\mathcal{D}$ cannot be an axiom. We apply induction on the depth of $\mathcal{D}$. If the final rule in $\mathcal{D}$ was a right rule, the premises are of the form $\Gamma \Rightarrow C'$ and the induction hypothesis applies. If the last rule applied in $\mathcal{D}$ is an application of L$\to$ (there is no other possibility), the final step is of the form

$$\frac{\Gamma \Rightarrow A_i \qquad \Gamma', B_i \Rightarrow C}{\Gamma', A_i \to B_i \Rightarrow C}$$

where $\Gamma \equiv \Gamma', A_i \to B_i$, and the result follows.

4.3.6A. We discuss the last formula of the exercise, abbreviated by not writing many implication arrows to:

$$(a)\ (PR)R \to (QR)R \Rightarrow (PQ)R \to R.$$

Let us write A for $(PR)R \to (QR)R$. We shall always tacitly reduce any problem by use of the invertible rule R$\to$. As a shortcut we also observe that (for all B, C, D, Γ) $DB, B, \Gamma \Rightarrow C$ is derivable iff $B, \Gamma \Rightarrow C$ is derivable. Our initial problem is replaced by (R$\to$):

$$(b)\ A, (PQ)R \Rightarrow R.$$

(1) First treat in (b) $(PQ)R$. This reduces with L$\to\to$ to

$$A, QR, P \Rightarrow PQ \ \text{ and } \ R \to R \text{ (Axiom)}$$

We then apply L$\to\to$ to A and obtain from the first problem:

$$R \to (QR)R, PR, QR, P \Rightarrow Q \ \text{ and } \ (QR)R, QR \Rightarrow R.$$

We continue with the first and apply L0$\to$:

$$R{\to}(QR)R, P, R, QR \Rightarrow Q.$$

Again with L0$\to$

$$R, (QR)R, P, QR \Rightarrow Q.$$

This is equivalent to $P, R \Rightarrow Q$ which is obviously underivable.

(2) Now treat in (b) first $(PQ)R$ by L$\to\to$:

$$R{\to}(QR)R, QR, PR, P \Rightarrow Q \ \text{ and } \ R{\to}P(QR)R, R, PR \Rightarrow R \text{ (Axiom).}$$

Apply L0$\to$ to the first of these formulas, then

$$R \to (QR)R, QR, P, R \Rightarrow Q.$$

Again with L0$\to$

$$(QR)R, QR, P, R \Rightarrow Q.$$

which is equivalent to $P, R \Rightarrow Q$ which is underivable.

Hence the formula to be tested is underivable.

5.1.13C. The proofs of properties (i) and (ii) are more or less routine and left to the reader; we concentrate on the proof of (iii). Let us write $\vdash_n$ for $\vdash_n^0$. Assume we have

$$\vdash_{n+1} \Gamma, u{:}\,(A{\to}B){\to}C \Rightarrow t{:}\,D.$$

Let the final step in the deduction be an application of L$\to$, with $(A\to B)\to C$ as principal formula. Then there are premises:

$$\vdash_n \Gamma, u\colon (A\to B)\to C \Rightarrow s_0\colon A\to B,$$
$$\vdash_n \Gamma, u\colon (A\to B)\to C, w\colon C \Rightarrow t_0\colon D.$$

By the inversion property (ii), the first line yields

$$\vdash_n \Gamma, u\colon (A\to B)\to C, y_0\colon A \Rightarrow s_1\colon B,$$

where $s_0 \equiv \lambda y_0^A.s_1$. By the IH applied to this result, we find

$$\vdash_n \Gamma, z\colon B\to C, y\colon A, y_0\colon A \Rightarrow s_2\colon B,$$

where

$$s_2 =_\beta s_1[u/\lambda x^{A\to B}.z(xy)].$$

From now on we write σ for the substitution $[u/\lambda x^{A\to B}.z(xy)]$. Apply the contraction property (i):

$$(*) \qquad \vdash_n \Gamma, z\colon B\to C, y\colon A \Rightarrow s_2[y_0/y]\colon B.$$

Put

$$s_3 \equiv s_2[y_0/y], \quad s_4 \equiv s_1[y_0/y],$$

and note that $s_2[y_0/y] =_\beta s_1[y_0/y]$, i.e., $s_3 =_\beta s_4\sigma$. Also, applying the IH to the other premise,

$$(**) \qquad \vdash_n \Gamma, z\colon B\to C, y\colon A, w\colon C \Rightarrow t_1\colon D,$$

where $t_1 =_\beta t_0\sigma$. Combining $(*)$ and $(**)$ we find

$$\vdash_{n+1} \Gamma, z\colon B\to C, y\colon A \Rightarrow t_1[w/zs_3]\colon D.$$

Since, with an appeal to the substitution lemma 1.2.4,

$$t_1[w/zs_4\sigma] =_\beta t_0\sigma[w/zs_4\sigma] \equiv t_0[w/zs_4]\sigma,$$

we can take in this case $t' \equiv t_0[w/zs_4]$. The other cases of the induction step are easier.

6.8.7C. The proof of C1 and C2 of lemma 6.8.3 for the extra case is easy; only C3 asks some extra attention. Assume

$$(\forall t' \prec_1 t)\mathrm{Comp}_{A\wedge B}(t').$$

We have to show $\mathrm{Comp}_{A\wedge B}(t)$, that is to say $\mathrm{Comp}_A(\mathbf{p}_0 t)$, $\mathrm{Comp}_B(\mathbf{p}_1 t)$; for this it suffices to show

$$(\forall t'' \prec_1 \mathbf{p}_0 t)\mathrm{Comp}_A(t''), \quad (\forall t''' \prec_1 \mathbf{p}_1 t)\mathrm{Comp}_B(t''').$$

Case 1. The term t is not of the form $\mathbf{p}t_0t_1$. Then if $t'' \prec_1 \mathbf{p}_0 t$, $t'' \equiv \mathbf{p}_0 t'$, $t' \prec_1 t$, and by hypothesis $\mathrm{Comp}_{A\wedge B}(t')$, hence $\mathrm{Comp}_A(\mathbf{p}_0 t')$, i.e., $\mathrm{Comp}_A(t'')$; similarly $\mathrm{Comp}_B(t''')$ for all $t''' \prec_1 \mathbf{p}_1 t$.

Case 2. If t is of the form $\mathbf{p}t_0t_1$, and $t'' \prec_1 \mathbf{p}_0 t$, t'' is either of the form $\mathbf{p}_0(\mathbf{p}t'_0t_1)$ with $t'_0 \prec_1 t_0$, or of the form $\mathbf{p}_0(\mathbf{p}t_0t'_1)$ with $t'_1 \prec_1 t_1$, or is equal to t_0. In the first subcase, since by assumption $\mathrm{Comp}_{A\wedge B}(\mathbf{p}t'_0t_1)$, also $\mathrm{Comp}_A(\mathbf{p}_0(\mathbf{p}t'_0t_1))$. In the second subcase, $\mathrm{Comp}_{A\wedge B}(\mathbf{p}t_0t'_1)$, hence $\mathrm{Comp}_A(\mathbf{p}_0(\mathbf{p}t_0t'_1))$. In the third subcase, we conclude that $\mathrm{Comp}_A(t_0)$, since as before we have $\mathrm{Comp}_A(\mathbf{p}_0(\mathbf{p}t'_0t_1))$ for all $t'_0 \prec t_0$, and since C2 holds for A, it also follows that $\mathrm{Comp}_A(t'_0)$. etc. In the same way we establish in this case $(\forall t''' \prec_1 \mathbf{p}_1 t)\mathrm{Comp}_B(t''')$.

But note that a better result may be obtained by strengthening the property defining $\mathrm{Comp}_{A\wedge B}$, see 8.3.1.

Bibliography

S. Abramsky, D. M. Gabbay, and T. S. E. Maibaum

[1992] eds., *Handbook of Logic in Computer Science, Vol. 1. Background: Mathematical Structures*, Clarendon Press, Oxford. Editor of the volume D. M. Gabbay.

P. H. G. Aczel

[1968] Saturated intuitionistic theories, in *Contributions to Mathematical Logic*, H. A. Schmidt, K. Schütte, and H.-J. Thiele, eds., Studies in Logic and the Foundations of Mathematics, North-Holland Publ. Co., Amsterdam, 1–11.

T. Altenkirch

[1993] *Constructions, Inductive Types and Strong Normalization*, PhD thesis, The University of Edinburgh, Department of Computer Science, Edinburgh.

Y. Andou

[1995] A normalization-procedure for the first-order natural deduction with full logical symbols, *Tsukuba Journal of Mathematics*, 19, 153–162.

J.-M. Andreoli

[1992] Logic programming with focusing proofs in linear logic, *Journal of Logic and Computation*, 2, 297–347.

J.-M. Andreoli and R. Pareschi

[1991] Linear objects: logical processes with built-in inheritance, *New Generation Computing*, 9, 445–473.

K. R. Apt

[1990] Logic programming, in *Handbook of Theoretical Computer Science, Volume B. Formal Methods and Semantics*, J. v. Leeuwen, ed., Elsevier Publ. Co., 493–574.

A. Avron

[1991] Hypersequents, logical consequence and intermediate logic for concurrency, *Annals of Mathematics and Artificial Intelligence*, 4, 225–248.

[1996] The method of hypersequents in the proof theory of propositional non-classical logics, in *Logic: from Foundations to Applications. European Logic Colloquium*, Clarendon Press, Oxford, 1–32.

[1998] Two types of multiple-conclusion systems, *Logic Journal of the IGPL Interest Group in Pure and Applied Logics*, 6, 695–717.

F. Baader and J. Siekmann

[1994] Unification theory, in *Gabbay et al. [1994]*, 41–125.

A. A. Babaev and S. V. Solovjov

[1979] A coherence theorem for canonical maps in CCC's (Russian, with English summary), *Zapiski Nauchnykh Seminarov Leningradskogo Otdeleniya Ordena Lenina Matematicheskogo Instituta imeni V. A. Steklova Akademii Nauk SSSR (LOMI)*, 88, 3–29. Translation in *Journal of Soviet Mathematics,* 20 (1982), 2263–2279.

[1990] On conditions of full coherence in biclosed categories: a new application of proof theory, in *COLOG-88*, P. Martin-Löf and G. E. Mints, eds., Lecture Notes in Computer Science 417, Springer-Verlag, Berlin, Heidelberg, New York, 3–8.

H. Bachmann

[1955] *Transfinite Zahlen*, Springer-Verlag, Berlin, Heidelberg, New York. 2nd, revised edition 1967.

H. P. Barendregt

[1984] *The Lambda Calculus*, North-Holland Publ. Co., Amsterdam. 2nd edition.

[1992] Lambda calculi with types, in *Handbook of Logic in Computer Science, Vol. 2*, S. Abramsky, D. M. Gabbay, and T. S. E. Maibaum, eds., Oxford University Press, Oxford, 118–309.

M. J. Beeson

[1985] *Foundations of Constructive Mathematics*, Springer-Verlag, Berlin, Heidelberg, New York.

N. D. Belnap

[1982] Display logic, *Journal of Philosophical Logic*, 11, 375–417.

[1990] Linear logic displayed, *The Notre Dame Journal of Formal Logic*, 31, 14–25.

N. Benton, G. Bierman, J. M. E. Hyland, and V. C. V. de Paiva

[1992] Term assignment for intuitionistic linear logic, Tech. Rep. 262, Computer Laboratory, University of Cambridge.

P. Bernays

[1970] On the original Gentzen consistency proof for number theory, in *Myhill et al. [1970]*, 409–417.

E. W. Beth

[1953] On Padoa's method in the theory of definition, *Indagationes Mathematicae*, 15, 330–339.

[1955] Semantic entailment and formal derivability, *Mededelingen der Koninklijke Nederlandse Akademie van Wetenschappen (Amsterdam), Afdeling Letterkunde. Nieuwe Reeks*, 18, 309–342.

[1956] Semantic construction of intuitionistic logic, *Mededelingen der Koninklijke Nederlandse Akademie van Wetenschappen (Amsterdam), Afdeling Letterkunde. Nieuwe Reeks*, 19, 357–388.

[1959] *The Foundations of Mathematics*, Studies in Logic and the Foundations of Mathematics, North-Holland Publ. Co., Amsterdam. 2nd edition, 1965.

[1962a] *Formal Methods*, D. Reidel Publ. Co., Dordrecht, Netherlands.

[1962b] Umformung einer abgeschlossenen deduktiven oder semantischen Tafel in eine natürliche Ableitung auf Grund der derivativen bzw. klassischen Implikationslogik, in *Logik und Logikkalkül*, M. Käsbauer and F. von Kutschera, eds., Verlag Karl Alber, Freiburg i. Br./München, Germany, 49–55.

M. N. Bezhanishvili

[1987] Notes on Wajsberg's proof of the separation theorem, in *Initiatives in Logic*, J. Srzednicki, ed., Martinus Nijhoff, Dordrecht, Netherlands, etc., 116–128.

W. Bibel and E. Eder

[1993] Methods and calculi for deduction, in *Gabbay et al. [1993]*, 68–182.

T. S. Blyth

[1986] *Categories*, Longman, London.

B. R. Boričić

[1985] On sequence-conclusion natural deduction systems, *Journal of Philosophical Logic*, 14, 359–377.

M. Borisavljević

[1999] A cut-elimination proof in intuitionistic predicate logic, *Annals of Pure and Applied Logic*, 99, 105–136.

N. G. de Bruijn

[1972] Lambda-calculus notation with nameless dummies, a tool for automatic formula manipulation, *Indagationes Mathematicae*, 34, 381–392.

W. Buchholz

[1991] Notation systems for infinitary derivations, *Archive for Mathematical Logic*, 30, 277–296.

W. Buchholz, S. Feferman, W. Pohlers, and W. Sieg

[1981] *Iterated Inductive Definitions and Subsystems of Analysis: Recent Proof-Theoretical Studies*, Lecture Notes in Mathematics 897, Springer-Verlag, Berlin, Heidelberg, New York.

W. Buchholz and S. S. Wainer

[1987] Provably computable functions and the fast growing hierarchy, in *Logic and Combinatorics. Proceedings of a Summer Research Conference held August 4–10, 1985*, S. G. Simpson, ed., Contemporary Mathematics 65, American Mathematical Society, Providence, RI, 179–198.

R. A. Bull and K. Segerberg

[1984] Basic modal logic, in *Handbook of Philosophical logic II. Extensions of Classical Logic*, D. Gabbay and F. Guenthner, eds., Reidel, Dordrecht, Netherlands, 1–88.

S. Buss and G. E. Mints

[1999] The complexity of the disjunction and existential properties in intuitionistic logic, *Annals of Pure and Applied Logic*, 99, 93–104.

D. H. C.

[1994] *From Logic to Logic Programming*, MIT Press, Cambridge, MA.

C. Cellucci

[1992] Existential instantiation and normalization in sequent natural deduction, *Annals of Pure and Applied Logic*, 58, 111–148.

C.-L. Chang and R. C.-T. Lee

[1973] *Symbolic Logic and Mechanical Theorem Proving*, Academic Press, New York.

A. Church

[1956] *Introduction to Mathematical Logic. Part I*, Princeton University Press, Princeton, NJ. 2nd edition.

T. Coquand and G. Huet

[1988] The calculus of constructions, *Information and Computation*, 76, 95–120.

W. Craig

[1957a] Linear reasoning. A new form of the Herbrand–Gentzen theorem, *The Journal of Symbolic Logic*, 22, 250–268.

[1957b] Three uses of the Herbrand–Gentzen theorem in relating model theory and proof theory, *The Journal of Symbolic Logic*, 22, 269–285.

P.-L. Curien

[1985] Typed categorical combinatory logic, in *Automata, Languages and Programming (ICALP 85)*, W. Brauer, ed., Lecture Notes in Computer Science 194, Springer-Verlag, Berlin, Heidelberg, New York, 130–139.

[1986] *Categorical Combinators, Sequential Algorithms and Functional Programming*, Pitman, London, and John Wiley and Sons, New York.

H. B. Curry

[1934] Functionality in combinatory logic, *Proceedings of the National Academy of the U.S.A.*, 20, 584–590.

[1942] The combinatory foundations of mathematical logic, *The Journal of Symbolic Logic*, 7, 49–64.

[1950] *A Theory of Formal Deducibility*, Notre Dame Mathematical Lectures 6, The University of Notre Dame Press, Notre Dame, IN.

[1952a] The elimination theorem when modality is present, *The Journal of Symbolic Logic*, 17, 249–265.

[1952b] The permutability of rules in the classical inferential calculus, *The Journal of Symbolic Logic*, 17, 245–248.

[1963] *Foundations of Mathematical Logic*, McGraw-Hill, New York. Also published by Dover, New York 1977.

H. B. Curry and R. Feys

[1958] *Combinatory Logic I*, Studies in Logic and the Foundations of Mathematics, North-Holland Publ. Co., Amsterdam. 2nd edition 1968.

D. van Dalen

[1994] *Logic and Structure*, Springer-Verlag, Berlin, Heidelberg, New York. 3rd edition.

D. van Dalen and R. Statman

[1979] Equality in the presence of apartness, in *Essays on Mathematical Logic. Proceedings of the Fourth Scandinavian Logic Symposium and of the 1st Soviet–Finnish Logic Conference*, J. Hintikka, I. Niiniluoto, and E. Saarinen, eds., Reidel, Dordrecht, Netherlands, 95–116.

V. Danos

[1990] *La logique linéaire appliquée à l'étude de divers processus de normalisation (principalement du λ-calcul)*, PhD thesis, Université Paris VII, Juin.

V. Danos, J.-B. Joinet, and H. A. J. M. Schellinx

[1995] On the linear decoration of intuitionistic derivations, *Archive for Mathematical Logic*, 33, 387–412. Slightly revised and condensed version of a technical report from 1993 with the same title.

[1997] A new deconstructive logic: classical logic, *The Journal of Symbolic Logic*, 62, 755–807.

[1999] Computational isomorphisms in classical logic. To appear in *Theoretical Computer Science*.

V. Danos and L. Regnier

[1989] The structure of multiplicatives, *Archive for Mathematical Logic*, 28, 181–203.

J. Diller

[1970] Zur Berechenbarkeit primitiv-rekursiver Funktionale endlicher Typen, in *Contributions to Mathematical Logic*, H. A. Schmidt, K. Schütte, and H.-J. Thiele, eds., North-Holland Publ. Co., Amsterdam, 109–120.

K. Došen

[1987] A note on Gentzen's decision procedure for intuitionistic propositional logic, *Zeitschrift für Mathematische Logik und Grundlagen der Mathematik*, 33, 453–456.

[1993] A historical introduction to substructural logics, in *Substructural Logics*, K. Dǒsen and P. Schroeder-Heister, eds., Clarendon Press, Oxford, 1–30.

A. G. Dragalin

[1979] *Mathematical Intuitionism. Introduction to Proof Theory* (Russian), Nauka, Moscow. Translated as Volume 67 in the series *Translations of Mathematical Monographs*, under the title *Mathematical Intuitionism*. American Mathematical Society, Providence, RI, 1988.

M. A. E. Dummett

[1959] A propositional calculus with denumerable matrix, *The Journal of Symbolic Logic*, 24, 97–106.

R. Dyckhoff

[1992] Contraction-free sequent calculi for intuitionistic logic, *The Journal of Symbolic Logic*, 57, 795–807.

[1996] Dragalin's proof of cut-admissibility for the intuitionistic sequent calculi **G3i** and **G3i′**, Tech. Rep. CS–96–9, Computer Science Division, St Andrews University.

R. Dyckhoff and S. Negri

[1999] Admissibility of structural rules for contraction-free systems of intuitionistic logic, *The Journal of Symbolic Logic*, 64, to appear.

R. Dyckhoff and L. Pinto

[1999] Permutability of proofs in intuitionistic sequent calculi, *Theoretical Computer Science*, 212, 141–155.

N. Eisinger and H. J. Ohlbach

[1993] Deduction systems based on resolution, in *Gabbay et al. [1993]*, 184–271.

S. Feferman

[1968] Lectures on proof theory, in *Proceedings of the Summer School in Logic*, M. H. Löb, ed., Lecture Notes in Mathematics 70, Springer-Verlag, Berlin, Heidelberg, New York, 1–107.

W. Felscher

[1975] Kombinatorische Konstruktionen mit Beweisen und Schnittelimination, in *ISILC Proof Theory Symposin, Kiel 1974*, J. Diller and G. H. Müller, eds., Springer-Verlag, Berlin, Heidelberg, New York, 119–151.

[1976] On interpolation when function symbols are present, *Archiv für Mathematische Logik und Grundlagenforschung*, 17, 145–157.

J. E. Fenstad

[1971] ed., *Proceedings of the Second Scandinavian Logic Symposium*, North-Holland Publ. Co., Amsterdam.

M. D. Fitting

[1969] *Intuitionistic Logic, Model Theory and Forcing*, North-Holland Publ. Co., Amsterdam.

[1983] *Proof Methods for Modal and Intuitionistic Logics*, Reidel, Dordrecht, Netherlands.

[1988] First-order modal tableaux, *Journal of Automated Reasoning*, 4, 191–213.

[1993] Basic modal logic, in *Gabbay et al. [1993]*, 365–448.

[1996] *First-Order Logic and Automated Theorem Proving*, Springer-Verlag, Berlin, Heidelberg, New York. 2nd edition.

R. C. Flagg and H. M. Friedman

[1986] Epistemic and intuitionistic formal systems, *Annals of Pure and Applied Logic*, 32, 53–60.

G. Frege

[1879] *Begriffschrift, eine der arithmetischen nachgebildete Formelsprache des reinen Denkens*, Louis Nebert, Halle. Reprinted in: Ignacio Angelelli (ed.) *Begriffschrift und andere Aufsätze*, Olms, Hildesheim 1964. Translation in van Heijenoort [1967], 5–82.

H. M. FRIEDMAN

[1978] Classically and intuitionistically provable functions, in *Higher Set Theory*, G. H. Müller and D. S. Scott, eds., Lecture Notes in Mathematics, Springer-Verlag, Berlin, Heidelberg, New York, 21–27.

H. M. FRIEDMAN AND A. SCEDROV

[1986] Intuitionistically provable recursive well-orderings, *Annals of Pure and Applied Logic*, 30, 165–171.

H. M. FRIEDMAN AND M. SHEARD

[1995] Elementary descent recursion and proof theory, *Annals of Pure and Applied Logic*, 71, 1–45.

T. FUJIWARA

[1978] A generalization of the Lyndon–Keisler theorem on homomorphism and its application to interpolation theorem, *Journal of the Mathematical Society of Japan*, 30, 278–302.

D. M. GABBAY, C. J. HOGGER, AND J. A. ROBINSON

[1993] eds., *Handbook of Logic in Artificial Intelligence and Logic Programming. Vol. 1, Logical Foundations*, Clarendon Press, Oxford.

[1994] eds., *Handbook of Logic in Artificial Intelligence and Logic Programming. Vol. 2, Deduction Methodologies*, Clarendon Press, Oxford.

J. GALLIER

[1991] What's so special about Kruskal's theorem and the ordinal γ_0? A survey of some results in proof theory, *Annals of Pure and Applied Logic*, 53, 199–260.

[1993] Constructive Logics. Part I: a tutorial on proof systems and typed λ-calculi, *Theoretical Computer Science*, 110, 249–339.

[1995] Proving properties of typed lambda-terms using realizability, covers and sheaves, *Theoretical Computer Science*, 142, 299–368.

R. O. GANDY

[1980] An early proof of normalization by A. M. Turing, in *Seldin and Hindley [1980]*, 453–455.

G. GENTZEN

[1933a] Über das Verhältnis zwischen intuitionistischer und klassischer Logik. Originally to appear in the *Mathematische Annalen*, reached the stage of galley proofs but was withdrawn. It was finally published in *Archiv für Mathematische Logik und Grundlagenforschung*, 16 (1974), 119–132. Translation in Gentzen [1969], 53–67.

[1933b] Über die Existenz unabhängiger Axiomensysteme zu unendlichen Satzsystemen, *Mathematische Annalen*, 107, 329–350.

[1935] Untersuchungen über das logische Schliessen I, II, *Mathematische Zeitschrift*, 39, 176–210, 405–431. Translation in Gentzen [1969], 68–131.

[1936] Die Widerspruchsfreiheit der reinen Zahlentheorie, *Mathematische Annalen*, 112, 493–565. Translation in Gentzen [1969], 132–170.

[1938] Neue Fassung des Widerspruchsfreiheitsbeweises für die reine Zahlentheorie, *Forschungen zur Logik und zur Grundlegung der exakten Wissenschaften. Neue Reihe*, 4, 19–44. Translation in Gentzen [1969], 252–286.

[1943] Beweisbarkeit und Unbeweisbarkeit von Anfangsfällen der transfiniten Induktion in der reinen Zahlentheorie, *Mathematische Annalen*, 119, 140–161. Translation in Gentzen [1969], 287–311.

[1969] *The Collected Papers of Gerhard Gentzen*, North-Holland Publ. Co., Amsterdam. English translation of Gentzen's papers, edited and introduced by M. E. Szabo.

H. GEUVERS

[1993] *Logics and Type Systems*, PhD thesis, Katholieke Universiteit Nijmegen.

[1994] Conservativity between logics and typed λ-calculi, in *Types for Proofs and Programs*, H. Barendregt and T. Nipkow, eds., Lecture Notes in Computer Science 806, Springer-Verlag, Berlin, Heidelberg, New York, 79–107.

H. GEUVERS AND M. J. NEDERHOF

[1991] A modular proof of strong normalization for the calculus of constructions, *Journal of Functional Programming*, 1, 155–189.

J.-Y. GIRARD

[1971] Une extension de l'interprétation de Gödel à l'analyse, et son application à l'élimination des coupures dans l'analyse et la théorie des types, in *Fenstad [1971]*, 63–92.

[1972] *Interprétation fonctionelle et élimination des coupures de l'arithmétique d'ordre supérieur*, PhD thesis, Université Paris VII.

[1976] Three-valued logic and cut elimination: the actual meaning of Takeuti's conjecture, *Dissertationes Mathematicae*, 136.

[1987a] Linear logic, *Theoretical Computer Science*, 50, 1–102.

[1987b] *Proof Theory and Logical Complexity*, Bibliopolis, Napoli.

[1991] Quantifiers in linear logic II, in *Nuovi problemi della logica e della filosofia della scienza, Volume II*, G. Corsi and G. Sambin, eds., CLUEB, Bologna (Italy). Proceedings of the conference with the same name, Viareggio, 8–13 gennaio 1990.

[1993] On the unity of logic, *Annals of Pure and Applied Logic*, 59, 201–217.

[1996] Proof-nets: the parallel syntax for proof theory, in *Logic and Algebra. Papers from the International Conference in memory of Roberto Magari, held in Pontignano, April 26–30, 1994*, A. Ursini and P. Aglianó, eds., Lecture Notes in Pure and Applied Mathematics 180, Marcel Dekker Inc., New York, 97–124.

J.-Y. GIRARD, Y. LAFONT, AND P. TAYLOR

[1988] *Proofs and Types*, Cambridge Tracts in Theoretical Computer Science 7, Cambridge University Press, Cambridge, UK.

V. GLIVENKO

[1929] Sur quelques points de la logique de M. Brouwer, *Académie Royale de Belgique. Bulletins de la Classe des Sciences, série 5*, 15, 183–188.

K. GÖDEL

[1933a] Eine Interpretation des intuitionistischen Aussagenkalküls, *Ergebnisse eines mathematischen Kolloquiums*, 4, 39–40. Also, with translation, in Gödel [1986], 300–303.

[1933b] Zur intuitionistischen Arithmetik und Zahlentheorie, *Ergebnisse eines mathematischen Kolloquiums*, 4, 34–38. Also, with translation, in Gödel [1986], pp. 286–295.

[1958] Über eine bisher noch nicht benützte Erweiterung des finiten Standpunktes, *Dialectica*, 12, 280–287. Also, with translation, in Gödel [1990], 240–251.

[1986] *Collected Works, Volume I*, Oxford University Press, Oxford.

[1990] *Collected Works, Volume II*, Oxford University Press, Oxford.

N. D. GOODMAN

[1984] Epistemic arithmetic is a conservative extension of intuitionistic arithmetic, *The Journal of Symbolic Logic*, 192–203.

L. GORDEEV

[1987] On Cut elimination in the presence of Peirce rule, *Archiv für Mathematische Logik und Grundlagenforschung*, 26, 147–164.

[1988] Proof-theoretic analysis: weak systems of functions and classes, *Annals of Pure and Applied Logic*, 38, 1–121.

R. P. GORÉ

[1992] Cut-free sequent and tableau systems for propositional normal modal logics, Tech. Rep. 257, Computer Laboratory, University of Cambridge.

C. A. GRABMAYER

[1999] *Cut-elimination in the implicative fragment* $\rightarrow$**G3mi** *of an intuitionistic G3-Gentzen system and its Computational Meaning*, Master's thesis, Institute for Logic, Language and Computation, University of Amsterdam.

T. HARDIN

[1989] Confluence results for the pure strong categorical logic, *Theoretical Computer Science*, 65, 291–342.

J. A. HARLAND

[1994] A proof-theoretic analysis of goal-directed provability, *Journal of Logic and Computation*, 4, 69–88.

L. A. HARRINGTON AND J. B. PARIS

[1977] A mathematical incompleteness in Peano arithmetic, in *Handbook of Mathematical Logic*, J. Barwise, ed., North-Holland Publ. Co., Amsterdam, 1133–1142.

R. HARROP

[1956] On disjunctions and existential statements in intuitionistic systems of logic, *Mathematische Annalen*, 132, 347–361.

[1960] Concerning formulas of the type $A \rightarrow B \vee C$, $A \rightarrow (Ex)B(x)$ in intuitionistic formal systems, *The Journal of Symbolic Logic*, 25, 27–32.

J. VAN HEIJENOORT

[1967] ed., *From Frege to Gödel. A Source Book in Mathematical Logic 1879–1931*, Harvard University Press, Cambridge, MA. Reprinted 1970.

L. HEINDORF

[1994] *Elementare Beweistheorie*, BI-Wissenschaftsverlag, Mannheim, Germany.

H. HERBELIN

[1995] A λ-calculus structure isomorphic to Gentzen-style sequent calculus structure, in *Computer Science Logic. 8th Workshop, CSL'94. Kazimierz, Poland, September 1994*, L. Pacholski and J. Tiuryn, eds., Lecture Notes in Computer Science 933, Springer-Verlag, Berlin, Heidelberg, New York, 61–75.

J. HERBRAND

[1928] Sur la théorie de la démonstration, *Académie des Sciences de Paris. Comptes Rendus Hebdomadaires des Séances*, 186, 1274–1276. Also in Herbrand [1968].

[1930] Recherches sur la théorie de la démonstration, *Société des Sciences et des Lettres de Varsovie. Comptes Rendus des Sciences. Classe III: Sciences Mathématiques et Physiques*, 33. Also in Herbrand [1968].

[1968] *Écrits Logiques*, Presses Universitaires de France, Paris. Translated as *Logical Writings*, Harvard University Press, Cambridge, MA, 1971.

P. HERTZ

[1929] Über Axiomensysteme für beliebige Satzsysteme, *Mathematische Annalen*, 101, 457–514.

A. HEYTING

[1930a] Die formalen Regeln der intuitionistischen Logik, *Sitzungsberichte der Preussischen Akademie von Wissenschaften. Physikalisch-mathematische Klasse*, 42–56.

[1930b] Die formalen Regeln der intuitionistischen Mathematik II, *Sitzungsberichte der Preussischen Akademie von Wissenschaften. Physikalisch-mathematische Klasse*, 57–71.

D. HILBERT

[1926] Über das Unendliche, *Mathematische Annalen*, 95, 161–190.

[1928] Die Grundlagen der Mathematik, *Abhandlungen aus dem mathematischen Seminar der Hamburgischen Universität*, 6, 65–85.

D. HILBERT AND W. ACKERMANN

[1928] *Grundzüge der theoretischen Logik*, Springer-Verlag, Berlin, Heidelberg, New York.

D. HILBERT AND P. BERNAYS

[1934] *Grundlagen der Mathematik, Bd. I*, Springer-Verlag, Berlin, Heidelberg, New York. 2nd edition 1968.

[1939] *Grundlagen der Mathematik, Bd. II*, Springer-Verlag, Berlin, Heidelberg, New York. 2nd edition 1970.

J. R. HINDLEY

[1993] BCK- and BCI-logics, condensed detachment and the 2-property, *Notre Dame Journal of Formal Logic*, 34, 231–250.

[1997] *Basic Simple Type Theory*, Cambridge University Press, Cambridge, UK.

J. R. HINDLEY AND D. MEREDITH

[1990] Principal type-schemes and condensed detachment, *The Journal of Symbolic Logic*, 55, 90–105.

K. J. J. HINTIKKA

[1955] Form and content in quantification theory. Two papers on symbolic logic, *Acta Philosophica Fennica*, 8, 7–55.

S. HIROKAWA

[1992] Balanced formulas, BCK-minimal formulas and their proofs, in *Logical Foundations of Computer Science (LFCS'92)*, A. Nerode and M. Taitslin, eds., Lecture Notes in Computer Science 620, Springer-Verlag, Berlin, Heidelberg, New York, 198–208.

J. HODAS AND D. MILLER

[1994] Logic programming in a fragment of intuitionistic linear logic, *Information and Computation*, 110, 327–365.

W. HODGES

[1993] Logical features of Horn clauses, in *Gabbay et al. [1993]*, 449–518.

W. A. HOWARD

[1970] Assignment of ordinals to terms for primitive recursive functionals of finite type, in *Myhill et al. [1970]*, 443–458.

[1980] The formulae-as-types notion of construction, in *Seldin and Hindley [1980]*, 480–490. Circulated as preprint since 1969.

J. HUDELMAIER

[1989] *Bounds for Cut Elimination in Intuitionistic Propositional Logic*, PhD thesis, Eberhard-Karls Universität, Tübingen, Germany.

[1992] Bounds for cut elimination in intuitionistic propositional logic, *Archive for Mathematical Logic*, 31, 331–354.

[1993] An $\mathcal{O}(n \log n)$-space decision procedure for intuitionistic propositional logic, *Journal of Logic and Computation*, 3, 63–75.

[1998] *Semantische Sequenzenkalküle*, habilitationsschrift, Fakultät für Informatik der Eberhard-Karls-Üniversität Tübingen, Germany.

G. E. HUGHES AND M. J. CRESSWELL

[1968] *An Introduction to Modal Logic*, Methuen, London.

J. M. E. HYLAND AND C. L. ONG

[1993] Modified realizability semantics and strong normalization proofs, in *Typed Lambda Calculi and Applications*, M. Bezem and J. Groote, eds., Springer Lecture Notes in Computer Science 664, 179–194.

G. JÄGER AND R. F. STÄRK

[1994] A proof-theoretic framework for logic programming. Draft for a chapter in the *Handbook of Proof Theory*, edited by S. Buss, to be published by North-Holland Publ. Co.

S. JAŚKOWSKI

[1934] On the rules of supposition in formal logic (Polish), *Studia Logica (old series)*, 1, 5–32. Translation in *Polish Logic 1920–39*, S. McCall, ed., Clarendon Press, Oxford, 1967, 232–258.

[1963] Über Tautologieen, in welchen keine Variabele mehr als zweimal vorkommt, *Zeitschrift für Mathematische Logik und Grundlagen der Mathematik*, 9, 231–250.

F. Joachimski and R. Matthes

[1999] Short proofs of nornalization for the simply-typed λ-calculus, permutative conversions and Gödel's T. Submitted.

I. Johansson

[1937] Der Minimalkalkül, ein reduzierter intuitionistischer Formalismus, *Compositio Mathematica*, 4, 119–136.

J.-B. Joinet, H. A. J. M. Schellinx, and L. Tortora de Falco

[1998] Linear decorations, simulations and normalization, Tech. Rep. Preprint nr.1067, Mathematisch Instituut, Universiteit Utrecht. Submitted.

A. Joyal and R. Street

[1991] The geometry of tensor calculus 1, *Advances in Mathematics*, 88, 55–112.

M. B. Kalsbeek

[1994] Gentzen systems for logic programming styles, Tech. Rep. CT–94–12, Institute for Logic, Language and Computation, University of Amsterdam.

[1995] *Meta-Logics for Logic Programming*, PhD thesis, Universiteit van Amsterdam.

S. Kanger

[1957] *Provability in Logic*, Acta Universitatis Stockholmiensis. Stockholm Studies in Philosophy, vol. 1, Almqvist and Wiksell, Stockholm.

G. M. Kelly

[1964] On Mac Lane's conditions for coherence of natural associativities, commutativities etc., *Journal of Algebra*, 1, 397–402.

[1972a] An abstract approach to coherence, in *Mac Lane [1972]*, 106–147.

[1972b] A cut-elimination theorem, in *Mac Lane [1972]*, 196–213.

G. M. Kelly and S. Mac Lane

[1971] Coherence in closed categories, *Journal of Pure and Applied Algebra*, 1, 97–140.

O. Ketonen

[1944] Untersuchungen zum Prädikatenkalkül, *Annales Academiae Scientiarum Fennicae, ser. A, I. Mathematica-physica*, 23. A detailed review by P. Bernays is in *The Journal of Symbolic Logic,* 10 (1945), 127–130.

S. C. Kleene

[1952a] *Introduction to Metamathematics*, North-Holland Publ. Co., Amsterdam.

[1952b] Permutability of inferences in Gentzen's calculi LK and LJ, *Memoirs of the American Mathematical Society*, 10, 1–26.

[1955] Hierarchies of number-theoretic predicates, *Bulletin of the American Mathematical Society*, 61, 193–213. Additions and corrections in *Proceedings of the American Mathematical Society* 8 (1957), p. 1006.

[1967] *Mathematical Logic*, Wiley and Sons, New York.

A. N. KOLMOGOROV

[1925] On the principle of the excluded middle (Russian), *Matematicheskij Sbornik. Akademiya Nauk SSSR i Moskovskoe Matematicheskoe Obshchestvo*, 32, 646–667. Translation in van Heijenoort [1967], 414–437.

R. A. KOWALSKI

[1974] Predicate logic as a programming language, in *Information Processing 74. Proceedings of the IFIP congress 74*, J. L. Rosenfeld, ed., North-Holland Publ. Co., Amsterdam, 569–574.

G. KREISEL

[1953] A variant to Hilbert's theory of the foundations of arithmetic, *British Journal for the Philosophy of Science*, 4, 107–127.

[1958] Elementary completeness properties of intuitionistic logic with a note on negations of prenex formulae, *The Journal of Symbolic Logic*, 23, 317–330.

[1968] Notes concerning the elements of proof theory. Course notes of a course on proof theory at U.C.L.A., 1967–1968.

[1977] Wie die Beweistheorie zu ihren Ordinalzahlen kam und kommt, *Jahresbericht der Deutschen Mathematiker-Vereinigung*, 78, 177–223.

G. KREISEL AND J.-L. KRIVINE

[1972] *Modelltheorie*, Springer-Verlag, Berlin, Heidelberg, New York.

G. KREISEL AND G. TAKEUTI

[1974] Formally self-referential propositions for cut free classical analysis and related systems, *Dissertationes Mathematicae*, 118.

S. A. KRIPKE

[1963] Semantical analysis of modal logic I, *Zeitschrift für Mathematische Logik und Grundlagen der Mathematik*, 9, 67–96.

[1965] Semantical analysis of intuitionistic logic I, in *Formal Systems and Recursive Functions*, J. N. Crossley and M. A. E. Dummett, eds., Studies in Logic and the Foundations of Mathematics, North-Holland Publ. Co., Amsterdam, 92–130.

S. KURODA

[1951] Intuitionistische Untersuchungen der formalistischen Logik, *Nagoya Mathematical Journal*, 2, 35–47.

J. LAMBEK

[1958] The mathematics of sentence structure, *The American Mathematical Monthly*, 65, 154–170.

[1968] Deductive systems and categories I: syntactic calculi and residuated categories, *Mathematical Systems Theory*, 2, 287–318.

[1969] Deductive systems and categories II: standard constructions and closed categories, in *Category Theory, Homology Theory and their Applications*, P. J. Hilton, ed., Lecture Notes in Mathematics 86, Springer-Verlag, Berlin, Heidelberg, New York, 76–122.

[1972] Deductive systems and categories III: cartesian closed categories, intuitionist propositional calculus, and combinatory logic, in *Toposes, Algebraic Geometry and Logic*, F. W. Lawvere, ed., Lecture Notes in Mathematics 274, Springer-Verlag, Berlin, Heidelberg, New York, 57–82.

[1974] Functional completeness of cartesian categories, *Annals of Pure and Applied Logic*, 6, 259–292.

J. LAMBEK AND P. J. SCOTT

[1986] *Introduction to Higher-Order Categorical Logic*, Cambridge University Press, Cambridge, UK.

D. LEIVANT

[1979] Assumption classes in natural deduction, *Zeitschrift für Mathematische Logik und Grundlagen der Mathematik*, 25, 1–4.

[1990] Contracting proofs to programs, in *Logic and Computer Science*, P. Odifreddi, ed., Academic Press, New York, 279–327.

[1994] Higher-order logic, in *Gabbay et al. [1994]*, 229–321.

C. I. LEWIS AND C. H. LANGFORD

[1932] *Symbolic Logic*, Appleton-Century-Crofts, New York. Reprinted Dover Publications, New York, 1951, 1959.

V. A. LIFSCHITZ

[1989] What is the inverse method?, *Journal of Automated Reasoning*, 5, 1–23.

P. LINCOLN, J. MITCHELL, A. SCEDROV, AND N. SHANKAR

[1992] Decision problems for propositional linear logic, *Annals of Pure and Applied Logic*, 56, 239–311.

J. W. LLOYD

[1987] *Foundations of Logic Programming*, Springer-Verlag, Berlin, Heidelberg, New York.

P. LORENZEN

[1950] Konstruktive Begründung der Mathematik, *Mathematische Zeitschrift*, 53, 162–202.

H. LUCKHARDT

[1989] Herbrand-Analysen zweier Beweise des Satzes von Roth: Polynomiale Anzahlschranken, *The Journal of Symbolic Logic*, 54, 234–263.

R. C. LYNDON

[1959] An interpolation theorem in the predicate calculus, *Pacific Journal of Mathematics*, 9, 129–142.

S. MAC LANE

[1963] Natural associativity and commutativity, *Rice University Studies*, 49, 28–46.

[1971] *Categories for the Working Mathematician*, Springer-Verlag, Berlin, Heidelberg, New York.

[1972] ed., *Coherence in Categories*, Springer-Verlag, Berlin, Heidelberg, New York.

[1976] Topology and logic as a source of algebra, *Bulletin of the American Mathematical Society*, 82, 1–40.

[1982] Why commutative diagrams coincide with equivalence proofs, *Contemporary Mathematics*, 13, 387–401.

S. MAEHARA

[1954] Eine Darstellung der intuitionistische Logik in der klassischen, *Nagoya Mathematical Journal*, 7, 45–64.

[1960] On the interpolation theorem of Craig (Japanese), *Sugaku*, 12, 235–237. Not seen by us.

S. MAEHARA AND G. TAKEUTI

[1961] A formal system of first-order predicate calculus with infinitely long expressions, *Journal of the Mathematical Society of Japan*, 13, 357–370.

P. E. MALMNÄS AND D. PRAWITZ

[1969] A survey of some connections between classical, intuitionistic and minimal logic, in *Contributions to Mathematical Logic*, H. Schmidt, K. Schütte, and H. Thiele, eds., North-Holland Publ. Co., Amsterdam, 215–229.

C. R. MANN

[1975] The connection between equivalence of proofs and Cartesian closed categories, *Proceedings of the London Mathematical Society. Third series*, 31, 289–310.

A. MARTELLI AND U. MONTANARI

[1982] An efficient unification algorithm, *ACM Transactions on Programming Languages and Systems*, 4, 258–282.

P. MARTIN-LÖF

[1971a] Hauptsatz for the intuitionistic theory of iterated inductive definitions, in *Proceedings of the Second Scandinavian Logic Symposium*, J. Fenstad, ed., North-Holland Publ. Co., Amsterdam, 179–216.

[1971b] Hauptsatz for the theory of species, in *Proceedings of the Second Scandinavian Logic Symposium*, J. Fenstad, ed., North-Holland Publ. Co., Amsterdam, 217–233.

S. MARTINI AND A. MASINI

[1993] A computational interpretation of modal proofs, Tech. Rep. TR–27/93, Dipartimento di Informatica, Università di Pisa.

[1994] A modal view of linear logic, *The Journal of Symbolic Logic*, 59, 888–899.

A. MASINI

[1992] 2-Sequent calculus: a proof theory of modalities, *Annals of Pure and Applied Logic*, 58, 229–246.

[1993] 2-Sequent calculus: intuitionism and natural deduction, *Journal of Language and Computation*, 3, 533–562.

R. MATTHES

[1998] *Extensions of System* F *by Iteration and Primitive Recursion on Monotone Inductive Types*, PhD thesis, Mathematisches Institut der Universität München, Germany.

J. C. C. McKinsey and A. Tarski

[1948] Some theorems about the sentential calculi of Lewis and Heyting, *The Journal of Symbolic Logic*, 13, 1–15.

C. McLarty

[1992] *Elementary Categories, Elementary Toposes*, Oxford Logic Guides 21, Clarendon Press, Oxford.

F. Metayer

[1994] Homology of proof nets, *Archive for Mathematical Logic*, 33, 169–188.

D. Miller

[1994] A multiple-conclusion meta-logic, in *Proceedings. Ninth Annual Symposium on Logic in Computer Science. July 1994, Paris*, S. Abramsky, ed., IEEE Computer Society Press, Los Alamitos, California, 272–281.

D. Miller, G. Nadathur, F. Pfenning, and A. Scedrov

[1991] Uniform proofs as a foundation for logic programming, *Annals of Pure and Applied Logic*, 51, 125–157.

G. E. Mints

[1971] Exact estimates of the provability of transfinite induction in the initial segments of arithmetic (Russian), *Zapiski Nauchnykh Seminarov Leningradskogo Otdeleniya Ordena Lenina Matematicheskogo Instituta imeni V. A. Steklova Akademii Nauk SSSR (LOMI)*, 20, 134–144. Translation in *Journal of Soviet Mathematics*, 1 (1973), 85–91.

[1975] Finite investigations of infinite derivations (Russian), *Zapiski Nauchnykh Seminarov Leningradskogo Otdeleniya Ordena Lenina Matematicheskogo Instituta imeni V. A. Steklova Akademii Nauk SSSR (LOMI)*, 49, 67–122. Translation in *Journal of Soviet Mathematics*, 10 (1978), 548–596.

[1977] Closed categories and the theories of proofs(Russian), *Zapiski Nauchnykh Seminarov Leningradskogo Otdeleniya Ordena Lenina Matematicheskogo Instituta imeni V. A. Steklova Akademii Nauk SSSR (LOMI)*, 68, 83–114. Translation in *Journal of Soviet Mathematics*, 15 (1981), 45–62; also revised translation in Mints [1992c], 183–212.

[1978] On Novikov's hypothesis (Russian). Photocopied proceedings. Translation in Mints [1992c], 147–151.

[1979] A coherence theorem for for cartesian closed categories (abstract), *The Journal of Symbolic Logic*, 44, 453–454.

[1990] Gentzen-type systems and resolution rules. Part I. Propositional logic, in *Colog-88*, G. E. Mints and P. Martin-Löf, eds., Lecture Notes in Mathematics 417, Springer-Verlag, Berlin, Heidelberg, New York, 198–231.

[1992a] Normalization of natural deduction and the effectivity of classical existence, in *Mints [1992c]*, 123–146. This is a translation of the Russian original in *Logicheskij Vyvod (Logical Inference). Proceedings of the All-Union Symposium on the Theory of Logical inference,* V. A. Smirnov, ed., Nauka, Moskva, 1979, 245–265.

[1992b] Proof theory and category theory, in *Mints [1992c]*, Bibliopolis, Napoli, and North-Holland Publ. Co., Amsterdam, 157–182.

[1992c] *Selected Papers in Proof Theory*, North-Holland Publ. Co., Amsterdam; Bibliopolis, Napoli.

[1992d] *A Short Introduction to Modal Logic*, CSLI Lecture Notes 30, Center for the Study of Language and Information, Stanford, California.

[1992e] A simple proof of the coherence theorem for CCC, in *Mints [1992c]*, Bibliopolis, Napoli, and North-Holland Publ. Co., Amsterdam, 213–220.

[1993] Resolution calculus for the first order linear logic, *Journal of Logic, Language and Information*, 2, 59–83.

[1994a] Cut-elimination and normal forms of sequent derivations, Tech. Rep. CSLI–94–193, CSLI, Stanford. Contains: Normal forms for sequent derivations; Indexed systems of sequents and cut-elimination; Normalization as an epsilon substitution process.

[1994b] Gentzen-type systems and resolution rule. Part II. Predicate Logic, in *Logic Colloquium '90*, J. Oikkonen and J. Väänänen, eds., Lecture Notes in Logic 2, Springer-Verlag, Berlin, Heidelberg, New York, 163–190.

[1994c] Resolution strategies for intuitionistic logic, in *Constraint Programming*, B. Mayoh, E. Tyugu, and J. Penjam, eds., Springer-Verlag, Berlin, Heidelberg, New York, 289–312.

[1995] Natural deduction in intuitionistic linear logic. Manuscript dated May 19, 1995.

[1996] Normal forms for sequent derivations, in *Kreiseliana*, P. Odifreddi, ed., A.K. Peters, Wellesley, MA., 469–492.

[1997] Indexed systems and cut-elimination, *Journal of Philosophical Logic*, 26, 671–696.

[1999] Axiomatization of a Skolem function in intuitionistic logic, in *Formalizing the Dynamics of Information*, M. Faller, S. Kaufmann, and M. Pauly, eds., CSLI, Stanford, CA. To appear.

N. Motohashi

[1984a] Approximation theory of uniqueness conditions by existence conditions, *Fundamenta Mathematicae*, 120, 127–142.

[1984b] Equality and Lyndon's interpolation theorem, *The Journal of Symbolic Logic*, 49, 123–128.

J. Myhill, A. Kino, and R. E. Vesley

[1970] eds., *Intuitionism and Proof Theory*, North-Holland Publ. Co., Amsterdam.

T. Nagashima

[1966] An extension of the Craig–Schütte interpolation theorem, *Annals of the Japan Association for the Philosophy of Science*, 3, 12–18.

R. P. Nederpelt, J. H. Geuvers, and R. C. de Vrijer

[1994] eds., *Selected Papers on Automath*, North-Holland Publ. Co., Amsterdam.

S. Negri

[1999] Sequent calculus proof theory of intuitionistic apartness and order relations, *Archive for Mathematical Logic*, 38, 521–547.

S. Negri and J. von Plato

[1998] Cut elimination in the presence of axioms, *The Bulletin of Symbolic Logic*, 4, 418–435.

[1999] Sequent calculus in natural deduction style. Manuscript.

J. von Neumann

[1927] Zur Hilbertschen Beweistheorie, *Mathematische Zeitschrift*, 26, 1–46.

M. H. A. Newman

[1942] On theories with a combinatorial definition of “equivalence”, *Annals of Mathematics, 2nd series*, 43, 223–243.

A. Oberschelp

[1968] On the Craig–Lyndon interpolation theorem, *The Journal of Symbolic Logic*, 33, 271–274.

H. Ono and Y. Komori

[1985] Logics without the contraction rule, *The Journal of Symbolic Logic*, 50, 169–201.

V. P. Orevkov

[1979] Lower bounds for the lengthening of proofs after cut-elimination (Russian), *Zapiski Nauchnykh Seminarov Leningradskogo Otdeleniya Ordena Lenina Matematicheskogo Instituta imeni V. A. Steklova Akademii Nauk SSSR (LOMI)*, 88, 137–162, 242–243. Translation *Journal of Soviet Mathematics*, 20 (1982), 2337–2350.

[1984] Upper bounds for the lengthening of proofs after cut-elimination (Russian), *Zapiski Nauchnykh Seminarov Leningradskogo Otdeleniya Ordena Lenina Matematicheskogo Instituta imeni V. A. Steklova Akademii Nauk SSSR (LOMI)*, 137, 87–98. Translation *Journal of Soviet Mathematics*, 34 (1986), 1810–1819.

[1987] Applications of Cut elimination to obtain estimates of proof lengths, *Doklady Akademii Nauk SSSR*, 296, 539–542. Translation *Soviet Mathematics Doklady*, 36 (1988), 292–295.

H. Osswald

[1973] Ein syntaktischer Beweis für die Zuverlässigkeit der Schnittregel im Kalkül von Schütte für die intuitionistische Typenlogik, *Manuscripta Mathematica*, 8, 243–249.

P. Päppinghaus

[1983] Completeness properties of classical theories of finite type and the normal form theorem, *Dissertationes Mathematicae*, 207.

J. B. Paris

[1978] Some independence results for Peano arithmetic, *The Journal of Symbolic Logic*, 43, 725–731.

C. Parsons

[1973] Transfinite induction in subsystems of number theory (abstract), *The Journal of Symbolic Logic*, 38, 544–545.

F. Pfenning

[1994] A structural proof of cut elimination and its representation in a logical framework, Tech. Rep. CMU–CS–94–218, School of Computer Science, Carnegie-Mellon University.

A. M. PITTS

[1992] On an interpretation of second-order quantification in first-order intuitionistic propositional logic, *The Journal of Symbolic Logic*, 57, 33–52.

J. VON PLATO

[1998] Structure of derivations in natural deduction. Manuscript.

[1999] A proof of Gentzen's *Hauptsatz* without multicut. To appear in the *Archive for Mathematical Logic.*

W. POHLERS

[1989] *Proof Theory. An Introduction*, Lecture Notes in Mathematics 1407, Springer-Verlag, Berlin, Heidelberg, New York.

A. POIGNÉ

[1992] Basic category theory, in *Abramsky et al. [1992]*, 413–640.

J. C. VAN DE POL AND H. SCHWICHTENBERG

[1995] Strict functionals for termination proofs, in *Proceedings of the Second International Conference on Typed Lambda Calculi and Applications, Edinburgh, Scotland*, M. Dezani-Ciancaglini and G. Plotkin, eds., Lecture Notes in Computer Science 902, Springer-Verlag, Berlin, Heidelberg, New York, 350–364.

G. POTTINGER

[1977] Normalization as a homomorphic image of cut-elimination, *Annals of Mathematical Logic*, 12, 323–357.

[1983] Uniform, cut-free formulations of T, S_4 and S_5, *The Journal of Symbolic Logic*, 48, 900.

D. PRAWITZ

[1965] *Natural Deduction. A Proof-Theoretical Study*, Almquist and Wiksell, Stockholm.

[1967] Completeness and Hauptsatz for second-order logic, *Theoria*, 33, 246–253.

[1968] Hauptsatz for higher-order logic, *The Journal of Symbolic Logic*, 33, 452–457.

[1970] Some results for intuitionistic logic with second-order quantification rules, in *Myhill et al. [1970]*, 259–269.

[1971] Ideas and results in proof theory, in *Proceedings of the Second Scandinavian Logic Symposium*, J. E. Fenstad, ed., North-Holland Publ. Co., Amsterdam, 235–307.

[1981] Validity and normalizability of proofs in 1st and 2nd order classical and intuitionistic logic, in *Atti de Congresso Nazionale di Logica*, S. Bernini, ed., Bibliopolis, Napoli, 11–36.

D. J. PYM AND J. A. HARLAND

[1994] A uniform proof-theoretic investigation of logic programming, *Journal of Logic and Computation*, 4, 175–207.

H. RASIOWA

[1954] Constructive theories, *Bulletin de l'Académie Polonaise des Sciences. Série des Sciences Mathématiques, Astronomiques et Physiques*, 2, 121–124.

[1955] Algebraic models of axiomatic theories, *Fundamentae Mathematicae*, 41, 291–310.

H. Rasiowa and R. Sikorski

[1953] Algebraic treatment of the notion of satisfiability, *Fundamentae Mathematicae*, 40, 62–95.

[1960] On the Gentzen theorem, *Fundamentae Mathematicae*, 48, 57–69.

[1963] *The Mathematics of Metamathematics*, PAN, Warszawa.

W. Rautenberg

[1983] Modal tableau calculi, *Journal of Philosophical Logic*, 12, 403–423.

J. C. Reynolds

[1974] Towards a theory of type structure, in *Programming Symposium, Proceedings. Colloque sur la Programmation*, B. Robinet, ed., Lecture Notes in Computer Science 19, Springer-Verlag, Berlin, Heidelberg, New York, 408–425.

A. Robinson

[1956] A result on consistency and its application to the theory of definition, *Indagationes Mathematicae*, 15, 330–339.

J. A. Robinson

[1965] A machine-oriented logic based on the resolution principle, *Journal of the Association for Computing Machinery*, 12, 23–41.

S. Ronchi della Rocca and L. Roversi

[1994] Lambda-calculus and intuitionistic linear logic, tech. rep., Department of Computer Science, Università di Torino, Torino, Italy.

D. Roorda

[1994] Interpolation in fragments of classical linear logic, *The Journal of Symbolic Logic*, 419–444.

G. F. Rose

[1953] Propositional calculus and realizability, *Transactions of the American mathematical Society*, 175, 1–19.

G. Sambin, G. Battilotti, and C. Faggian

[1997] Basic logic: reflection, symmetry, visibility. To appear.

G. Sambin and C. Faggian

[1998] From basic logic to quantum logics with cut-elimination, *International Journal of Theoretical Physics*, 37, 31–37.

L. E. Sanchis

[1967] Functionals defined by recursion, *The Notre Dame Journal of Formal Logic*, 8, 161–174.

[1971] A generalization of the Gentzen Hauptsatz, *The Notre Dame Journal of Formal Logic*, 12, 499–504.

H. A. J. M. Schellinx

[1991] Some syntactical observations on linear logic, *Journal of Logic and Computation*, 1, 537–559.

[1994] *The Noble Art of Linear Decorating*, PhD thesis, Universiteit van Amsterdam.

P. SCHROEDER-HEISTER

[1984] A natural extension of natural deduction, *The Journal of Symbolic Logic*, 49, 1284–1300.

J. SCHULTE-MÖNTING

[1976] Interpolation formulas for predicates and terms which carry their own history, *Archiv für Mathematische Logik und Grundlagenforschung*, 17, 159–170.

K. SCHÜTTE

[1950a] Beweistheoretische Erfassung der unendliche Induktion in der Zahlentheorie, *Mathematische Annalen*, 122, 369–389.

[1950b] Schlussweisen-Kalküle der Prädikatenlogik, *Mathematische Annalen*, 122, 47–65.

[1951] Die Eliminierbarkeit des bestimmten Artikels, *Mathematische Annalen*, 123, 166–186.

[1956] Ein System des verknüpfenden Schliessens, *Archiv für Mathematische Logik und Grundlagenforschung*, 2, 34–67.

[1960] *Beweistheorie*, Springer-Verlag, Berlin, Heidelberg, New York.

[1962] Der Interpolationssatz der intuitionistischen Prädikatenlogik, *Mathematische Annalen*, 148, 192–200.

[1977] *Proof Theory*, Springer-Verlag, Berlin, Heidelberg, New York.

H. SCHWICHTENBERG

[1976] Definierbare Funktionen im λ-Kalkül mit Typen, *Archiv für Mathematische Logik und Grundlagenforschung*, 17, 113–114.

[1977] Proof theory: some applications of cut-elimination, in *Handbook of Mathematical Logic*, J. Barwise, ed., North-Holland Publ. Co., Amsterdam, 867–895.

[1991] Normalization, in *Logic, Algebra and Computation*, F. Brauer, ed., Springer-Verlag, Berlin, Heidelberg, New York, 201–235.

[1992] Minimal from classical proofs, in *Computer Science Logic. 5th Workshop, CSL'91. Berne, Switzerland, October 1991. Proceedings*, E. Börger, G. Jäger, H. Kleine Büning, and M. Richter, eds., Lecture Notes in Computer Science 626, Springer-Verlag, Berlin, Heidelberg, New York, 326–328.

[1998] Finite notations for infinite terms, *Annals of Pure and Applied Logic*, 94, 201–222.

[1999] Termination of permutative conversions in intuitionistic Gentzen calculi, *Theoretical Computer Science*, 212, 247–260.

D. S. SCOTT

[1979] Identity and existence in intuitionistic logic, in *Applications of Sheaves*, M. P. Fourman, C. J. Mulvey, and D. S. Scott, eds., Lecture Notes in Mathematics 753, Springer-Verlag, Berlin, Heidelberg, New York, 660–669.

J. P. Seldin and J. R. Hindley

[1980] eds., *To H. B. Curry: Essays on Combinatory logic, Lambda Calculus and Formalism*, Academic Press, New York.

D. J. Shoesmith and T. J. Smiley

[1978] *Multiple-Conclusion Logic*, Cambridge University Press, Cambridge, UK.

T. Skolem

[1923] Begründung der elementaren Arithmetik durch die rekurrierende denkweise ohne Anwendung scheinbarer Veränderlichen mit unendlichen Ausdehnungsbereich, *Videnskaps Selskapet i Kristiania, Skrifter Utgit (1)*, 6, 1–38. Translation van Heijenoort [1967], 303–333.

R. M. Smullyan

[1965] Analytic natural deduction, *The Journal of Symbolic Logic*, 30, 123–139.

[1966] Trees and nest structures, *The Journal of Symbolic Logic*, 31, 303–321.

[1968] *First-order Logic*, Springer-Verlag, Berlin, Heidelberg, New York.

R. Socher-Ambrosius

[1994] *Deduktionssysteme*, BI-Wissenschaftsverlag, Mannheim, Germany.

S. V. Solovjov

[1979] Derivation of equivalence of proofs under reduction of formula depth (Russian, with English summary), *Zapiski Nauchnykh Seminarov Leningradskogo Otdeleniya Ordena Lenina Matematicheskogo Instituta imeni V. A. Steklova Akademii Nauk SSSR (LOMI)*, 88, 197–207. Translation *Journal of Soviet Mathematics*, 20 (1982), 2370–2376.

[1997] Proof of a conjecture of S. Mac Lane, *Annals of Pure and Applied Logic*, 90, 101–162.

C. Spector

[1962] Provably recursive functions of analysis: a consistency proof of analysis by an extension of principles formulated in current intuitionistic mathematics, in *Recursive Function Theory*, J. C. E. Dekker, ed., Symposia in Pure Mathematics V, American Mathematical Society, Providence, RI, 1–27.

G. Stålmarck

[1991] Normalization theorems for full first order classical natural deduction, *The Journal of Symbolic Logic*, 56, 129–149.

R. F. Stärk

[1990] A direct proof for the completeness of SLD-resolution, in *Computer Science Logic, Selected Papers from CSL '89*, E. Börger, H. Kleine Büning, and M. M. Richter, eds., Lecture Notes in Computer Science 440, Springer-Verlag, Berlin, Heidelberg, New York, 382–383.

R. Statman

[1978] Bounds for proof-search and speed-up in the predicate calculus, *Annals of Pure and Applied Logic*, 15, 225–287.

E. TAHHAN BITTAR

[1999] Strong normalization proofs for cut elimination in Gentzen's sequent calculi, *Banach Center Publications*, 46, 179–225. The title of the volume is *Logic, Algebra, and Computer Science*, published by the Polish Academy of Sciences, Warszawa.

W. W. TAIT

[1965] Infinitely long terms of transfinite type, in *Formal Systems and Recursive Functions*, J. N. Crossley and M. A. E. Dummett, eds., Studies in Logic and the Foundations of Mathematics, North-Holland Publ. Co., Amsterdam, 176–185.

[1966] A nonconstructive proof of Gentzen's Hauptsatz for second-order predicate logic, *Bulletin of the American Mathematical Society*, 72, 980–988.

[1967] Intensional interpretation of functionals of finite type, I, *The Journal of Symbolic Logic*, 32, 198–212.

[1968] Normal derivability in classical logic, in *The Syntax and Semantics of Infinitary Languages*, K. J. Barwise, ed., Lecture Notes in Mathematics 72, Springer-Verlag, Berlin, Heidelberg, New York, 204–236.

[1975] A realizability interpretation of the theory of species, in *Proceedings of Logic Colloquium*, R. J. Parikh, ed., Lecture Notes in Mathematics 453, Springer-Verlag, Berlin, Heidelberg, New York, 240–251.

M. TAKAHASHI

[1967] A proof of cut-elimination in simple type theory, *Journal of the Mathematical Society of Japan*, 19, 399–410.

[1970] A system of simple type theory of Gentzen style with inference on extensionality and the cut-elimination in it, *Commentarii Mathematici Universitatis Sancti Pauli*, 18, 129–147.

G. TAKEUTI

[1978] *Two Applications of Logic to Mathematics*, Princeton University Press, Princeton NJ.

[1987] *Proof Theory*. 2nd edition, North-Holland Publ. Co., Amsterdam.

A. TARSKI

[1956] *Logic, Semantics, Metamathematics. Papers from 1923 to 1938*, Clarendon Press, Oxford, UK.

A. S. TROELSTRA

[1973] *Metamathematical Investigation of Intuitionistic Arithmetic and Analysis. Chapters 1–4*, Lecture Notes in Mathematics 344, Springer-Verlag, Berlin, Heidelberg, New York.

[1983] Logic in the writings of Brouwer and Heyting, in *Atti del Convegno Internazionale di Storia della Logica, San Gimignano, 4–8 dicembre 1982*, V. M. Abrusci, E. Casari, and M. Mugnai, eds., Cooperativa Libraria Universitaria Editrice Bologna, Bologna, Italy, 193–210.

[1986] Strong normalization for typed terms with surjective pairing, *The Notre Dame Journal of Formal Logic*, 27, 547–550.

[1990] On the early history of intuitionistic logic, in *Mathematical Logic*, P. P. Petkov, ed., Plenum Press, New York, 3–27. Proceedings of the Heyting '88 Summer School and Conference on Mathematical Logic, September 13–23, 1988 in Chaika, Bulgaria.

[1992a] *Lectures on Linear Logic*, CSLI-Lecture Notes 29, Center for the Study of Language and Information, Stanford, California.

[1992b] Realizability, Tech. Rep. ML–92–09, Institute for Logic, Language and Computation, University of Amsterdam. A revised version is to appear in *Handbook of Proof Theory*, S. Buss, ed., North-Holland Publ. Co., Amsterdam.

[1995] Natural deduction for intuitionistic logic, *Annals of Pure and Applied Logic*, 73, 79–108.

[1999] Marginalia on sequent calculi, *Studia Logica*, 62, 291–303.

A. S. Troelstra and D. van Dalen

[1988] *Constructivism in Mathematics*, Studies in Logic and the Foundations of Mathematics, North-Holland Publ. Co., Amsterdam. Two vols.

T. Uesu

[1984] An axiomatization of the apartness fragment of the theory DLO^+ of dense linear order, in *Logic Colloquium '84*, Lecture Notes in Mathematics 1104, Springer-Verlag, Berlin, Heidelberg, New York, 453–475.

A. M. Ungar

[1992] *Normalization, Cut-Elimination, and the Theory of Proofs*, Center for the Study of Language and Information, Stanford, California. CSLI-Lecture Notes 28.

A. Urquhart

[1995] The complexity of propositional proofs, *The Bulletin of Symbolic Logic*, 1, 425–467.

S. Valentini

[1986] A syntactic proof of cut-elimination of GL_{lin}, *Zeitschrift für Mathematische Logik und Grundlagen der Mathematik*, 32, 137–144.

R. Vestergaard

[1998a] A computational anomaly in the Troelstra-Schwichtenberg $G3i(m)$ system. Manuscript.

[1998b] The cut rule and explicit substitutions (author's cut), Tech. Rep. TR-1998-9, Department of Computer Science, University of Glasgow.

A. Visser

[1996] Uniform interpolation and layered bisimulation, in *Gödel '96 (Brno, 1996)*, Lecture Notes in Logic 6, Springer-Verlag, Berlin, Heidelberg, New York, 139–164.

R. Voreadou

[1977] Coherence and non-commutative diagrams in closed categories, *Memoirs of the American Mathematical Society*, 182.

N. N. Vorob'ev

[1964] A new algorithm for derivability in a constructive propositional calculus (Russian), *Trudy Ordena Matematicheskogo Instituta imeni V. A. Steklova. Akademiya Nauk SSSR*, 72, 195–227. Translation in *American Mathematical Society Translations. Series 2*, 94 (1970), 37–71.

R. C. DE VRIJER

[1987] *Surjective Pairing and Strong Normalization: two Themes in Lambda Calculus*, PhD thesis, University of Amsterdam.

M. WAJSBERG

[1938] Untersuchungen über den Aussagenkalkül von A. Heyting, *Wiadomości Matematyczne*, 46, 45–101. Translation in Wajsberg [1977], 132–171.

[1977] *Logical Works*, Zakład Narodowy Imiena Ossolińskich., Wydawnictwo Polskiej Akademii Nauk. Wrocław, Poland. Edited by S. J. Surma.

H. WANSING

[1994] Sequent calculi for normal modal propositional logics, *Journal of Logic and Computation*, 4, 125–142.

[1998] Translation of hypersequents into display sequents, *Journal of the IGPL Interest Group in Pure and Applied Logics*, 6, 719–733.

J. I. ZUCKER

[1974] The correspondence between cut-elimination and normalization I, II, *Annals of Mathematical Logic*, 7, 1–156.

Symbols and notations

Below we list symbols and notations which either appear in the text more than just locally, or are important for other reasons. The more important notations and conventions in use throughout the book are found in section 1.1. For conventions on prooftrees, see 1.3.2, 3.1.4, 4.1.2.

Logical operators

$\wedge, \vee, \to$	2	(primitive propositional operators)
$\forall, \exists$	2, 345	(quantifiers)
$\neg, \leftrightarrow$	3	(defined operators)
$\bot$	2, 294	(falsity)
$\top$	3, 294	(truth)
$\Rightarrow$	7	(sequent arrow)
$\Box, \Diamond$	284	(necessary, possible)
$\bigwedge, \bigvee$	3, 7	(iterated conjunction and disjunction)
$\sqcap, \sqcup, \star, +, \multimap, \rightsquigarrow$	294	(binary linear logic operators)
$\sim, \mathbf{0}, \mathbf{1}$	294	(negation, zero, unit of linear logic)
$!, ?$	295	(modalities of linear logic)

Substitution

$\mathcal{E}[x/t]$	4, 11	(substitution in expression)
$\mathcal{E}[\vec{x}/\vec{t}]$	4, 11	(simultaneous substitution)
$A[X^n/\lambda\vec{x}.B]$	4	(second-order substitution)
$A(t)$	4	(convention on substitution)
$F[*], F[A]$	6	(context, substitution in context)

Measures on trees and formulas

$\lvert\mathcal{T}\rvert$	9	(depth of a tree $\mathcal{T}$)
$\mathrm{s}(\mathcal{T})$	9	(size of a tree $\mathcal{T}$)
$\mathrm{ls}(\mathcal{T})$	9	(leafsize of a tree $\mathcal{T}$)
$\lvert A\rvert$	10	(depth of a formula A)
$\mathrm{s}(A)$	10	(size of a formula A)
$\mathrm{w}(A)$	113	(weight of a formula A)

$\mathbf{p}(t,s), \langle t,s\rangle$	46, 47	(pair of t,s)
$\mathbf{p}_0, \mathbf{p}_1$	46	(unpairing functions, left and right)
$\mathbf{k}_0, \mathbf{k}_1$	46	(injections, left and right)
$\mathrm{E}^{\vee}_{x,y}$	46	(disjunction elimination operator)
$\mathrm{E}^{\exists}_{y,z}$	47	(existence elimination operator)
$\mathrm{E}^{\perp}_{A}$	47	(operator corresponding to $\perp_{\mathrm{i}}$)

Rules and axioms. Let $\dagger \in \{\wedge, \vee, \to, \forall, \exists\}$, $\ddagger \in \{\wedge, \vee, \to, \forall, \exists, \Box, \Diamond, \mathrm{W}, \mathrm{C}\}$.

$\dagger$I, $\dagger$E	36–37	(introduction and elimination rules)
$\perp_{\mathrm{i}}$,$\perp_{\mathrm{c}}$	37	(absurdity rules)
$\Box$I	284	(neccessitation rule)
Ax, L$\perp$		axioms; see the respective G-systems.
R$\ddagger$, L$\ddagger$		see the respective G-systems.
$\forall^2$I, $\forall^2$E, $\exists^2$I, $\exists^2$E	204, 345	(second-order quantifier-rules)
Cut	66,86	(context-free Cut rule)
$\mathrm{Cut}_{\mathrm{cs}}$	67	(context-sharing Cut)
R,R_{u}	236	(resolution, unrestricted resolution)
Ind	317	(induction schema)
TI	317	(transfinite induction)

Reductions and conversions

In chapter 10 $\prec$, $\preceq$, $\succ$, $\succeq$ are used for orderings of the natural numbers, in particular for the standard ordering of ordertype ε_0.

$\succ, \succeq, \succ_1$	12	(reduction, general)
$\prec, \preceq, \prec_1$	12	(converse reduction)
$\prec$	320	(in chapter 10: standard well-ordering)
$\succ_\beta, \succeq_\beta, \succ_{\beta,1}, \succ_{\beta\eta}$ etc.	13	(β-reduction, etc.)
$=_\beta$	15	(β-equality)
$=_{\beta\eta}$	15	($\beta\eta$-equality)
$\succeq_{\mathrm{w}}, =_{\mathrm{w}}$	19	(weak reduction, equality)
$\succ_{\mathrm{g}\beta}, \succeq_{\mathrm{g}\beta},$	217	(generalized β-reduction)

Translations and embeddings

^{g}	49	(Gödel–Gentzen negative translation)
^{k}	50	(Kolmogorov's negative translation)
^{q}	50	(Kuroda's negative translation)
$^{\circ}, ^{\Box}$	288	(modal embeddings)
*	300	(Girard's embedding)

Resolution (chapter 7)

dom, ranv	232	(domain and variable-range)
$\mathcal{H}$, $\mathcal{H}'$,...	236	(metavariables for Horn clauses)
$\mathcal{P}$, $\mathcal{P}'$,...	236	(metavariables for programs)
$\mathcal{H}^{\forall}$, $\mathcal{P}^{\forall}$,	236	(universal closures)
$[\,]$	236	(empty goal)
$\mathrm{R},\mathrm{R_u}$	236	(resolution rules)
$\Gamma' \xrightarrow{\theta}_{\mathrm{u}} \Gamma$	237	(unrestricted resolution step)
$\Gamma' \xrightarrow{\theta} \Gamma$	237	(resolution step)
$\Gamma \mid \mathcal{P}$	239	(computed answer)

Categorical logic (chapter 8)

$f{:}\,A \Rightarrow B$, $A \overset{f}{\Rightarrow} B$	259	(notation for arrows)
$g \circ f$	259	(composition of f and g)
$A \wedge B$, $A \to B$	259	(product and exponent objects)
id^A	259	(identity arrow)
$\pi_0^{A,B}$, $\pi_1^{A,B}$	259	(projection arrows)
$\langle f, g\rangle$	259	(pairing of arrows)
$\top$	259	(terminal object)
tr^A	259	(truth arrow)
$\mathrm{ev}^{A,B}$	259	(evaluation arrow)
cur	259	(currying operator)
$\mathrm{CCC}(\mathcal{PV})$	263	(free category over $\mathcal{PV}$)
$\simeq$	263	(isomorphism of objects)

Other notations

$\boxtimes$	2	(end-of-proof symbol)
$\equiv$	2	(literal identity)
$\mathbb{N}$	2	(natural numbers)
$\mathcal{PV}$	2	(propositional variables)
FV	4, 46	(free variables)
$\lvert\Delta\rvert$	6	(number of elements in multiset)
$\Gamma \cup \Delta$, $\Gamma\Delta$	6	(join of multisets)
$\mathrm{Set}(\Gamma)$	7	(set associated with multiset Γ)
$\mathrm{hyp}(t, t', t'')$	149	(hyperexponentiation)
$2^i_k, 4^i_k, 2_k, 4_k$	149	(abbreviations for hyperexponentiation)
$\bar{n}_A$, N_A	20	(Church numerals on basis A)
$\bar{n}$, N	350	(second-order Church numerals)
Comp	210	(computability predicate)

Index